T0317498

ROBUST CONTROL

ROBUST CONTROL
THEORY AND APPLICATIONS

Kang-Zhi Liu
Professor, Chiba University, Japan
Yu Yao
Professor, Harbin Institute of Technology, China

Registered office
John Wiley & Sons Singapore Pte. Ltd., 1 Fusionopolis Walk, #07-01 Solaris South Tower, Singapore 138628.

For details of our global editorial offices, for customer services and for information about how to apply for permission to reuse the copyright material in this book please see our website at www.wiley.com.

Library of Congress Cataloging-in-Publication Data applied for

ISBN: 9781118754375

A catalogue record for this book is available from the British Library.

Typeset in 10/12pt TimesLTStd by Laserwords Private Limited, Chennai, India

1 2016

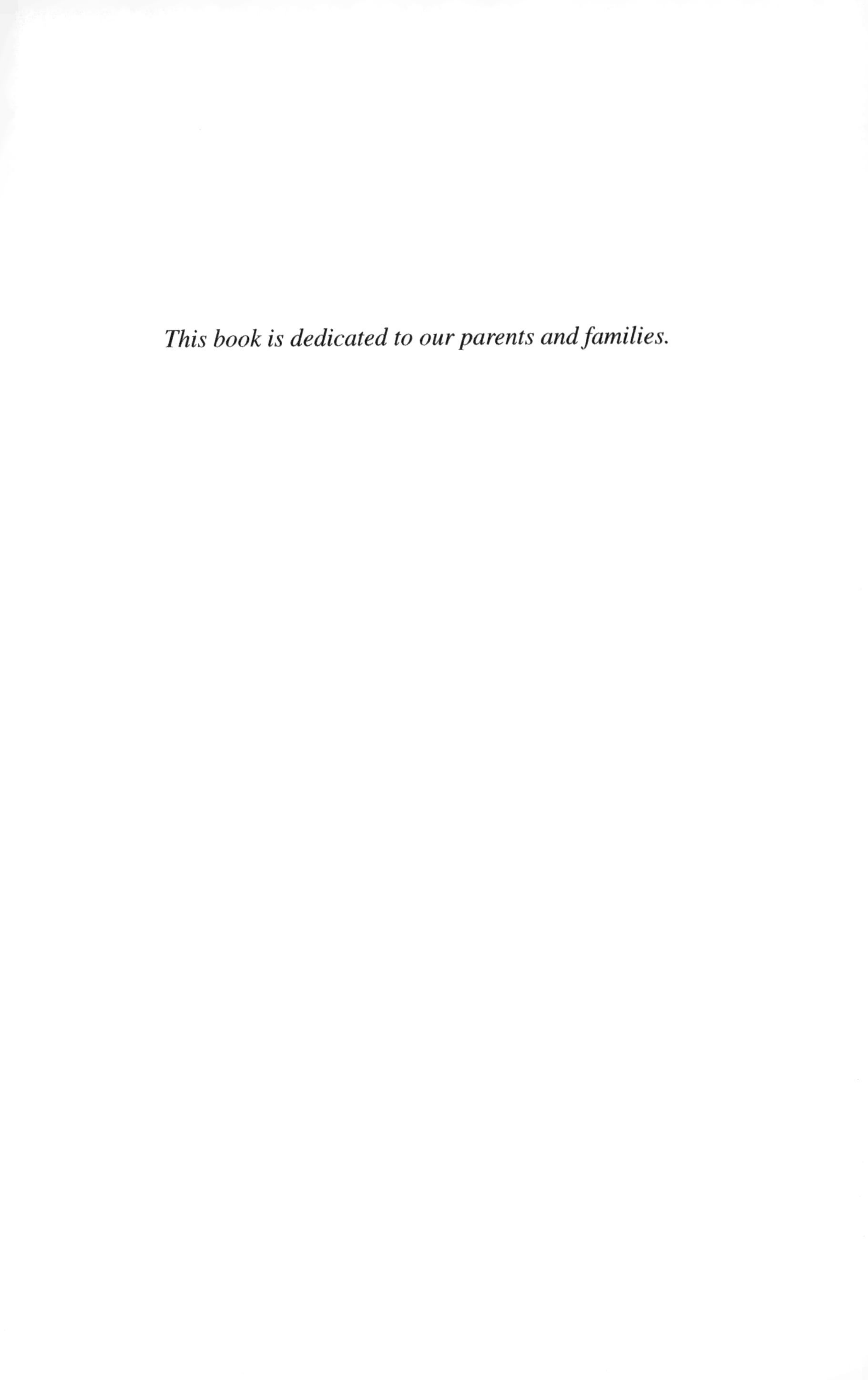

This book is dedicated to our parents and families.

Contents

Preface xvii

List of Abbreviations xix

Notations xxi

1 Introduction 1
1.1 Engineering Background of Robust Control 1
1.2 Methodologies of Robust Control 4
1.2.1 Small-Gain Approach 5
1.2.2 Positive Real Method 5
1.2.3 Lyapunov Method 6
1.2.4 Robust Regional Pole Placement 6
1.2.5 Gain Scheduling 7
1.3 A Brief History of Robust Control 8

2 Basics of Linear Algebra and Function Analysis 10
2.1 Trace, Determinant, Matrix Inverse, and Block Matrix 10
2.2 Elementary Linear Transformation of Matrix and Its Matrix Description 12
2.3 Linear Vector Space 14
2.3.1 Linear Independence 15
2.3.2 Dimension and Basis 16
2.3.3 Coordinate Transformation 18
2.4 Norm and Inner Product of Vector 18
2.4.1 Vector Norm 19
2.4.2 Inner Product of Vector 20
2.5 Linear Subspace 22
2.5.1 Subspace 22

2.6 Matrix and Linear Mapping 23
2.6.1 Image and Kernel Space 23
2.6.2 Similarity Transformation of Matrix 25
2.6.3 Rank of Matrix 26
2.6.4 Linear Algebraic Equation 27
2.7 Eigenvalue and Eigenvector 28
2.8 Invariant Subspace 30
2.8.1 Mapping Restricted in Invariant Subspace 32
2.8.2 Invariant Subspace over $\mathbb{R}^n$ 32
2.8.3 Diagonalization of Hermitian/Symmetric Matrix 33
2.9 Pseudo-Inverse and Linear Matrix Equation 34
2.10 Quadratic Form and Positive Definite Matrix 35
2.10.1 Quadratic Form and Energy Function 35
2.10.2 Positive Definite and Positive Semidefinite Matrices 36
2.11 Norm and Inner Product of Matrix 37
2.11.1 Matrix Norm 37
2.11.2 Inner Product of Matrices 39
2.12 Singular Value and Singular Value Decomposition 40
2.13 Calculus of Vector and Matrix 43
2.13.1 Scalar Variable 43
2.13.2 Vector or Matrix Variable 43
2.14 Kronecker Product 44
2.15 Norm and Inner Product of Function 45
2.15.1 Signal Norm 45
2.15.2 Inner Product of Signals 47
2.15.3 Norm and Inner Product of Signals in Frequency Domain 48
2.15.4 Computation of 2-Norm and Inner Product of Signals 49
2.15.5 System Norm 50
2.15.6 Inner Product of Systems 53
Exercises 53
Notes and References 56

3 Basics of Convex Analysis and LMI 57
3.1 Convex Set and Convex Function 57
3.1.1 Affine Set, Convex Set, and Cone 57
3.1.2 Hyperplane, Half-Space, Ellipsoid, and Polytope 60
3.1.3 Separating Hyperplane and Dual Problem 64
3.1.4 Affine Function 67
3.1.5 Convex Function 68
3.2 Introduction to LMI 72
3.2.1 Control Problem and LMI 72
3.2.2 Typical LMI Problems 73
3.2.3 From BMI to LMI: Variable Elimination 74
3.2.4 From BMI to LMI: Variable Change 79
3.3 Interior Point Method* 81
3.3.1 Analytical Center of LMI 81

3.3.2 Interior Point Method Based on Central Path 82
Exercises 83
Notes and References 84

4 Fundamentals of Linear System **85**
4.1 Structural Properties of Dynamic System 85
4.1.1 Description of Linear System 85
4.1.2 Dual System 87
4.1.3 Controllability and Observability 87
4.1.4 State Realization and Similarity Transformation 90
4.1.5 Pole 90
4.1.6 Zero 91
4.1.7 Relative Degree and Infinity Zero 96
4.1.8 Inverse System 97
4.1.9 System Connections 97
4.2 Stability 100
4.2.1 Bounded-Input Bounded-Output Stability 100
4.2.2 Internal Stability 103
4.2.3 Pole–Zero Cancellation 106
4.2.4 Stabilizability and Detectability 107
4.3 Lyapunov Equation 108
4.3.1 Controllability Gramian and Observability Gramian 111
4.3.2 Balanced Realization 113
4.4 Linear Fractional Transformation 114
Exercises 117
Notes and References 118

5 System Performance **119**
5.1 Test Signal 120
5.1.1 Reference Input 120
5.1.2 Persistent Disturbance 121
5.1.3 Characteristic of Test Signal 121
5.2 Steady-State Response 122
5.2.1 Analysis on Closed-Loop Transfer Function 122
5.2.2 Reference Tracking 124
5.2.3 Disturbance Suppression 127
5.3 Transient Response 130
5.3.1 Performance Criteria 130
5.3.2 Prototype Second-Order System 131
5.3.3 Impact of Additional Pole and Zero 134
5.3.4 Overshoot and Undershoot 136
5.3.5 Bandwidth and Fast Response 139
5.4 Comparison of Open-Loop and Closed-Loop Controls 140
5.4.1 Reference Tracking 140
5.4.2 Impact of Model Uncertainty 142
5.4.3 Disturbance Suppression 144

	Exercises	146
	Notes and References	147
6	**Stabilization of Linear Systems**	**148**
6.1	State Feedback	148
	6.1.1 Canonical Forms	150
	6.1.2 Pole Placement of Single-Input Systems	154
	*6.1.3 Pole Placement of Multiple-Input Systems**	156
	6.1.4 Principle of Pole Selection	159
6.2	Observer	160
	6.2.1 Full-Order Observer	161
	6.2.2 Minimal Order Observer	163
6.3	Combined System and Separation Principle	167
	6.3.1 Full-Order Observer Case	167
	6.3.2 Minimal Order Observer Case	168
	Exercises	170
	Notes and References	172
7	**Parametrization of Stabilizing Controllers**	**173**
7.1	Generalized Feedback Control System	174
	7.1.1 Concept	174
	7.1.2 Application Examples	175
7.2	Parametrization of Controllers	178
	7.2.1 Stable Plant Case	178
	7.2.2 General Case	181
7.3	Youla Parametrization	184
7.4	Structure of Closed-Loop System	186
	7.4.1 Affine Structure in Controller Parameter	186
	7.4.2 Affine Structure in Free Parameter	187
7.5	2-Degree-of-Freedom System	188
	7.5.1 Structure of 2-Degree-of-Freedom Systems	188
	7.5.2 Implementation of 2-Degree-of-Freedom Control	191
	Exercises	193
	Notes and References	196
8	**Relation between Time Domain and Frequency Domain Properties**	**197**
8.1	Parseval's Theorem	197
	8.1.1 Fourier Transform and Inverse Fourier Transform	197
	8.1.2 Convolution	198
	8.1.3 Parseval's Theorem	199
	8.1.4 Proof of Parseval's Theorem	200
8.2	KYP Lemma	200
	8.2.1 Application in Bounded Real Lemma	201
	8.2.2 Application in Positive Real Lemma	204
	*8.2.3 Proof of KYP Lemma**	209

	Exercises	214
	Notes and References	214
9	**Algebraic Riccati Equation**	**215**
9.1	Algorithm for Riccati Equation	215
9.2	Stabilizing Solution	218
9.3	Inner Function	223
	Exercises	224
	Notes and References	224
10	**Performance Limitation of Feedback Control**	**225**
10.1	Preliminaries	226
	10.1.1 Poisson Integral Formula	226
	10.1.2 All-Pass and Minimum-Phase Transfer Functions	227
10.2	Limitation on Achievable Closed-loop Transfer Function	228
	10.2.1 Interpolation Condition	228
	10.2.2 Analysis of Sensitivity Function	229
10.3	Integral Relation	231
	10.3.1 Bode Integral Relation on Sensitivity	231
	10.3.2 Bode Phase Formula	234
10.4	Limitation of Reference Tracking	237
	10.4.1 1-Degree-of-Freedom System	237
	10.4.2 2-Degree-of-Freedom System	243
	Exercises	244
	Notes and References	244
11	**Model Uncertainty**	**245**
11.1	Model Uncertainty: Examples	245
	11.1.1 Principle of Robust Control	247
	11.1.2 Category of Model Uncertainty	247
11.2	Plant Set with Dynamic Uncertainty	248
	11.2.1 Concrete Descriptions	248
	11.2.2 Modeling of Uncertainty Bound	251
11.3	Parametric System	253
	11.3.1 Polytopic Set of Parameter Vectors	255
	11.3.2 Matrix Polytope and Polytopic System	257
	11.3.3 Norm-Bounded Parametric System	258
	11.3.4 Separation of Parameter Uncertainties	262
11.4	Plant Set with Phase Information of Uncertainty	264
11.5	LPV Model and Nonlinear Systems	266
	11.5.1 LPV Model	266
	11.5.2 From Nonlinear System to LPV Model	267
11.6	Robust Stability and Robust Performance	269
	Exercises	270
	Notes and References	271

12 Robustness Analysis 1: Small-Gain Principle **272**
12.1 Small-Gain Theorem 272
12.2 Robust Stability Criteria 276
12.3 Equivalence between $\mathcal{H}_\infty$ Performance and Robust Stability 277
12.4 Analysis of Robust Performance 279
12.4.1 Sufficient Condition for Robust Performance 280
12.4.2 Introduction of Scaling 281
12.5 Stability Radius of Norm-Bounded Parametric Systems 282
Exercises 283
Notes and References 287

13 Robustness Analysis 2: Lyapunov Method **288**
13.1 Overview of Lyapunov Stability Theory 288
13.1.1 Asymptotic Stability Condition 289
13.1.2 Condition for State Convergence Rate 290
13.2 Quadratic Stability 290
13.2.1 Condition for Quadratic Stability 291
13.2.2 Quadratic Stability Conditions for Polytopic Systems 292
13.2.3 Quadratic Stability Condition for Norm-Bounded Parametric Systems 294
13.3 Lur'e System 296
13.3.1 Circle Criterion 300
13.3.2 Popov Criterion 304
13.4 Passive Systems 307
Exercises 310
Notes and References 311

14 Robustness Analysis 3: IQC Approach **312**
14.1 Concept of IQC 312
14.2 IQC Theorem 314
14.3 Applications of IQC 316
14.4 Proof of IQC Theorem* 319
Notes and References 321

15 $\mathcal{H}_2$ Control **322**
15.1 $\mathcal{H}_2$ Norm of Transfer Function 322
15.1.1 Relation with Input and Output 323
15.1.2 Relation between Weighting Function and Dynamics of Disturbance/Noise 324
15.1.3 Computing Methods 325
15.1.4 Condition for $\|G\|_2 < \gamma$ 327
15.2 $\mathcal{H}_2$ Control Problem 329
15.3 Solution to Nonsingular $\mathcal{H}_2$ Control Problem 331
15.4 Proof of Nonsingular Solution 332
15.4.1 Preliminaries 332
15.4.2 Proof of Theorems 15.1 and 15.2 334
15.5 Singular $\mathcal{H}_2$ Control 335

15.6 Case Study: $\mathcal{H}_2$ Control of an RTP System 337

15.6.1 Model of RTP 337

15.6.2 Optimal Configuration of Lamps 340

15.6.3 Location of Sensors 340

15.6.4 $\mathcal{H}_2$ Control Design 340

15.6.5 Simulation Results 341

Exercises 342

Notes and References 345

16 $\mathcal{H}_\infty$ Control **346**

16.1 Control Problem and $\mathcal{H}_\infty$ Norm 346

16.1.1 Input–Output Relation of Transfer Matrix's $\mathcal{H}_\infty$ Norm 346

16.1.2 Disturbance Control and Weighting Function 347

16.2 $\mathcal{H}_\infty$ Control Problem 348

16.3 LMI Solution 1: Variable Elimination 349

16.3.1 Proof of Theorem 16.1 350

16.3.2 Computation of Controller 351

16.4 LMI Solution 2: Variable Change 351

16.5 Design of Generalized Plant and Weighting Function 352

16.5.1 Principle for Selection of Generalized Plant 352

16.5.2 Selection of Weighting Function 353

16.6 Case Study 354

16.7 Scaled $\mathcal{H}_\infty$ Control 355

Exercises 358

Notes and References 359

17 μ Synthesis **360**

17.1 Introduction to μ 360

17.1.1 Robust Problems with Multiple Uncertainties 360

17.1.2 Robust Performance Problem 363

17.2 Definition of μ and Its Implication 364

17.3 Properties of μ 365

17.3.1 Special Cases 365

17.3.2 Bounds of $\mu_\Delta(M)$ 366

17.4 Condition for Robust $\mathcal{H}_\infty$ Performance 368

17.5 D–K Iteration Design 369

17.5.1 Convexity of the Minimization of the Largest Singular Value 370

17.5.2 Procedure of D–K Iteration Design 370

17.6 Case Study 371

Exercises 373

Notes and References 374

18 Robust Control of Parametric Systems **375**

18.1 Quadratic Stabilization of Polytopic Systems 375

18.1.1 State Feedback 375

18.1.2 Output Feedback 376

18.2 Quadratic Stabilization of Norm-Bounded Parametric Systems 379
18.3 Robust $\mathcal{H}_\infty$ Control Design of Polytopic Systems 379
18.4 Robust $\mathcal{H}_\infty$ Control Design of Norm-Bounded Parametric Systems 382
Exercises 382

19 Regional Pole Placement 384
19.1 Convex Region and Its Characterization 384
19.1.1 Relationship between Control Performance and Pole Location 384
19.1.2 LMI Region and Its Characterization 385
19.2 Condition for Regional Pole Placement 387
19.3 Composite LMI Region 392
19.4 Feedback Controller Design 394
19.4.1 Design Method 394
19.4.2 Design Example: Mass–Spring–Damper System 396
19.5 Analysis of Robust Pole Placement 396
19.5.1 Polytopic System 396
19.5.2 Norm-Bounded Parametric System 398
19.6 Robust Design of Regional Pole Placement 402
19.6.1 On Polytopic Systems 402
19.6.2 Design for Norm-Bounded Parametric System 403
19.6.3 Robust Design Example: Mass–Spring–Damper System 405
Exercises 405
Notes and References 406

20 Gain-Scheduled Control 407
20.1 General Structure 407
20.2 LFT-Type Parametric Model 408
20.2.1 Gain-Scheduled $\mathcal{H}_\infty$ Control Design with Scaling 410
20.2.2 Computation of Controller 413
20.3 Case Study: Stabilization of a Unicycle Robot 414
20.3.1 Structure and Model 414
20.3.2 Control Design 418
20.3.3 Experiment Results 420
20.4 Affine LPV Model 422
20.4.1 Easy-to-Design Structure of Gain-Scheduled Controller 423
20.4.2 Robust Multiobjective Control of Affine Systems 425
20.5 Case Study: Transient Stabilization of a Power System 428
20.5.1 LPV Model 429
20.5.2 Multiobjective Design 430
20.5.3 Simulation Results 430
20.5.4 Robustness 432
20.5.5 Comparison with PSS 432
Exercises 434
Notes and References 435

21 Positive Real Method 436
21.1 Structure of Uncertain Closed-Loop System 436
21.2 Robust Stabilization Based on Strongly Positive Realness 438
21.2.1 Variable Change 439
21.2.2 Variable Elimination 440
21.3 Robust Stabilization Based on Strictly Positive Realness 441
21.4 Robust Performance Design for Systems with Positive Real Uncertainty 442
21.5 Case Study 445
Exercises 448
Notes and References 449

References 450

Index 455

The sections and chapters marked by * require relatively high level mathematics and may be skipped in the first reading, which does not affect the understanding of the main body.

Preface

This textbook evolved from our course teachings both at Chiba University and at Harbin Institute of Technology (HIT). It is intended for use in graduate courses on advanced control method. Many case studies are included in the book, hence it is also useful for practicing control engineers.

Robust control aims at conducting realistic system design in the face of model uncertainty. Since its advent in the 1980s, it has received extensive attention and have been applied to many real-world systems such as steel making process, vehicles, mass storage devices, and so on.

In this book, our ultimate goal is to summarize comprehensively the major methods and results on robust control, and provide a platform for readers to study and apply robust control. We also intend to provide a shortcut to the frontier of robust control research for those who are interested in theoretical research. Further, we hope to achieve the following goals:

1. Easy to understand: presenting profound contents in the most concise manner;
2. Showing the essence of robust control clearly and thoroughly;
3. Emphasizing engineering implications while keeping the theoretic rigor;
4. Self-contained: reducing to the lowest level the necessity of referring to other sources.

In order to realize the preceding objectives, whenever we introduce a new content, we first use a simple example or the concept that the readers are familiar with to illustrate the engineering background and idea clearly. Then we show the simplest proof known to the authors. All examples in this book stem from engineering practice and have clear backgrounds. It is hoped that this will help the readers learn how to apply these knowledge. For each important result, a rigorous proof is provided. To be self-contained, a quite complete coverage on linear algebra and convex analysis is provided, including proofs.

The book falls into three parts: mathematical preliminaries, fundamentals of linear systems and robust control. Chapters 2 and 3 summarize the knowledge of linear algebra, convex analysis, and LMI, which are required in order to understand the robust control theory completely. Chapter 4 reviews the basics of linear systems. Chapter 5 analyzes the performance specifications of reference tracking, disturbance rejection and fast response. Chapter 6 treats the stabilization of a given plant by using state feedback and observer. Further, Chapter 7 shows how to use a simple formula to parameterize all controllers that can stabilize a given plant, and analyze the structure of the closed-loop system, including the two-degree-of-freedom system. Chapter 8 provides several important results that bridge the time domain and frequency domain characteristics of signals/systems, especially the KYP lemma that lays the foundation for solving the robust control problem in the time domain. Chapter 9 describes some of the main results on the Riccati equation. Chapter 10 discusses the limitation of feedback control, so as to provide a guide for setting up reasonable performance specifications as well as the design of easy-to-control systems. Chapter 11 discusses the model uncertainty, including its classification, the features of different types, and the modeling methods. Chapter 12 is about the robust control analysis based on the gain information of uncertainty. The central issue is how to use the nominal model and the bounding function of uncertainty gain to derive the conditions for robust stability and robust performance. Based on Lyapunov stability theory and quadratic common Lyapunov functions, Chapter 13 establishes the robust stability conditions for systems with parameter uncertainty, and discusses Lur'e systems as well as passive systems. Chapter 14 presents the so-called IQC (integral quadratic constraint) theory. Chapters 15 and 16 handle, respectively, the $\mathcal{H}_2$ and $\mathcal{H}_\infty$ control theory and their applications. Chapter 17 is devoted to the robustness analysis and design of systems with multiple uncertainties. Chapter 18 covers systems with parameter uncertainty. Chapter 19 focuses on the parametric systems and shows how to improve the transient response by placing the closed-loop poles in a suitable region. Chapter 20 treats systems with known time-varying parameters. A higher performance can be achieved through the gain-scheduling of the controller, that is, letting the controller parameters vary together with the plant parameters. Finally, Chapter 21 describes how to use the positive real property of an uncertain system to realize robust control. Its essence is to make use of the phase information of uncertainty.

This book is an extended version of our textbook written in Chinese and published by Science Press, Beijing. Many staffs and students at HIT involved in the translation. Meanwhile, students at Chiba University helped us in doing the application researches and numerical designs. All case studies included in the book were conducted by our students. Without their efforts, this book will not be as it is.

We are indebted to Prof. Stephen Boyd and Prof. Kemin Zhou. Their books on convex analysis and small-gain approach to robust control have a significant influence on the writing of this book. Further, we would like to thank our supervisors, Prof. Tsutomu Mita and Prof. Zicai Wang, who guided us into the wonderful field of control engineering.

Kang-Zhi Liu
Yu Yao

List of Abbreviations

For simplicity, frequently used words are abbreviated as follows.

ARE	algebraic Riccati equation
BMI	bilinear matrix inequality
BIBO	bounded-input bounded-output
EVP	eigenvalue problem
GEVP	generalized eigenvalue problem
HDD	hard disk drive
IQC	Integral quadratic constraint
LFT	linear fractional transformation
LMI	linear matrix inequality
LPV	linear parameter varying
MIMO	multi-input multi-output
PSS	power system stabilizer
RTP	rapid thermal processing
SISO	single-input single-output
SVD	singular value decomposition
1-DOF	1-degree-of-freedom
2-DOF	2-degree-of-freedom
i.e.	that is
w.r.t.	with respect to
iff	if and only if

Notations

The notations used in this book are listed below.

$:=$	defined as
$\Leftrightarrow$	equivalent to
$\in$	belong to
$\forall$	for all
$\subset$	included in
$\emptyset$	empty set
δ_{ij}	Dirac index, satisfying $\delta_{ii} = 1, \delta_{ij} = 0 (i \neq j)$
$\mathbb{R}$	field of real numbers
$\mathbb{C}$	field of complex numbers
$\mathbb{R}^n$	n-dimensional space of real vectors
$\mathbb{C}^n$	n-dimensional space of complex vectors
$\mathbb{R}^{m \times p}$	set of real matrices with m rows and p columns
$\mathbb{C}^{m \times p}$	set of complex matrices with m rows and p columns
$\mathbb{F}$	In stating a result which is true for both real and complex variables, $\mathbb{F}$ is used to present both.
sup	Supremum (i.e., least upper bound), may be regarded as the maximum max in engineering.
inf	Infimum (i.e., greatest lower bound), may be regarded as the minimum min.
$\max\limits_i$	maximum w.r.t. all variables i
$\bar{x}$	conjugate of $x = a + jb$, namely $\bar{x} = a - jb$.
$\Re(x)$, $\Im(x)$	real and imaginary parts of a complex number x.
arg (x)	phase angle of a complex number x
$\hat{u}(s) = \mathcal{L}[u(t)]$	Laplace transform of time function $u(t)$
$\hat{u}(j\omega) = \mathcal{F}[u(t)]$	Fourier transform of time function $u(t)$
$\dot{u}(t), \ddot{u}(t)$	First and second derivatives of $u(t)$ about time t
$A = (a_{ij})$	matrix with a_{ij} as its element in the ith row and jth column
$X^* = \bar{X}^T$	Conjugate transpose of complex matrix X

$\text{He}(X)$	$\text{He}(X) = X + X^*$
$F^{\sim}(s)$	$F^{\sim}(s) = F^T(-s)$
$\text{diag}(a_1, \cdots, a_n)$	diagonal matrix with a_i as its ith diagonal element
$\det(A)$	determinant of square matrix A
$\lambda_i(A)$	the ith eigenvalue of matrix A
$\sigma(A)$	set of eigenvalues $\{\lambda_1, \cdots, \lambda_n\}$ of matrix A
$\rho(A)$	spectral radius $\max_i \lvert\lambda_i(A)\rvert$ of matrix A
$\sigma_i(A)$	the ith singular value of matrix A. Be cautious not to confuse with the eigenvalue set $\sigma(A)$.
$\text{rank}(A)$	rank of matrix A
$A_\perp$	orthogonal matrix of matrix A ($m \times n$), i.e., a matrix with the maximum rank among all matrices U satisfying $AU = 0\ (m \le n)$ or $UA = 0\ (m \ge n)$.
$\text{Tr}(A)$	trace of $A = (a_{ij}) \in \mathbb{C}^{n \times n}$, i.e., the sum of its diagonal elements $\sum_{i=1}^n a_{ii}$
$\text{Im}A$	$\{y \in \mathbb{C}^n \mid y = Ax, x \in \mathbb{C}^m\}$, image of A.
$\text{Ker}\ A$	$\{x \mid Ax = 0, x \in \mathbb{C}^n\}$, kernel space of A.
$A \otimes B$	Kronecker product
$A \oplus B$	Kronecker sum
$\text{vec}(A)$	vector formed by sequentially aligning the columns of matrix A starting from the first column
$A^\dagger$	pseudo-inverse of matrix A
$A > 0 (\ge 0)$	positive definite (positive semidefinite) matrix
$\lVert x\rVert$	norm of vector x
$\lVert A\rVert$	norm of matrix A
$\langle u, v\rangle$	inner product of vectors u, v.
$\langle A, B\rangle$	inner product of matrices A, B.
$\text{span}\{u_1, \cdots, u_k\}$	space spanned by the vector set $u_1, \cdots, u_k$
$\textbf{conv}\ C$	convex hull of set C
$x \prec y$	inequality of vectors (x, y), satisfying $x_i < y_i$.
$\frac{\partial f(x)}{\partial x}$	first-order partial derivative of function $f(x)$ whose transpose is the gradient $\nabla f(x)$ of $f(x)$.
$\frac{\partial^2 f(x)}{\partial x^2}$	second-order partial derivative of $f(x)$ whose transpose is the Hessian matrix $\nabla^2 f(x)$ of $f(x)$.
$\mathcal{F}_\ell(M, X)$	lower linear fractional transformation
$\mathcal{F}_u(M, X)$	upper linear fractional transformation
$f_1(t) * f_2(t)$	convolution integral of $f_1(t)$ and $f_2(t)$
$\textbf{dom}\ f$	domain of function f
$f'(x)$	derivative of function $f(x)$
$\text{Res}_{s_i} \hat{f}(s)$	residue of complex function $\hat{f}(s)$ at the point s_i

1

Introduction

Mathematical model is indispensable in the simulation of physical systems or the design of control systems. The so-called natural science is, in fact, the systematic and categorized study of various kinds of physical, chemical, and other natural phenomena. Various models are used to describe and reproduce the observed phenomena. The models used in engineering are usually described as differential equations, difference equations, or statistical data. It may be said that the design of modern control systems is essentially based on the models of systems.

However, a mathematical model cannot describe the physical phenomenon of a system perfectly. Even if it could, the model would be unnecessarily complicated which makes it difficult to capture the main characteristic of the system. In particular, this is often the case in engineering practice. In almost all cases, a system is not isolated from the outside world. Instead, it is constantly influenced by the surrounding environment. It is difficult to describe the external influence by models. This means that there is inevitably a gap between the actual system and its mathematical model. This gap is called as the *model uncertainty*. The purpose of robust control is to extract the characteristics of model uncertainty and apply this information to the design of control system, so as to enhance the performance of the actual control system to the limit.

1.1 Engineering Background of Robust Control

Here, we give some specific examples to illustrate the model uncertainty.

Example 1.1 *Hard disk (refer to Figure 1.1(a)) is commonly used as the data storage device for computers. The frequency response of a hard disk drive (HDD in short hereafter) is shown in Figure 1.1(b) as the dotted line. The approximate model only considers the rigid body and is described as a double integrator (the solid line). Since the arm is very thin, the drive has numerous resonant modes in the high-frequency band, and these resonant modes vary with the manufacturing error in mass production. So it is difficult to obtain the exact model. Therefore, we have to design the control system based on the model of rigid body.*

Robust Control: Theory and Applications, First Edition. Kang-Zhi Liu and Yu Yao.

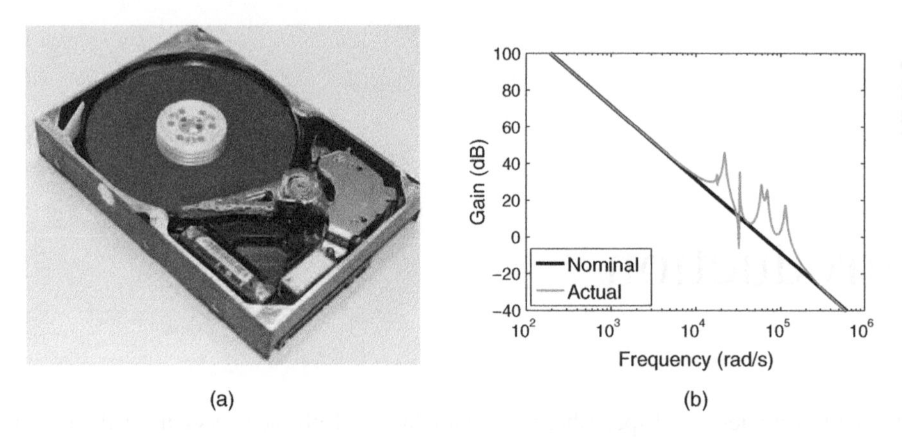

Figure 1.1 Hard disk drive and its frequency response (a) Photo, (b) Bode plot

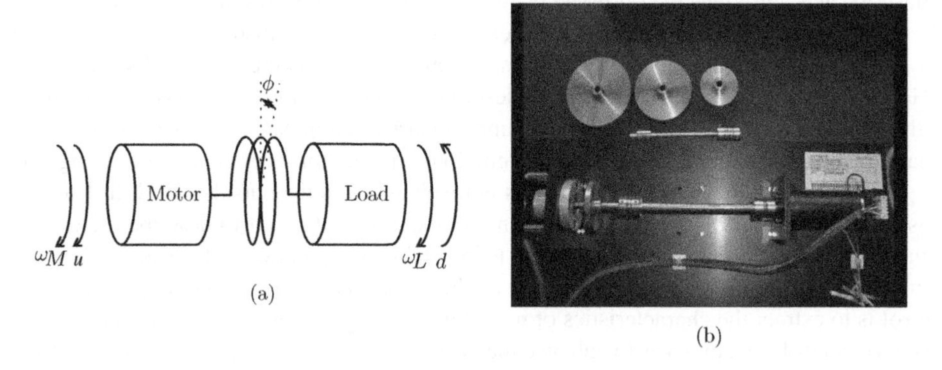

Figure 1.2 Motor drive system (a) Two-mass–spring system, (b) Motor drive

Example 1.2 *The system in Figure 1.2(a) is called* two-mass–spring system. *As shown in Figure 1.2(b), this is essentially a motor drive system in which the motor and load are connected through a shaft. The purpose is to control the load speed indirectly by controlling the motor inertia moment. There are many such systems around us. Typical examples are the DVD drive in home electronics, the rolling mill in steel factory, and so on.*

Let J_M be the inertia moment of the motor, k the spring constant of the shaft, J_L the inertia moment of the load, and D_M and D_L the viscous friction coefficients of the motor and the load. In addition, the motor speed is ω_M, the load speed is ω_L, the torsional angle of the shaft is ϕ, and the motor torque is u. As the direct measurement of load speed is difficult, we usually measure the motor speed ω_M only. The equations of moment balance as well as the speed equation are

$$\begin{aligned} J_L\dot{\omega}_L + D_L\omega_L &= k\phi + d \\ \dot{\phi} &= \omega_M - \omega_L \\ J_M\dot{\omega}_M + D_M\omega_M + k\phi &= u. \end{aligned}$$

Here, d denotes the torque disturbance acting on the load. Select the state vector as $x = [\omega_L \ \phi \ \omega_M]^T$, *the following state equation is obtained easily:*

$$\dot{x} = \begin{bmatrix} -\dfrac{D_L}{J_L} & \dfrac{k}{J_L} & 0 \\ -1 & 0 & 1 \\ 0 & -\dfrac{k}{J_M} & -\dfrac{D_M}{J_M} \end{bmatrix} x + \begin{bmatrix} \dfrac{1}{J_L} \\ 0 \\ 0 \end{bmatrix} d + \begin{bmatrix} 0 \\ 0 \\ \dfrac{1}{J_M} \end{bmatrix} u \tag{1.1}$$

$$y = [0 \ \ 0 \ \ 1]x. \tag{1.2}$$

However, in practice the motor load is rather diverse. Specifically, the load inertia moment J_L *and the spring constant k of the shaft change over a wide range. This will undoubtedly affect the performance of the motor drive system.*

As illustrated by these examples, the mathematical model of a system always contains some uncertain part. Nevertheless, we still hope that a control system, designed based on a model with uncertainty, can run normally and has high performance. To achieve this goal, it is obvious that we should make use of all available information about the uncertainty. For example, in the loop shaping design of the classical control, a very important principle is to ensure that the high-frequency gain of controller rolls off sufficiently to avoid exciting the unmodeled high-frequency dynamics of the plant, so that the controller can be successfully applied to the actual system. In addition, the open-loop transfer function needs also to have sufficient gain margin and phase margin so as to ensure that the closed-loop system is insensitive to the model uncertainty in the low- and middle-frequency domain. This implies that the information about model uncertainty has already been used *indirectly* in system designs based on the classical control methods. Of course, indirect usage of model uncertainty has only a limited effect. So, the mission of robust control theory is to develop an effective, high-performance design method that makes use of the uncertainty information *fully and directly* for systems with uncertainty.

Next, we use an example to show the influence of model uncertainty. It will be illustrated that if the model uncertainty is neglected in the system design, the controller may have trouble when applied to the actual system. In the worst case, even the stability of the closed-loop system may be lost.

Example 1.3 *Consider a system containing both rigid body and a resonant mode (a simplified dynamics of HDD)*

$$\tilde{P}(s) = \frac{1}{s^2} - \frac{0.1 \times \omega_n^2}{s^2 + 2\zeta\omega_n s + \omega_n^2}, \quad \zeta = 0.02, \omega_n = 5.$$

In the control design, only the rigid body model $P(s) = 1/s^2$ *is considered, and the resonant mode* $-0.1 \times \omega_n^2/(s^2 + 2\zeta\omega_n s + \omega_n^2)$ *is ignored. In the stabilization design of the rigid body model, we hope that the closed-loop system response is fast enough while the overshoot is as small as possible. So the damping ratio of the closed-loop system is set as* $1/2$ *and the natural*

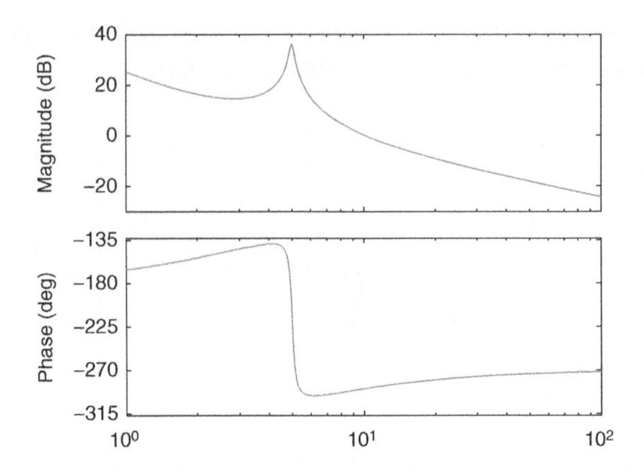

Figure 1.3 Bode plot of open-loop transfer function

frequency set as 4 [rad/s] (slightly lower than the natural frequency of the resonant mode). The corresponding characteristic polynomial is

$$p(s) = s^2 + 4s + 16.$$

The following proportional derivative (PD) compensator is able to achieve the goal:

$$K(s) = 4(s + 4).$$

However, when this controller is applied on the actual system $\tilde{P}(s)$, we find that the characteristic polynomial of the actual closed-loop becomes

$$\tilde{p}(s) = s^4 - 5.8s^3 + 1.8s^2 + 103.2s + 400$$

which obviously has unstable roots. In fact, the closed-loop poles are xx

$$5.2768 \pm j3.8875, \quad -2.3768 \pm j1.9357,$$

in which two of them are unstable. That is, the closed-loop system not only fails to achieve performance improvement but also losses the stability. This phenomenon is known as spillover in vibration control. The reason for the spillover is clear from Figure 1.3, the Bode plot of the open-loop transfer function $\tilde{P}K$. Since the response speed of the closed loop is designed too fast for the rigid body model, the roll-off of the controller gain is not sufficient near $\omega_n = 5$ rad/s, the natural frequency of the resonant mode, which excites the resonant mode.

1.2 Methodologies of Robust Control

In this section, various descriptions of model uncertainty and the corresponding robust control methods are briefly illustrated for single-input single-output systems.

1.2.1 Small-Gain Approach

Suppose that the system model is $P(s)$ while the actual system is $\tilde{P}(s)$. If no information is known about the structure of the actual system $\tilde{P}(s)$, the simplest method is to regard the difference between them as the model uncertainty $\Delta(s)$, that is,

$$\tilde{P}(s) = P(s) + \Delta(s). \tag{1.3}$$

Then, the closed-loop system can be transformed equivalently into the system of Figure 1.4 in which

$$M(s) = \frac{K}{1 + PK}$$

is the closed-loop transfer function about $P(s)$ and should be stable.

As for the description of the characteristics of $\Delta(s)$, a straightforward method is to use the frequency response. In general, the frequency response of $\Delta(s)$ varies within a certain range, such as

$$0 \leq |\Delta(j\omega)| \leq |W(j\omega)| \ \ \forall\omega. \tag{1.4}$$

Here $W(s)$ is a known transfer function whose gain specifies the boundary of the uncertainty $\Delta(s)$. When $\Delta(s)$ is a stable transfer function,

$$|M(j\omega)\Delta(j\omega)| < 1 \ \ \forall\omega \ \ \Leftrightarrow \ \ |M(j\omega)W(j\omega)| < 1 \ \ \forall\omega \tag{1.5}$$

guarantees the stability of the closed-loop system for all uncertainties $\Delta(s)$. This condition is known as the *small-gain condition*. The reason for the stability of the closed-loop system is clear. The open-loop transfer function in Figure 1.4 is $L(s) = M(s)\Delta(s)$ and is stable. $L(j\omega)$ does not encircle the critical point $(-1, j0)$ when the small-gain condition is satisfied, so the Nyquist stability condition holds. Here, the information of uncertainty gain is used.

Chapter 12 describes in detail the robust analysis based on the small-gain condition, while Chapters 16 and 17 provide the corresponding robust design methods.

1.2.2 Positive Real Method

On the contrary, in (1.3) if the phase angle $\arg\Delta(j\omega)$ of the uncertainty $\Delta(s)$ changes in a finite range, particularly when

$$-90^\circ \leq \arg\Delta(j\omega) \leq 90^\circ \ \ \forall\omega \ \ \Leftrightarrow \ \ \Re[\Delta(j\omega)] \geq 0 \ \ \forall\omega, \tag{1.6}$$

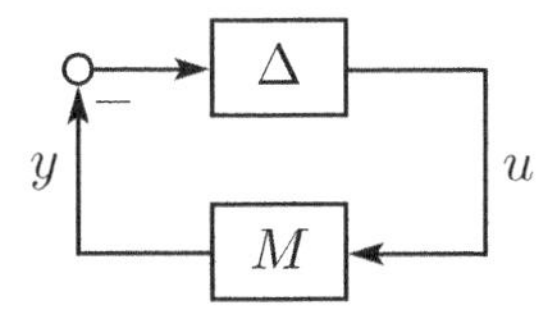

Figure 1.4 Closed-loop system with uncertainty

the following condition guarantees the stability of the closed-loop system for all the uncertainties $\Delta(s)$:

$$-90^\circ < \arg M(j\omega) < 90^\circ \ \forall\omega \ \Leftrightarrow \ \Re[M(j\omega)] > 0 \ \forall\omega. \tag{1.7}$$

This is because the open-loop transfer function $L(s) = M(s)\Delta(s)$ is stable and the phase angle is never equal to $\pm 180^\circ$. As such, $L(j\omega)$ does not encircle the critical point $(-1, j0)$, and the Nyquist stability condition is met.

For all frequencies ω, a transfer function satisfying $\Re[G(j\omega)] \geq 0$ is called a *positive real function*. So the previous condition is named as the *positive real condition*. General positive real condition is described detailedly in Section 13.4, and the robust design based on positive real condition is discussed in Chapter 21.

1.2.3 Lyapunov Method

For systems with known dynamics but uncertain parameters, such as the two-mass–spring system, methods based on quadratic Lyapunov function are quite effective. For example, consider an autonomous system

$$\dot{x} = Ax, \quad x(0) \neq 0.$$

If the quadratic Lyapunov function

$$V(x) = x^T Px, \quad P > 0$$

satisfies the strictly decreasing condition

$$\dot{V}(x) < 0 \quad \forall x \neq 0,$$

then $V(x)$ converges to zero eventually since it is bounded below. As $V(x)$ is positive definite, $V(x) = 0$ implies that $x = 0$. Further, the strictly decreasing condition is equivalent to the inequality on the matrix A:

$$A^T P + PA < 0.$$

Even if some parameters of the matrix A take value in a certain range, the stability of the system is secured so long as there exists a matrix $P > 0$ satisfying this inequality.

Refer to Chapter 13 for the analysis on such uncertain systems and Section 18.3, Chapter 19, and Chapter 20 for the design methods.

1.2.4 Robust Regional Pole Placement

For a system with parameter uncertainty, we can place the poles of the closed-loop system in some region of the complex plane, such as the region shown in Figure 1.5, so as to guarantee the quality of transient response. Detailed design conditions and design methods are given in Chapter 19.

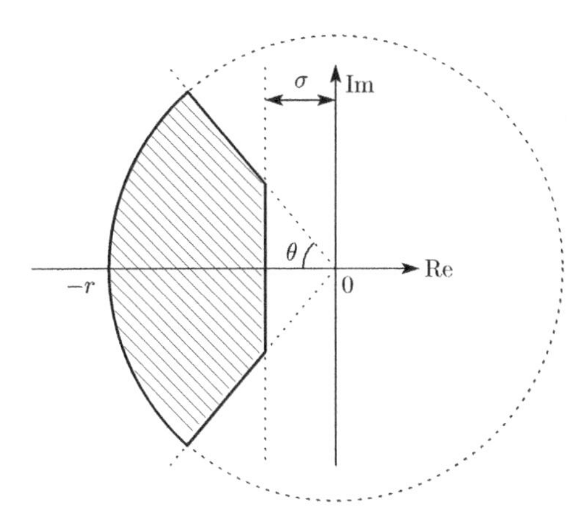

Figure 1.5 Regional pole placement

1.2.5 Gain Scheduling

Many nonlinear systems can be described as linear systems with time-varying parameters, in which the time-varying parameters are bounded functions of some states. Such a system is called as a *linear parameter-varying* (LPV) system . When the states contained in the LPV system can be measured online, the time-varying parameters can be computed. Then, if the controller parameters vary in accordance with the time-varying parameters, it is possible to obtain a better control performance. This method is called as *gain scheduling*. The details can be found in Chapter 20.

Example 1.4 *Let us look at a very simple example: a one-link arm. Assume that the angle between the arm and the vertical line is* θ*, the mass and inertia moment are* m*,* J*, the distance between the center of gravity and the joint is* l*, and the control torque is* u*. Then the motion equation is*

$$J\ddot{\theta} + mgl\ \sin\theta = u.$$

If we set $p(t) = \sin\theta/\theta$*, the motion equation can be rewritten as*

$$J\ddot{\theta}(t) + mglp(t)\theta(t) = u(t).$$

This can be regarded as an LPV system with a time-varying parameter $p(t)$*. Clearly, the time-varying parameter satisfies*

$$|p(t)| = \left|\frac{\sin\theta(t)}{\theta(t)}\right| \leq 1$$

and is bounded. In the control design, a better performance may be obtained by using a controller containing $p(t)$*, or the control design is made easier. For example, when we use the control input*

$$u(t) = mglp(t)\theta(t) - 2\zeta\omega_n J\dot{\theta}(t) - \omega_n^2 J\theta(t),$$

the closed-loop system becomes

$$\ddot{\theta} + 2\zeta\omega_n\dot{\theta} + \omega_n^2\theta = 0.$$

By adjusting the damping ratio ζ *and the natural frequency* ω_n*, we can easily achieve a high control performance. In this designed input, the first term about the time-varying parameter is in fact a nonlinear term* $mgl \sin\theta$.

1.3 A Brief History of Robust Control

As mentioned earlier, in the classical control design, the model uncertainty in the high-frequency domain is taken into account, and effective feedback control is realized by tuning controller parameters online. However, model uncertainty is excluded from the modern control theory. So, it is rare that a controller designed by pole placement or least quadratic Gaussian (LQG) optimal control theory can be directly applied to the actual system. This is because the actual control effect is quite different from the design specification and simulation. In successful applications of modern control, filters are always used to filter out the high-frequency component of the output signal before feeding it to the controller. This in fact considers the neglected high-frequency characteristic of the plant indirectly.

For example, although the state feedback least quadratic control has a gain margin ranging from $1/2$ to infinity and a phase margin of ±60° [82], Doyle [23] proved that in the output feedback of LQG control, even a small perturbation in the gain may destabilize the closed-loop system. Facing this flaw of the modern control theory, researchers, mainly from North America, began to consider how to deal with model uncertainty. First of all, inspired by the *perturbation theory* in mathematics, in 1964 Cruz *et al.* [19] analyzed the rate of relative change of the closed-loop transfer function $H(s) = L(s)/(1 + L(s))$, that is,

$$\frac{\Delta H/H}{\Delta L/L} = \frac{\dfrac{L+\Delta L}{1+L+\Delta L} - \dfrac{L}{1+L}}{\Delta L}\frac{L}{H}$$

when the open-loop transfer function $L(s)$ has a small perturbation. It is easy to know that when the perturbation ΔL approaches zero,

$$\lim_{\Delta L \to 0} \frac{\Delta H/H}{\Delta L/L} = \frac{1}{1+L} := S(s)$$

holds. This is the reason why the transfer function $S(s)$ is called as the *sensitivity*. Unfortunately, this relationship is valid only for very small perturbations and has nothing to do with the dynamic property of uncertainty. So, it makes no sense in handling uncertainty.

In the 1980s, Zames [99] and Doyle-Stein [28] for the first time challenged the issue of model uncertainty and discussed how to introduce the information of model uncertainty into the feedback control design. Both of them believe that the model uncertainty should be described by the range of the gain of its frequency response. The former stressed that the disturbance should be treated as a set and the control performance of the disturbance control measured by the $\mathcal{H}_\infty$ norm of the closed-loop transfer function. Meanwhile, the latter proposed the small-gain principle. In the following decade, robust control was developed along this line. Because the uncertainty and the control performance are characterized in the form of frequency response, robust control research commenced in the frequency domain, and the main mathematical tools were operator theory and Nevanlinna–Pick interpolation theory [48]. Later, Doyle advocated the use of state space in $\mathcal{H}_\infty$ control problems and completed the Riccati equation solution [36] with Glover in 1988. Particularly, the famous paper published in 1989, known as DGKF paper [26], has an immeasurable impact on the robust control research thereafter. This is because it reveals that, to solve the $\mathcal{H}_\infty$ control problem, one does not need to use advanced mathematical tools in function spaces which is formidable for engineers and the familiar state-space tool is sufficient.

In a roughly same period, Boyd [9, 10] advocated the application of numerical method and optimization approach to robust control. In particular, the book [10] had a great impact. It happened in the end of 1980s that Russian mathematicians Nesterov and Nemirovski [72] completed the numerical approach, the interior point method, to convex programming, thus paving the way for solving the robust problems by optimization methods. Since 1990s, numerous robust control methods based on LMI tools have sprung up. In particular, the French school, represented by Gahinet, not only proposed the LMI solution to $\mathcal{H}_\infty$ control but also solved a series of robust control problems, such as regional pole placement and gain scheduling [33, 16, 3]. Moreover, they developed the LMI toolbox [35] which helped promoting the LMI approach. So far, the mainstream of robust control is LMI. This book presents the major contents of robust control mainly based on the LMI approach.

The aforementioned studies are all robust control methods that use the information of uncertainty gain or parameter uncertainty. When the phase information of uncertainty is known and its range is not big, it is effective to use the phase information to design robust control system. Haddad-Berstein [37] and Tits [88] have done some works in this direction. More new results on this approach will be shown in this book.

2

Basics of Linear Algebra and Function Analysis

In this chapter, the fundamentals of linear algebra and function analysis to be used in this book will be briefly summarized, and several new mathematical tools will be introduced.

2.1 Trace, Determinant, Matrix Inverse, and Block Matrix

First, the trace $\text{Tr}(A)$ of a square matrix A is the sum of all diagonal elements and has the following properties:

1. $\text{Tr}(\alpha A) = \alpha \text{Tr}(A)\ \forall \alpha \in \mathbb{F}, A \in \mathbb{F}^{n\times n}$.
2. $\text{Tr}(A+B) = \text{Tr}(A) + \text{Tr}(B)\ \forall A, B \in \mathbb{F}^{n\times n}$.
3. $\text{Tr}(AB) = \text{Tr}(BA)\ \forall A \in \mathbb{F}^{n\times m}, B \in \mathbb{F}^{m\times n}$.

These properties can be easily verified based on the definition of trace.

The properties of determinant are summarized in the following. For square matrices A, B with the same size, the determinant of their product satisfies

$$\det(AB) = \det(A)\det(B), \tag{2.1}$$

that is, the determinant of a matrix product is equal to the product of matrix determinants. From this equation and the identity $A^{-1}A = I$, it is easy to see that

$$\det(A^{-1}) = \frac{1}{\det(A)}. \tag{2.2}$$

Further, for a block triangular matrix with square diagonal blocks A, B (the sizes may be different), its determinant can be calculated by

$$\det \begin{bmatrix} A & * \\ 0 & B \end{bmatrix} = \det(A)\det(B), \tag{2.3}$$

Robust Control: Theory and Applications, First Edition. Kang-Zhi Liu and Yu Yao.

$$\det \begin{bmatrix} A & 0 \\ * & B \end{bmatrix} = \det(A)\det(B). \tag{2.4}$$

When square matrices A, B are both nonsingular, the inverse of their product satisfies

$$(AB)^{-1} = B^{-1}A^{-1}. \tag{2.5}$$

Partition a square matrix A as

$$A := \begin{bmatrix} A_{11} & A_{12} \\ A_{21} & A_{22} \end{bmatrix} \tag{2.6}$$

in which A_{11} and A_{22} are square. If A_{11} is nonsingular, then A can be decomposed as

$$\begin{bmatrix} A_{11} & A_{12} \\ A_{21} & A_{22} \end{bmatrix} = \begin{bmatrix} I & 0 \\ A_{21}A_{11}^{-1} & I \end{bmatrix} \begin{bmatrix} A_{11} & 0 \\ 0 & A_{22} - A_{21}A_{11}^{-1}A_{12} \end{bmatrix} \begin{bmatrix} I & A_{11}^{-1}A_{12} \\ 0 & I \end{bmatrix}. \tag{2.7}$$

Similarly, when A_{22} is nonsingular, we have

$$\begin{bmatrix} A_{11} & A_{12} \\ A_{21} & A_{22} \end{bmatrix} = \begin{bmatrix} I & A_{12}A_{22}^{-1} \\ 0 & I \end{bmatrix} \begin{bmatrix} A_{11} - A_{12}A_{22}^{-1}A_{21} & 0 \\ 0 & A_{22} \end{bmatrix} \begin{bmatrix} I & 0 \\ A_{22}^{-1}A_{21} & I \end{bmatrix}. \tag{2.8}$$

Based on these decompositions, it is easy to derive the following formulae:

$$\det(A) = \det(A_{11})\det(A_{22} - A_{21}A_{11}^{-1}A_{12}), \text{when } \det(A_{11}) \neq 0; \tag{2.9}$$

$$\det(A) = \det(A_{22})\det(A_{11} - A_{12}A_{22}^{-1}A_{21}), \text{when } \det(A_{22}) \neq 0. \tag{2.10}$$

Further, for matrices $B \in \mathbb{F}^{m\times n}$ and $C \in \mathbb{F}^{n\times m}$,

$$\det \begin{bmatrix} I_m & B \\ -C & I_n \end{bmatrix} = \det(I_n + CB) = \det(I_m + BC) \tag{2.11}$$

holds. Particularly, for vectors $x, y \in \mathbb{C}^n$, there holds

$$\det(I_n + xy^*) = 1 + y^*x. \tag{2.12}$$

In addition, it is also easy to verify the following inversion formula about block triangular matrices[1]:

$$\begin{bmatrix} A_{11} & 0 \\ A_{21} & A_{22} \end{bmatrix}^{-1} = \begin{bmatrix} A_{11}^{-1} & 0 \\ -A_{22}^{-1}A_{21}A_{11}^{-1} & A_{22}^{-1} \end{bmatrix}, \tag{2.13}$$

$$\begin{bmatrix} A_{11} & A_{12} \\ 0 & A_{22} \end{bmatrix}^{-1} = \begin{bmatrix} A_{11}^{-1} & -A_{11}^{-1}A_{12}A_{22}^{-1} \\ 0 & A_{22}^{-1} \end{bmatrix}. \tag{2.14}$$

[1] The nondiagonal block on the right-hand side of the equation is obtained by multiplying clockwise the inverses of diagonal blocks and the nondiagonal block of the original matrix.

It is worth noting that for a general nonsingular matrix A partitioned as (2.6), its inverse can be calculated using the decompositions (2.7) and (2.8), the inversion formulae (2.13) and (2.14) for block triangular matrices, and the property $(RST)^{-1} = T^{-1}S^{-1}R^{-1}$. So, there is no need to remember the rather complicated inversion formulae.

Moreover, from the identities

$$A - ABA = A(I - BA) = (I - AB)A, (I - AB)^{-1}(I - AB) = I,$$

we obtain the following properties of matrix inverse:

$$A(I - BA)^{-1} = (I - AB)^{-1}A, \tag{2.15}$$

$$(I - AB)^{-1} = I + (I - AB)^{-1}AB = I + A(I - BA)^{-1}B. \tag{2.16}$$

In deriving the second equation of (2.16), we have used the property of (2.15). Based on them, we can deduce a very useful identity. Assume that A, B, C, D are matrices of appropriate dimensions, with A and D being invertible. Then,

$$(A - BD^{-1}C)^{-1} = A^{-1} + A^{-1}B(D - CA^{-1}B)^{-1}CA^{-1} \tag{2.17}$$

holds. This identity is proved as follows. First of all, according to (2.15), (2.16) the left-hand side of (2.17) can be written as

$$(I - A^{-1}BD^{-1}C)^{-1}A^{-1} = [I + A^{-1}B(I - D^{-1}CA^{-1}B)^{-1}D^{-1}C]A^{-1}$$

since $A - BD^{-1}C = A(I - A^{-1}BD^{-1}C)$. Then, substitution of $(I - D^{-1}CA^{-1}B)^{-1} = (D - CA^{-1}B)^{-1}D$ gives the right-hand side of (2.17).

2.2 Elementary Linear Transformation of Matrix and Its Matrix Description

Elementary linear transformation of block matrices will be frequently used when we perform similarity transformation on the state-space realization of transfer function in subsequent chapters. Here, we illustrate the elementary transformation of matrix and its matrix description briefly. Matrix $A \in \mathbb{R}^{m\times n}$ is used as an example.

(1) Interchange of row i and row j

This is realized by premultiplying matrix A with the matrix T obtained from performing the same transformation on the identity matrix I_n.

For example, to exchange row 1 and row 3 of matrix $A = \begin{bmatrix} a_1 \\ a_2 \\ a_3 \end{bmatrix}$ (a_i is a row vector), we transform I_3 in the same way so that

$$I_3 = \begin{bmatrix} 1 & 0 & 0 \\ 0 & 1 & 0 \\ 0 & 0 & 1 \end{bmatrix} \to T = \begin{bmatrix} 0 & 0 & 1 \\ 0 & 1 & 0 \\ 1 & 0 & 0 \end{bmatrix} \Rightarrow TA = \begin{bmatrix} a_3 \\ a_2 \\ a_1 \end{bmatrix}.$$

Obviously, by exchanging column j and column i of T, we can restore it back to the identity matrix. So T^{-1} is obtained by the same transformation on the identity matrix.

(2) Interchange of column i and column j

This is achieved by transforming the identity matrix I_n in the same way to obtain a matrix T and then postmultiplying matrix A by T.

For example, to exchange column 1 and column 3 of matrix $A = [a_1\ a_2\ a_3]$ (a_i is a column vector), we transform I_3 in the same way so that

$$T = \begin{bmatrix} 0 & 0 & 1 \\ 0 & 1 & 0 \\ 1 & 0 & 0 \end{bmatrix} \Rightarrow AT = [a_3\ a_2\ a_1].$$

Further, exchanging row j and row i of T, the identity matrix is recovered. So T^{-1} is obtained by the same transformation on the identity matrix.

(3) Adding α times row i to row j

This is realized by transforming the identity matrix I_n in the same way to matrix T and then premultiplying matrix A with T.

For example, when adding 2 times row 1 of matrix $A = \begin{bmatrix} a_1 \\ a_2 \\ a_3 \end{bmatrix}$ to row 3, we transform I_3 in the same way so that

$$T = \begin{bmatrix} 1 & 0 & 0 \\ 0 & 1 & 0 \\ 2 & 0 & 1 \end{bmatrix} \Rightarrow TA = \begin{bmatrix} a_1 \\ a_2 \\ a_3 + 2a_1 \end{bmatrix}.$$

Moreover, the identity matrix is recovered by multiplying column j of T by $-\alpha$ and adding it to column i. So, T^{-1} is obtained by the same transformation on the identity matrix.

(4) Adding α times column i to column j

This is achieved by transforming the identity matrix I_n in the same way to matrix T and then postmultiplying matrix A with T.

For example, when adding 2 times column 1 to column 3 of matrix $A = [a_1\ a_2\ a_3]$, we perform the same transformation on I_3 so that

$$T = \begin{bmatrix} 1 & 0 & 2 \\ 0 & 1 & 0 \\ 0 & 0 & 1 \end{bmatrix} \Rightarrow AT = [a_1\ a_2\ 2a_1 + a_3].$$

Multiplying column j by $(-\alpha)$ and adding it to column i recover the identity matrix. So T^{-1} is obtained by the same transformation on the identity matrix.

(5) Multiplying row i by α

This is realized by premultiplying matrix A with the matrix T obtained from the same transformation on the identity matrix I_n.

When we want to multiply row 1 of matrix $A = \begin{bmatrix} a_1 \\ a_2 \\ a_3 \end{bmatrix}$ by 3, we perform the same transformation on I_3 so that

$$T = \begin{bmatrix} 3 & 0 & 0 \\ 0 & 1 & 0 \\ 0 & 0 & 1 \end{bmatrix} \Rightarrow TA = \begin{bmatrix} 3a_1 \\ a_2 \\ a_3 \end{bmatrix}.$$

When $\alpha \neq 0$, multiplying column i of T by α^{-1} yields the identity matrix. So T^{-1} can be obtained by the same transformation on the identity matrix.

(6) Multiplying column i by α

This is done by performing the same transformation on the identity matrix I_n to get a matrix T and then postmultiplying matrix A by T.

For example, multiplying column 3 of matrix $A = [a_1\ a_2\ a_3]$ by 2 is carried out as follows: transforming I_3 in the same way, then we have

$$T = \begin{bmatrix} 1 & 0 & 0 \\ 0 & 1 & 0 \\ 0 & 0 & 2 \end{bmatrix} \Rightarrow AT = [a_1\ a_2\ 2a_3].$$

Obviously, when $\alpha \neq 0$, multiplying row j of T by α^{-1} yields the identity matrix. So T^{-1} is obtained by the same transformation on the identity matrix.

These rules of elementary transformation also apply to block matrices. Note that *in multiplying a matrix to a block matrix, row blocks must be multiplied from the left and column blocks be multiplied from the right.*

2.3 Linear Vector Space

A *vector space* is a nonempty set composed of vectors with the same dimension whose elements share the same character. When all elements are real numbers, the n-dimensional vector space is denoted by $\mathbb{R}^n$. When all elements are complex numbers, the n-dimensional vector space is written as $\mathbb{C}^n$. $\mathbb{F}^n$ is used to denote both $\mathbb{R}^n$ and $\mathbb{C}^n$. The real vector space $\mathbb{R}^n$ has the following properties:

1. If $x, y \in \mathbb{R}^n$, then $x + y \in \mathbb{R}^n$.
2. If $x \in \mathbb{R}^n$ and $a \in \mathbb{R}$, then $ax \in \mathbb{R}^n$.

These two properties can be condensed into one (see Exercise 2.2):

$$ax + by \in \mathbb{R}^n \text{ for any } x, y \in \mathbb{R}^n \text{ and } a, b \in \mathbb{R}.$$

Similarly, the complex vector space $\mathbb{C}^n$ has the following property:

$$ax + by \in \mathbb{C}^n \text{ for any } x, y \in \mathbb{C}^n \text{ and } a, b \in \mathbb{C}.$$

$ax + by$ is a linear operation on the vectors x, y. The vector space is closed w.r.t. this linear operation, so it is called a *linear vector space*.

As a house is supported by pillars, a vector space is formed by a set of special vectors called *basis*. Moreover, as a wall is placed in the plane formed by some of the pillars, any vector lies in a plane spanned by some base vectors. Abstraction of this master and servant relationship leads to the concepts of linear independence and linear dependence. Further, just as a building can be separated into several rooms, a vector space can be decomposed into several subspaces. In addition, the features such as the length of vector and the angle between two vectors can be expressed by norm and inner product, respectively. In the following subsections, we describe these basic properties of vector space in detail.

2.3.1 Linear Independence

Let $x_1, x_2, \ldots, x_k \in \mathbb{F}^n$ and $\alpha_i \in \mathbb{F}$ $(i = 1, \ldots, k)$. Then, the vector

$$\alpha_1 x_1 + \cdots + \alpha_k x_k \tag{2.18}$$

is called a *linear combination* of $x_1, x_2, \ldots, x_k$[2], $\alpha_1, \alpha_2, \ldots, \alpha_k$ are called the combination coefficients.

A set of vectors $x_1, x_2, \ldots, x_k$ is said to be *linearly dependent* if there exists at least a nonzero scalar in $\alpha_1, \alpha_2, \ldots, \alpha_k$ satisfying

$$\alpha_1 x_1 + \alpha_2 x_2 + \cdots + \alpha_k x_k = 0. \tag{2.19}$$

Conversely, when the equation holds only for $\alpha_1 = \cdots = \alpha_k = 0$, the vectors $x_1, x_2, \ldots, x_k$ are said to be *linearly independent.*

When vectors $x_1, x_2, \ldots, x_k$ are linearly dependent, there must be a nonzero scalar α_i in the combination coefficients satisfying (2.19). Without loss of generality, we may set $\alpha_1 \neq 0$. Then, x_1 can be written as

$$x_1 = -\frac{\alpha_2}{\alpha_1} x_2 - \cdots - \frac{\alpha_k}{\alpha_1} x_k. \tag{2.20}$$

That is, x_1 is a linear combination of $x_2, \ldots, x_k$. Therefore, the so-called linear dependence of $x_1, x_2, \ldots, x_k$ means that there exists a vector that can be expressed as a linear combination of other vectors. On the contrary, in a set of linearly independent vectors, no vector can be expressed as a linear combination of other vectors. The concepts of linear independence and linear dependence are illustrated intuitively in Figure 2.1. This picture shows the linear dependence of u and $\{x_1, x_2\}$ as well as the linear independence of v and $\{x_1, x_2\}$.

Let us imagine picking up vectors one by one from a set of linearly independent vectors and doing linear combination. For one vector, its linear combinations form a straight line in the space. For two vectors, their linear combinations form a plane; the linear combinations of three vectors form a three-dimensional space (see Figure 2.2). As such, whenever a new linearly independent vector is added, the space formed by linear combination expands into a new direction[3].

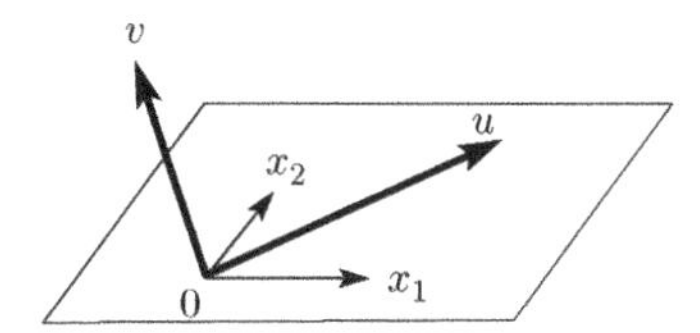

Figure 2.1 Linear independence and linear dependence

[2] Note that the elements of vector x_j and the coefficient α_i share the same nature. For example, when the elements of x_j are real numbers, the coefficient α_i must also be real.

[3] However, as shown in the next section, the number of linearly independent vectors cannot exceed the dimension of the linear space. Therefore, the expansion of space is constrained by the dimension.

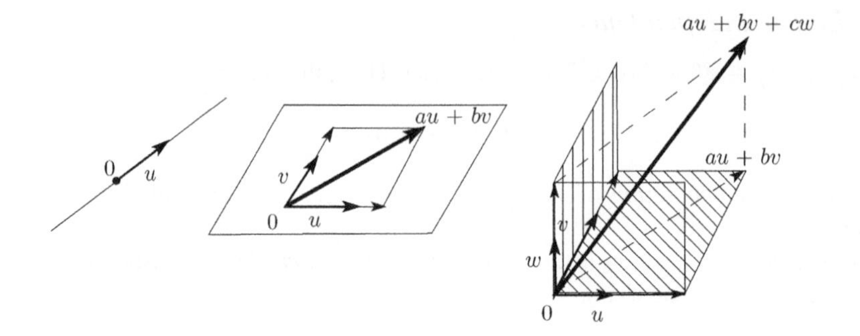

Figure 2.2 Number of linearly independent vectors and expansion of space

Example 2.1 *For vectors* $u = [1\ 0\ 2]^T$, $v = [0\ 2\ 1]^T$ *and* $w = [2\ 2\ 5]^T$, *there holds*

$$2u + v - w = 2\begin{bmatrix}1\\0\\2\end{bmatrix} + \begin{bmatrix}0\\2\\1\end{bmatrix} - \begin{bmatrix}2\\2\\5\end{bmatrix} = 0.$$

So they are linearly dependent. However, for u *and* v

$$au + bv = \begin{bmatrix}a\\2b\\2a+b\end{bmatrix} = 0 \Rightarrow a = b = 0,$$

so they are linearly independent. Similarly, u *and* w, *and* v *and* w *are also linearly independent, respectively. Geometrically,* u *and* v *are vectors pointing in different directions, their linear combination forming a plane.* w *is located in this plane. This is why* w *is linearly dependent on* $\{u, v\}$. *Moreover, the linear combination of* u *and* w *as well as the linear combination of* v *and* w *also forms a plane, respectively.*

2.3.2 Dimension and Basis

The next question is, how many linearly independent vectors there are in $\mathbb{F}^n$? Before answering this question, let us look at an example. Here, e_i denotes a vector whose elements are all 0 except that the ith one is 1.

Example 2.2 *In the three-dimensional space* $\mathbb{R}^3$, *any vector* u *can be expressed as*

$$u = \begin{bmatrix}x\\y\\z\end{bmatrix} = x\begin{bmatrix}1\\0\\0\end{bmatrix} + y\begin{bmatrix}0\\1\\0\end{bmatrix} + z\begin{bmatrix}0\\0\\1\end{bmatrix} = xe_1 + ye_2 + ze_3. \tag{2.21}$$

That is, u *can be expressed as a linear combination of* (e_1, e_2, e_3). *Clearly,* (e_1, e_2, e_3) *are linearly independent. Therefore, the number of linearly independent vectors in* $\mathbb{R}^3$ *is 3, equal to the number of elements in a vector.*

In fact, vectors (e_1, e_2, e_3) are unit vectors along the x-axis, y-axis, and z-axis, respectively, also known as the natural basis of $\mathbb{R}^3$.

Extending this example, we see that any vector in $\mathbb{F}^n$ must be a linear combination of the following n vectors:

$$e_1, \ldots, e_i, \ldots, e_n.$$

Thus, linearly independent vectors contained in an n-dimensional linear space are no more than n. Further, since these vectors are linearly independent, the number of linearly independent vectors is n. In summary, we get the following theorem.

Theorem 2.1 *The dimension of linear vector space $\mathbb{F}^n$ equals the largest number of linearly independent vectors in it.*

Secondly, a set formed by the linear combinations of $\{u_1, u_2, \ldots, u_k\}$ is called a space spanned by them, denoted by

$$\text{span}\{u_1, u_2, \ldots, u_k\} := \{x \mid x = \alpha_1 u_1 + \cdots + \alpha_k u_k, \alpha_i \in \mathbb{F}\}. \tag{2.22}$$

Moreover, the set of vectors $\{u_1, u_2, \ldots, u_n\} \subset \mathbb{F}^n$ is called a *basis* of $\mathbb{F}^n$ if its elements are all linearly independent. The following theorem shows that $\mathbb{F}^n$ can be spanned by this basis.

Theorem 2.2 *Let $\{u_1, u_2, \ldots, u_n\} \subset \mathbb{F}^n$ be a basis of $\mathbb{F}^n$. Then, $\mathbb{F}^n = \text{span}\{u_1, u_2, \ldots, u_n\}$.*

Proof. $x \in \mathbb{F}^n$ whenever $x \in \text{span}\{u_1, u_2, \ldots, u_n\}$. Then, apparently $\text{span}\{u_1, u_2, \ldots, u_n\} \subset \mathbb{F}^n$. Set a square matrix as $[u_1 \ u_2 \ \cdots u_n] = U$. Due to the linear independence of base vectors, the algebraic equation

$$0 = [u_1 \ u_2 \ \cdots u_n]c = Uc, \quad c \in \mathbb{F}^n$$

only has a trivial solution $c = 0$. So U must be nonsingular. Then, for any $x \in \mathbb{F}^n$, by setting $\alpha = U^{-1}x$, we have $x = U\alpha = \sum \alpha_i u_i$ ($\alpha_i \in \mathbb{F}$ is an element of α). So $\mathbb{F}^n \subset \text{span}\{u_1, u_2, \ldots, u_n\}$, which ends the proof. •

According to this theorem, any vector $x \in \mathbb{F}^n$ can be described as

$$x = \alpha_1 u_1 + \cdots + \alpha_n u_n \tag{2.23}$$

by using a basis $\{u_1, \ldots, u_n\}$ and scalars α_i $(i = 1, \ldots, n)$. Here, $(\alpha_1, \ldots, \alpha_n)$ is called the *coordinate* of vector x on the basis $\{u_1, \ldots, u_n\}$. Geometrically, α_i is the projection of x in the direction of base vector u_i (see Example 2.2).

2.3.3 Coordinate Transformation

Let us consider the relation of two bases $\{v_1, \dots, v_n\}$, $\{u_1, \dots, u_n\}$ of $\mathbb{F}^n$. First, according to the definition of basis, there is $t_{ij} \in \mathbb{F}$ such that v_j is described as

$$v_j = t_{1j}u_1 + \cdots + t_{nj}u_n = [u_1 \ \cdots \ u_n] \begin{bmatrix} t_{1j} \\ \vdots \\ t_{nj} \end{bmatrix}.$$

Aligning all v_j horizontally, we get

$$[v_1 \ \cdots \ v_n] = [u_1 \ \cdots \ u_n]T, \quad T = (t_{ij}).$$

Owing to the linear independence of basis, matrices $[v_1 \ \cdots \ v_n]$, $[u_1 \ \cdots \ u_n]$ must be nonsingular. Therefore, the matrix T transforming $\{u_1, \dots, u_n\}$ to $\{v_1, \dots, v_n\}$ must also be nonsingular. For any vector $x \in \mathbb{F}^n$, let its coordinates on basis $\{u_1, \dots, u_n\}$ be $(\alpha_1, \dots, \alpha_n)$, and the coordinates on basis $\{v_1, \dots, v_n\}$ be $(\beta_1, \dots, \beta_n)$. Then

$$x = [u_1 \ \cdots \ u_n] \begin{bmatrix} \alpha_1 \\ \vdots \\ \alpha_n \end{bmatrix} = [v_1 \ \cdots \ v_n]T^{-1} \begin{bmatrix} \alpha_1 \\ \vdots \\ \alpha_n \end{bmatrix} = [v_1 \ \cdots \ v_n] \begin{bmatrix} \beta_1 \\ \vdots \\ \beta_n \end{bmatrix}.$$

Therefore, between the coordinates on these two bases, there holds

$$\begin{bmatrix} \beta_1 \\ \vdots \\ \beta_n \end{bmatrix} = T^{-1} \begin{bmatrix} \alpha_1 \\ \vdots \\ \alpha_n \end{bmatrix}. \tag{2.24}$$

This equation is called a *coordinate transformation.*

Example 2.3 *Consider a vector* $x = [1 \ 1]^T = e_1 + e_2$ *whose coordinate is* $(1, 1)$ *on the natural basis* $e_1 = [1 \ 0]^T$, $e_2 = [0 \ 1]^T$. *In a new coordinate system that rotates 45° counterclockwise, the base vectors become* $u_1 = [\cos 45° \sin 45°]^T$, $u_2 = [-\sin 45° \cos 45°]^T$. *The coordinate of* x *turns out to be* $(\sqrt{2}, 0)$ *(show it by drawing a figure). That is, this vector can be expressed as* $x = \sqrt{2}u_1$.

From the viewpoint of coordinate transformation, since $u_1 = \cos 45° e_1 + \sin 45° e_2$, $u_2 = -\sin 45° e_1 + \cos 45° e_2$, *the transformation matrix* T *from* (e_1, e_2) *to* (u_1, u_2) *is*

$$T = \begin{bmatrix} \cos 45° & -\sin 45° \\ \sin 45° & \cos 45° \end{bmatrix} \Rightarrow T^{-1} = \begin{bmatrix} \frac{1}{\sqrt{2}} & \frac{1}{\sqrt{2}} \\ -\frac{1}{\sqrt{2}} & \frac{1}{\sqrt{2}} \end{bmatrix}.$$

Therefore, the coordinate of x *in the new coordinate system becomes* $T^{-1}[1 \ 1]^T = [\sqrt{2} \ \ 0]^T$.

2.4 Norm and Inner Product of Vector

In this section, we discuss the issues of vector size as well as the relation of directions between vectors.

2.4.1 Vector Norm

The size of real numbers or complex numbers is measured by their absolute values. Now, how should we measure the size of a vector? To answer this question, let us review the notion of distance in Euclidean space. The distance between a point $P(x, y, z)$ and the origin in the three-dimensional Euclidean space is defined as

$$d(P) = \sqrt{x^2 + y^2 + z^2}. \tag{2.25}$$

As is well known, this denotes the length of vector $u = [x\ y\ z]^T$. The size of vector (as well as matrix and function described later) is called the *norm* and denoted by $\|\cdot\|$ (Figure 2.3). So $d(P)$ can also be written as $d(P) = \|u\|$. After investigating the property of function $\|u\| = d(P)$ carefully, we see that it satisfies the following conditions (refer to Exercise 2.5):

1. $\|u\| \geq 0$ (positivity)
2. $\|u\| = 0$ iff $u \equiv 0$ (positive definiteness)
3. $\|\alpha u\| = |\alpha|\|u\|$ for any scalar $\alpha \in \mathbb{R}$ (homogeneity)
4. $\|u + v\| \leq \|u\| + \|v\|$ for any vectors u, v (triangle inequality)

These properties of Euclidean distance form the starting point for treating the norms of vector or function. Hereafter, a *scalar real-valued function* defined in any vector space (as well as matrix space, function space) is called a norm of the corresponding space so long as it satisfies all the aforementioned properties, and is used to measure the size of vector (matrix, function). It is worth noting that for real-valued vector (matrix, function), the property (3) holds for $\alpha \in \mathbb{R}$. However, for complex-valued vector (matrix, function), the property (3) holds for $\alpha \in \mathbb{C}$.

Norm is simply an extension of the distance notion in three-dimensional Euclidean space, and its property is identical to that of the distance in Euclidean space. Therefore, we may imagine intuitively any norm in terms of Euclidean distance.

For a vector $u \in \mathbb{F}^n$, the frequently used norms are listed as follows:

1-norm $\|u\|_1 = \sum_{i=1}^n |u_i|$

2-norm $\|u\|_2 = \sqrt{\sum_{i=1}^n |u_i|^2}$

Infinity-norm $\|u\|_\infty = \max_{1\leq i\leq n} |u_i|$

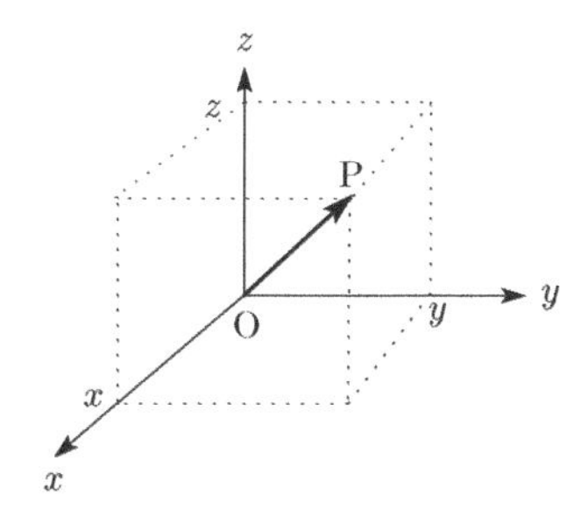

Figure 2.3 Distance in Euclidean space

Example 2.4 *Let us prove that the function* $f(u) = \sum_{i=1}^{n} |u_i|$ *meets the condition of norm.*

Obviously, $f(u) \geq 0$. *Secondly*

$$f(u) = 0 \Leftrightarrow |u_i| = 0 \ \forall i \Leftrightarrow u_i = 0 \ \forall i \Leftrightarrow u = 0$$

holds. It is also easy to see that

$$f(\alpha u) = \sum_{i=1}^{n} |\alpha u_i| = |\alpha| \sum_{i=1}^{n} |u_i| = |\alpha| f(u).$$

Further,

$$f(u+v) = \sum_{i=1}^{n} |u_i + v_i| \leq \sum_{i=1}^{n} (|u_i| + |v_i|) = f(u) + f(v)$$

is true because the triangle inequality $|u_i + v_i| \leq |u_i| + |v_i|$ *holds for scalars. So, this* $f(u)$ *is indeed a norm.*

2.4.2 Inner Product of Vectors

In vector space, there is a direction relation between vectors, that is, the angle between them, in addition to their sizes. Here, we consider how to describe this direction relation.

In the two-dimensional Euclidean space $\mathbb{R}^2$, the angle between two vectors is defined as the geometric angle between them (refer to Figure 2.4). Suppose that the coordinates of two vectors are $u_i = [x_i \ y_i]^T$ $(i = 1, 2)$, respectively, in Figure 2.4. Then, the angle between them can be calculated based on the cosine rule

$$\|u_1 - u_2\|_2^2 = \|u_1\|_2^2 + \|u_2\|_2^2 - 2\|u_1\|_2\|u_2\|_2 \cos\theta. \tag{2.26}$$

Expanding the left-hand side according to the definition of 2−norm, we get

$$\cos\theta = \frac{x_1x_2 + y_1y_2}{\|u_1\|_2\|u_2\|_2} = \frac{u_1^T u_2}{\|u_1\|_2\|u_2\|_2}. \tag{2.27}$$

$u_1^T u_2$ is a function mapping two vectors into a scalar, called *inner product* and denoted by

$$\langle u_1, u_2 \rangle := u_1^T u_2. \tag{2.28}$$

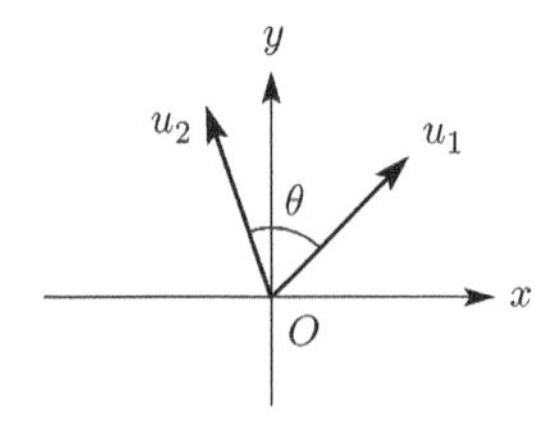

Figure 2.4 Inner product and angle

Then, we have

$$\cos\theta = \frac{\langle u_1, u_2\rangle}{\|u_1\|_2\|u_2\|_2}, \quad \theta \in [0, \pi]. \tag{2.29}$$

Therefore, inner product and angle have a one-to-one relationship. In vector spaces with higher dimensions as well as matrix and function spaces to be described later on, the angle cannot be drawn. So, it is necessary to use the inner product to define the angle between the elements of each space.

Generalizing the preceding example, we define the inner product between real-valued vectors $u, v \in \mathbb{R}^n$ as

$$\langle u, v\rangle := u^T v. \tag{2.30}$$

Similarly, the inner product between complex-valued vectors $u, v \in \mathbb{C}^n$ is defined as

$$\langle u, v\rangle := u^* v \tag{2.31}$$

which takes complex value. Further, the angle between two vectors u, v is defined via[4]

$$\cos\theta = \frac{\Re(\langle u, v\rangle)}{\|u\|_2\|v\|_2}, \theta \in [0, \pi]. \tag{2.32}$$

According to this definition of inner product, when $\langle u, v\rangle = 0$ the angle between u, v is 90°. In this case, we say that u, v are *orthogonal*, denoted as $u \perp v$.

Example 2.5 *Given vectors*

$$u = \begin{bmatrix}1\\1\end{bmatrix}, \quad v = \begin{bmatrix}-1\\1\end{bmatrix}, \quad w = \begin{bmatrix}1\\0\end{bmatrix}.$$

Let the angle between u, v be ϕ and the angle between u, w be θ. Calculation based on inner product yields

$$\cos\phi = \frac{u^T v}{\|u\|_2\|v\|_2} = 0 \;\Rightarrow\; \phi = 90^\circ,$$

$$\cos\theta = \frac{u^T w}{\|u\|_2\|w\|_2} = \frac{1}{\sqrt{2}} \;\Rightarrow\; \theta = 45^\circ.$$

We can verify the correctness of the calculation by drawing a figure.

The inner product defined in vector space $\mathbb{F}^n$ has the following properties. They can be validated according to the definition:

1. $\langle x, \alpha y + \beta z\rangle = \alpha\langle x, y\rangle + \beta\langle x, z\rangle$ holds for any scalars $\alpha, \beta \in \mathbb{F}$.

[4] As is well known, in geometry a complex number $a + jb$ can be expressed as a vector $[a \; b]^T$ in two-dimensional real space. Therefore, a complex-valued vector u and a real-valued vector $\begin{bmatrix}\Re(u)\\ \Im(u)\end{bmatrix}$ are one to one. It is easy to verify that $\Re(\langle u, v\rangle) = [\Re(u) \; \Im(u)] \begin{bmatrix}\Re(v)\\ \Im(v)\end{bmatrix}$. This is why the angle between complex vectors u, v is defined as such.

2. $\langle x, y\rangle = \overline{\langle y, x\rangle}$.
3. $\langle x, x\rangle \geq 0$, and $\langle x, x\rangle = 0$ iff $x = 0$.

$\sqrt{\langle u, u\rangle}$ satisfies all conditions of norm and is called an *induced norm*. In fact, there holds $\sqrt{\langle u, u\rangle} = \|u\|_2$. The inner product and induced norm have the following properties.

Theorem 2.3 *For any $u, v \in \mathbb{F}^n$, the following statements are true.*

1. $|\langle u, v\rangle| \leq \|u\|_2\|v\|_2$ *(Cauchy–Schwarz inequality). The equality holds only when $u = \alpha v$ (α is a constant), $u = 0$ or $v = 0$.*
2. $\|u + v\|_2^2 + \|u - v\|_2^2 = 2\|u\|_2^2 + 2\|v\|_2^2$ *(parallelogram law).*
3. $\|u + v\|_2^2 = \|u\|_2^2 + \|v\|_2^2$ *when $u \perp v$ (Pythagoras' law).*

2.5 Linear Subspace

2.5.1 Subspace

All linear combinations of vectors $x_1, x_2, \ldots, x_k \in \mathbb{F}^n$

$$\text{span}\{x_1, x_2, \ldots, x_k\} := \{x \mid x = \alpha_1 x_1 + \cdots + \alpha_k x_k, \alpha_i \in \mathbb{F}\} \tag{2.33}$$

is called the *subspace* spanned by $x_1, x_2, \ldots, x_k$. In the subspace $S := \text{span}\{x_1, x_2, \ldots, x_k\}$, any $u, v \in S$ can be written as

$$u = \alpha_1 x_1 + \cdots + \alpha_k x_k, v = \beta_1 x_1 + \cdots + \beta_k x_k \tag{2.34}$$

with combination coefficients $\alpha_i, \beta_i \in \mathbb{F}(i = 1, \ldots, k)$. So, for any $a, b \in \mathbb{F}$, there holds

$$au + bv = (a\alpha_1 + b\beta_1)x_1 + \cdots + (a\alpha_k + b\beta_k)x_k. \tag{2.35}$$

Further, since $a\alpha_i + b\beta_i \in \mathbb{F}$, we have $au + bv \in S$. In other words, the subspace S satisfies the property of linear space. Therefore, a subspace itself is also a linear space. Note that a subspace must contain the *origin*.

Moreover, in the subspace S of $\mathbb{F}^n$, a set of linearly independent vectors with the largest possible number is called a *basis* of S[5]. The number of vectors contained in the basis of S is called the *dimension* of the subspace S, denoted as $\dim(S)$. In other words, the dimension of subspace S is equal to the number of linearly independent vectors contained in S.

Example 2.6 *In Example 2.1, vectors $u = [1\ 0\ 2]^T$ and $v = [0\ 2\ 1]^T$ are linearly independent, so they span a two-dimensional subspace in $\mathbb{R}^3$. This can be verified from the fact that* $\text{span}\{u, v\}$ *forms the gray plane in Figure 2.5.*

On the other hand, the vector $w = [2\ 2\ 5]^T$ is linearly dependent on u and v so that $\text{span}\{u, v, w\} = \text{span}\{u, v\}$. *This can also be validated from the fact that w locates in the gray plane shown in Figure 2.5.*

[5] Note that the vectors contained in a basis must be linearly independent. On the other hand, a vector set $\{x_1, \ldots, x_k\}$ that spans $S = \text{span}\{x_1, \ldots, x_k\}$ may have linearly dependent vectors.

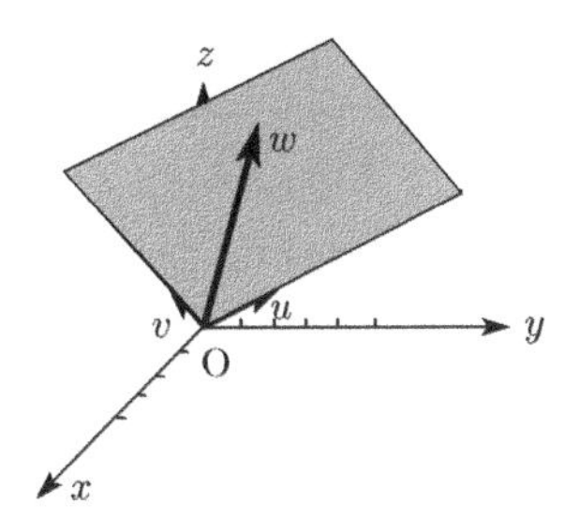

Figure 2.5 Linear subspace and basis

2.6 Matrix and Linear Mapping

2.6.1 Image and Kernel Space

Multiplying vector $x \in \mathbb{F}^n$ to matrix $A \in \mathbb{F}^{m \times n}$, we get a new vector

$$y = Ax \in \mathbb{F}^m. \tag{2.36}$$

That is, the matrix A can be regarded as a mapping from linear space $\mathbb{F}^n$ to linear space $\mathbb{F}^m$[6]

$$A : \ \mathbb{F}^n \mapsto \mathbb{F}^m. \tag{2.37}$$

This mapping obviously has the following linearity property:

$$A(ax + by) = a(Ax) + b(Ay) \ \forall a, \ \ b \in \mathbb{F}, x, y \in \mathbb{F}^n. \tag{2.38}$$

So it is called a *linear mapping*.

The *image* (range) of linear mapping A is denoted by

$$\mathrm{Im}A := \{y \in \mathbb{F}^m \ | \ y = Ax, x \in \mathbb{F}^n\}. \tag{2.39}$$

$\mathrm{Im}A$ is a subspace of $\mathbb{F}^m$ (prove it). Also, it is conceivable that some nonzero vector $x \neq 0$ in the domain $\mathbb{F}^n$ may be mapped to the origin of the image $\mathbb{F}^m$, that is, $Ax = 0$. The set of such vectors is called the *kernel space* (null space) of A and is denoted as

$$\mathrm{Ker}A := \{x \in \mathbb{F}^n \ | \ Ax = 0\}. \tag{2.40}$$

It is easy to see that $\mathrm{Ker}A$ *is a subspace of the domain* $\mathbb{F}^n$ (refer to Exercise 2.6). The relationship between these subspaces is illustrated in Figure 2.6.

Example 2.7 *Mapping vector* $x = [x_1 \ x_2 \ x_3]^T$ *with matrix*

$$A = \begin{bmatrix} 1 & 1 & 0 \\ 0 & 0 & 1 \end{bmatrix},$$

[6] In fact, A is the matrix description of this linear mapping.

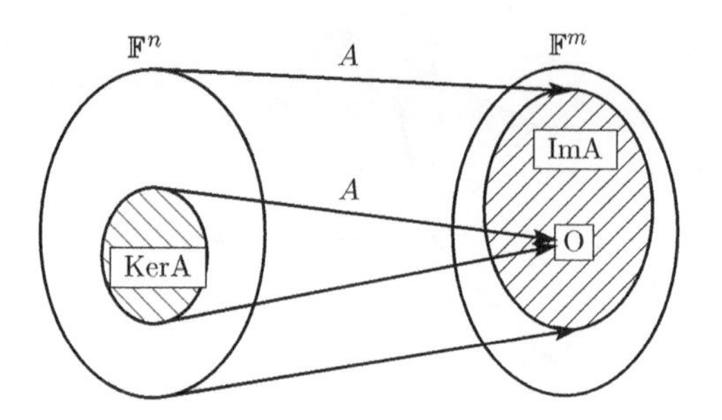

Figure 2.6 Image and kernel space

the map y is

$$y = Ax = \begin{bmatrix} x_1 + x_2 \\ x_3 \end{bmatrix} = (x_1 + x_2) \begin{bmatrix} 1 \\ 0 \end{bmatrix} + x_3 \begin{bmatrix} 0 \\ 1 \end{bmatrix}.$$

So the image becomes

$$\text{Im}A = \text{span}\left\{ \begin{bmatrix} 1 \\ 0 \end{bmatrix}, \begin{bmatrix} 0 \\ 1 \end{bmatrix} \right\} = \mathbb{R}^2. \tag{2.41}$$

Secondly, since

$$Ax = 0 \Rightarrow x_2 = -x_1, x_3 = 0 \Rightarrow x = x_1[1 \quad -1 \quad 0]^T,$$

the kernel space is given by

$$\text{Ker}A = \text{span}\{[1 \quad -1 \quad 0]^T\}. \tag{2.42}$$

Further, the following relationships between dimensions are true:

$$\dim(\text{Ker}A) + \dim(\text{Im}A) = n, \quad \dim(\text{Im}A) = \dim\,[(\text{Ker}A)^{\perp}]. \tag{2.43}$$

Note that $(\text{Ker}A)^{\perp}$ is also a subspace of $\mathbb{F}^n$.

Lemma 2.1 *Let $a_i (i = 1, 2, \ldots, n)$ be a column vector of matrix $A \in \mathbb{F}^{m \times n}$. Then*

$$\text{Im}A = \text{span}\{a_1, a_2, \ldots, a_n\}$$

holds and

$$\begin{aligned} \dim(\text{Im}A) &= \text{maximum number of linearly independent columns of } A \\ &= \text{maximum number of linearly independent rows of } A. \end{aligned}$$

Clearly, the dimension of image Im A is no more than the dimension of domain $\mathbb{F}^n$ (Figure 2.7). That is,

$$\dim(\text{Im}A) \leq n = \dim(\text{domain}) = \text{number of columns of } A. \tag{2.44}$$

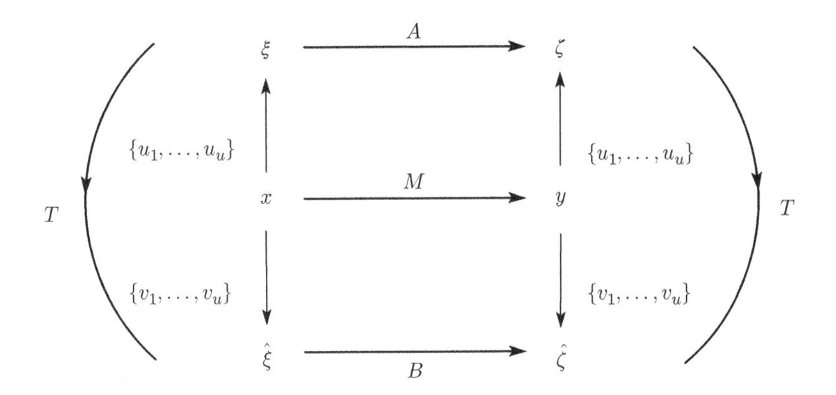

Figure 2.7 Linear transformation described by mapping matrix

2.6.2 *Similarity Transformation of Matrix*

Consider a linear mapping M that maps vector $x \in \mathbb{F}^n$ into $y \in \mathbb{F}^n$. Let the coordinate of x in the basis $\{u_i\}$ be ξ, the coordinate of y be ζ, and the matrix description of mapping M be A. Then, $\zeta = A\xi$ holds. Similarly, when the coordinate of x in the basis $\{v_i\}$ is $\hat{\xi}$, the coordinate of y is $\hat{\zeta}$, and the matrix description of mapping M is B, then $\hat{\zeta} = B\hat{\xi}$. Further, assume that the matrix transforming basis $\{u_i\}$ into basis $\{v_i\}$ is T. Then, according to (2.24), the relation between the coordinates on each basis, we have

$$\hat{\xi} = T^{-1}\xi, \quad \hat{\zeta} = T^{-1}\zeta \;\Rightarrow\; T\hat{\zeta} = \zeta = A\xi = AT\hat{\xi} \;\Rightarrow\; \hat{\zeta} = T^{-1}AT\hat{\xi}.$$

Therefore, we obtain the following relation:

$$B = T^{-1}AT. \tag{2.45}$$

Matrices A, B both express the same linear mapping; only the coordinate systems are different. For this reason, the preceding equation is called *similarity transformation.*

Example 2.8 *Consider the mapping that rotates vector u an angle of θ counterclockwise in $\mathbb{R}^2$. On the natural basis $\{e_1, e_2\}$, let the coordinate of vector u be $\xi = [r\cos\phi \;\; r\sin\phi]^T$. Then, the coordinate ζ of its image v is obtained as*

$$\zeta = \begin{bmatrix} r\cos(\phi+\theta) \\ r\sin(\phi+\theta) \end{bmatrix} = r\begin{bmatrix} \cos\phi\cos\theta - \sin\phi\sin\theta \\ \sin\phi\cos\theta + \cos\phi\sin\theta \end{bmatrix} = \begin{bmatrix} \cos\theta & -\sin\theta \\ \sin\theta & \cos\theta \end{bmatrix}\xi$$

from Figure 2.8. So, the matrix description of this mapping is

$$A = \begin{bmatrix} \cos\theta & -\sin\theta \\ \sin\theta & \cos\theta \end{bmatrix}.$$

Considering the same mapping on another basis $\{u_1, u_2\} = \{2e_1, 3e_2\}$, the transformation matrix between the bases is obviously

$$T = \begin{bmatrix} 2 & 0 \\ 0 & 3 \end{bmatrix}.$$

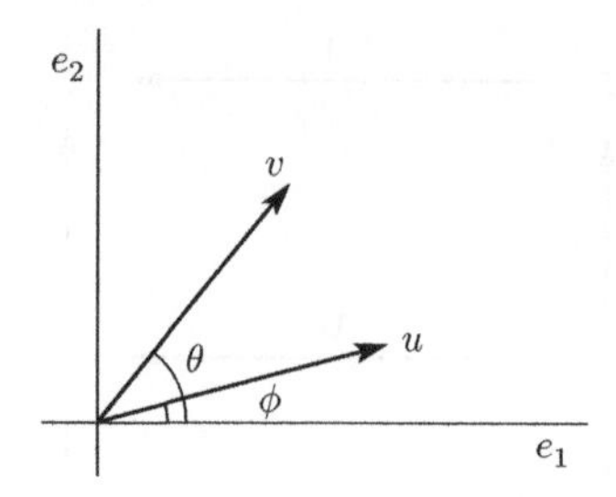

Figure 2.8 Rotation of vector

Therefore, the matrix description of this mapping on the basis $\{u_1, u_2\}$ *is equal to* $B = T^{-1}AT$.

2.6.3 Rank of Matrix

The rank of matrix A is defined as the dimension of its image, that is,

$$\text{rank}(A) = \dim\ (\text{Im}A). \tag{2.46}$$

So

$$\text{rank}(A) = \text{number of linearly independent columns of } A \tag{2.47}$$

$$= \text{number of linearly independent rows of } A. \tag{2.48}$$

For instance, in Example 2.7 $\text{rank}(A) = 2$.

When matrix $A \in \mathbb{F}^{m\times n}$ satisfies $m \leq n$ (wide matrix) and $\text{rank}(A) = m$, it has *full row rank*. Similarly, when it satisfies $n \leq m$ (tall matrix) and $\text{rank}(A) = n$, it has *full column rank*. Also, a square matrix with full rank is called a *nonsingular matrix*.

The following lemma holds w.r.t. matrix multiplication.

Lemma 2.2 *The following statements hold:*

1. *For matrix* $A \in \mathbb{F}^{m\times n}$, $\text{rank}(A) = \text{rank}(AT) = \text{rank}(PA)$ *when T and P are nonsingular matrices with appropriate dimensions.*
2. *(Sylvester's inequality) Assume that* $A \in \mathbb{F}^{m\times n}$, $B \in \mathbb{F}^{n\times k}$. *The following relationship is true:*

$$\text{rank}(A) + \text{rank}(B) - n \leq \text{rank}(AB) \leq \min\{\text{rank}(A),\ \text{rank}(B)\}.$$

Moreover, there hold the following relations between the image, kernel space, and determinant of a matrix.

Theorem 2.4 *For square matrix* $A \in \mathbb{F}^{n\times n}$, *the following statements are equivalent:*

1. $\text{Ker}A \neq \{0\}$.

2. $\text{Im}A \neq \mathbb{F}^n$.
3. $\det(A) = 0$.

From this theorem, we see that $\det(A) = 0$ is equivalent to the existence of linearly dependent column/row in A.

Example 2.9 *In matrix*

$$A = \begin{bmatrix} 1 & 0 & 2 \\ 1 & 0 & 2 \\ 0 & 1 & 1 \end{bmatrix} := [a_1 \ a_2 \ a_3],$$

as $a_3 = 2a_1 + a_2$, *we have*

$$\begin{aligned} \text{Im}A &= \text{span}\{a_1, a_2\} \neq \mathbb{R}^3, \quad \det(A) = 0 \\ \text{Ker}A &= \text{span}\{[2 \ \ 1 \ \ -1]^T\} \neq \{0\}. \end{aligned}$$

2.6.4 Linear Algebraic Equation

Let us consider the following linear algebraic equation:

$$Ax = b \tag{2.49}$$

where $A \in \mathbb{F}^{n\times m}$ and $b \in \mathbb{F}^n$ are given matrix and vector, respectively, $x \in \mathbb{F}^m$ is the unknown vector. The left-hand side of this equation can be interpreted as a linear combination of the columns of matrix A with elements of vector x as the coefficients. Therefore, if this equation has a solution, the vector b can be written as a linear combination of the columns of A.

The following theorem is very famous.

Theorem 2.5 *For the linear equation (2.49), the following statements are equivalent:*

1. *There exists a solution* $x \in \mathbb{F}^m$.
2. $b \in \text{Im}A$.
3. $\text{rank}[A \ \ b] = \text{rank}(A)$.
4. $\text{Ker}A^* \subset \text{Ker}b^*$.

   Further, the following conclusions are true when a solution exists:
5. *Given a special solution* x_0, *all solutions are characterized by the set*

$$x_0 + \text{Ker}A = \{x_0 + y \mid y \in \text{Ker}A\}. \tag{2.50}$$

6. *A unique solution exists if the matrix A has full column rank.*

Example 2.10 *Given the following matrix and vectors*

$$A = \begin{bmatrix} 1 & 0 & 2 \\ 1 & 0 & 2 \\ 0 & 1 & 1 \end{bmatrix}, \quad b_1 = \begin{bmatrix} 0 \\ 1 \\ 0 \end{bmatrix}, \quad b_2 = \begin{bmatrix} 1 \\ 1 \\ 2 \end{bmatrix}.$$

Expanding $Ax = b_1$, we obtain

$$x_1 + 2x_3 = 0, \quad x_1 + 2x_3 = 1, \quad x_2 + x_3 = 0.$$

The first two equations are contradictory, so no solution exists. This corresponds to $b_1 \notin \mathrm{Im}A$. On the other hand, expansion of $Ax = b_2$ about vector $b_2 \in \mathrm{Im}A$ yields

$$\begin{cases} x_1 + 2x_3 = 1 \\ x_1 + 2x_3 = 1 \\ x_2 + x_3 = 2 \end{cases} \Rightarrow \begin{bmatrix} x_1 \\ x_2 \\ x_3 \end{bmatrix} = \begin{bmatrix} 1 - 2x_3 \\ 2 - x_3 \\ x_3 \end{bmatrix}, x_3 \neq 0. \tag{2.51}$$

Meanwhile, $x_0 = [-1\ 1\ 1]^T$ is a special solution of $Ax = b_2$. Together with $\mathrm{Ker}A = \mathrm{span}\{[2\ 1\ \ -1]^T\}$, we obtain the general solution

$$x = x_0 + y = \begin{bmatrix} -1 \\ 1 \\ 1 \end{bmatrix} + \alpha \begin{bmatrix} 2 \\ 1 \\ -1 \end{bmatrix} = \begin{bmatrix} -1 + 2\alpha \\ 1 + \alpha \\ 1 - \alpha \end{bmatrix} \tag{2.52}$$

according to Theorem 2.5 where α is an arbitrary real number. This agrees with (2.51) with $x_3 = 1 - \alpha$ in the solution.

2.7 Eigenvalue and Eigenvector

When square matrix $A \in \mathbb{F}^{n \times n}$, scalar $\lambda \in \mathbb{C}$, and vector $u \in \mathbb{C}^n$ satisfy

$$Au = \lambda u, \quad u \neq 0, \tag{2.53}$$

λ is called the *eigenvalue* of A, and u the *eigenvector* of A. It is worth noting that even if the matrix A is real, its eigenvalue and eigenvector are not necessarily real.

Further, when (2.53) is true, there also holds

$$\det(\lambda I - A) = 0. \tag{2.54}$$

Therefore, the eigenvalues can be equivalently regarded as the roots of the determinant $\det(\lambda I - A)$. $\det(\lambda I - A)$ is called the *characteristic polynomial* of A, and its roots are called the *characteristic roots*. Since the degree of the characteristic polynomial is n, there are n eigenvalues. The set of all eigenvalues of A is denoted by

$$\sigma(A) = \{\lambda_1, \ldots, \lambda_n\}. \tag{2.55}$$

Example 2.11 *Compute the eigenvalues and eigenvectors of real matrix*

$$A = \begin{bmatrix} 0 & 1 \\ 3 & 2 \end{bmatrix}.$$

Solution The characteristic polynomial of A is

$$\det(\lambda I - A) = \begin{vmatrix} \lambda & -1 \\ -3 & \lambda - 2 \end{vmatrix} = (\lambda + 1)(\lambda - 3).$$

Therefore, its eigenvalues are $\lambda = -1, 3$. The eigenvectors can be calculated by the following method. For $\lambda_1 = -1$, set the corresponding eigenvector as $u = [\alpha\ \beta]^T$ which satisfies

$$(\lambda_1 I - A)u = \begin{bmatrix} -1 & -1 \\ -3 & -3 \end{bmatrix} \begin{bmatrix} \alpha \\ \beta \end{bmatrix} = 0.$$

Solving this simultaneous equation, we get $\beta = -\alpha$. So, when $\alpha = 1$ is selected, one eigenvector is obtained as $u = [1\ \ -1]^T$. Obviously, if α is nonzero, αu is also an eigenvector. That is, eigenvector is not unique.

Similarly, eigenvector $v = [\gamma\ \delta]^T$ corresponding to the eigenvalue $\lambda_2 = 3$ is obtained by solving simultaneous equation $(\lambda_2 I - A)v = 0$. One eigenvector is $v = [1/3\ \ 1]^T$. The eigenvectors in this example are shown in Figure 2.9.

From the calculations, we have

$$A[u\ v] = [Au\ Av] = [u\ v] \begin{bmatrix} -1 & 0 \\ 0 & 3 \end{bmatrix}.$$

Therefore, when a vector $x = c_1 u + c_2 v$ in the subspace

$$\text{span}\{u\ v\} := \{x \mid x = c_1 u + c_2 v,\ c_i \in \mathbb{R}\}$$

is mapped by matrix A, there holds $Ax = c_1 Au + c_2 Av = -c_1 u + 3c_2 v \in \text{span}\{u\ v\}$. This property is also noteworthy. ▽

Viewing the matrix A from the angle of mapping[7], an eigenvector is a special vector in the domain of A. Being mapped by A, this vector remains on the same straight line; only the length changes $|\lambda|$ times. In other words, the absolute value of an eigenvalue of A may be interpreted as the amplification rate of the corresponding eigenvector. When the eigenvalue is a real number, its sign indicates the directional relation between the image and the eigenvector. Figure 2.9 shows such a relation in the forgoing example.

When there exist multiple eigenvalues and the number of eigenvectors is less than that of multiple eigenvalues, we need to extend the notion of eigenvector, that is, to consider *generalized eigenvector*. For example, when matrix A has r multiple eigenvalues λ and there are

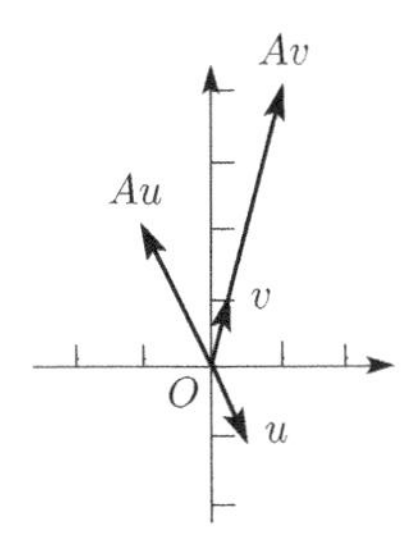

Figure 2.9 Eigenvalues and eigenvectors

[7] From the viewpoint of system engineering, A can be regarded as a signal amplifier with u as its input signal.

nonzero vectors $u_1, u_2, \ldots, u_r$ satisfying

$$A[u_1\ u_2\ \cdots\ u_r] = [u_1\ u_2\ \cdots\ u_r]\begin{bmatrix} \lambda & 1 & & & \\ & \lambda & 1 & & \\ & & \ddots & \ddots & \\ & & & \lambda & 1 \\ & & & & \lambda \end{bmatrix}, \tag{2.56}$$

only u_1 is an eigenvector, while $u_2, \ldots, u_r$ are not. $u_2, \ldots, u_r$ are called the generalized eigenvectors. Expanding this equation, we see that the generalized eigenvectors satisfy

$$Au_i = \lambda u_i + u_{i-1}, \quad i \geq 2. \tag{2.57}$$

This property shows that when we map the subspace $\text{span}\{u_1, \ldots, u_r\} := \{x \mid x = \sum c_i u_i, c_i \in \mathbb{C}\ \forall i\}$ spanned by the eigenvectors and generalized eigenvectors by matrix A, its image still returns to the same subspace. A subspace with such property is called an *invariant subspace* of matrix A.

In general, when there exist $n \times r$ matrix U and $r \times r$ matrix Λ satisfying

$$AU = U\Lambda, \tag{2.58}$$

the subspace $S = \text{Im}U = \{y \in \mathbb{C}^n \mid y = Ux, x \in \mathbb{C}^r\}$ with the columns of U as its basis becomes an invariant subspace of A. Obviously, any subspace formed by eigenvectors is A-invariant. It should be noted that the so-called invariant subspace must correspond to a specific mapping and a subspace alone does not have such property. For details about invariant subspace, refer to Section 2.8.

The following Cayley–Hamilton theorem plays an important role in the analysis of linear system structure.

Theorem 2.6 (Cayley–Hamilton) *Assume that $A \in \mathbb{C}^{n\times n}$ and*

$$\det(\lambda I - A) = \lambda^n + a_1\lambda^{n-1} + \cdots + a_n. \tag{2.59}$$

Then, the following matrix equation holds:

$$A^n + a_1A^{n-1} + \cdots + a_nI = 0. \tag{2.60}$$

This implies that A^n (and higher power of A) is a linear combination of $I, \ldots, A^{n-1}$.

2.8 Invariant Subspace

We regard square matrix A as a linear mapping $A : \mathbb{F}^n \mapsto \mathbb{F}^n$. In subspace S of $\mathbb{F}^n$, if any vector returns to S after being mapped by A, that is,

$$Ax \in S \quad \forall x \in S, \tag{2.61}$$

we say that S is an invariant subspace of the mapping A, or simply A-invariant. This property of invariant subspace can be succinctly expressed as

$$AS \subset S. \tag{2.62}$$

The concept of invariant subspace of linear mapping plays a very important role in the analysis of system structure.

Several examples are used to illustrate the invariant subspace. For example, if λ is an eigenvalue of matrix A and x is the corresponding eigenvector, then $Ax = \lambda x$ holds. This implies that the one-dimensional subspace span$\{x\}$ spanned by the eigenvector is A-invariant.

Example 2.12 *Prove that* $\{0\}$, $\mathbb{C}^n$, $\mathrm{Ker}A$, *and* $\mathrm{Im}A$ *are A-invariant subspaces.*

Solution Obviously, $A\{0\} \subset \{0\}$, $A\mathbb{C}^n \subset \mathbb{C}^n$ hold. $Ax = 0$ is true for any $x \in \mathrm{Ker}A$. So, $\mathrm{Ker}A$ is A-invariant because $0 \in \mathrm{Ker}A$. Also, the matrix A considered in this section is square, so naturally we have $\mathrm{Im}A \subset \mathbb{C}^n$, that is, the image is contained in the domain. Thus, mapping $\mathrm{Im}A$ again with A, the image still returns to $\mathrm{Im}A$. ∇

In general, subspaces spanned by all the eigenvectors corresponding to eigenvalues $\lambda_1, \ldots, \lambda_k$ of A (not necessarily distinct) as well as all eigenvectors and generalized eigenvectors are A-invariant. Let us see the following example.

Example 2.13 *Suppose that matrix* $A \in \mathbb{C}^{3\times 3}$ *has a Jordan canonical form*

$$A\begin{bmatrix} x_1 & x_2 & x_3 \end{bmatrix} = \begin{bmatrix} x_1 & x_2 & x_3 \end{bmatrix}\begin{bmatrix} \lambda_1 & 1 & \\ & \lambda_1 & \\ & & \lambda_2 \end{bmatrix}.$$

Let us consider the following subspaces:

$$S_1 = \mathrm{span}\{x_1\}, S_2 = \mathrm{span}\{x_2\}, S_3 = \mathrm{span}\{x_3\},$$
$$S_{12} = \mathrm{span}\{x_1, x_2\}, S_{13} = \mathrm{span}\{x_1, x_3\}.$$

Since x_1, x_3 *are eigenvectors of matrix A, both* S_1 *and* S_3 *are invariant subspaces. As for* S_{12}, *for any* $x \in S_{12}$, *we have* $x = ax_1 + bx_2 \Rightarrow Ax = aAx_1 + bAx_2 = a\lambda_1 x_1 + b(x_1 + \lambda_1 x_2) = (a\lambda_1 + b)x_1 + (b\lambda_1)x_2 \in S_{12}$. *Therefore,* S_{12} *is also A-invariant. Similarly,* S_{13} *is A-invariant. However,* $S_2 = \mathrm{span}\{x_2\}$ *is not an invariant subspace because* $Ax_2 = \lambda_1 x_2 + x_1$, *and* x_1 *(independent of* x_2*) is not contained in* S_2. *So* $Ax_2 \in S_2$ *does not hold.*

The following theorem shows that, when performing similarity transformation on matrix A with the basis of A-invariant subspace, A can be transformed into a block triangular matrix.

Theorem 2.7 *Assume that* $S \subset \mathbb{C}^n$ *is A-invariant and set its basis as* $\{t_1, \ldots, t_k\}$ $(k < n)$. *Then*

1. *The following equation holds for some matrix* $A_{11} \in \mathbb{C}^{k\times k}$

$$AT_1 = T_1 A_{11}, \quad T_1 = [t_1 \ \cdots \ t_k].$$

2. *There exist vectors* $t_{k+1}, \ldots, t_n \in \mathbb{C}^n$ *such that the matrix*

$$T = [t_1 \ \cdots \ t_k \mid t_{k+1} \ \cdots \ t_n] = [T_1 \ \ T_2]$$

is nonsingular and satisfies

$$AT = T \begin{bmatrix} A_{11} & A_{12} \\ 0 & A_{22} \end{bmatrix}$$

in which A_{21}, A_{22} *are compatible matrices.*

2.8.1 Mapping Restricted in Invariant Subspace

Let us consider restricting the mapping A in its invariant subspace S and denote this mapping by $A|S$. According to Theorem 2.7, for matrix T constructed by the basis of S, there must be a square matrix $\underline{A}$ satisfying

$$AT = T\underline{A}. \tag{2.63}$$

This matrix $\underline{A}$ is exactly the matrix description of $A|S$, the mapping restricted in subspace S. Clearly, elements of $\sigma(\underline{A})$, the eigenvalue set of $\underline{A}$, are also eigenvalues of A. They are called the eigenvalues of mapping A restricted to subspace S, denoted as

$$\sigma(A|S) = \sigma(\underline{A}). \tag{2.64}$$

2.8.2 Invariant Subspace over $\mathbb{R}^n$

As mentioned earlier, real matrix may have complex eigenvalues and complex eigenvectors. In such case, the subspace with eigenvectors as the basis is naturally a complex subspace. This is why we have been discussing invariant subspaces over $\mathbb{C}^n$ up to now. However, a real matrix can also have real invariant subspace. For example, when the matrix $A \in \mathbb{R}^{n\times n}$ has complex eigenvalue $a + jb$ and eigenvector $x + jy$, by comparing the real and imaginary parts of the two sides of $A(x + jy) = (a + jb)(x + jy)$, we get $Ax = ax - by$, $Ay = bx + ay$. Thus,

$$A[x \ y] = [x \ y] \begin{bmatrix} a & b \\ -b & a \end{bmatrix}$$

holds. Next, by the assumption that the eigenvalue is complex (i.e., $b \neq 0$), we can prove that x, y, are linearly independent[8]. Therefore, when setting $U = [x \ y]$, we have

$$AU = U\Lambda, \quad \Lambda := \begin{bmatrix} a & b \\ -b & a \end{bmatrix}. \tag{2.65}$$

Then, Im$U \subset \mathbb{R}^n$ is a real invariant subspace of A. The background is that the characteristic polynomial of a real matrix has real coefficients, so when there is a complex eigenvalue, its

[8] First, $a - jb$ is also an eigenvalues of A, its corresponding eigenvector is $x - jy$ which is linearly independent of $x + jy$ as the eigenvalues are different. If x, y are not linearly independent, there must be a nonzero scalar $c \in \mathbb{R}$ such that $y = cx$. This means that $x + jy = (1 + jc)x$ and $x - jy = (1 - jc)x$ are linearly dependent, which is contradictory.

conjugate must also be an eigenvalue. In fact, the eigenvalues of Λ are $a \pm jb$ which are also eigenvalues of A.

Although a real invariant subspace of matrix A does not contain the complex eigenvectors of A, they however can be generated from the basis of the subspace. For example, in the preceding example, the eigenvectors of Λ are $v_{1,2} = [1 \quad \pm j]^T$, and the eigenvectors of A can be generated by $[x \quad y]v_{1,2} = x \pm jy$. Repetition of this argument leads to Theorem 2.8.

Theorem 2.8 *Assume that $S = \text{Im}\, T$ is an n-dimensional invariant subspace of $A \in \mathbb{R}^{n\times n}$ and the matrix description of the mapping $A|S$ restricted in S is $\underline{A}$. Then, $S \subset \mathbb{R}^n$ iff $\underline{A}$ is a real matrix.*

Moreover, from this theorem, we immediately get the following corollary.

Corollary 2.1 *For $A \in \mathbb{R}^{n\times n}$, let $S \subset \mathbb{R}^n$ be an invariant subspace of A with basis $\{t_1, \ldots, t_k\}$ $(k < n)$. Then, the following statements are true:*

1. *There exists a matrix $A_{11} \in \mathbb{R}^{k\times k}$ satisfying*

$$AT_1 = T_1 A_{11}, \quad T_1 = [t_1 \ \cdots \ t_k].$$

2. *There exist vectors $t_{k+1}, \ldots, t_n \in \mathbb{R}^n$ such that the matrix*

$$T = [t_1 \ \cdots \ t_k \mid t_{k+1} \ \cdots \ t_n] = [T_1 \quad T_2]$$

is nonsingular and satisfies

$$AT = T \begin{bmatrix} A_{11} & A_{12} \\ 0 & A_{22} \end{bmatrix}.$$

Here A_{21} and A_{22} are real matrices with suitable dimensions.

This corollary shows that any real square matrix can always be transformed into a real block triangular matrix by a real transformation matrix. Its difference from Theorem 2.7 is that Theorem 2.7 only says that any matrix can be transformed into a complex block triangular matrix.

2.8.3 Diagonalization of Hermitian/Symmetric Matrix

When $A \in \mathbb{F}^{n\times n}$ satisfies

$$A^* = A, \tag{2.66}$$

it is called an *Hermitian matrix*. Particularly, when A is a real square matrix, this relation becomes

$$A^T = A. \tag{2.67}$$

Such matrix A is called a *symmetric matrix*. Here, we show a result about the diagonalization of Hermitian matrix which is derived from Theorem 2.7 and Corollary 2.1. The conclusions are particularly useful in discussing the singular value decomposition (SVD) in Section 2.12.

Theorem 2.9 *For an Hermitian matrix A, the following statements are true:*

1. *All eigenvalues of A are real.*
2. *There exists a unitary matrix $U \in \mathbb{F}^{n\times n}$ which diagonalizes A:*

$$U^*AU = \begin{bmatrix} \lambda_1 & & \\ & \ddots & \\ & & \lambda_n \end{bmatrix}, \quad U^*U = I.$$

Note that for a real matrix, the transformation matrix for diagonalization is an orthonormal matrix.

Example 2.14 *Diagonalize the matrix* $A = \begin{bmatrix} 1 & 2 \\ 2 & 1 \end{bmatrix}$.

First of all, $\det(\lambda I - A) = (\lambda - 1)^2 - 4 = 0$ *leads to the eigenvalues* $\lambda_1 = 3, \lambda_2 = -1$. *Let the eigenvector corresponding to* $\lambda_1 = 3$ *be* x_1. *Then*

$$(\lambda_1 I - A)x_1 = \begin{bmatrix} 2 & -2 \\ -2 & 2 \end{bmatrix} x_1 = 0 \Rightarrow x_1 = \begin{bmatrix} a \\ a \end{bmatrix}, \quad a \neq 0.$$

Normalizing x_1, *we get the eigenvector* $u_1 = [1\ 1]^T/\sqrt{2}$. *Similarly, the eigenvector corresponding to* $\lambda_2 = -1$ *is* $x_2 = [b\ \ -b]^T$, *and the normalized eigenvector is* $u_2 = [1\ \ -1]^T/\sqrt{2}$. *Clearly,* $u_1 \perp u_2$. *Setting the similarity transformation matrix as* $U = [u_1\ u_2]$, *it is orthonormal, and the transformed matrix is a diagonal matrix*

$$U^*AU = \frac{1}{2}\begin{bmatrix} 1 & 1 \\ 1 & -1 \end{bmatrix}^T \begin{bmatrix} 1 & 2 \\ 2 & 1 \end{bmatrix} \begin{bmatrix} 1 & 1 \\ 1 & -1 \end{bmatrix} = \begin{bmatrix} 3 & 0 \\ 0 & -1 \end{bmatrix}.$$

2.9 Pseudo-Inverse and Linear Matrix Equation

For a nonsquare or singular matrix, there is no inverse in the normal sense. However, we may define the so-called *pseudo-inverse*. The pseudo-inverse is denoted as $A^\dagger$ and satisfies all the following conditions:

1. $AA^\dagger A = A$.
2. $A^\dagger AA^\dagger = A^\dagger$.
3. $(AA^\dagger)^* = AA^\dagger$.
4. $(A^\dagger A)^* = A^\dagger A$.

$A^\dagger$ can be calculated via SVD (refer to Section 2.12). Suppose that the SVD of A is

$$A = U\Sigma V^*, \quad \Sigma = \begin{bmatrix} \Sigma_r & 0 \\ 0 & 0 \end{bmatrix}, \quad \det(\Sigma_r) \neq 0.$$

Then, $A^\dagger$ is given by

$$A^\dagger = V\Sigma^\dagger U^*, \quad \Sigma^\dagger = \begin{bmatrix} \Sigma_r^{-1} & 0 \\ 0 & 0 \end{bmatrix}.$$

It can be proven that the pseudo-inverse exists and is unique [8]. The pseudo-inverse is very useful in solving linear vector (or matrix) equations.

Lemma 2.3 *[8] The linear matrix equation*

$$AXB = C \tag{2.68}$$

is solvable iff

$$AA^\dagger CB^\dagger B = C. \tag{2.69}$$

Further, its general solution is given by

$$X = A^\dagger CB^\dagger + Y - A^\dagger AYBB^\dagger \tag{2.70}$$

where Y is an arbitrary compatible matrix.

2.10 Quadratic Form and Positive Definite Matrix

2.10.1 Quadratic Form and Energy Function

The scalar function $ax_1^2 + 2bx_1x_2 + cx_2^2$ of vector $x = [x_1\ x_2]^T$ is called a quadratic form because all its terms are second-order functions of variables x_1, x_2 [9]. The energy function of a physical system is usually described by second-order function of physical variables (states). For example, the kinetic energy $mv^2/2$ of a mass m is a second-order function of the speed, and the rotational energy $J\omega^2/2$ of a rigid body with inertia J is also a second-order function of the angular velocity ω. System stability or control performance is closely related to energy, so we often encounter quadratic forms in system analysis and design. More details will be provided in the Lyapunov stability analysis of Chapter 13 and Section 4.3.

In general, the quadratic form of an n-dimensional real vector x has the form of

$$\begin{aligned} V(x) =& \sum_{i=1}^{n}\sum_{j=1}^{n} b_{ij}x_ix_j \\ =&(b_{11}x_1^2 + b_{12}x_1x_2 + \cdots + b_{1n}x_1x_n) + \cdots \\ &+ (b_{n1}x_nx_1 + b_{n2}x_nx_2 + \cdots + b_{nn}x_n^2) \end{aligned} \tag{2.71}$$

in which $b_{ij} \in \mathbb{R}$. Using $x_ix_j = x_jx_i$ and setting

$$a_{ii} = b_{ii}, \quad a_{ij} = a_{ji} = \frac{b_{ij} + b_{ji}}{2}, \quad i \neq j, \tag{2.72}$$

[9] The cross term x_1x_2 is also regarded as being second order.

we can always write $V(x)$ as

$$V(x) = x^T A x \tag{2.73}$$

with a symmetric matrix $A = (a_{ij})$. For example, the quadratic form in the preceding example can be written as

$$ax_1^2 + 2bx_1x_2 + cx_2^2 = [x_1 \ x_2] \begin{bmatrix} a & b \\ b & c \end{bmatrix} \begin{bmatrix} x_1 \\ x_2 \end{bmatrix}.$$

Similarly, the quadratic form about complex vector $x \in \mathbb{C}^n$ can be defined as

$$V(x) = x^* A x \tag{2.74}$$

with an Hermitian matrix[10] $A^* = A$. As energy is real valued, to fit this fact, the quadratic form is also restricted to real number. This is why the quadratic form of complex vector is defined as such ($V^*(x) = V(x)$ holds).

Energy is always positive. So the quadratic form describing energy should also be positive. This naturally leads to the notion of positive definite function. For any nonzero vector $x \in \mathbb{F}^n$, quadratic form satisfying $V(x) = x^* A x \geq 0$ is called a *positive semidefinite function*, and quadratic form satisfying $V(x) = x^* A x > 0$ is called a *positive definite function*. By definition, whether a quadratic form is positive definite or not does not depend on the vector x, but is determined solely by the coefficient matrix A. Matrix with such property is the positive definite matrix/positive semidefinite matrix to be discussed in the next subsection.

2.10.2 Positive Definite and Positive Semidefinite Matrices

When an Hermitian matrix $A = A^*$ satisfies $x^* A x > 0$ for any $x \neq 0$, it is called a *positive definite matrix*, denoted by $A > 0$. Similarly, if $x^* A x \geq 0$ for any $x \neq 0$, A is called a *positive semidefinite matrix*, denoted by $A \geq 0$. For example, $B^* B \geq 0$ since

$$x^* B^* B x = \|Bx\|_2^2 \geq 0$$

for matrix B and arbitrary vector x.

The following lemma gives conditions for an Hermitian matrix A to be positive (semi) definite.

Theorem 2.10 *When $A \in \mathbb{F}^{n \times n}$ is Hermitian, the following statements hold:*

1. *$A \geq 0$ iff its eigenvalues are all nonnegative.*
2. *$A > 0$ iff its eigenvalues are all positive.*
3. *When $A \geq 0$, there exists $B \in \mathbb{F}^{n \times r}$ such that A is decomposed as $A = BB^*$ where $r \geq \mathrm{rank}(A)$.*

For a semidefinite matrix A, we define its square root as the positive semidefinite matrix $A^{1/2} = (A^{1/2})^* \geq 0$ satisfying

$$A = A^{1/2} A^{1/2}. \tag{2.75}$$

[10] Note that all eigenvalues of an Hermitian matrix are real (refer to Exercise 2.16).

From Theorem 2.9, we see that $A^{1/2}$ can be computed by

$$A^{1/2} = U\ \mathrm{diag}(\sqrt{\lambda_1}, \ldots, \sqrt{\lambda_n})U^*. \tag{2.76}$$

Schur's lemma is shown as follows. It describes the relation between the whole matrix and its blocks for positive (semi)definite matrices.

Lemma 2.4 (Schur's Lemma) *Partition Hermitian matrix $X = X^*$ as*

$$X = \begin{bmatrix} X_{11} & X_{12} \\ X_{12}^* & X_{22} \end{bmatrix}$$

in which X_{11}, X_{22} are square. Then, the following statements are true:

1. $X > 0$ *iff one of the following conditions is satisfied:*
   (a) $X_{22} > 0,\ X_{11} - X_{12}X_{22}^{-1}X_{12}^* > 0.$
   (b) $X_{11} > 0,\ X_{22} - X_{12}^*X_{11}^{-1}X_{12} > 0.$
2. $X \geq 0$ *iff one of the following conditions holds:*
   (a) $X_{22} \geq 0,\ \mathrm{Ker}X_{22} \subset \mathrm{Ker}X_{12},\ X_{11} - X_{12}X_{22}^{\dagger}X_{12}^* \geq 0.$
   (b) $X_{11} \geq 0,\ \mathrm{Ker}X_{11} \subset \mathrm{Ker}X_{12}^*,\ X_{22} - X_{12}^*X_{11}^{\dagger}X_{12} \geq 0.$

When the elements of matrix X are variables, matrix inequality such as $X_{11} - X_{12}X_{22}^{-1}X_{12}^* > 0$ is nonlinear and difficult to solve numerically. With Schur's lemma, it is possible to change it into a linear matrix inequality by extending the size of matrix inequality. This is the most important application of Schur's lemma.

2.11 Norm and Inner Product of Matrix

2.11.1 Matrix Norm

Assume that $A = (a_{ij}) \in \mathbb{F}^{m \times n}$. The mapping of vectors using this matrix is shown in Figure 2.10. From the viewpoint of system, the matrix can be regarded as an amplifier and the vector as a signal. So, u corresponds to the input and Au the output.

Then, a matrix norm can be regarded as the amplification rate of signal. Therefore, the matrix norm can be defined by the ratio of input and output vector norms. A norm defined like this is called an *induced norm*. But it should be noted that the ratio of input and output vector norms is not a fixed number. It varies with the change in the direction of input vector.

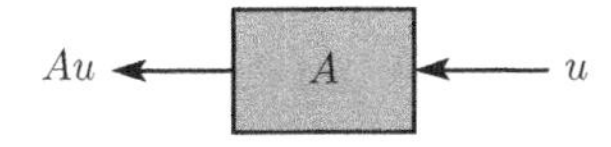

Figure 2.10 Mapping vector with matrix

Example 2.15 *Multiplying input vectors* $u_1 = [1\ 0]^T$, $u_2 = [0\ 1]^T$, *and* $u_3 = [1\ 1]^T/\sqrt{2}$ *to matrix* $A = \begin{bmatrix} 1 & 2 \\ 3 & 4 \end{bmatrix}$, *we get output vectors* $y_1 = [1\ 3]^T$, $y_2 = [2\ 4]^T$, *and* $y_3 = [3\ 7]^T/\sqrt{2}$, *respectively. Therefore, the 2-norm ratios of input and output are, respectively,* $\sqrt{10}, 2\sqrt{5}, \sqrt{29}$. *Apparently, they are not equal.*

Therefore, we should use the supremum of the ratio of input and output vector norms as the norm of matrix. For example,

$$\|A\|_1 := \sup_{u \neq 0} \frac{\|Au\|_1}{\|u\|_1} \tag{2.77}$$

$$\|A\|_2 := \sup_{u \neq 0} \frac{\|Au\|_2}{\|u\|_2} \tag{2.78}$$

$$\|A\|_\infty := \sup_{u \neq 0} \frac{\|Au\|_\infty}{\|u\|_\infty} \tag{2.79}$$

are all norms of matrix A.

Secondly, from the geometric viewpoint, matrix A is a mapping that maps the vector space $\mathbb{F}^n$ into the vector space $\mathbb{F}^m$. The induced norm of matrix indicates the maximum amplification rate of vector length after the mapping.

The matrix norms $\|A\|_1, \|A\|_2, \|A\|_\infty$ do not depend on the input vector u, but are determined solely by its elements:

1-norm $\|A\|_1 = \max_{1 \le j \le n} \sum_{i=1}^{m} |a_{ij}|$ (column sum)

2-norm $\|A\|_2 = \sqrt{\lambda_{\max}(A^*A)}$

Infinity-norm $\|A\|_\infty = \max_{1 \le i \le m} \sum_{j=1}^{n} |a_{ij}|$ (row sum)

Example 2.16 *Prove the formula of 1-norm. First, according to the definition of vector's 1-norm,*

$$\|Au\|_1 = \sum_{i=1}^{m} \left| \sum_{j=1}^{n} a_{ij} u_j \right| \le \sum_{i=1}^{m} \sum_{j=1}^{n} |a_{ij}||u_j| = \sum_{j=1}^{n} \left(\sum_{i=1}^{m} |a_{ij}| \right) |u_j|$$

$$\le \max_{1 \le j \le n} \sum_{i=1}^{m} |a_{ij}| \sum_{j=1}^{n} |u_j| = \max_{1 \le j \le n} \sum_{i=1}^{m} |a_{ij}| \|u\|_1$$

$$\Rightarrow \frac{\|Au\|_1}{\|u\|_1} \le \max_{1 \le j \le n} \sum_{i=1}^{m} |a_{ij}|$$

holds. This inequality is true for arbitrary vector u. So when the left-hand side takes the supremum w.r.t to u, the inequality is still satisfied. That is, $\|A\|_1 \leq \max_j \sum_{i=1}^{m} |a_{ij}|$. *Next, we prove the opposite inequality. Assume that the column sum takes the maximum at the* j^**th column, that is,*

$$\sum_{i=1}^{m} |a_{ij^*}| = \max_{1\leq j\leq n} \sum_{i=1}^{m} |a_{ij}|.$$

Set $u_* = e_{j^*}$ *(it is the column vector whose all elements are zero except that the* j^**th element is 1). Then* $\|u_*\| = 1$ *and*

$$\|Au_*\|_1 = \sum_{i=1}^{m} |a_{ij^*}| = \max_{1\leq j\leq n} \sum_{i=1}^{m} |a_{ij}| = \max_{1\leq j\leq n} \sum_{i=1}^{m} |a_{ij}| \|u_*\|_1$$

$$\Rightarrow \frac{\|Au_*\|_1}{\|u_*\|_1} = \max_{1\leq j\leq n} \sum_{i=1}^{m} |a_{ij}| \Rightarrow \|A\|_1 \geq \frac{\|Au_*\|_1}{\|u_*\|_1} = \max_{1\leq j\leq n} \sum_{i=1}^{m} |a_{ij}|.$$

So, we have

$$\|A\|_1 = \max_{1\leq j\leq n} \sum_{i=1}^{m} |a_{ij}|.$$

In Example 2.15, $\|A\|_1 = 2 + 4 = 6$, the input direction that yields the maximum amplification is $[0\ 1]^T$. In addition, $\|A\|_2 = \sqrt{15 + \sqrt{221}}$ and $\|A\|_\infty = 7$.

As another important feature of induced norm, the following so-called submultiplicative property holds

$$\|AB\| \leq \|A\| \|B\|. \tag{2.80}$$

It can be derived easily as follows. Assume that

$$y = Av, \quad v = Bu.$$

Then

$$\frac{\|y\|}{\|u\|} = \frac{\|y\|}{\|v\|}\frac{\|v\|}{\|u\|} \leq \sup \frac{\|y\|}{\|v\|} \sup \frac{\|v\|}{\|u\|}$$

$$\Rightarrow \sup \frac{\|y\|}{\|u\|} \leq \sup \frac{\|y\|}{\|v\|} \sup \frac{\|v\|}{\|u\|}$$

$$\Rightarrow \|AB\| \leq \|A\| \|B\|.$$

2.11.2 Inner Product of Matrix

The inner product of matrices $A, B \in \mathbb{F}^{m\times n}$ is defined by the trace of their product. That is,

$$\langle A, B\rangle = \text{Tr}(A^* B). \tag{2.81}$$

It is easy to see that it satisfies all properties of inner product. To understand why the inner product of matrix is defined as such, we denote the ith columns of A, B by a_i, b_i. Then, we have

$$\mathrm{Tr}(A^*B) = \sum_{i=1}^{n} a_i^* b_i = \mathrm{vec}(A)^*\mathrm{vec}(B) = \langle \mathrm{vec}(A), \mathrm{vec}(B) \rangle. \tag{2.82}$$

That is, the inner product of matrices is equal to the inner product of the vectors formed, respectively, by the columns of these matrices. Furthermore, the scalar

$$\|A\|_F := \sqrt{\mathrm{Tr}(A^*A)} \tag{2.83}$$

induced by the matrix inner product is called *Frobenius norm* of matrix A. Via the SVD to be introduced in the following section, it is easy to prove that $\|A\|_F$ satisfies

$$\|A\|_F^2 = \sum_{i=1}^{n} \sigma_i^2(A) \geq \|A\|_2^2. \tag{2.84}$$

2.12 Singular Value and Singular Value Decomposition

What the matrix norm introduced in Section 2.11.1 represents is the maximal possible amplification rate for input vectors in all directions. However, a vector has its own direction. The amplification rate is different when the direction of vector differs. However, the matrix norm cannot express this property. On the other hand, although the absolute value of eigenvalue of a square matrix indicates the degree of amplification in the corresponding eigenvector direction, for nonsquare matrix, eigenvalue cannot be defined. To overcome this difficulty, we note that for a matrix A of arbitrary size, A^*A is always square and positive semidefinite[11]. Owing to this property, we introduce a nonnegative real number called *singular value*.

The singular value of matrix $A \in \mathbb{F}^{m\times n}$ is defined as

$$\sigma_i(A) := \sqrt{\lambda_i(A^*A)} \tag{2.85}$$

in which $\lambda_i(A^*A)$ stands for the ith largest eigenvalue of A^*A. So, $\sigma_i(A)$ is the ith largest singular value of A. Further, there exists a nonzero vector $v_i \neq 0$ satisfying

$$A^*Av_i = \sigma_i^2 v_i, \tag{2.86}$$

which is called a *singular vector*. Obviously, by premultiplying both sides with v^* and taking square root, we obtain

$$\frac{\|Av_i\|_2}{\|v_i\|_2} = \sigma_i. \tag{2.87}$$

Hence, in the sense of vector 2-norm, a singular value indicates the amplification rate of input vector in the direction of the corresponding singular vector.

The largest and the smallest singular values of matrix A are denoted, respectively, by $\sigma_{\max}(A), \sigma_{\min}(A)$. Clearly,

$$\sigma_{\max}(A) = \|A\|_2. \tag{2.88}$$

[11] All eigenvalues of a positive semidefinite matrix are nonnegative real numbers (refer to Exercise 2.16).

Next, we show a quite useful result, known as singular value decomposition (SVD).

Theorem 2.11 *For any matrix $A \in \mathbb{F}^{m\times n}$, there exist* unitary matrices

$$U = [u_1 \; u_2 \; \cdots \; u_m] \in \mathbb{F}^{m\times m}, \quad V = [v_1 \; v_2 \; \cdots \; v_n] \in \mathbb{F}^{n\times n}$$

such that A is decomposed into

$$A = U\Sigma V^*, \quad \Sigma = \begin{bmatrix} \Sigma_1 & 0 \\ 0 & 0 \end{bmatrix} \tag{2.89}$$

where

$$\Sigma_1 = \begin{bmatrix} \sigma_1 & 0 & \cdots & 0 \\ 0 & \sigma_2 & \cdots & 0 \\ \vdots & \vdots & \ddots & \vdots \\ 0 & 0 & \cdots & \sigma_p \end{bmatrix},$$

$$\sigma_1 \geq \sigma_2 \geq \cdots \geq \sigma_p \geq 0, \quad p = \min\{m, n\}. \tag{2.90}$$

σ_i is the ith *singular value* of A, u_i and v_j are, respectively, the ith *left singular vector* and the jth *right singular vector*. First, from $AV = U\Sigma$ we have

$$Av_i = \sigma_i u_i. \tag{2.91}$$

Next, after taking conjugate transpose on both sides of $AV = U\Sigma$, we premultiply V and postmultiply u_i to the equation and get

$$A^* u_i = \sigma_i v_i. \tag{2.92}$$

These two equations can be further written as

$$A^*Av_i = \sigma_i^2 v_i,$$
$$AA^*u_i = \sigma_i^2 u_i.$$

So, σ_i^2 is an eigenvalue of AA^* or A^*A, u_i the eigenvector of AA^*, and v_i the eigenvector of A^*A. These relationships are illustrated in Figure 2.11.

Example 2.17 *We consider mapping the unit disk $\{x \in \mathbb{R}^2 \mid \|x\|_2 \leq 1\}$ with a 2×2 real matrix A. Note that a two-dimensional unitary matrix U corresponds to a rotation matrix and can always be written as*[12]

$$U = \begin{bmatrix} \cos\theta & -\sin\theta \\ \sin\theta & \cos\theta \end{bmatrix}$$

[12] Assume that $U = (u_{ij})$. Expanding $U^TU = I$, we get $u_{11}^2 + u_{21}^2 = 1$, $u_{12}^2 + u_{22}^2 = 1$, and $u_{11}u_{12} + u_{21}u_{22} = 0$. From the first two equations, we have $u_{11} = \cos\theta, u_{21} = \sin\theta$, $u_{22} = \cos\phi, u_{12} = \sin\phi$. From the third equation, we get $\sin(\theta + \phi) = 0 \Rightarrow \phi = -\theta$. Therefore, $\sin\phi = -\sin\theta, \cos\phi = \cos\theta$.

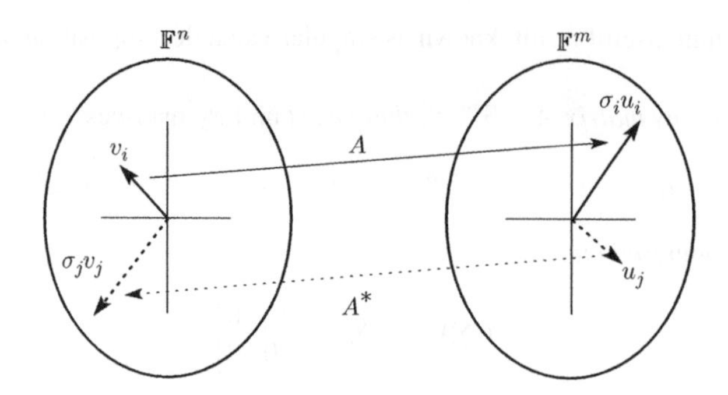

Figure 2.11 Relationships between singular vectors

with a variable θ. If the singular values σ_1, σ_2 of A are both nonzero, then

$$y = Ax = U \begin{bmatrix} \sigma_1 & \\ & \sigma_2 \end{bmatrix} V^T x \Rightarrow \hat{y} := U^T y = \begin{bmatrix} \sigma_1 & \\ & \sigma_2 \end{bmatrix} V^T x$$

$$\Rightarrow \frac{\hat{y}_1^2}{\sigma_1^2} + \frac{\hat{y}_2^2}{\sigma_2^2} = x^T V V^T x = x^T x \leq 1$$

holds. In the coordinate $\hat{y}$, this inequality represents an ellipsoid whose long-axis length is $2\sigma_1$ and short-axis length is $2\sigma_2$. Moreover, because of

$$y = U\hat{y} = \begin{bmatrix} \cos\theta & -\sin\theta \\ \sin\theta & \cos\theta \end{bmatrix} \hat{y},$$

it becomes an ellipsoid rotating an angle of θ counterclockwise in the coordinate system y. From relations $Av_1 = \sigma_1 u_1$, $Av_2 = \sigma_2 u_2$, we see that the long-axis direction of the image is u_1 and the short-axis direction is u_2. Finally, v_1 in the domain is mapped to the long-axis direction and v_2 mapped to the short-axis direction.

As shown by this example, geometrically the singular values of matrix A are equal to the half-lengths of the ellipsoid's axes:

$$E = \{y \mid y = Ax, x \in \mathbb{F}^n, \|x\|_2 \leq 1\}.$$

So, v_1 is the direction that maximizes $\|y\|_2$ among all vectors x. On the contrary, v_n is the direction that minimizes $\|y\|_2$.

By the SVD, we see that

$$\sigma_{\max}^2(A)I - A^*A = \sigma_{\max}^2(A)I - V\Sigma^2 V^* = V(\sigma_{\max}^2(A)I - \Sigma^2)V^* \geq 0. \quad (2.93)$$

Obviously, $\gamma^2 I - A^*A \geq 0$ fails for $\gamma < \sigma_{\max}(A)$. Therefore, the largest singular value is equivalent to the minimum among all positive real number γ satisfying

$$\gamma^2 I - A^*A \geq 0. \quad (2.94)$$

This inequality may be used as an equivalent definition of the largest singular value. This relationship will play a role in solving $\mathcal{H}_\infty$ control problem and other issues later.

2.13 Calculus of Vector and Matrix

2.13.1 *Scalar Variable*

When the variable of vector/matrix function is a scalar, the calculus is defined by those of its elements about the scalar. For example, when the variable is t

$$\dot{x}(t) := \begin{bmatrix} x_1(t) \\ \vdots \\ x_n(t) \end{bmatrix}, \quad \dot{A}(t) := [\dot{a}_{ij}(t)], \quad \int A(t)\,dt := \left[\int a_{ij}(t)\,dt\right]. \tag{2.95}$$

Here, $\dot{a}(t)$ denotes the first derivative $\frac{d}{dt}a(t)$ of $a(t)$ about t. Further, similar to scalar functions, the derivative formula of product and the partial integration formula hold. Namely,

$$\frac{d}{dt}(AB) = \frac{dA}{dt}B + A\frac{dB}{dt} \tag{2.96}$$

$$\int_a^b \frac{dA}{dt}B\,dt = AB\Big|_a^b - \int_a^b A\frac{dB}{dt}\,dt. \tag{2.97}$$

These properties are easy to verify.

2.13.2 *Vector or Matrix Variable*

For a scalar function $f(x)$ of column vector $x = [x_1, \ldots, x_n]^T$, its partial derivatives are defined as follows:

$$\frac{\partial f}{\partial x} := \left[\frac{\partial f}{\partial x_1}, \ldots, \frac{\partial f}{\partial x_n}\right], \tag{2.98}$$

$$\frac{\partial^2 f}{\partial x^2} := \begin{bmatrix} \frac{\partial^2 f}{\partial x_1^2} & \cdots & \frac{\partial^2 f}{\partial x_n \partial x_1} \\ \vdots & & \vdots \\ \frac{\partial^2 f}{\partial x_1 \partial x_n} & \cdots & \frac{\partial^2 f}{\partial x_n^2} \end{bmatrix}. \tag{2.99}$$

According to the definition, the following equations hold for $b \in \mathbb{R}^n$, $A^T = A \in \mathbb{R}^{n \times n}$:

$$\frac{\partial}{\partial x}b^T x = b^T, \quad \frac{\partial}{\partial x}x^T Ax = 2x^T A, \quad \frac{\partial^2}{\partial x^2}x^T Ax = 2A. \tag{2.100}$$

Similar to the partial derivatives about column vector, for a scalar function, its first-order partial derivative about a matrix is given by differentiating the scalar function with matrix elements one by one and putting them in the same order as the transpose of the original matrix.

Specifically, for $X \in \mathbb{F}^{m\times n}$, we have

$$\frac{\partial f(X)}{\partial X} = \begin{bmatrix} \frac{\partial f}{\partial x_{11}} & \cdots & \frac{\partial f}{\partial x_{1n}} \\ \vdots & & \vdots \\ \frac{\partial f}{\partial x_{m1}} & \cdots & \frac{\partial f}{\partial x_{mn}} \end{bmatrix}^T = \begin{bmatrix} \frac{\partial f}{\partial x_{11}} & \cdots & \frac{\partial f}{\partial x_{m1}} \\ \vdots & & \vdots \\ \frac{\partial f}{\partial x_{1n}} & \cdots & \frac{\partial f}{\partial x_{mn}} \end{bmatrix}. \tag{2.101}$$

When A, B, X are matrices with appropriate dimensions, the following differentiation formulae hold:

$$\begin{aligned}
&\frac{\partial}{\partial X}\mathrm{Tr}(AXB) = A^T B^T \\
&\frac{\partial}{\partial X}\mathrm{Tr}(AX^T B) = BA \\
&\frac{\partial}{\partial X}\mathrm{Tr}(AX^{-1}B) = -(X^{-1}BAX^{-1})^T \\
&\frac{\partial}{\partial X}\det(X) = \frac{\partial}{\partial X}\det(X^T) = \det(X)(X^T)^{-1} \\
&\frac{\partial}{\partial X}\log\det(X) = (X^T)^{-1}.
\end{aligned}$$

2.14 Kronecker Product

The *Kronecker product* of matrices $A \in \mathbb{F}^{m\times n}$ and $B \in \mathbb{F}^{p\times q}$ is an $mp \times nq$ matrix defined by

$$A \otimes B = \begin{bmatrix} a_{11}B & \cdots & a_{1n}B \\ \vdots & & \vdots \\ a_{m1}B & \cdots & a_{mn}B \end{bmatrix}. \tag{2.102}$$

Meanwhile, the *Kronecker sum* of square matrices $A \in \mathbb{F}^{n\times n}$ and $B \in \mathbb{F}^{m\times m}$ is defined as

$$A \oplus B = A \otimes I_m + I_n \otimes B. \tag{2.103}$$

Putting the columns of matrix A into a vector sequentially from the first column, we get

$$\mathrm{vec}(A) = [a_{11} \ \cdots \ a_{m1} \ \cdots \ a_{1n} \ \cdots \ a_{mn}]^T. \tag{2.104}$$

Kronecker product and Kronecker sum have the following properties:

1. For arbitrary scalar α, there holds

$$\alpha(A \otimes B) = (\alpha A) \otimes B = A \otimes (\alpha B).$$

2. For $A \in \mathbb{F}^{l\times m}, B \in \mathbb{F}^{p\times q}$ and $C \in \mathbb{F}^{m\times n}, D \in \mathbb{F}^{q\times r}$, we have

$$(AC) \otimes (BD) = (A \otimes B)(C \otimes D).$$

3. Let λ_i, μ_j $(i = 1, \ldots, n, j = 1, \ldots, m)$ be the eigenvalues of square matrices $A \in \mathbb{F}^{n\times n}, B \in \mathbb{F}^{m\times m}$, respectively, and x_i, y_j be their corresponding eigenvectors. The following statements are true:
   (a) The eigenvalues of Kronecker product $A \otimes B$ are $\lambda_i\mu_j$ $(i = 1, \ldots, n, j = 1, \ldots, m)$ with corresponding eigenvectors $x_i \otimes y_j$.
   (b) The eigenvalues of Kronecker sum $A \oplus B$ are $\lambda_i + \mu_j$ $(i = 1, \ldots, n, j = 1, \ldots, m)$ with corresponding eigenvectors $x_i \otimes y_j$.
4. For matrices A, B, C with appropriate dimensions, there holds

$$\text{vec}(ABC) = (C^T \otimes A)\text{vec}(B).$$

5. When the elements of matrices $A(t), B(t)$ are differentiable w.r.t. scalar t, the following differentiation formula holds:

$$\frac{d}{dt}(A(t) \otimes B(t)) = \left(\frac{d}{dt}A(t)\right) \otimes B(t) + A(t) \otimes \left(\frac{d}{dt}B(t)\right).$$

All these properties can be verified by using simple matrices such as 2×2 matrices.

2.15 Norm and Inner Product of Function

2.15.1 Signal Norm

Consider the disturbance attenuation problem shown in Figure 2.12. Typical disturbance response is shown in Figure 2.12(b). In order to quantify the effect of disturbance attenuation (the size of disturbance response), as well as the quality of reference tracking (the size of tracking error), we need an objective measure on signal size. The measure of signal size is called *signal norm*. On one hand, as a signal is a function of time, it does not make sense considering only the response amplitude at a particular time instant t_0. We must investigate the whole time response. On the other hand, the size of signal is used to compare the relation between different signals. When we observe two signals in time sequence, the size order of signals changes at different moments, making it impossible to conduct an objective comparison. Therefore, we need to find a measure of signal size that is independent of time. This means that the signal norm must be independent of specific time instant.

Commonly used signal norms are shown as follows.

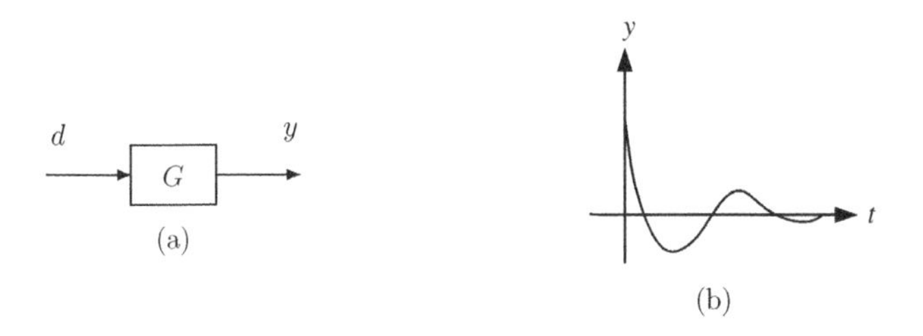

Figure 2.12 Disturbance attenuation (a) System (b) Disturbance response

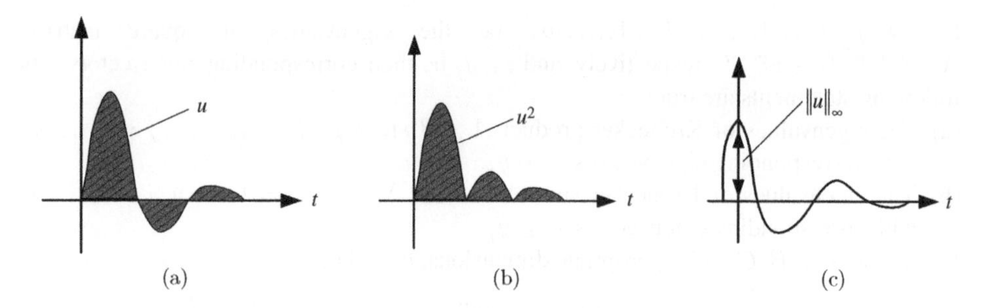

Figure 2.13 Signal norms (a) 1-norm (b) 2-norm (c) Infinity-norm

2.15.1.1 Scalar Signal

1. 1-norm (area of absolute value; refer to Figure 2.13(a))

$$\|u\|_1 = \int_0^\infty |u(t)|\,dt \tag{2.105}$$

2. 2-norm (square root of squared area; refer to Figure 2.13(b))

$$\|u\|_2 = \sqrt{\int_0^\infty u^2(t)\,dt} \tag{2.106}$$

3. Infinity-norm (maximum amplitude; refer to Figure 2.13(c))

$$\|u\|_\infty = \sup_{t\in[0,\infty)} |u(t)| \tag{2.107}$$

All these functions are independent of specific time instants; their values indicate different aspects of the response. It is easy to prove that they all satisfy the condition of norm and thus can be used as a norm.

It should be noted that in comparing two signals, we must use the same norm. This is because different norms have different values even for the same signal.

Example 2.18 *For the signal below, calculate the norms defined earlier.*

$$u(t) = e^{-3t}, \quad t \geq 0.$$

Solution Calculation based on the definitions yields

$$\|u\|_1 = \int_0^\infty e^{-3t}\,dt = \frac{1}{3}, \quad \|u\|_2 = \sqrt{\int_0^\infty e^{-6t}\,dt} = \frac{\sqrt{6}}{6},$$

$$\|u\|_\infty = \max_{t\geq 0} |e^{-3t}| = 1.$$

Clearly, the values of various norms are different. ▽

2.15.1.2 Vector Signal

Let a vector signal $u(t)$ be

$$u(t) = \begin{bmatrix} u_1(t) \\ \vdots \\ u_n(t) \end{bmatrix}. \tag{2.108}$$

Its various kinds of norms are defined as follows:

$$\|u\|_1 = \int_0^\infty \sum_{i=1}^n |u_i(t)|\,dt, \tag{2.109}$$

$$\|u\|_2 = \sqrt{\int_0^\infty \sum_{i=1}^n u_i^2(t)\,dt}, \tag{2.110}$$

$$\|u\|_\infty = \max_{1\le i\le n} \sup_{t\in[0,\infty)} |u_i(t)|. \tag{2.111}$$

2.15.2 *Inner Product of Signals*

For quadratically integrable scalar signals $u(t), v(t)$, their inner product is defined as

$$\langle u, v\rangle = \int_0^\infty u(t)v(t)\,dt. \tag{2.112}$$

And for quadratically integrable vector signals $u(t), v(t) \in \mathbb{R}^n$, their inner product is

$$\langle u, v\rangle = \int_0^\infty u^T(t)v(t)\,dt = \sum_{i=1}^n \int_0^\infty u_i(t)v_i(t)\,dt. \tag{2.113}$$

In circuit theory, in addition to the difference of amplitude between sine waves with the same frequency, there also exists the phase difference. Let us consider how to describe this characteristic. Intuitively, the sine wave $\sin(\omega t + \varphi)$ is the projection on the vertical axis of a unit vector rotating counterclockwise with an angular velocity ω from an initial angle φ in the two-dimensional vector space. Therefore, the phase difference of the sine wave signals $\sin(\omega t)$ and $\sin(\omega t - \varphi)$ can be thought of as the angle between the two unit vectors rotating at the same speed. The next question is, how to use the inner product to express the phase angle? Noting that the integrals of the product of trigonometric functions are equal in each cycle, we define the inner product and norm of trigonometric functions (vectors) $u(t), v(t)$ with the same frequency as

$$\langle u, v\rangle = \frac{2}{T}\int_0^T u^T(t)v(t)\,dt, \tag{2.114}$$

$$\|u\| = \sqrt{\langle u, u\rangle} = \sqrt{\frac{2}{T}\int_0^T u^T(t)u(t)\,dt}. \tag{2.115}$$

For example, the inner product of $u(t) = A\sin(\omega t), v(t) = B\sin(\omega t - \varphi)$ is

$$\begin{aligned}\langle u, v\rangle &= AB\frac{2}{T}\int_0^T \sin(\omega t)\sin(\omega t - \varphi)dt \\ &= AB\frac{1}{T}\int_0^T [\cos(\varphi) - \cos(2\omega t - \varphi)]dt = AB\cos\varphi.\end{aligned}$$

Based on this equation and the definition of norm, it is easy to know that $\|u\| = A, \|v\| = B$ by setting $v(t)$ as $u(t)$ and $u(t)$ as $v(t)$, respectively. Thus,

$$\cos\varphi = \frac{\langle u, v\rangle}{\|u\|\|v\|}. \tag{2.116}$$

This confirms that the phase difference between two trigonometric signals with the same frequency indeed has the same meaning as the angle between two vectors in vector space. In fact, the inner product and norm defined earlier apply to any periodic functions with the same period.

2.15.3 Norm and Inner Product of Signals in Frequency Domain

In the frequency domain, the 2-norm of a quadratically integrable vector signal $\hat{u}(s) = \mathcal{L}[u(t)]$ is defined as

$$\|\hat{u}\|_2 = \sqrt{\frac{1}{2\pi}\int_{-\infty}^{\infty} \hat{u}^*(j\omega)\hat{u}(j\omega)d\omega}. \tag{2.117}$$

Moreover, the inner product between quadratically integrable vector signals $\hat{f}(s), \hat{g}(s)$ is defined as

$$\langle \hat{f}, \hat{g}\rangle = \frac{1}{2\pi}\int_{-\infty}^{\infty} \hat{f}^*(j\omega)\hat{g}(j\omega)d\omega. \tag{2.118}$$

The inner product defined as such has the following properties, which can be simply proved based on the previous definition (Exercise 2.31).

Lemma 2.5 *Assume that vector functions $\hat{f}(s)$, $\hat{g}(s)$ are quadratically integrable. Then, the following statements hold:*

1. $\langle a\hat{f}, b\hat{g}\rangle = \bar{a}b\langle \hat{f}, \hat{g}\rangle$ *for arbitrary* $a, b \in \mathbb{C}$.
2. $\langle \hat{f}, \hat{f}\rangle = \|\hat{f}\|_2^2$.
3. $\langle \hat{f}, H\hat{g}\rangle = \langle H^{\sim}\hat{f}, \hat{g}\rangle$ *holds when matrix* $H(s)$ *has no poles on the imaginary axis.*
4. *If* $A^*(j\omega)A(j\omega) = I\ (\forall\omega)$, *then* $\|A\hat{f}\|_2 = \|\hat{f}\|_2$.

The statement (4) means that the 2-norm is invariant w.r.t. all-pass transfer function (matrix)[13]. This property will be frequently used hereafter. According to Parseval's theorem in Section 8.1, when the norm/inner product is bounded in the time domain, the time domain and frequency domain norms/inner products defined earlier are equal to each other.

[13] A transfer matrix satisfying $A^*(j\omega)A(j\omega) = I\ (\forall\omega)$ is known as the all-pass transfer matrix. See Section 9.3 and Section 10.1.2 for the detail.

2.15.4 Computation of 2-Norm and Inner Product of Signals

Assume that $\hat{f}(s), \hat{g}(s)$ are both strictly proper, rational functions (vectors) with real coefficients and have no purely imaginary poles[14]. Then, $\lim_{R\to\infty} R\cdot \hat{f}^T(-Re^{j\,\theta})\hat{g}(Re^{j\,\theta}) = 0$ is true. Therefore, the integral

$$\lim_{R\to\infty}\int_{\pi/2}^{3\pi/2} \hat{f}^T(-Re^{j\theta})\hat{g}(Re^{j\theta})d(Re^{j\theta})$$

along a semicircle with infinite radius is also zero. Hence, we have

$$\langle \hat{f},\hat{g}\rangle = \frac{1}{2\pi}\int_{-\infty}^{\infty} \hat{f}^*(j\omega)\hat{g}(j\omega)d\omega = \frac{1}{2\pi j}\oint_{\partial\Omega} \hat{f}^T(-s)\hat{g}(s)ds. \tag{2.119}$$

The path of the final closed integral is composed of the imaginary axis and a semicircle on the left half-plane with infinite radius. According to the residue theorem, the inner product $\langle \hat{f},\hat{g}\rangle$ equals the sum of residues of all poles of $\hat{f}^T(-s)\hat{g}(s)$ on the *left* half-plane. That is, let s_i be an arbitrary left-half- plane pole of $\hat{f}^T(-s)\hat{g}(s)$; then

$$\langle \hat{f},\hat{g}\rangle = \sum_i \operatorname*{Res}_{\Re(s_i)<0} \hat{f}^T(-s)\hat{g}(s) \tag{2.120}$$

holds in which Res_{s_i} denotes the residue at the point s_i. Therefore, the calculation of inner product boils down to the calculation of residue.

Further, for a stable function $\hat{g}(s)$, its poles are the same as the left-half-plane poles of $\hat{g}^T(-s)\hat{g}(s)$. Substituting $\hat{f}(s) = \hat{g}(s)$ into (2.120), we see that the 2-norm $\|g\|_2 = \|\hat{g}\|_2$ of stable function $\hat{g}(s)$ can be calculated as follows by using residue

$$\|g\|_2 = \|\hat{g}\|_2 = \sqrt{\sum_i \operatorname*{Res}_{s_i} \hat{g}^T(-s)\hat{g}(s)}. \tag{2.121}$$

Here, s_i denotes a pole of $\hat{g}(s)$.

Example 2.19 *Let $u(t)$ be the input of stable transfer function $G(s) = 1/(s+1)$ and $y(t)$ be the output.*

1. *Calculate the 2-norm of the unit impulse response $g(t)$.*
2. *For $u(t) = e^{-5t}$, compute $\|y\|_2$.*

Solution (1) First, $G(-s)G(s) = 1/(1-s)(1+s)$ has a pole $p = -1$ on the left half-plane. The residue at the pole is

$$\lim_{s\to -1}(s+1)G(-s)G(s) = \lim_{s\to -1}(s+1)\frac{1}{(1-s)(s+1)} = \lim_{s\to -1}\frac{1}{1-s} = \frac{1}{2}.$$

So, $\|g\|_2 = 1/\sqrt{2}$.

[14] Refer to Chapter 4 for the concepts of strict properness, pole, and stability.

(2) Since $\hat{u}(s) = 1/(s+5)$, $\hat{y}(s)$ is equal to $1/(s+1)(s+5)$ and has two poles $p = -1, -5$. The residues at the stable poles of $\hat{y}(-s)\hat{y}(s) = 1/(1-s)(5-s)(1+s)(5+s)$ are

$$\lim_{s\to -1}(s+1)\hat{y}(-s)\hat{y}(s) = \frac{1}{48}, \quad \lim_{s\to -5}(s+5)\hat{y}(-s)\hat{y}(s) = -\frac{1}{240}.$$

So, $\|y\|_2 = 1/\sqrt{60}$. ▽

The following lemma shows that stable function is orthogonal to antistable function without purely imaginary poles.

Lemma 2.6 *Assume that $\hat{f}(s)$ is a stable (vector) function and $\hat{g}(s)$ is an antistable (vector) function without purely imaginary poles, both being strictly proper. Then, there hold*

$$\langle \hat{f}, \hat{g} \rangle = 0, \quad \|\hat{f} + \hat{g}\|_2^2 = \|\hat{f}\|_2^2 + \|\hat{g}\|_2^2.$$

Proof. As $\hat{f}, \hat{g}$ are strictly proper and have no imaginary poles, their inner product exists. Also, since $\hat{f}^T(-s)\hat{g}(s)$ has no poles on the left half-plane, according to (2.120), $\langle \hat{f}, \hat{g} \rangle = 0$ is true. The second equation comes from Pythagoras' law (refer to Theorem 2.3(3)). •

The following simple results will be used in the proofs of Chapter 10 (Exercise 2.32).

Lemma 2.7 *Let $\Re(\lambda) > 0$, $\Re(\eta) > 0$. Then, we have*

$$\left\langle \frac{1}{\lambda - s}, \frac{1}{\eta - s} \right\rangle = \frac{1}{\bar{\lambda} + \eta}, \quad \left\| \frac{1}{\lambda - s} \right\|_2^2 = \frac{1}{2\Re(\lambda)}. \tag{2.122}$$

2.15.5 System Norm

The transfer function of a system is essentially used for quantifying its amplification or attenuation ability of input signal. When analyzing the amplification or attenuation rate (i.e., gain) of a system from the angle of frequency response, since

$$\hat{y}(j\omega) = G(j\omega)\hat{u}(j\omega) \tag{2.123}$$

we see that the gain at frequency ω is $|G(j\omega)|$. However, in comparing the gains of two systems, if we use $|G(j\omega)|$, then we will get completely different conclusions at different frequencies. For example, in Figure 2.14 $|G_1(j\omega_1)| > |G_2(j\omega_1)|$ at frequency ω_1, while $|G_1(j\omega_2)| < |G_2(j\omega_2)|$ at frequency ω_2. In order to avoid such situation, a measure independent of specific frequency is necessary.

The norms of the system shown in Figure 2.15 are defined as follows.

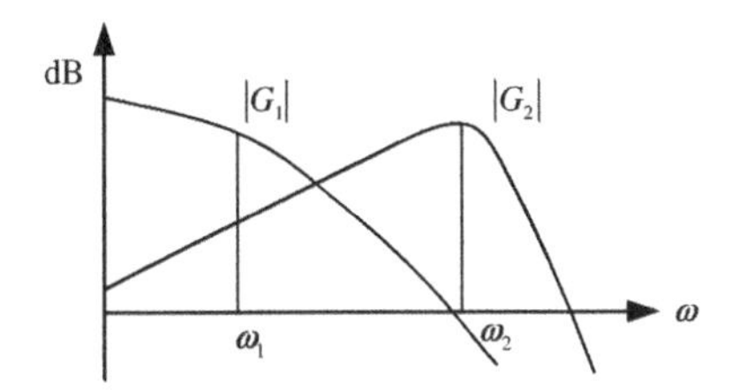

Figure 2.14 Comparison of gains of transfer functions

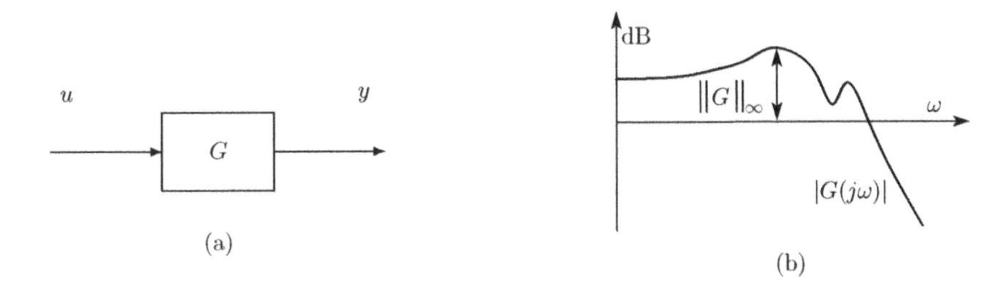

Figure 2.15 System and frequency response (a) Transfer function (b) $\mathcal{H}_\infty$ norm

2.15.5.1 Transfer Function

$\mathcal{H}_2$ norm:

$$\|G\|_2 = \sqrt{\frac{1}{2\pi}\int_{-\infty}^{\infty}|G(j\omega)|^2 d\omega} = \sqrt{\int_0^{\infty} g^2(t)\,dt}. \tag{2.124}$$

Here, $g(t) = \mathcal{L}^{-1}[G(s)]$ denotes the unit impulse response of transfer function $G(s)$. $\mathcal{H}_2$ norm is the square root of the squared area of the frequency response gain, which is also equal to the square root of the squared area of unit impulse response. The second equality in the equation comes from Parseval's theorem (refer to Section 8.1).

$\mathcal{H}_\infty$ norm:

$$\|G\|_\infty = \sup_{\omega\in(-\infty,\infty)} |G(j\omega)| \tag{2.125}$$

As shown in Figure 2.15(b), the $\mathcal{H}_\infty$ norm is the maximal amplitude of the frequency response of transfer function.

Example 2.20 *Calculate the $\mathcal{H}_2$ norm and $\mathcal{H}_\infty$ norm of stable transfer function*

$$G(s) = \frac{10}{(s+1)(s+10)}.$$

Solution First, we calculate the unit impulse response of $G(s)$. By partial fraction expansion, we get

$$G(s) = \frac{10}{9}\left(\frac{1}{s+1} - \frac{1}{s+10}\right).$$

So, the unit impulse response is

$$g(t) = \frac{10}{9}(e^{-t} - e^{-10t}), \quad t \geq 0.$$

Thus, we get

$$\|G\|_2 = \sqrt{\int_0^\infty |g(t)|^2 dt} = \sqrt{\frac{5}{11}}.$$

On the other hand,

$$|G(j\omega)|^2 = \frac{100}{(\omega^2+1)(\omega^2+100)}$$

is a continuous function. So the slope is zero at its maximum point. The solutions of $0 = \frac{d|G(j\omega)|^2}{d\omega} = \frac{d|G(j\omega)|^2}{d\omega^2}\frac{d\omega^2}{d\omega} = 2\omega\frac{d|G(j\omega)|^2}{d\omega^2}$ are $\omega = 0$ and $\omega = \infty$. Since $|G(j\infty)| = 0$, we finally get

$$\|G\|_\infty = |G(j0)| = 1.$$

In fact, for this example, there is no need to calculate the $\mathcal{H}_\infty$ norm. The conclusion can be obtained simply by drawing the Bode plot. ▽

2.15.5.2 Transfer Matrix

$\mathcal{H}_2$ norm:

$$\begin{aligned}\|G\|_2 &= \sqrt{\frac{1}{2\pi}\int_{-\infty}^{\infty} \mathrm{Tr}(G^*(j\omega)G(j\omega))d\omega} \\ &= \sqrt{\int_0^\infty \mathrm{Tr}(g^T(t)g(t))\,dt}, \quad g(t) = \mathcal{L}^{-1}[G(s)] \end{aligned} \tag{2.126}$$

The relationship between input/output sizes and the $\mathcal{H}_2$ norm of transfer matrix, as well as the calculation method of the $\mathcal{H}_2$ norm, will be explained detailedly in Chapter 15.

$\mathcal{H}_\infty$ norm:

$$\|G\|_\infty = \sup_{\omega\in[0,\infty)} \sigma_{\max}(G(j\omega)) \tag{2.127}$$

Here, $\sigma_{\max}(G(j\omega))$ denotes the largest singular value of matrix $G(j\omega)$. The calculation method of $\mathcal{H}_\infty$ norm will be introduced in Example 8.2 of Section 8.2.1, while its relation with the input and output will be introduced in Chapter 16.

2.15.6 Inner Product of Systems

For quadratically integrable transfer matrices $\hat{F}(s), \hat{G}(s)$ and unit impulse response matrices $F(t), G(t)$, their inner products are defined, respectively, as

$$\langle \hat{F}(j\omega), \hat{G}(j\omega)\rangle = \frac{1}{2\pi}\int_{-\infty}^{\infty} \mathrm{Tr}(\hat{F}^*(j\omega)\hat{G}(j\omega))d\omega, \tag{2.128}$$

$$\langle F(t), G(t)\rangle = \int_{0}^{\infty} \mathrm{Tr}(F^T(t)G(t))\,dt. \tag{2.129}$$

And they are equal (refer to Section 8.1).

Exercises

2.1 Prove that the following equation holds for any matrices $B \in \mathbb{R}^{n\times m}, C \in \mathbb{R}^{m\times n}$

$$\det(I_n + BC) = \det(I_m + CB).$$

In particular, when b is a column vector and c is a row vector, $\det(I + bc) = 1 + cb$.

2.2 Prove that the statement
(a) If $x, y \in \mathbb{R}^n$, then $x + y \in \mathbb{R}^n$.
(b) If $x \in \mathbb{R}^n$ and $a \in \mathbb{R}$, then $ax \in \mathbb{R}^n$.
is equivalent to the following statement:

For any $x, y \in \mathbb{R}^n$ and $a, b \in \mathbb{R}$, there holds $ax + by \in \mathbb{R}^n$.

2.3 Calculate the 1-norm, 2-norm, and infinity-norm of vector $x = [1\ 2\ 3]^T$.

2.4 Given vectors $u = [0\ 1]^T$ and $v = [1\ 1]^T$, calculate the 2-norm, inner product, and the angle between them, respectively.

2.5 The distance between the origin and the point $P(x, y, z)$ in the three-dimensional Euclidean space is

$$d(P) = \sqrt{x^2 + y^2 + z^2}.$$

Prove that this distance function $d(P)$ satisfies the norm condition.

2.6 Prove that the image $\mathrm{Im}A$ of matrix $A \in \mathbb{F}^{m\times n}$ is a subspace in the range $\mathbb{F}^m$ and the kernel space $\mathrm{Ker}A$ is a subspace in the domain $\mathbb{F}^n$.

2.7 Given a matrix

$$A = \begin{bmatrix} 1 & 0 & 0 \\ 0 & 0 & 1 \end{bmatrix},$$

find $\mathrm{rank}(A)$ and the bases of $\mathrm{Im}A = \{y \in R^2 \mid y = Ax, x \in R^3\}$, $\mathrm{Ker}A = \{x \in R^3 \mid Ax = 0\}$.

2.8 Discuss the existence and uniqueness of the solution of linear algebraic equation

$$\begin{bmatrix} 2 & -1 \\ -3 & 3 \\ -1 & 2 \end{bmatrix} x = b, \quad b = \begin{bmatrix} 1 \\ 0 \\ 1 \end{bmatrix}.$$

If b changes to $b = [1\ 1\ 1]^T$, whether there is a solution or not?

2.9 Find the general solution for algebraic equation

$$\begin{bmatrix} 1 & 2 & 3 & 4 \\ 0 & -1 & -2 & 2 \\ 0 & 0 & 0 & 1 \end{bmatrix} x = \begin{bmatrix} 3 \\ 2 \\ 1 \end{bmatrix}.$$

How many free parameters are there in the solution?

2.10 Consider the equation

$$x[n] = A^n x[0] + A^{n-1} bu[0] + A^{n-2} bu[1] + \cdots + Abu[n-2] + bu[n-1]$$

where $A \in \mathbb{R}^{n \times n}, b \in \mathbb{R}^n$ are known. To ensure that this equation has a solution $u[0], \ldots, u[n-1]$ for arbitrary $x[n], x[0]$, what condition (A, b) must satisfy?

2.11 Given matrix $A \in \mathbb{R}^{m \times n}$, $\text{rank}(A) = \min(m, n)$, consider the linear algebraic equation $Ax = b$.

(a) If $m > n$, no solution exists except the case of $b \in \text{Im}(A)$. However, there exists a solution that minimizes the norm $\|e\|^2 = e^T e$ of error $e = Ax - b$. Prove that the minimum solution is $x = (A^T A)^{-1} A^T b$.

(b) If $m < n$, there exist an infinite number of solutions. Prove that the solution minimizing the norm $\|x\|^2 = x^T x$ is $x = A^T (AA^T)^{-1} b$. (Hint: Use the Lagrange multiplier method.)

2.12 Calculate the eigenvalues and eigenvectors of matrices

$$A = \begin{bmatrix} 0 & 1 \\ 3 & -2 \end{bmatrix}, \quad B = \begin{bmatrix} 0 & 1 \\ 3 & 0 \end{bmatrix}.$$

2.13 Calculate the eigenvalues and eigenvectors of matrix

$$A = \begin{bmatrix} 1 & 1 & 0 \\ 0 & 0 & 1 \\ 0 & 0 & 1 \end{bmatrix}$$

and convert it into the Jordan canonical form. Then, compute A^k using the Jordan canonical form (k is a natural number).

2.14 Let λ and x be the eigenvalue and eigenvector of matrix A, respectively. Prove that the matrix function $f(A)$ has the eigenvalue $f(\lambda)$ and the eigenvector x. (Hint: Use Taylor expansion of $f(\lambda)$.)

2.15 Assume that the eigenvalues of matrix A are distinct and that λ_i, q_i are, respectively, the eigenvalue and *right eigenvector* satisfying $Aq_i = \lambda_i q_i$.
(a) Set $Q = [q_1\ q_2\ \cdots q_n]$. Prove that Q is nonsingular.
(b) Set a row vector p_i as the ith row of the inverse of Q, that is,

$$P = Q^{-1} := \begin{bmatrix} p_1 \\ \vdots \\ p_n \end{bmatrix}.$$

Prove that p_i is a *left eigenvector* of A, that is, it satisfies $p_i A = \lambda_i p_i$.

2.16 Prove that all eigenvalues of an Hermitian matrix $A \in \mathbb{F}^{n\times n}$ are real numbers, and then prove that all eigenvalues of a positive semidefinite matrix are nonnegative.

2.17 Prove, based on Theorem 2.7, that when a subspace S (except the zero vector) is an invariant subspace of A, there must be a nonzero vector $x \in S$ and a scalar λ satisfying $Ax = \lambda x$. (This means that the subspace S must contain some eigenvectors of A.)

2.18 Check whether the symmetric matrix

$$A_1 = \begin{bmatrix} 2 & 1 \\ 1 & 1 \end{bmatrix}, \quad A_2 = \begin{bmatrix} 2 & 1 & 1 \\ 1 & 1 & 1 \\ 1 & 1 & 2 \end{bmatrix}$$

is positive definite using Schur's lemma.

2.19 Find the singular values and singular vectors of the matrices in Exercise 2.12.

2.20 When matrix A is an Hermitian matrix, discuss the relationship between its eigenvalues and singular values.

2.21 Calculate the 1-norm, 2-norm, and infinity-norm of the matrices in Exercise 2.12.

2.22 Mimicking Example 2.16 to prove that the induced norms defined by $\|A\|_p = \sup \frac{\|Au\|_p}{\|u\|_p}$ $(p = 2, \infty)$ are equal to

$$\|A\|_2 = \sqrt{\lambda_{\max}(A^*A)}, \quad \|A\|_\infty = \max_{1\le i\le m} \sum_{j=1}^{n} |a_{ij}|.$$

2.23 For the matrix

$$A = \begin{bmatrix} 1+\sin^2\theta & \sin\theta\cos\theta \\ \sin\theta\cos\theta & 1+\cos^2\theta \end{bmatrix},$$

calculate the mapping in Example 2.17 and verify it by drawing a figure.

2.24 For matrix $A(t) \in \mathbb{R}^{n\times n}$, prove the following equation using $A^{-1}A = I$

$$\frac{dA^{-1}}{dt} = -A^{-1}\frac{dA}{dt}A^{-1}.$$

2.25 Prove the differentiation equation (2.100) about matrix.

2.26 Among the following functions, which can be used as a norm?
(a) $\sup_t |\dot{u}(t)|$
(b) $|u(0)| + \sup_t |\dot{u}(t)|$
(c) $\lim_{T\to\infty} \sqrt{\frac{1}{T}\int_{-T}^{T} u^2(t)dt}$

2.27 Calculate the $\mathcal{H}_2$ norm and $\mathcal{H}_\infty$ norm of stable transfer function

$$G(s) = \frac{s+1}{(s+2)(s+5)}.$$

2.28 Calculate the $\mathcal{H}_2$ norm of the following stable transfer function by using, respectively, the residue and impulse response methods

$$G(s) = \frac{s+5}{(s+1)(s+10)}.$$

2.29 Prove the following equations. If necessary, it may be assumed that the $\mathcal{H}_2$ norm or $\mathcal{H}_\infty$ norm of transfer function $G(s)$ exists.
(a) $\|DG\|_2 = \|G\|_2$, $\|DG\|_\infty = \|G\|_\infty$ for the time delay $D(s) = e^{-sT}$.
(b) $\|AG\|_2 = \|G\|_2$, $\|AG\|_\infty = \|G\|_\infty$ for the all-pass filter $A(s) = \frac{s-a}{s+a}$ $(a > 0)$.
(These properties indicate that the $\mathcal{H}_2$ and $\mathcal{H}_\infty$ norms of transfer function are invariant to time delay and all-pass filter.)

2.30 Prove that the function defined in (2.118) is an inner product.

2.31 Prove Lemma 2.5.

2.32 Prove Lemma 2.7.

Notes and References

The fundamentals of complex analysis can be found in Ref. [1]. Bellman [7] and Kodama and Suda [49] are classical references for linear algebra. Boullion and Odell [8] provides a quite complete summary on the generalized inverse matrices. For more details about the norms and inner products of signals and systems, refer to Refs [9], [10] and [100]. Liu *et al.* [62] investigated the engineering implication of system norms and revealed the essence of weighting function.

Due to the page limitation, some proofs are omitted. They are provided on the supplementary web site of this book.

3

Basics of Convex Analysis and LMI

In the analysis and design of robust control systems, linear matrix inequality (LMI) method plays a vital role. Meanwhile, convex analysis is an essential knowledge for the understanding of LMI approach and other associated optimization methods. So, this chapter introduces both of them briefly.

3.1 Convex Set and Convex Function

3.1.1 Affine Set, Convex Set, and Cone

3.1.1.1 Affine Set

Given two points (vectors) x_1, x_2 pointing to different directions in $\mathbb{R}^n$ space, a new point

$$y = \theta x_1 + (1-\theta)x_2 = x_2 + \theta(x_1 - x_2) \tag{3.1}$$

may be created using an arbitrary real number θ. The set of all these points forms a straight line passing through x_1, x_2 (imagine the two-dimensional case and refer to Figure 3.1[1]). As long as x_1, x_2 are not zero vectors and the angle between them is neither 0 [rad] nor π [rad], this line does not pass through the origin. $y = x_2$ when $\theta = 0$, and $y = x_1$ when $\theta = 1$. In particular, when θ takes value in the interval $[0, 1]$, this set becomes the line segment between x_1 and x_2. The feature of Eq. (3.1) is that coefficients of the linear combination of vectors x_1, x_2 sum to 1. This is a special case of linear combination, called *affine combination*.

For any two points x_1, x_2 in the set $C \subset \mathbb{R}^n$, if their affine combination still belongs to C, then C is called an *affine set*. That is, $\theta x_1 + (1-\theta)x_2 \in C$ holds for any $\theta \in \mathbb{R}$. It is worth

[1] Originally both x_1 and x_2 are vectors, so they should be drawn as vectors connecting the origin and end point x_1, x_2. However, the sets to be studied later do not necessarily contain the origin. It is rather cumbersome to draw the vector from the origin every time. For this reason, we usually omit the origin and only use the vertex of a vector to represent it. One more benefit of this convention is that it is very easy to describe a set in multidimensional spaces.

Robust Control: Theory and Applications, First Edition. Kang-Zhi Liu and Yu Yao.

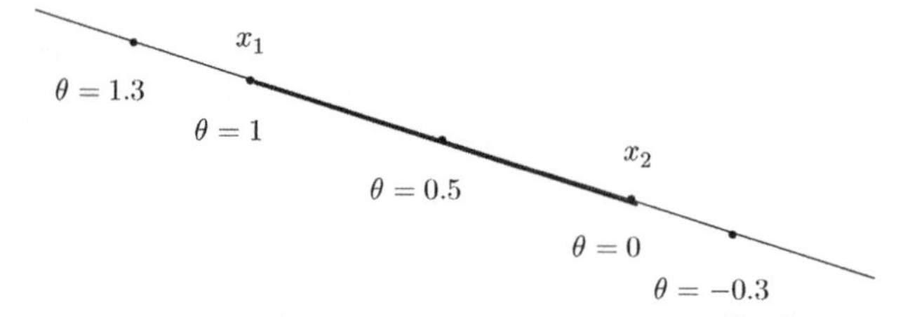

Figure 3.1 Affine set

noting that an affine set can be extended to the infinity along certain direction, so it is not bounded (refer to Figure 3.1).

Example 3.1 *The solution set of linear equation $Ax = b$ is an affine set. This can be easily understood as follows. Assume that x_1, x_2 are two arbitrary solutions. Mapping their affine combination $\theta x_1 + (1-\theta)x_2$ with A, we get*

$$\begin{aligned} A(\theta x_1 + (1-\theta)x_2) &= \theta A x_1 + (1-\theta)A x_2 \\ &= \theta b + (1-\theta)b \\ &= b. \end{aligned}$$

So, $\theta x_1 + (1-\theta)x_2$ is again a solution.

The notion of affine set can be extended to affine combinations of more than two points. For example, using coefficients $\theta_i \in \mathbb{R}$ satisfying $\theta_1 + \cdots + \theta_k = 1$ to do affine combination of k points $x_1, \ldots, x_k \in C$, we obtain a new point $\theta_1 x_1 + \cdots + \theta_k x_k$. As long as the set C is affine, we may use the induction method to show that this new point $\theta_1 x_1 + \cdots + \theta_k x_k$ is also contained in the set C (Exercise 3.1).

Affine combinations of three points x_1, x_2, x_3 pointing in different directions form a plane. In general, affine combinations of several points form a hyperplane[2]. This means that an affine set is a hyperplane. However, an affine set does not necessarily contain the origin, so it is not necessarily a subspace. Shifting one point to the origin via translation, an affine set turns into a subspace. Specifically, assume that the set C is affine and $x_0 \in C$. Then, the set

$$V = C - x_0 = \{x - x_0 \mid x \in C\} \tag{3.2}$$

becomes a subspace. That is, the set V is closed w.r.t the manipulation of linear combination. To see this, let $v_1, v_2 \in V$ and α, β be any real numbers; then $x_1 = v_1 + x_0, x_2 = v_2 + x_0$ are also points in C. The point

$$\alpha v_1 + \beta v_2 + x_0 = \alpha(x_1 - x_0) + \beta(x_2 - x_0) + x_0 = \alpha x_1 + \beta x_2 + (1 - \alpha - \beta)x_0$$

is an affine combination of point x_0, x_1, x_2, so it belongs to C. Therefore, $(\alpha x_1 + \beta x_2 + x_0) - x_0 = \alpha v_1 + \beta v_2 \in V$ holds. This shows that V is indeed a subspace.

[2] For details on the hyperplane, refer to Subsection 3.1.2.

From (3.2) we see that the affine set C in turn can be expressed by a subspace V and a point $x_0 \in C$ as

$$C = V + x_0 = \{x = v + x_0 \mid v \in V\}. \tag{3.3}$$

3.1.1.2 Convex Set

Further, restricting the combination coefficients to nonnegative real numbers in the affine combination, that is, $\theta_i \geq 0$ and $\theta_1 + \cdots + \theta_k = 1$. If any point obtained from the combination of $x_i \in C$ still belongs to C, that is

$$\theta_1 x_1 + \cdots + \theta_k x_k \in C, \tag{3.4}$$

then the set C is called a *convex set*. Such combination is called *convex combination*. The convex combination of two points x_1, x_2 is a line segment, and the convex combination of three points x_1, x_2, x_3 is a triangle. In general, a convex set is not necessarily a compact set. For example, any quadrant of the two-dimensional space is a convex set, but unbounded. In addition, it should be remembered that the combination coefficients of a convex combination are all nonnegative, and their sum is 1. Due to these constraints, the coefficient θ_i is limited in the interval $[0, 1]$.

The characteristic of convex set is that all points in the segment between two points of the set are again in the convex set (see (a) of Figure 3.2). Why? This is because for any $\theta \in [0, 1]$, $\theta x_1 + (1 - \theta)x_2 \in C$ as long as $x_1, x_2 \in C$. Since an affine set contains all points of the line connecting any two points of the set, it naturally contains the segment between these two points. Therefore, *an affine set automatically becomes a convex set*.

Moreover,

$$\textbf{conv}\ C = \{\theta_1 x_1 + \cdots + \theta_k x_k \mid x_i \in C, \theta_i \geq 0, \theta_1 + \cdots + \theta_k = 1\}, \tag{3.5}$$

a set formed by all convex combinations of finite points $x_i \quad (i = 1, \ldots, k)$ in a set C that is not necessarily convex is called a *convex hull* of the set C. Here, k, the number of combined points is arbitrary. This set is a closed convex set, and it is also the minimum set that contains the set C. For example, the convex hull of the nonconvex set in (b) of Figure 3.2 is equal to the set shown in Figure 3.3.

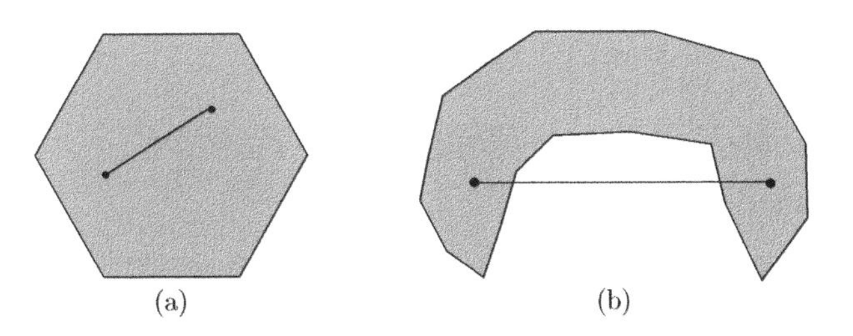

Figure 3.2 Convex set (a) and nonconvex set (b)

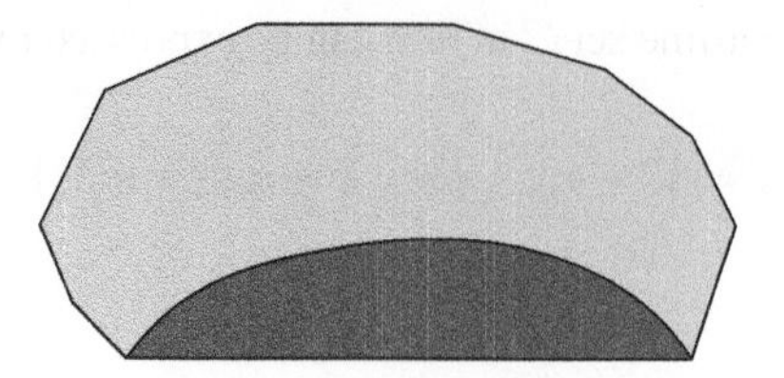

Figure 3.3 Convex hull of nonconvex set

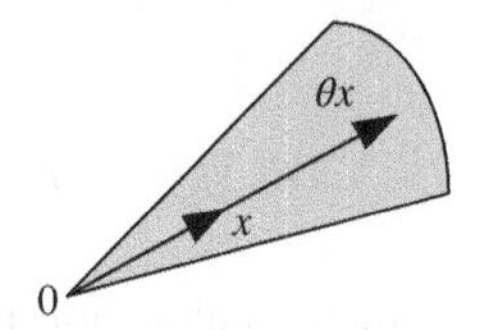

Figure 3.4 Cone

3.1.1.3 Cone

Next, we consider cones. A set C is called a *cone* if $\theta x \in C$ for any vector $x \in C$ and coefficient $\theta \geq 0$. Because the coefficient θ is limited to nonnegative real number, θx can stretch arbitrarily in the x direction, but not along the $-x$ direction (refer to Figure 3.4). Obviously, a cone is not bounded. In particular, a cone with convexity is called a *convex cone*. For a convex cone C, due to its convexity, a point formed by a convex combination of any two points $x_1, x_2 \in C$ belongs to C; further, multiplication of this point by a positive constant still belongs to C. Therefore, for any $\theta_1, \theta_2 \geq 0$ there holds

$$\theta x_1 + \theta_2 x_2 \in C. \tag{3.6}$$

The shape of a two-dimensional cone looks like a cut of pie (Figure 3.4). Moreover, the point $\theta x_1 + \cdots + \theta_k x_k$ formed by linear combination with coefficients $\theta_1, \ldots, \theta_k \geq 0$ is called a *conic combination*.

The so-called *conic hull* of a set C is the set of all conic combinations of points in C, that is,

$$\{\theta_1 x_1 + \cdots + \theta_k x_k \mid x_i \in C,\ \theta_i > 0,\ i = 1, \ldots, k\} \tag{3.7}$$

in which k is an arbitrary natural number. It is the smallest one in all convex cones that contain the set C.

3.1.2 *Hyperplane, Half-Space, Ellipsoid, and Polytope*

3.1.2.1 Hyperplane

The so-called *hyperplane* refers to a set that includes all points $x \in \mathbb{R}^n$ satisfying the equation $a^T x = b$ w.r.t. vector $a \in \mathbb{R}^n$ and scalar $b \in \mathbb{R}$. In other words, a hyperplane is the set

$$\{x \in \mathbb{R}^n \mid a^T x = b\}. \tag{3.8}$$

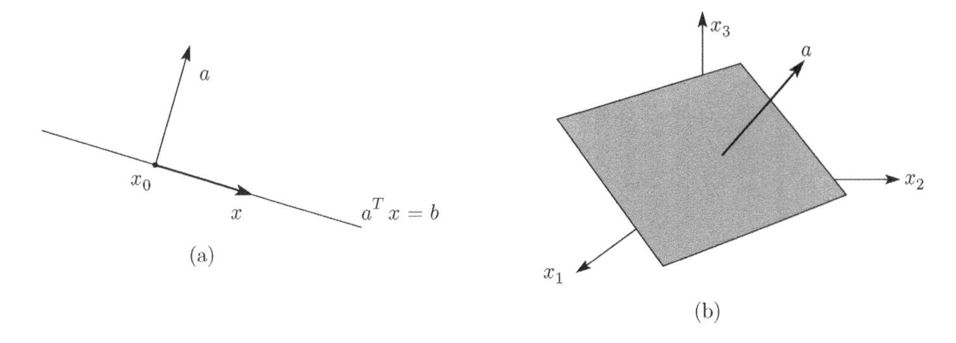

Figure 3.5 Hyperplanes (a) One-dimensional case (b) Two-dimensional case

In the two-dimensional space, this set is a straight line with a normal vector a (Figure 3.5(a)); the set becomes a plane with a normal vector a in the three-dimensional space (Figure 3.5(b)). This statement may be understood as follows. If x_0 is a point on the hyperplane, all points on the hyperplane satisfy

$$a^T(x - x_0) = b - b = 0.$$

This shows that the vector a is orthogonal to the vector $x - x_0$ connecting two points on the hyperplane. Therefore, a is a normal vector of the hyperplane.

Example 3.2 *A hyperplane about vector a and scalar b below*

$$a = [1\ 1\ 1]^T, \quad b = 1$$

is given by

$$x_1 + x_2 + x_3 = 1 \Rightarrow x_3 = 1 - (x_1 + x_2).$$

There are two free variables (x_1, x_2) here. So, it is a plane and shown in Figure 3.5(b). It is seen from this figure that a is indeed the normal vector of the hyperplane.

Later on, we will need to find the intersection point of the normal and hyperplane. So its calculation method is described here. If we extend the normal vector a by multiplying it with a constant β, it will eventually intersect the hyperplane. Let the point of intersection be $x_0 = \beta a$. Then, we have

$$b = a^T(\beta a) = \beta \|a\|^2 \Rightarrow \beta = \frac{b}{\|a\|^2}.$$

So, the point of intersection is $x_0 = \frac{b}{\|a\|^2} a$.

3.1.2.2 Half-Space

A hyperplane separates a space into two half-spaces. These (closed) *half-space* are the set $\{x | a^T x \leq b\}$ and the set $\{x | a^T x \geq b\}$ (refer to Figure 3.6). They are the solution sets of linear inequalities. Although a half-space is a convex set (think why), it is however not affine. Figure 3.6 shows that it cannot be extended to the opposite side of the hyperplane $a^T x = b$.

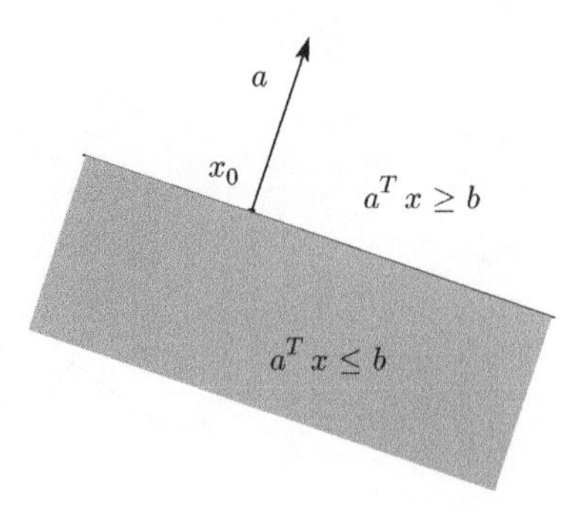

Figure 3.6 Half-space

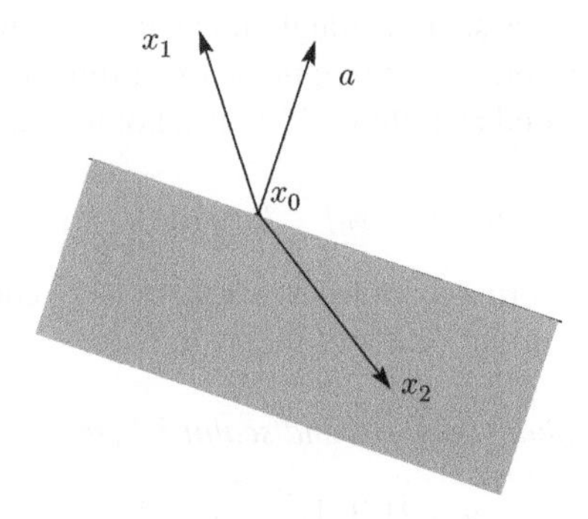

Figure 3.7 Vector $x_1 - x_0$ in the upper half-space $a^T(x - x_0) \geq 0$ has an acute angle with a, vector $x_2 - x_0$ in the lower half-space $a^T(x - x_0) \leq 0$ has an obtuse angle with a

In addition, when a point x_0 satisfies $a^T x_0 = b$, that is, it is a point on the hyperplane $\{x \mid a^T x = b\}$, the half-space $\{x | a^T x \leq b\}$ can be expressed as follows:

$$\{x | a^T(x - x_0) \leq 0\}. \tag{3.9}$$

Geometrically, this means that the vector from a point x_0 on the hyperplane to any point x in the half-space $\{x | a^T x \leq b\}$ always form an obtuse angle with the normal vector a of the hyperplane (Figure 3.7). Therefore, the half-space $\{x \in \mathbb{R}^n \mid a^T x \leq b\}$ is on the side opposite to a while the half-space $\{x \in \mathbb{R}^n \mid a^T x \geq b\}$ is on the same side of a. Clearly, the boundary of these two half-spaces is the hyperplane $\{x \mid a^T x = b\}$. It is worth noting that a half-space is not closed w.r.t the operation of linear combination, so it is not a subspace.

3.1.2.3 Ellipsoid

The so-called *ellipsoid* is a set of vectors (see Figure 3.8):

$$\mathcal{E} = \{x \mid (x - x_c)^T P^{-1}(x - x_c) \leq 1\} \tag{3.10}$$

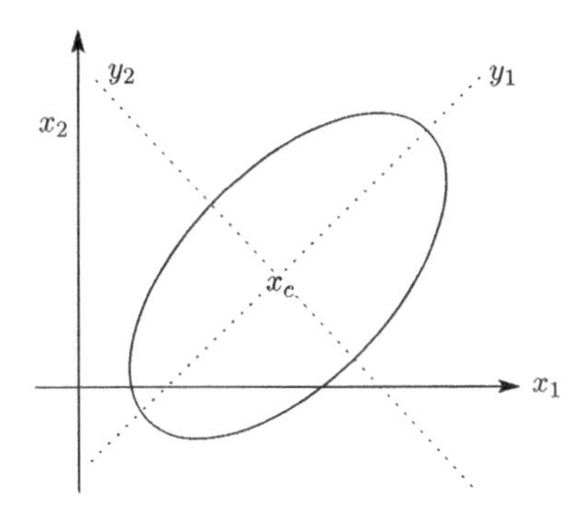

Figure 3.8 Ellipsoid in the two-dimensional space: it becomes symmetric in the y coordinate after translation and rotation

in which x_c is the center of ellipsoid and matrix $P = P^T$ is positive definite. Let λ_i be an eigenvalue of P; then $\sqrt{\lambda_i}$ is the length of a semiaxis of the ellipsoid. For example, in the case of three-dimensional space, P can be diagonalized as

$$UPU^T = \begin{bmatrix} \lambda_1 & & \\ & \lambda_2 & \\ & & \lambda_3 \end{bmatrix}$$

by using a unitary matrix U. Therefore, after a translation $\hat{x} = x - x_c$ and a rotation $y = U\hat{x}$, this set satisfies

$$\begin{aligned}(U\hat{x})^T \begin{bmatrix} \frac{1}{\lambda_1} & & \\ & \frac{1}{\lambda_2} & \\ & & \frac{1}{\lambda_3} \end{bmatrix} (U\hat{x}) &= y^T \begin{bmatrix} \frac{1}{\lambda_1} & & \\ & \frac{1}{\lambda_2} & \\ & & \frac{1}{\lambda_3} \end{bmatrix} y \\ &= \frac{y_1^2}{\lambda_1} + \frac{y_2^2}{\lambda_2} + \frac{y_3^2}{\lambda_3} \\ &\leq 1\end{aligned}$$

in the y coordinate. This confirms that $\sqrt{\lambda_1}, \sqrt{\lambda_2}, \sqrt{\lambda_3}$ are indeed the lengths of semiaxes. Further, the volume of a three-dimensional ellipsoid is proportional to

$$\lambda_1\lambda_2\lambda_3 = \det(P)$$

(Exercise 3.3). In general, the volume $\mathbf{vol}(\mathcal{E})$ of an n-dimensional ellipsoid is equal to

$$\mathbf{vol}(\mathcal{E}) = \det(P). \tag{3.11}$$

Here, we have omitted the factor which depends on the dimension of ellipsoid.

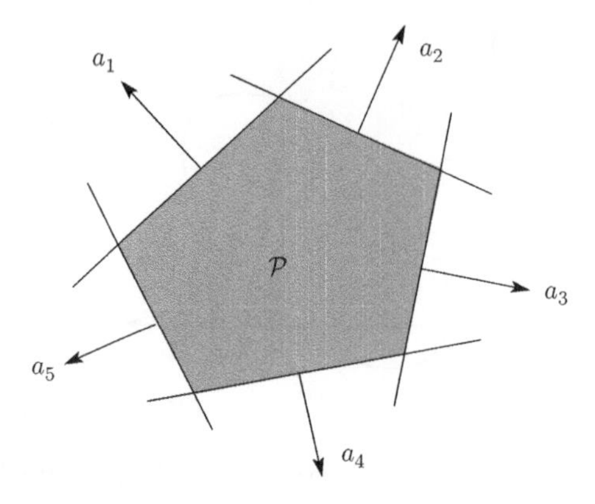

Figure 3.9 Polyhedron $\mathcal{P}$ formed by the intersections of half-spaces with normal a_i

3.1.2.4 Polyhedron

A *polyhedron* is the following set of vectors:

$$\mathcal{P} = \{x | a_i^T x \leq b_i, \ i = 1, \ldots, m; c_j^T x = d_j, j = 1, \ldots, p\}. \tag{3.12}$$

By definition, a polyhedron is the intersection of several half-spaces and hyperplanes (Figure 3.9). Affine sets, segments, and half-space are all polyhedra. It is easy to know that a polyhedron is convex (Exercise 3.4).

For example, the first quadrant

$$\{x \in \mathbb{R}^3 \mid x_i \geq 0, \ i = 1, 2, 3\}$$

of the three-dimensional space is a polyhedron. It is the intersection of three half-spaces $x_1 \geq 0$, $x_2 \geq 0$, and $x_3 \geq 0$. Incidentally, this set is also a cone.

Next, a bounded polyhedron is called a *polytope*. For example, the cube

$$0 \leq x_1 \leq 1, \quad 0 \leq x_2 \leq 1, \quad 0 \leq x_3 \leq 1$$

in the three-dimensional space is the intersection of six half-spaces, which is obviously bounded.

3.1.3 Separating Hyperplane and Dual Problem

3.1.3.1 Separating Hyperplane

The two-dimensional plane shown in Figure 3.10 has two convex sets C, D. Consider the case when they are disjoint. Intuitively, there should be a straight line passing through the area between these two convex sets. That is, we can use a straight line to separate the convex sets C, D. This is proved in the sequel.

First, we assume that the convex sets C and D are *compact sets* (i.e., closed and bounded) for simplicity. Since they are disjoint, $C \cap D = \emptyset$ holds. The distance between points $u \in C$

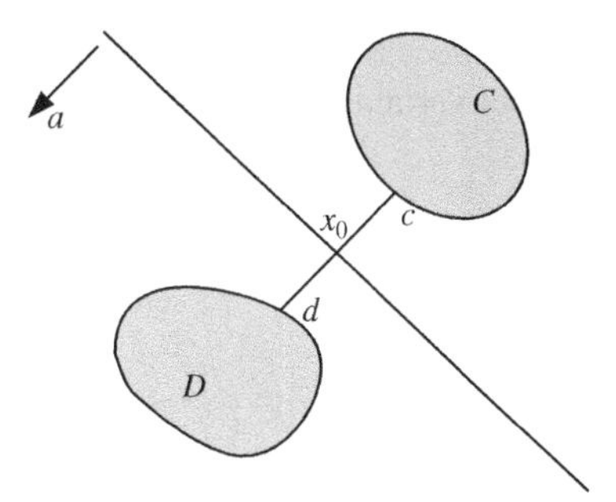

Figure 3.10 Separating hyperplane

and $v \in D$ is $\|u - v\|_2$. Meanwhile, the distance between two sets is defined as the infimum of point-wise distance, denoted by

$$\mathrm{dist}(C, D) := \inf\{\|u - v\|_2 \mid u \in C,\ v \in D\}. \tag{3.13}$$

It is the shortest distance between any two points in these two sets. Due to $C \cap D = \emptyset$ and their compactness, this distance is not zero. In addition, owing to the compactness of C, D, there must be points $c \in C, d \in D$ satisfying

$$\|c - d\|_2 = \mathrm{dist}(C, D) > 0.$$

Figure 3.10 shows that a line, which passes through the midpoint of the segment connecting the points c, d and is perpendicular to the vector $d - c$, can separate these two sets. The normal vector of this line $\{x \mid a^T x = b\}$ is $a = d - c$. Lastly we need only to calculate the offset b. To this end, we note that the midpoint of the segment connecting c and d is $x_0 = (c + d)/2$[3]; as a result, we have

$$b = a^T x_0 = \frac{1}{2}(d - c)^T (c + d) = \frac{\|d\|_2^2 - \|c\|_2^2}{2}.$$

Now, we prove that this line separates these two sets. Note that the feature of the open half-plane on the a side is $a^T x > b$, while the open half-plane on the opposite side satisfies $a^T x < b$. We prove that D is contained in the open half-plane $a^T x > b$ by contradiction. Namely, we suppose that there is a point $u \in D$ satisfying $a^T u \leq b$ and then deduce a contradiction. By assumption, we have

$$\begin{aligned} 0 \geq a^T u - b &= (d - c)^T u - \frac{(d - c)^T (d + c)}{2} \\ &= (d - c)^T \left(u - \frac{1}{2}(d + c)\right) = (d - c)^T \left(u - d + \frac{1}{2}(d - c)\right) \\ &= (d - c)^T (u - d) + \frac{1}{2}\|d - c\|_2^2. \end{aligned}$$

[3] This can be understood easily by setting the origin and drawing the vector sum $c + d$ in Figure 3.10

It means that $(d-c)^T(u-d)<0$ since $\|d-c\|_2>0$. Further, the derivative of the distance function[4] $\|d+t(u-d)-c\|_2^2$ about variable $t\in[0,1]$ satisfies

$$\begin{aligned}\frac{d}{dt}\|d+t(u-d)-c\|_2^2\Big|_{t=0} &= 2(u-d)^T(d+t(u-d)-c)\Big|_{t=0}\\ &= 2(d-c)^T(u-d)<0.\end{aligned}$$

Therefore, for sufficiently small t, there holds

$$\|d+t(u-d)-c\|_2<\|d-c\|_2.$$

However, from $u,d\in D$ and the convexity of D, we have $d+t(u-d)=tu+(1-t)d\in D$. This inequality implies that the distance between this point and c is even shorter than the shortest distance, which is obviously contradictory. By a similar argument, C belongs to the open half-plane $a^Tx<0$.

As such, we have proved that the line $a^Tx=b$ separates the convex sets C,D. That is,

$$C\subset\{x\mid a^Tx<b\},\quad D\subset\{x\mid a^Tx>b\}. \tag{3.14}$$

When the convex sets C,D are compact, it is clear that these two sets are disjoint with the separating line $a^Tx=b$ in between.

In a space with a higher dimension, the line that separates two convex sets turns into a hyperplane and is called a *separating hyperplane*. The preceding proof is independent of the space dimension, so it applies to any space.

However, when one of C and D is open or unbounded, it may approach arbitrarily close to the separating hyperplane $a^Tx=b$. Therefore, in this case, we can only conclude that

$$C\subset\{x\mid a^Tx\le b\},\quad D\subset\{x\mid a^Tx\ge b\}. \tag{3.15}$$

A major application of separating hyperplane is that if a problem can be transformed into a problem that two sets are disjoint, then it can further be converted equivalently to the existence of a separating hyperplane. In this way, it is possible to transform a problem which is difficult to solve directly into an easier one. This new problem is known as the *dual problem.*

The duality result to be shown in Example 3.3 is very effective in testing the *feasibility* of LMI which will be described in the next section.

Example 3.3 *Prove that the following statements are equivalent:*

1. *There is no $x\in\mathbb{R}^m$ satisfying LMI*

$$F(x)=F_0+x_1F_1+\cdots+x_mF_m<0,\quad F_i=F_i^T,\quad i=0,1,\ldots,m. \tag{3.16}$$

   (The LMI $F(x)<0$ is said to be infeasible.)
2. *There is a nonzero matrix $W=W^T\ge 0$ satisfying*

$$\mathrm{Tr}(F_0W)\ge 0,\quad \mathrm{Tr}(F_iW)=0,\quad i=0,1,\ldots,m. \tag{3.17}$$

[4] This function is the distance from a point on the segment between two points u,d in the set D to a point c in the set C.

Proof. (2)⇒(1): As long as the statement (2) is true,

$$\mathrm{Tr}(W^{1/2}F(x)W^{1/2}) = \mathrm{Tr}(F(x)W) = \mathrm{Tr}(F_0W) \geq 0$$

holds for any $x \in \mathbb{R}^m$. Owing to $W \neq 0$ and $W \geq 0$, if $F(x) < 0$, then $\mathrm{Tr}(W^{1/2}F(x)W^{1/2}) < 0$. However, it is contradictory to the statement (2). This shows that $F(x) < 0$ has no solution.

(1)⇒(2): In this case, the set $F(\mathbb{R}^m) = \{F(x) \mid x \in \mathbb{R}^m\}$ and the set S_- of negative definite Hermitians are disjoint. As these two sets are both convex, there must be a separating hyperplane. Further, since both $F(\mathbb{R}^m)$ and S_- are sets of Hermitians, the space under consideration is that of Hermitians. So, the normal W of the separating hyperplane also becomes an Hermitian. The inner product of matrix A, B is $\mathrm{Tr}(A^TB)$; thus there exist a matrix $W = W^T$ and $a \in \mathbb{R}$ satisfying

$$\mathrm{Tr}(F(x)W) \geq a \ \forall x \in \mathbb{R}^m; \quad \mathrm{Tr}(HW) \leq a \ \forall H < 0.$$

The first inequality can be expanded into

$$\begin{aligned}\mathrm{Tr}(F_0W) - a &\geq -x_1\mathrm{Tr}(F_1W) - \cdots - x_m\mathrm{Tr}(F_mW)\\ &= -[x_1 \ \cdots \ x_m]\begin{bmatrix}\mathrm{Tr}(F_1W)\\ \vdots\\ \mathrm{Tr}(F_mW)\end{bmatrix}.\end{aligned}$$

To ensure that the right-hand side is bounded for any $x \in \mathbb{R}^m$, there must be $\mathrm{Tr}(F_iW) = 0 \quad (i = 1, \ldots, m)$. So $\mathrm{Tr}(F_0W) \geq a$. Moreover, to ensure that the left-hand side of the second inequality $\mathrm{Tr}(HW) \leq a$ is bounded from above, there should be $W \geq 0$ and $a \geq 0$. This completes the proof. •

3.1.3.2 Converse Separating Hyperplane Theorem

In general, the existence of a separating hyperplane may not be able to guarantee that the two separated sets are disjoint. For example, although $x = \{0\}$ separates two sets, $C = D = \{0\}$, but both are the same. However, if two sets C, D are both convex and at least one of them is open, then the existence of a separating hyperplane guarantees that they are disjoint. This is one of the so-called *converse separating hyperplane theorems*.

This converse separating hyperplane theorem can be explained as follows. Here, assume that the separating hyperplane is $\{x \mid a^Tx = b\}$ and the set C is open and located in the half-space $a^Tx \leq b$. If C intersects the separating hyperplane $a^Tx = b$ at point x_0, then since C is open, there must be a $x \in C$ satisfying $a^Tx > b$ in the neighborhood of x_0. This contradicts the assumption that C is located in the half-space $a^Tx \leq b$. Therefore, C must be located in the half-space $a^Tx < b$. On the other hand, the set D is located in the half-space $a^Tx \geq b$. So these two sets are disjoint.

3.1.4 Affine Function

For a scalar variable $x \in \mathbb{R}$, a map in the form of

$$f(x) = ax + b$$

is called an *affine function*, in which $a, b \in \mathbb{R}$ are scalar constants. From the viewpoint of set, it presents a line on the plane $(x, f(x))$ with an offset b. This line passes through the origin when $b = 0$ and becomes a linear function. An affine function about vector variable $x \in \mathbb{R}^n$ is the map

$$f(x) = Ax + b, \quad A \in \mathbb{R}^{m \times n}, \quad b \in \mathbb{R}^m. \tag{3.18}$$

It represents a hyperplane in the augmented space $\mathbb{R}^m \times \mathbb{R}^n$ (the direct sum space).

The translation

$$S + a = \{x + a \mid x \in S\}, \quad S \subset \mathbb{R}^n$$

is the simplest example of affine functions.

One of the important properties of affine function is that it maps a convex set into another convex set. The image of convex set S about affine mapping $f(x) = Ax + b$ is denoted by

$$f(S) = \{f(x) \mid x \in S\}.$$

$\theta x + (1 - \theta)y \in S$ holds for any $\theta \in [0, 1]$ as long as $x, y \in S$. Further, the convex combination of their images $f(x), f(y) \in f(S)$ satisfies

$$\theta f(x) + (1 - \theta)f(y) = \theta(Ax + b) + (1 - \theta)(Ay + b) = A[\theta x + (1 - \theta)y] + b \in f(S).$$

So, $f(S)$ is also convex. Similarly, if S is a convex set and f is an affine function, then the inverse map of f

$$f^{-1}(S) = \{x \mid f(x) = Ax + b \in f(S)\}$$

is also a convex set (prove it).

Example 3.4 *Mapping a ball* $\{u \mid \|u\|_2 \leq 1\}$ *using the affine map*

$$f(u) = P^{1/2}u + x_c$$

about positive definite matrix P *and vector* x_c, *the image becomes an ellipsoid as shown here. Assume that the image is* $x = f(u)$; *then* $u = P^{-1/2}(x - x_c)$ *and*

$$1 \geq u^T u = [P^{-1/2}(x - x_c)]^T [P^{-1/2}(x - x_c)] = (x - x_c)^T P^{-1}(x - x_c)$$

holds. It represents an ellipsoid centered at x_c.

3.1.5 Convex Function

Consider a function $f : \mathbb{R}^n \mapsto \mathbb{R}$ defined in a convex set $\mathbf{dom}f$. For any $x, y \in \mathbf{dom}f$ and $\theta \in [0, 1]$, if f has the property

$$f(\theta x + (1 - \theta)y) \leq \theta f(x) + (1 - \theta)f(y), \tag{3.19}$$

then f is called a *convex function*. Geometrically, this implies that the line segment (chord) between $(x, f(x))$ and $(y, f(y))$ is above the graph of f (Figure 3.11). When the inequality is strict except the two ends of the line segment as shown in Figure 3.11, f is called a *strictly convex function*.

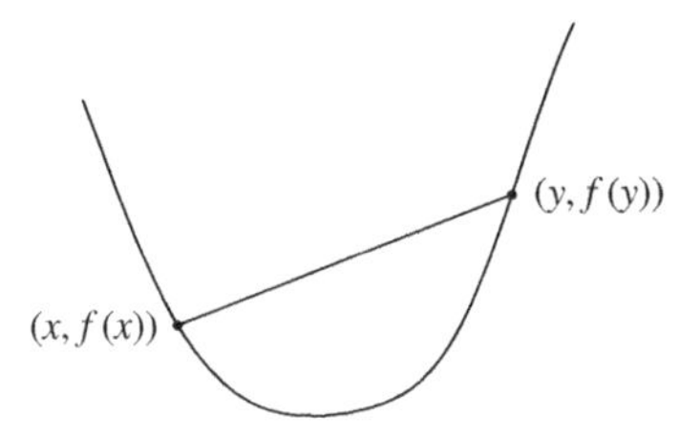

Figure 3.11 Geometric implication of convex function

On the other hand, when $-f$ is a convex function, f is called a *concave function*. That is, a concave function f satisfies the following inequality for any $x, y \in \mathbf{dom} f$ and $\theta \in [0, 1]$ (consider its geometric interpretation):

$$f(\theta x + (1 - \theta)y) \geq \theta f(x) + (1 - \theta)f(y). \tag{3.20}$$

3.1.5.1 First-Order Condition for Convex Function

When $f(x)$ is differentiable, the necessary and sufficient condition for $f(x)$ to be a convex function is that

$$f(y) \geq f(x) + \nabla f(x)^T (y - x) \tag{3.21}$$

holds for any $x, y \in \mathbf{dom} f$ (its geometric implication is shown in Figure 3.12.) Obviously, the right-hand side is the first-order approximation of Taylor expansion of f in the neighborhood of x. This means that the convexity of f is equivalent to that the first-order approximation of Taylor expansion of f is below f.

Now, we prove (3.21). We start from the scalar case. By the convexity of f,

$$f(ty + (1 - t)x) \leq tf(y) + (1 - t)f(x)$$

holds for $0 < t < 1$. Dividing both sides by t, we get

$$f(y) \geq \frac{f(x + t(y - x)) - (1 - t)f(x)}{t} = f(x) + \frac{f(x + t(y - x)) - f(x)}{t}.$$

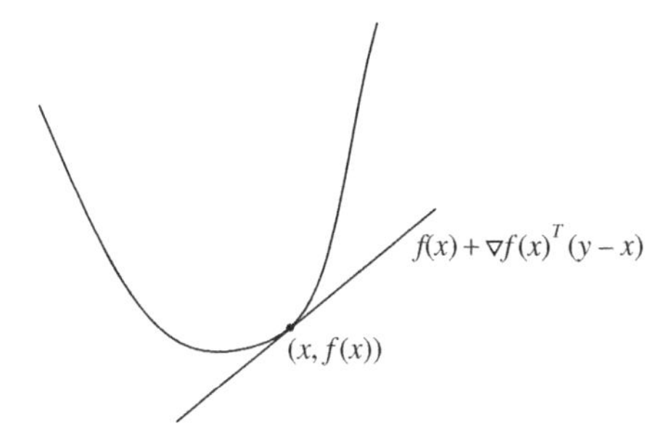

Figure 3.12 First-order condition for convex function

(3.21) is obtained by taking the limit for $t \to 0$. Conversely, when (3.21) holds, we choose two different points (x, y), a combination coefficient $\theta \in [0, 1]$, and define $z = \theta x + (1 - \theta)y$. Applying (3.21) to (z, x) and (z, y), respectively, we get

$$f(x) \geq f(z) + f'(z)(x - z), \quad f(y) \geq f(z) + f'(z)(y - z).$$

Multiplying the first inequality by θ and the second inequality by $1 - \theta$ and then adding them together, we get

$$\theta f(x) + (1 - \theta)f(y) \geq f(z) + f'(z)[\theta(x - z) + (1 - \theta)(y - z)] = f(z).$$

Thus, the convexity of f is proved.

Next, we prove the vector case. The proof is carried out by transforming the vector problem into a scalar one. Assume that $f : \mathbb{R}^n \mapsto \mathbb{R}$. Given arbitrary vectors $x, y \in \mathbf{dom} f$, we set a function $g(t) = f(ty + (1 - t)x)$ about $t \in [0, 1]$. It is easy to know that $g'(t) = \nabla f(ty + (1 - t)x)^T(y - x)$. The key of proof is to use the equivalence between the convexity of f and that of g[5]. Consequently, we only need to prove the equivalence between the convexity of g and (3.21).

When $g(t)$ is a convex function, from the convexity condition of scalar function, we know that $g(1) \geq g(0) + g'(0)(1 - 0)$ holds. This inequality is (3.21) itself. On the contrary, according to the convexity of $\mathbf{dom} f$, if $x, y \in \mathbf{dom} f$, then $t_1 y + (1 - t_1)x, t_2 y + (1 - t_2)x \in \mathbf{dom} f$ holds for $t_1, t_2 \in [0, 1]$. Therefore, when (3.21) holds, we have

$$\begin{aligned} f(t_2 y + (1 - t_2)x) &\geq f(t_1 y + (1 - t_1)x) + \nabla f(t_1 y + (1 - t_1)x)^T \\ &\quad \times [(t_2 y + (1 - t_2)x) - (t_1 y + (1 - t_1)x)] \\ &= f(t_1 y + (1 - t_1)x) + \nabla f(t_1 y + (1 - t_1)x)^T(y - x)(t_2 - t_1) \\ &\Rightarrow g(t_2) \geq g(t_1) + g'(t_1)(t_2 - t_1). \end{aligned}$$

Hence, g is a convex function.

Strictly convex condition can also be derived by the same argument. That is, when $x, y \in \mathbf{dom} f$ and $x \neq y$, there holds the following inequality:

$$f(y) > f(x) + \nabla f(x)^T(y - x). \tag{3.22}$$

3.1.5.2 Second-Order Condition for Convex Function

Second-order condition guaranteeing the convexity of a function is also known. Assume that the domain of f is a convex set and f has second-order derivative in its domain. Further, denote its second-order derivative (Hessian) by $\nabla^2 f$. Then, the function f is convex iff, for all $x \in \mathbf{dom} f$, the Hessian satisfies the inequality

$$\nabla^2 f(x) \geq 0. \tag{3.23}$$

[5] This relationship is as shown in Figure 3.13. $g(t) \leq tg(1) + (1 - t)g(0)$ and $f(ty + (1 - t)x) \leq tf(y) + (1 - t)f(x)$ are the same inequality, so they certainly are equivalent. Further, this equivalence is true for any $x \neq y$.

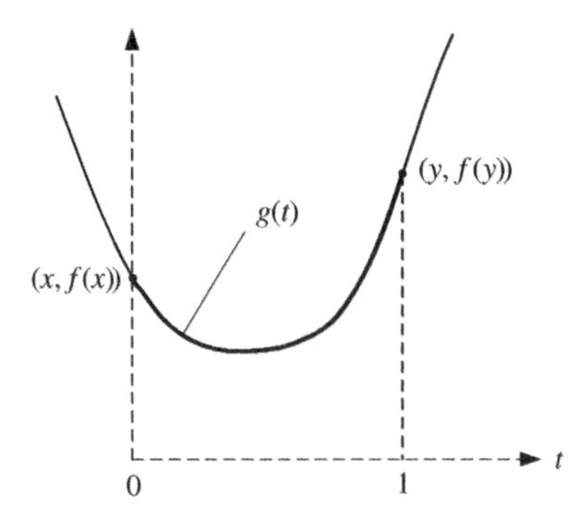

Figure 3.13 Reducing vector problem to scalar problem

Its proof is left to the readers (Exercise 3.11).

Example 3.5 *Consider a quadratic function*

$$V(x) = x^T Px + q^T x + r$$

defined in $\mathbb{R}^n$. *Since* $\nabla^2 V(x) = 2P$, *we know that the condition of convexity is* $P \geq 0$, *the semipositive definiteness of the matrix* P. *In fact, when* $P > 0$, *all points* x *satisfying* $V(x) \leq c$ *form an ellipsoid.*

Example 3.6 *Any norm is a convex function. This is because the following inequality*

$$\|ty + (1-t)x\| \leq \|ty\| + \|(1-t)x\| = t\|y\| + (1-t)\|x\|$$

holds for any $t \in [0, 1]$. *Moreover, for the logarithmic function* $f(x) = \log x (x > 0)$, *we have*

$$f'(x) = \frac{1}{x}, \quad f''(x) = -\frac{1}{x^2} < 0.$$

Therefore, it is a strictly concave function.

Example 3.7 *The following special function, called* barrier function, *plays a vital role in solving the optimization problem with constraints:*

$$f(X) = \log\det(X), \quad X > 0. \tag{3.24}$$

Although this function is a concave function, it is however very difficult to prove directly. For this reason, we prove it by converting it into a scalar problem. Here, set a scalar function $g(t) = f(Z + tV)$ *about* t *in which the matrices* $Z > 0, V \geq 0$ *are arbitrary and* t *is limited to* $t \geq 0$. *Further, when the eigenvalues of symmetric matrix* $Z^{-1/2}VZ^{-1/2}$ *are* $\lambda_1, \ldots, \lambda_n$, *the eigenvalues of* $I + tZ^{-1/2}VZ^{-1/2}$ *will be* $1 + t\lambda_1, \ldots, 1 + t\lambda_n$. *Noting that* $Z + tV = Z^{1/2}(I + tZ^{-1/2}VZ^{-1/2})Z^{1/2}$, *we obtain*

$$g(t) = \log\det(Z + tV)$$

$$= \log\ \det(Z^{1/2})\det(I + tZ^{-1/2}VZ^{-1/2})\det(Z^{1/2})$$

$$= \log\ \prod_{i=1}^{n}(1 + t\lambda_i) + \log\ \det(Z)$$

$$= \sum_{i=1}^{n}\log\ (1 + t\lambda_i) + \log\ \det(Z).$$

Therefore, its derivatives are

$$g'(t) = \sum_{i=1}^{n}\frac{\lambda_i}{1 + t\lambda_i},\quad g''(t) = -\sum_{i=1}^{n}\frac{\lambda_i^2}{(1 + t\lambda_i)^2}.$$

Since $g''(t) \leq 0$ for all t, $g(t)$ is a concave function. Therefore, $f(X)$ is a concave function too.

3.2 Introduction to LMI

Matrix inequality in the form of

$$F(x) = F_0 + \sum_{i=1}^{m} x_i F_i > 0 \tag{3.25}$$

is called a *linear matrix inequality*, or *LMI* for short. Here, $x \in \mathbb{R}^m$ is an unknown vector, and $F_i = F_i^T \in \mathbb{R}^{n\times n}\,(i = 1, \ldots, m)$ is a constant matrix. $F(x)$ is an affine function of variable x. The inequality means that $F(x)$ is positive definite, that is, $u^T F(x) u > 0$ for all nonzero vector u. LMI can be solved numerically by methods such as the interior point method introduced later. MATLAB has an LMI toolbox tailored for solving control problems.

3.2.1 Control Problem and LMI

In control problems, it is often the case that the variables are matrices. This is different from the LMI of (3.25) in form. However, it can always be converted to (3.25) equivalently via the introduction of matrix basis.

Example 3.8 *According to the Lyapunov stability theory in Chapter 13, the necessary and sufficient condition for the stability of a two-dimensional linear system*

$$\dot{x}(t) = Ax(t),\quad x(0) \neq 0$$

is that there exists a positive definite matrix $P = P^T \in \mathbb{R}^{2\times 2}$ satisfying

$$AP + PA^T < 0.$$

This 2×2 symmetric matrix P has the following symmetric basis:[6]

$$P_1 = \begin{bmatrix} 1 & 0 \\ 0 & 0 \end{bmatrix},\quad P_2 = \begin{bmatrix} 0 & 1 \\ 1 & 0 \end{bmatrix},\quad P_3 = \begin{bmatrix} 0 & 0 \\ 0 & 1 \end{bmatrix}.$$

[6] A matrix basis is a set composed of all linearly independent matrices with the same dimension. The linear combination of its elements can describe any matrix with the same dimension.

By using this symmetric basis, P can be expressed as

$$P = \begin{bmatrix} x_1 & x_2 \\ x_2 & x_3 \end{bmatrix} = x_1 P_1 + x_2 P_2 + x_3 P_3.$$

Substituting this equation into the inequality about P, we get an inequality about the vector $x = [x_1 \quad x_2 \quad x_3]^T$ *in the form of (3.25). Specifically, the result obtained is*

$$x_1(AP_1 + P_1 A^T) + x_2(AP_2 + P_2 A^T) + x_3(AP_3 + P_3 A^T) < 0.$$

In addition, when the dimension of symmetric matrix P is n, the number of its base matrices is $n(n+1)/2$.

Therefore, linear inequality with matrix variables is also called LMI hereafter.

3.2.2 Typical LMI Problems

Now, we introduce several types of LMI problems (LMIP) which will be frequently encountered later.

3.2.2.1 Feasibility Problem: LMIP

Given an LMIP, $F(x) < 0$. The so-called *feasibility problem* is to seek a vector x^* satisfying $F(x^*) < 0$, or make the judgment that the LMI has no solution. In judging whether $F(x) < 0$ has a solution, we may solve the dual problem (3.17) shown in Example 3.3. That is, $F(x) < 0$ has no solution iff there is a nonzero matrix $W = W^T \geq 0$ satisfying

$$\mathrm{Tr}(F_0 W) \geq 0, \quad \mathrm{Tr}(F_i W) = 0, \quad i = 0, 1, \ldots, m. \tag{3.26}$$

Example 3.9 *In the stability problem for the two-dimensional system of Example 3.8,* $F(x) = \mathrm{diag}(AP + PA^T, -P)$. *Since* $F_0 = 0$, $F_i = \mathrm{diag}(AP_i + P_i A^T, -P_i)$ $(i = 1, 2, 3)$, *the system is unstable iff there exists a nonzero matrix* $W = W^T = \mathrm{diag}(W_1, W_2) \geq 0$ *satisfying*

$$\begin{aligned} 0 &= \mathrm{Tr}(F_i W) \\ &= \mathrm{Tr}(AP_i W_1) + \mathrm{Tr}(P_i A^T W_1) + \mathrm{Tr}(-P_i W_2) \\ &= \mathrm{Tr}(P_i W_1 A) + \mathrm{Tr}(P_i A^T W_1) + \mathrm{Tr}(-P_i W_2) \\ &= \mathrm{Tr}(P_i(W_1 A + A^T W_1 - W_2)). \end{aligned}$$

Due to the special structure of symmetric basis P_i, *when* $i = 1, 3$, *the previous equation implies that the diagonal elements of* $W_1 A + A^T W_1 - W_2$ *are zeros; when* $i = 2$, *the nondiagonal elements of* $W_1 A + A^T W_1 - W_2$ *are zeros. Eventually, the instability condition reduces to the existence of nonzero matrices* $W_1, W_2 \geq 0$ *satisfying the linear matrix equation*

$$W_1 A + A^T W_1 - W_2 = 0$$

which can be solved easily.

3.2.2.2 Eigenvalue Problem: EVP

The following optimization problem with constraints is called an *eigenvalue problem* (EVP):

$$\begin{aligned}&\text{minimize } \lambda\\&\text{subject to } \lambda I - A(x) > 0,\ B(x) > 0.\end{aligned} \tag{3.27}$$

Here, $A(x)$ and $B(x)$ are symmetric and affine matrix functions of vector x. Minimizing λ subject to $\lambda I - A(x) > 0$ can be interpreted as minimizing the largest eigenvalue of matrix $A(x)$. This is because the largest λ that does not meet $\lambda I - A(x) > 0$ is the maximal λ satisfying $\det(\lambda I - A(x)) = 0$.

We may also regard λ as a variable and augment the vector x accordingly. Then, λ can be expressed as $c^T x$, and the two inequalities can be merged into one. Thus, we arrive at the following equivalent EVP (Exercise 3.13):

$$\begin{aligned}&\text{minimize } c^T x\\&\text{subject to } F(x) > 0.\end{aligned} \tag{3.28}$$

3.2.2.3 Generalized Eigenvalue Problem: GEVP

When λI in the EVP is generalized to $\lambda B(x)$, this problem becomes a *generalized eigenvalue problem* (GEVP). The detail is as follows:

$$\begin{aligned}&\text{minimize } \lambda\\&\text{subject to } \lambda B(x) - A(x) > 0, \quad B(x) > 0, \quad C(x) > 0\end{aligned} \tag{3.29}$$

in which $A(x), B(x)$, and $C(x)$ are symmetric and affine matrix functions of vector x. This is equivalent to the problem of minimizing the largest generalized eigenvalue satisfying $\det(\lambda B(x) - A(x)) = 0$. So it is called a generalized EVP.

This problem can also be described equivalently as

$$\begin{aligned}&\text{minimize } \lambda_{\max}(A(x), B(x))\\&\text{subject to } B(x) > 0, \quad C(x) > 0\end{aligned} \tag{3.30}$$

in which $\lambda_{\max}(A(x), B(x))$ denotes the largest generalized eigenvalue of matrices $(A(x), B(x))$.

3.2.3 From BMI to LMI: Variable Elimination

Let us look at the following stabilization problem.

Example 3.10 *Consider the problem of stabilizing linear system*

$$\dot{x} = Ax + Bu$$

via state feedback $u = Fx$. *The closed-loop system is given by*

$$\dot{x} = (A + BF)x.$$

Therefore, the stability condition is that there exist matrices $P > 0$ and F satisfying

$$(A + BF)P + P(A + BF)^T < 0.$$

In this inequality, a product of unknown matrices F, P appears. This kind of matrix inequality is called a bilinear matrix inequality[7], *or* BMI *for short. BMI problem is nonconvex and very difficult to solve numerically. From the viewpoint of system control, whether a system can be stabilized depends on the controllability of (A, B), not the control gain F. In other words, there must be a stabilization condition independent of F. In order to obtain this condition, we need to apply the following theorem.*

Theorem 3.1 *Given real matrices E, F, G with G being symmetric. There is a matrix X satisfying the inequality*

$$E^T XF + F^T X^T E + G < 0 \tag{3.31}$$

iff the following two inequalities hold simultaneously:

$$E_\perp^T GE_\perp < 0, \quad F_\perp^T GF_\perp < 0. \tag{3.32}$$

Proof. Set $E_\perp = [P_0, P_1]$, $F_\perp = [P_0, P_2]$ in which P_0 is the largest common part of $E_\perp$ and $F_\perp$. Then, $[P_0, P_1, P_2]$ has full column rank. So there is a matrix Q such that $T = [P_0, P_1, P_2, Q]$ is nonsingular. Thus, there hold

$$ET = [0, 0, E_1, E_2], \quad FT = [0, F_1, 0, F_2].$$

Moreover, both $[E_1, E_2]$ and $[F_1, F_2]$ have full column rank. Now, when

$$\begin{bmatrix} E_1^T \\ E_2^T \end{bmatrix} X[F_1 \ F_2] = \begin{bmatrix} X_{11} & X_{12} \\ X_{21} & X_{22} \end{bmatrix}$$

is known, X may be calculated from X_{ij} as follows:

$$X = \begin{bmatrix} E_1^T \\ E_2^T \end{bmatrix}^\dagger \begin{bmatrix} X_{11} & X_{12} \\ X_{21} & X_{22} \end{bmatrix} [F_1 \ F_2]^\dagger.$$

Therefore, we need only consider the existence of X_{ij}. Partition $T^T GT$ according to the decomposition of T as

$$T^T GT = \begin{bmatrix} G_{11} & G_{12} & G_{13} & G_{14} \\ G_{12}^T & G_{22} & G_{23} & G_{24} \\ G_{13}^T & G_{23}^T & G_{33} & G_{34} \\ G_{14}^T & G_{24}^T & G_{34}^T & G_{44} \end{bmatrix}. \tag{3.33}$$

[7] The so-called bilinearity means that if one variable is fixed, the function becomes a linear function of another variable.

Multiplying (3.31) with T^T and T from left and right, respectively, we get an equivalent inequality:

$$\left[\begin{array}{ccc|c} G_{11} & G_{12} & G_{13} & G_{14} \\ G_{12}^T & G_{22} & G_{23}+X_{11}^T & G_{24}+X_{21}^T \\ G_{13}^T & G_{23}^T+X_{11} & G_{33} & G_{34}+X_{12} \\ \hline G_{14}^T & G_{24}^T+X_{21} & G_{34}^T+X_{12}^T & G_{44}+X_{22}+X_{22}^T \end{array}\right] := \begin{bmatrix} S_{11} & S_{12} \\ S_{12}^T & S_{22} \end{bmatrix} < 0. \tag{3.34}$$

According to Schur's lemma, (3.34) is equivalent to

$$S_{11} < 0 \tag{3.35}$$

$$G_{44} + X_{22} + X_{22}^T - S_{12}^T S_{11}^{-1} S_{12} < 0. \tag{3.36}$$

But whenever X_{11}, X_{12}, X_{21} are given, there always exists a matrix X_{22} satisfying (3.36). Consequently, (3.35) becomes the equivalent condition of (3.34). Applying Schur's lemma once again to S_{11}, we obtain the following equivalent condition of (3.35):

$$\begin{bmatrix} G_{11} & 0 & 0 \\ 0 & G_{22} - G_{12}^T G_{11}^{-1} G_{12} & X_{11}^T + H^T \\ 0 & X_{11} + H & G_{33} - G_{13}^T G_{11}^{-1} G_{13} \end{bmatrix} < 0 \tag{3.37}$$

in which $H = G_{23}^T - G_{13}^T G_{11}^{-1} G_{12}$. Obviously, to guarantee (3.37), there must be

$$G_{11} < 0, \ \ G_{22} - G_{12}^T G_{11}^{-1} G_{12} < 0, \ \ G_{33} - G_{13}^T G_{11}^{-1} G_{13} < 0. \tag{3.38}$$

Conversely, when the preceding inequalities hold, it yields (3.37) by taking $X_{11} = -H$. Therefore, (3.38) is equivalent to (3.37). That is, (3.38) is the solvability condition for (3.31). Finally, these inequalities can be arranged as follows via an application of Schur's lemma:

$$0 > \begin{bmatrix} G_{11} & G_{12} \\ G_{12}^T & G_{22} \end{bmatrix} = E_\perp^T G E_\perp, \quad 0 > \begin{bmatrix} G_{11} & G_{13} \\ G_{13}^T & G_{33} \end{bmatrix} = F_\perp^T G F_\perp.$$

•

Example 3.11 *Let us use the preceding theorem to derive a stabilizability condition for Example 3.10. The stability condition can be written as*

$$(AP + PA^T) + PF^T B^T + BFP < 0.$$

For this inequality to have a solution F,

$$(B^T)_\perp^T (AP + PA^T)(B^T)_\perp < 0$$

must hold. This condition only depends on matrix P, so it is an LMI. Since $P_\perp$ does not exist, we do not need to consider the second inequality of Theorem 3.1. After obtaining P, we may substitute it back into the first inequality. Then, the inequality becomes an LMI about F and can be solved numerically.

We often need to construct a new matrix satisfying certain property from two given matrices in transforming a control problem into LMI. The required condition is given by the following lemma.

Lemma 3.1 *Given two positive definite matrices $X, Y \in \mathbb{R}^{n\times n}$ and a positive integer r, there is a positive definite matrix $P \in \mathbb{R}^{(n+r)\times(n+r)}$ satisfying*

$$P = \begin{bmatrix} Y & * \\ * & * \end{bmatrix}, \quad P^{-1} = \begin{bmatrix} X & * \\ * & * \end{bmatrix}$$

iff

$$\begin{bmatrix} X & I \\ I & Y \end{bmatrix} \geq 0, \quad \text{rank}\begin{bmatrix} X & I \\ I & Y \end{bmatrix} \leq n + r.$$

Further, when matrix $F \in \mathbb{R}^{n\times r}$ satisfies

$$FF^T = Y - X^{-1},$$

One of such P is given by

$$P = \begin{bmatrix} Y & F \\ F^T & I \end{bmatrix}.$$

Proof. Sufficiency: according to the inversion formula, when the matrix $P = \begin{bmatrix} Y & P_{12} \\ P_{21} & P_{22} \end{bmatrix}$ is nonsingular, its inverse is given by

$$P^{-1} = \begin{bmatrix} (Y - P_{12}P_{22}^{-1}P_{21})^{-1} & * \\ * & * \end{bmatrix}.$$

Under the given conditions, a congruent transformation on the positive semidefinite matrix $\begin{bmatrix} X & I \\ I & Y \end{bmatrix}$ leads to a positive semidefinite matrix $\begin{bmatrix} X & 0 \\ 0 & Y-X^{-1} \end{bmatrix}$, whose rank is the same and less than $n + r$. Hence, $\text{rank}(Y - X^{-1}) \leq r$ holds. Further, it is easy to see that there is a matrix $F \in \mathbb{R}^{n\times r}$ satisfying $FF^T = Y - X^{-1}$ via the singular value decomposition of $Y - X^{-1}$. Moreover, it can be confirmed that the (1, 1) block is equal to X by calculating the inverse of P given in the lemma. Finally, the positive definiteness of P can be verified easily by Schur's lemma.

Necessity: When such a positive definite matrix P exists, we know from the inversion formula that

$$X^{-1} = Y - P_{12}P_{22}^{-1}P_{12}^T.$$

So, both $Y - X^{-1} = P_{12}P_{22}^{-1}P_{12}^T \geq 0$ and $\text{rank}(Y - X^{-1}) = \text{rank}(P_{12}P_{22}^{-1}P_{12}^T) \leq r$ hold. Applying Schur's lemma again, we obtain the conditions on $\begin{bmatrix} X & I \\ I & Y \end{bmatrix}$. In other words, the conditions are necessary. •

Example 3.12 *Let us take the system stabilization problem as an example to illustrate the application of variable elimination method in the output feedback control. Suppose that the plant is*

$$\dot{x}_P = Ax_P + Bu \tag{3.39}$$

$$y = Cx_P + Du \tag{3.40}$$

and the dynamic output feedback controller is

$$\dot{x}_K = A_K x_K + B_K y \tag{3.41}$$

$$u = C_K x_K + D_K y. \tag{3.42}$$

Computing the state equation of the closed-loop system, we get

$$\begin{bmatrix} \dot{x}_P \\ \dot{x}_K \end{bmatrix} = A_c \begin{bmatrix} x_P \\ x_K \end{bmatrix}, \quad A_c = \begin{bmatrix} A + BD_K C & BC_K \\ B_K C & A_K \end{bmatrix}. \tag{3.43}$$

The condition for the stability of closed-loop system is that there is a matrix $\mathbb{P}$ satisfying inequalities

$$A_c^T \mathbb{P} + \mathbb{P} A_c < 0, \quad \mathbb{P} > 0. \tag{3.44}$$

To apply the variable elimination method, we need to collect all controller coefficient matrices into a single matrix. To this end, we note that the A_c can be written as

$$A_c = \begin{bmatrix} A & 0 \\ 0 & 0 \end{bmatrix} + \begin{bmatrix} B & 0 \\ 0 & I \end{bmatrix} \begin{bmatrix} D_K & C_K \\ B_K & A_K \end{bmatrix} \begin{bmatrix} C & 0 \\ 0 & I \end{bmatrix} = \overline{A} + \overline{B}\mathcal{K}\overline{C}$$

in which

$$\mathcal{K} = \begin{bmatrix} D_K & C_K \\ B_K & A_K \end{bmatrix}, \quad \overline{A} = \begin{bmatrix} A & 0 \\ 0 & 0 \end{bmatrix}, \quad \overline{B} = \begin{bmatrix} B & 0 \\ 0 & I \end{bmatrix}, \quad \overline{C} = \begin{bmatrix} C & 0 \\ 0 & I \end{bmatrix}.$$

Hence, the stability condition can be rewritten as

$$\mathbb{P}\overline{A} + \overline{A}^T \mathbb{P} + \mathbb{P}\overline{B}\mathcal{K}\overline{C} + (\mathbb{P}\overline{B}\mathcal{K}\overline{C})^T < 0.$$

It is easy to verify that

$$\overline{C}_\perp = \begin{bmatrix} C_\perp \\ 0 \end{bmatrix}, \quad (\overline{B}^T \mathbb{P})_\perp = \mathbb{P}^{-1} \begin{bmatrix} (B^T)_\perp \\ 0 \end{bmatrix}.$$

By Theorem 3.1, we see that an equivalent condition of $A_c^T \mathbb{P} + \mathbb{P} A_c < 0$ is

$$(\overline{C}_\perp)^T (\mathbb{P}\overline{A} + \overline{A}^T \mathbb{P}) \overline{C}_\perp < 0, \quad (\overline{B}^T)_\perp^T (\overline{A}\mathbb{P}^{-1} + \mathbb{P}^{-1}\overline{A}^T)(\overline{B}^T)_\perp < 0.$$

As in Lemma 3.1, we set

$$\mathbb{P} = \begin{bmatrix} Y & * \\ * & * \end{bmatrix}, \quad \mathbb{P}^{-1} = \begin{bmatrix} X & * \\ * & * \end{bmatrix}.$$

Concrete calculation with these partitions substituted shows that the equivalent condition for $A_c^T \mathbb{P} + \mathbb{P} A_c < 0$ is as follows: there exist symmetric matrices X, Y meeting LMIs

$$(B^T)_\perp^T (AX + XA^T)(B^T)_\perp < 0 \tag{3.45}$$

$$(C_\perp)^T (YA + A^T Y) C_\perp < 0. \tag{3.46}$$

Finally, according to Lemma 3.1 $\mathbb{P} > 0$ *iff matrices* X, Y *satisfy*

$$\begin{bmatrix} X & I \\ I & Y \end{bmatrix} \geq 0, \quad \text{rank} \begin{bmatrix} X & I \\ I & Y \end{bmatrix} \leq n + r. \tag{3.47}$$

3.2.4 From BMI to LMI: Variable Change

Consider again the state feedback stabilization problem of Example 3.10. The stability condition for the closed-loop system $\dot{x} = (A + BF)x$ is that there are matrices $P > 0$ and F satisfying

$$(A + BF)P + P(A + BF)^T < 0.$$

In dealing with product terms such as FP, another effective approach is to make variable change in addition to the variable elimination method described in the previous subsection. That is, we treat FP as a new variable M such that the stability condition turns into

$$A^T P + PA + BM + M^T B^T < 0.$$

Due to the positive definiteness of matrix P, once we obtain (P, M), the unique solution of F can be calculated from $F = MP^{-1}$.

In the case of output feedback, the variable change method gets much more complicated. The detail is described here. As shown in Example 3.12, the closed-loop system subject to output feedback is

$$\begin{bmatrix} \dot{x}_P \\ \dot{x}_K \end{bmatrix} = A_c \begin{bmatrix} x_P \\ x_K \end{bmatrix}, \quad A_c = \begin{bmatrix} A + BD_K C & BC_K \\ B_K C & A_K \end{bmatrix}. \tag{3.48}$$

Its stability condition is described by

$$A_c^T \mathbb{P} + \mathbb{P} A_c < 0, \quad \mathbb{P} > 0. \tag{3.49}$$

The current question is how to transform it to an LMI problem via a suitable variable change.

Let us take a look at the structure of $\mathbb{P}$ at first. For this purpose, we partition it as

$$\mathbb{P} = \begin{bmatrix} Y & N \\ N^T & * \end{bmatrix}, \quad \mathbb{P}^{-1} = \begin{bmatrix} X & M \\ M^T & * \end{bmatrix}. \tag{3.50}$$

Since $\mathbb{P}\mathbb{P}^{-1} = I$, obviously there hold

$$\mathbb{P} \begin{bmatrix} X \\ M^T \end{bmatrix} = \begin{bmatrix} I \\ 0 \end{bmatrix}, \quad \mathbb{P} \begin{bmatrix} I \\ 0 \end{bmatrix} = \begin{bmatrix} Y \\ N^T \end{bmatrix}.$$

This implies that $\mathbb{P}$ satisfies

$$\mathbb{P}\Pi_1 = \Pi_2, \quad \Pi_1 = \begin{bmatrix} X & I \\ M^T & 0 \end{bmatrix}, \quad \Pi_2 = \begin{bmatrix} I & Y \\ 0 & N^T \end{bmatrix} \tag{3.51}$$

and

$$\mathbb{P}^{-1}\Pi_2 = \Pi_1. \tag{3.52}$$

In other words, both $\mathbb{P}$ and $\mathbb{P}^{-1}$ can be described as the quotients of triangular matrices using four matrices (X, Y, M, N). The advantage of this description is that we can obtain an equivalent inequality $\Pi_1^T A_c^T \mathbb{P}\Pi_1 + \Pi_1^T \mathbb{P} A_c \Pi_1 < 0$ by multiplying (3.49) with Π_1^T and Π_1 from the left and right. A detailed calculation shows that one of the terms reduces to

$$\begin{aligned}\Pi_1^T \mathbb{P} A_c \Pi_1 &= \Pi_2^T A_c \Pi_1\\ &= \begin{bmatrix} I & Y \\ 0 & N^T \end{bmatrix}^T \begin{bmatrix} A+BD_KC & BC_K \\ B_KC & A_K \end{bmatrix} \begin{bmatrix} X & I \\ M^T & 0 \end{bmatrix}\\ &= \begin{bmatrix} AX+B\mathbb{C} & A+B\mathbb{D}C \\ \mathbb{A} & YA+\mathbb{B}C \end{bmatrix}\end{aligned} \tag{3.53}$$

in which the new unknown matrices $\mathbb{A}, \mathbb{B}, \mathbb{C}, \mathbb{D}$ are defined as

$$\begin{aligned}&\mathbb{A} = NA_KM^T + NB_KCX + YBC_KM^T + Y(A+BD_KC)X\\ &\mathbb{B} = NB_K + YBD_K, \quad \mathbb{C} = C_KM^T + D_KCX, \quad \mathbb{D} = D_K.\end{aligned} \tag{3.54}$$

Therefore, the stability condition turns into an LMI about $\mathbb{A}, \mathbb{B}, \mathbb{C}, \mathbb{D}$:

$$\begin{bmatrix} AX + XA^T + B\mathbb{C} + \mathbb{C}^TB^T & A + B\mathbb{D}C + \mathbb{A}^T \\ A^T + C^T\mathbb{D}^TB^T + \mathbb{A} & YA + A^TY + \mathbb{B}C + C^T\mathbb{B}^T \end{bmatrix} < 0.$$

The remaining question is whether there is a one-to-one relationship between the new unknown matrices and the controller coefficient matrices. The answer is that this correspondence is true when M, N have full row ranks. In this case, the controller coefficient matrices are

$$\begin{aligned}&D_K = \mathbb{D}, \quad C_K = (\mathbb{C} - D_KCX)(M^\dagger)^T, \quad B_K = N^\dagger(\mathbb{B} - YBD_K) \\ &A_K = N^\dagger(\mathbb{A} - NB_KCX - YBC_KM^T - Y(A+BD_KC)X)(M^\dagger)^T.\end{aligned} \tag{3.55}$$

Note that $NN^\dagger = I, MM^\dagger = I$.

Finally, according to Lemma 3.1 the condition for $\mathbb{P} > 0$ is

$$\begin{bmatrix} X & I \\ I & Y \end{bmatrix} \geq 0, \quad \text{rank} \begin{bmatrix} X & I \\ I & Y \end{bmatrix} \leq n + r.$$

If we strengthen this condition to

$$\begin{bmatrix} X & I \\ I & Y \end{bmatrix} > 0 \Rightarrow X - Y^{-1} > 0, \tag{3.56}$$

then $I - XY$ is nonsingular (Exercises 3.14). Therefore, there are nonsingular matrices M, N satisfying

$$MN^T = I - XY. \tag{3.57}$$

This equation comes from the (1,1) block of $\mathbb{P}^{-1}\mathbb{P} = I$.

The variable change method can be applied to many of the problems concerning matrix inequality in this book.

3.3 Interior Point Method*

In this section, we introduce the most efficient numerical method for LMI problems at present: the interior point method. To this end, we need to know some basic concepts such as the analytical center and the central path.

3.3.1 *Analytical Center of LMI*

The concept of analytical center plays a significant role in the interior point method. The idea is to transform the feasibility of an LMI into the minimization of a scalar barrier function. Consider the following LMI:

$$F(x) = F_0 + \sum_{k=1}^{n} x_i F_i > 0, \quad F_i = F_i^T \, \forall i.$$

Obviously, the function

$$\phi(x) = \begin{cases} \log\det F^{-1}(x), & F(x) > 0 \\ \infty, & \text{other} \end{cases} \tag{3.58}$$

is bounded iff $F(x) > 0$. Further, $\phi(x)$ tends to infinity when the vector x approaches the boundary of the feasible set $\{x \mid F(x) > 0\}$. In this sense, the function $\phi(x)$ can be regarded as a wall separating the feasible set from other points. This is why it is called a *barrier function*.

We further consider the case where the feasible set is nonempty and bounded[8]. Since ϕ is a strictly convex function (refer to Example 3.7), there is a solution

$$x^\star := \arg\min_x \phi(x) \tag{3.59}$$

minimizing ϕ. This $x^\star$ is called the *analytic center* of LMI $F(x) > 0$. It should be noted that this optimization problem has no constraint, so it can be solved easily by using a variety of methods.

For example, in applying Newton's method, let the gradient and Hessian of ϕ be

$$g(x) = \nabla\phi(x) = \left[\frac{\partial\phi(x)}{\partial x}\right]^T, \quad H(x) = \nabla^2\phi(x) = \left[\frac{\partial^2\phi(x)}{\partial x^2}\right]^T, \tag{3.60}$$

respectively. Then the analytical center $x^\star$ can be computed iteratively based on the algorithm

$$x^{(k+1)} = x^{(k)} - \alpha^{(k)} H^{-1}(x^{(k)}) g(x^{(k)}). \tag{3.61}$$

Here, $\alpha^{(k)}$ is the step length of the kth iteration.

---

[8] If $F_1, \dots, F_n$ is not linearly independent, there is a nonzero vector $x^\circ = [x_1, \dots, x_n]^T$ satisfying $\alpha \sum x_i F_i = 0$ for any scalar α. That is, the vector obtained by adding the straight line αx° to a feasible solution is always feasible. Then, the solution set is unbounded. Therefore, $F_1, \dots, F_n$ must be linearly independent when the feasible set is bounded.

3.3.2 Interior Point Method Based on Central Path

Now, how should we solve the following EVP?

$$\begin{array}{ll} \text{minimize} & c^T x \\ \text{subject to} & F(x) > 0 \end{array}$$

The idea is as follows. Suppose that the optimum of the objective function is λ^{opt}. For $\lambda > \lambda^{\text{opt}}$, the LMI

$$F(x) > 0, \quad c^T x < \lambda \Leftrightarrow \begin{bmatrix} F(x) & 0 \\ 0 & \lambda - c^T x \end{bmatrix} > 0 \tag{3.62}$$

is feasible. That is, a constrained optimization problem can be converted into an LMI feasibility problem. We can approach arbitrarily close to the optimum λ^{opt} by gradually reducing λ until (3.62) gets unfeasible. As shown in the previous subsection, this problem can be further converted to an optimization problem about a barrier function.

The analytical center of LMI (3.62) certainly depends on λ, so it is denoted by $x^\star(\lambda)$. By

$$\det \begin{bmatrix} F(x) & \\ & \lambda - c^T x \end{bmatrix}^{-1} = \det F^{-1}(x) \times \frac{1}{\lambda - c^T x},$$

we know that

$$x^\star(\lambda) = \arg\min_x \left(\log\det F^{-1}(x) + \log \frac{1}{\lambda - c^T x} \right). \tag{3.63}$$

Varying λ, then $x^\star(\lambda)$ forms a trajectory called *central path* of the EVP. The central path is analytic about λ, and the limit exists when $\lambda \to \lambda^{\text{opt}}$. We denote this limit as x^{opt}. Clearly, x^{opt} is the optimal solution for the EVP.

In computing the optimal solution x^{opt}, we need to gradually reduce λ to λ^{opt}. $\lambda > c^T x$ and $\lambda^{(k+1)} < \lambda^{(k)}$ must be satisfied in building an update law for λ. Here, focusing on $\lambda^{(k)} > c^T x^{(k)}$, we update λ using a convex combination of $\lambda^{(k)}$ and $c^T x^{(k)}$:

$$\lambda^{(k+1)} = (1-\theta) c^T x^{(k)} + \theta \lambda^{(k)}, \quad 0 < \theta < 1. \tag{3.64}$$

This $\lambda^{(k+1)}$ satisfies

$$\begin{aligned} \lambda^{(k+1)} - \lambda^{(k)} &= (1-\theta)(c^T x^{(k)} - \lambda^{(k)}) < 0 \\ \lambda^{(k+1)} &= c^T x^{(k)} + \theta(\lambda^{(k)} - c^T x^{(k)}) > c^T x^{(k)} \end{aligned}$$

simultaneously. In addition, the update law for vector x is

$$x^{(k+1)} = x^\star(\lambda^{(k+1)}). \tag{3.65}$$

Finally, we summarize the algorithm for the EVP:

1. Calculate a solution $x^{(0)}$ for LMI $F(x) > 0$ by the optimization of barrier function, and then determine a real number $\lambda^{(0)}$ such that $\lambda^{(0)} > c^T x^{(0)}$.
2. Update $\lambda^{(k+1)}$ for each $k = 1, 2, \ldots$ according to (3.64).
3. Minimize the barrier function (3.63) to get $x^{(k+1)}$.

4. Repeat this procedure until $\lambda^{(k)} - c^T x^{(k)}$ achieves the required accuracy.

Other constrained optimization problems can be solved in the same way.

Exercises

3.1 Suppose that C is an affine set and constant θ_i $(i = 1, 2, 3)$ satisfies $\theta_1 + \theta_2 + \theta_3 = 1$. Prove that

$$y = \theta_1 x_1 + \theta_2 x_2 + \theta_3 x_3 \in C \quad \forall x_i \in C.$$

3.2 Prove that the half-space $\{x|a^T x \le b\}$ is a convex set.

3.3 Prove that the volume of the following ellipsoid is proportional to $\lambda_1\lambda_2\lambda_3$:

$$\frac{y_1^2}{\lambda_1} + \frac{y_2^2}{\lambda_2} + \frac{y_3^2}{\lambda_3} \le 1.$$

3.4 Prove that the following polyhedron is a convex set:

$$\mathcal{P} = \{x|a_1^T x \le b_1,\ a_2^T x \le b_2,\ c^T x = d\}.$$

3.5 The solvability condition for linear equation $Ax = b$ is $b \in \text{Im } A$. Discuss the existence condition for the positive solution $x \succ 0$.

3.6 Prove that the set of positive definite matrices below is convex

$$S_+ = \{X \in \mathbb{R}^{n \times n} \mid X = X^T > 0\}.$$

3.7 Given a real matrix M, prove that the following statements are equivalent by using the separating hyperplane theorem:
(a) $M + M^T \ge 0$;
(b) $\text{Tr}(M^T X) \le 0$ holds for all $X < 0$.

3.8 Given a square matrix A, derive a necessary and sufficient condition for the existence of a positive definite matrix X satisfying LMI

$$XA + X^T A < 0$$

by applying the separating hyperplane theorem. (Hint: Use the result of Exercise 3.8.)

3.9 Prove that the largest eigenvalue $f(X) = \lambda_{\max}(X)$ of real symmetric matrix X is a convex function.

3.10 Suppose the domain $\mathbf{dom} f$ of scalar function f is a convex set and f is twice differentiable in its domain. Prove that f is a convex function iff

$$\nabla^2 f(x) \ge 0 \quad \forall x \in \mathbf{dom} f.$$

3.11 Let X be a complex Hermitian. Prove that $X \geq 0$ iff

$$\begin{bmatrix} \Re(X) & -\Im(X) \\ \Im(X) & \Re(X) \end{bmatrix} \geq 0.$$

3.12 In the eigenvalue problem (3.27), augment the unknown vector x to $\begin{bmatrix} \lambda \\ x \end{bmatrix}$. Find the constant vector c and matrix $F(x)$ in the equivalent problem (3.28).

3.13 Prove the following equivalent conditions:

$$\begin{bmatrix} X & I \\ I & Y \end{bmatrix} > 0 \Leftrightarrow X > 0, \quad Y > 0, \; \lambda_{\min}(XY) > 1.$$

Notes and References

Boyd and Vandenberghe [11] gives a complete and easy-to-read exposition on optimization theory and is worth reading. Boyd *et al.* [10] describes LMI and its applications in depth. Further, Refs [35, 42] are also useful references. The variable change method presented here comes from Ref. [32] and the variable elimination method is from Ref. [33]. Readers who are interested in the theory of interior-point method may consult [72].

4

Fundamentals of Linear System

In this chapter, we make a concise review on the basic knowledge of linear systems such as controllability, observability, pole and zero, stability, as well as associated balanced realization and linear fractional transformation (LFT).

4.1 Structural Properties of Dynamic System

4.1.1 Description of Linear System

The motion of a linear system with state $x(t) \in \mathbb{R}^n$, input $u(t) \in \mathbb{R}^m$, and output $y(t) \in \mathbb{R}^p$ is governed by the following *state equation*:

$$\dot{x}(t) = Ax(t) + Bu(t) \tag{4.1}$$

$$y(t) = Cx(t) + Du(t). \tag{4.2}$$

Here, (A, B, C, D) are the coefficient matrices. This state equation can further be described compactly by the following equation:

$$\begin{bmatrix} \dot{x} \\ y \end{bmatrix} = \begin{bmatrix} A & B \\ C & D \end{bmatrix} \begin{bmatrix} x \\ u \end{bmatrix}. \tag{4.3}$$

When we only consider the relationship between the input and output of linear system, we may use the *transfer matrix*

$$\hat{y}(s) = G(s)\hat{u}(s) = [C(sI - A)^{-1}B + D]\hat{u}(s) \tag{4.4}$$

to describe the system. This transfer matrix is obtained by taking Laplace transform on the state equation (4.1) subject to zero initial state $x(0) = 0$ and eliminating the state variables. Each element of the transfer matrix is a transfer function between the corresponding input and output, that is, its (i, j) element is a transfer function between the jth input and ith output. For single-input single-output (SISO) system, the transfer matrix reduces to a transfer function.

Robust Control: Theory and Applications, First Edition. Kang-Zhi Liu and Yu Yao.

Further, in order to facilitate the computation between transfer matrices, we often use the notation

$$\left[\begin{array}{c|c} A & B \\ \hline C & D \end{array}\right] := C(sI - A)^{-1}B + D \tag{4.5}$$

to represent a transfer matrix. The following expression is also used frequently:

$$(A, B, C, D) := C(sI - A)^{-1}B + D. \tag{4.6}$$

Example 4.1 *In the two-mass–spring system of Example 1.2, we have derived the state equation*

$$\dot{x} = Ax + B_1 d + B_2 u, \quad y = Cx$$

about state vector $x = [\omega_L \quad \phi \quad \omega_M]^T$ in which the coefficient matrices are given by

$$A = \begin{bmatrix} -\dfrac{D_L}{J_L} & \dfrac{k}{J_L} & 0 \\ -1 & 0 & 1 \\ 0 & -\dfrac{k}{J_M} & -\dfrac{D_M}{J_M} \end{bmatrix}, \quad B_1 = \begin{bmatrix} \dfrac{1}{J_L} \\ 0 \\ 0 \end{bmatrix}, \quad B_2 = \begin{bmatrix} 0 \\ 0 \\ \dfrac{1}{J_M} \end{bmatrix}$$

$$C = [0 \quad 0 \quad 1].$$

Therefore, the transfer function $P_d(s)$ from the load torque disturbance d to the output y and the transfer function $P_u(s)$ from the input u to the output y are

$$\begin{aligned} P_d(s) &= C(sI - A)^{-1}B_1 \\ &= \frac{\frac{k}{J_L^2}}{s^3 + \left(\frac{D_L}{J_L} + \frac{D_M}{J_M}\right)s^2 + \left(\frac{D_L D_M}{J_L J_M} + \frac{k}{J_L} + \frac{k}{J_M}\right)s + \frac{k(D_L + D_M)}{J_L J_M}} \\ P_u(s) &= C(sI - A)^{-1}B_2 \\ &= \frac{\frac{1}{J_M}\left(s^2 + \frac{D_L}{J_L}s + \frac{k}{J_L}\right)}{s^3 + \left(\frac{D_L}{J_L} + \frac{D_M}{J_M}\right)s^2 + \left(\frac{D_L D_M}{J_L J_M} + \frac{k}{J_L} + \frac{k}{J_M}\right)s + \frac{k(D_L + D_M)}{J_L J_M}}, \end{aligned}$$

respectively. They are consistent with the results obtained by directly taking Laplace transform on the motion equations:

$$J_L\dot{\omega}_L + D_L\omega_L = k\phi + d, \ \dot{\phi} = \omega_M - \omega_L, \ J_M\dot{\omega}_M + D_M\omega_M + k\phi = u.$$

Particularly, when friction is neglected, the transfer functions become

$$P_d(s) = \frac{\frac{k}{J_L^2}}{\left(s^2 + \frac{k}{J_L} + \frac{k}{J_M}\right)s}, \quad P_u(s) = \frac{\frac{1}{J_M}\left(s^2 + \frac{k}{J_L}\right)}{\left(s^2 + \frac{k}{J_L} + \frac{k}{J_M}\right)s}.$$

4.1.2 Dual System

By taking *transpose* on the transfer matrix $G(s)$, we obtain a new transfer matrix known as the *dual system* of $G(s)$:

$$G^T(s) = B^T(sI - A^T)^{-1}C^T + D^T$$
$$= \left[\begin{array}{c|c} A^T & C^T \\ \hline B^T & D^T \end{array}\right]. \tag{4.7}$$

That is, the dual system of (A, B, C, D) is (A^T, C^T, B^T, D^T). In system properties to be described later on, many have a duality relation. That is, when a system has two different physical properties A and B, if property B is the same as property A of the dual system mathematically, we say that property A and property B are dual to each other. Such situation often appears in the physical world. Therefore, by the notion of duality, the characteristics of property B will be clear if we know the mathematical characteristics of property A. This is very smart mathematically. A typical example is the duality between controllability and observability to be introduced next.

4.1.3 Controllability and Observability

The purpose of control is to make the physical variables of plant move as desired. Before the design of a control system, it is necessary to check in advance whether the system can meet such a requirement. Whether a plant can be controlled is determined by the structural property of the plant itself. Since all signals in the system are generated by the state, it is necessary to be able to control the state arbitrarily. By abstraction, such engineering requirement is described as the notion of controllability precisely.

Definition 4.1 *Given arbitrary initial state $x(0) = x_0$, terminal state x_f, and finite time instant $t_f > 0$, if there is a bounded input $u(t)$ such that the solution of (4.1) satisfies $x(t_f) = x_f$, then the dynamic system (4.1) or the pair (A, B) is called* controllable. *Otherwise, it is called* uncontrollable.

As shown in the phase plane[1] of Figure 4.1, the controllability means that the state can be moved from any initial state x_0 to any assigned terminal state x_f in finite time by using a bounded input. However, the path from x_0 to x_f can be arbitrary.

The controllability of a system can be tested by the following algebraic criteria.

Theorem 4.1 *The following statements are equivalent:*

1. *(A, B) is controllable;*
2. *The controllability matrix*

$$\mathcal{C} = \begin{bmatrix} B & AB & \cdots & A^{n-1}B \end{bmatrix}$$

has full row rank.

[1] The Cartesian coordinate with states as axes is called a phase plane. The time sequence (trajectory) of state vector can be drawn on the phase plane.

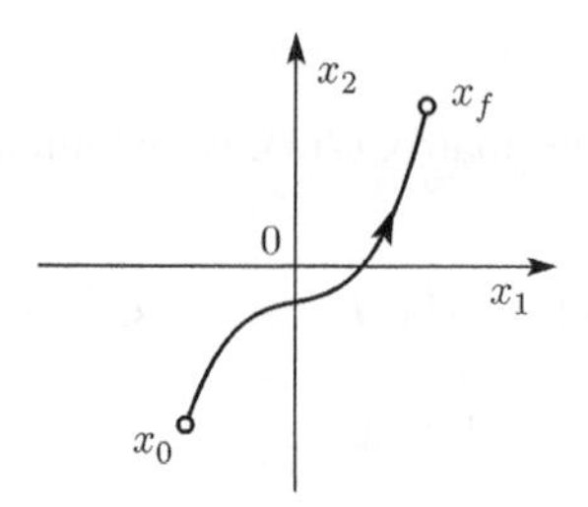

Figure 4.1 State trajectory

3. *For any $\lambda \in \mathbb{C}$, the matrix $[A - \lambda I \quad B]$ has full row rank.*
4. *For any $\lambda \in \sigma(A)$, the matrix $[A - \lambda I \quad B]$ has full column rank.*
5. *By choosing a suitable matrix F, the eigenvalue of $A + BF$ can be placed arbitrarily (for a complex eigenvalue, its conjugate must also be designated as an eigenvalue).*

As seen from this theorem, the controllability only depends on the coefficient matrices (A, B) of the state equation. This conclusion is a matter of course because the controllability is related only to the input and state, which has nothing to do with the output.

Example 4.2 *In the rigid body model of hard disk drive, when the states are chosen as the rotational angle (x_1) and angular velocity (x_2), the coefficient matrices of state equation are given by*

$$A = \begin{bmatrix} 0 & 1 \\ 0 & 0 \end{bmatrix}, \quad B = \begin{bmatrix} 0 \\ k \end{bmatrix}, \quad C = [1, 0]$$

respectively. Via a simple calculation, we get

$$\mathcal{C} = \begin{bmatrix} 0 & k \\ k & 0 \end{bmatrix}, \quad [A - \lambda I \quad B] = \begin{bmatrix} -\lambda & 1 & 0 \\ 0 & -\lambda & k \end{bmatrix}.$$

Both have full row rank, so this system is controllable. In physics, this corresponds to the fact that the drive arm can be moved to any angle by supplying appropriate current to the motor.

Next, we check if there is a matrix $F = [f_1, f_2]$ that places the eigenvalues of $A + BF$ to two arbitrary points (a, b). The identity

$$\det(\lambda I - (A + BF)) = (\lambda - a)(\lambda - b)$$
$$\Rightarrow \lambda^2 - kf_2\lambda - kf_1 = \lambda^2 - (a + b)\lambda + ab$$

certainly holds. This is an identity about polynomials of variable λ, so the coefficients of terms with the same degree on both sides must be equal. From it, we obtain simultaneous equations

$$-kf_1 = ab, \quad kf_2 = a + b.$$

The unique solution is

$$F = \frac{1}{k}[-ab, \ a + b].$$

As such, we have verified that as long as (A, B) is controllable, there exists a feedback gain F that can place the eigenvalues $(A + BF)$ arbitrarily.

Basically, to control all states freely, information about the system state is indispensable. However, in engineering practice the signals measured by sensors are not the states, but the outputs of system only. In other words, we need to compute the state based on the input and output information. The mathematical property on this possibility is clearly described by the notion of observability in Definition 4.2.

Definition 4.2 *For any finite time instant $t_1 > 0$, if the initial state $x(0) = x_0$ can be uniquely determined via the input $u(t)$ and output $y(t)$ over the time interval $[0, t_1]$, we say that the system (4.1) or (C, A) is* observable. *Otherwise, the system is* unobservable.

Observability conditions are given by the following theorem.

Theorem 4.2 *The following statements are equivalent:*

1. *(C, A) is observable.*
2. *The observability matrix*

$$\mathcal{O} = \begin{bmatrix} C \\ CA \\ \vdots \\ CA^{n-1} \end{bmatrix}$$

has full column rank.

3. *For all $\lambda \in \mathbb{C}$, the matrix $\begin{bmatrix} A-\lambda I \\ C \end{bmatrix}$ has full column rank.*
4. *For all $\lambda \in \sigma(A)$, the matrix $\begin{bmatrix} A-\lambda I \\ C \end{bmatrix}$ has full column rank.*
5. *By choosing a suitable matrix L, the eigenvalues of $A + LC$ can be placed freely (for a complex eigenvalue, its conjugate must also be designated as an eigenvalue).*
6. *(A^T, C^T) is controllable.*

The last condition shows that the controllability and observability are in a dual relation. In addition, from this theorem we see that the observability is only related to the coefficient matrices (C, A).

Example 4.3 *Let us examine the observability for the hard disk drive in Example 4.2. Since*

$$\mathcal{O} = \begin{bmatrix} 1 & 0 \\ 0 & 1 \end{bmatrix}, \quad \begin{bmatrix} A-\lambda I \\ C \end{bmatrix} = \begin{bmatrix} -\lambda & 1 \\ 0 & -\lambda \\ 1 & 0 \end{bmatrix}$$

have full column rank, this system is observable. Similarly, the matrix L placing the eigenvalues of $A + LC$ to two points (c, d) can be computed by comparing the coefficients of the following identity:

$$\det(\lambda I - (A + LC)) = (\lambda - c)(\lambda - d)$$
$$\Rightarrow \lambda^2 - l_1\lambda - l_2 = \lambda^2 - (c + d)\lambda + cd.$$

The solution obtained is $L = [c+d, -cd]^T$.

4.1.4 State Realization and Similarity Transformation

Definition 4.3 *The state-space description* (A, B, C, D) *corresponding to transfer matrix* $G(s)$ *is called its* realization. *Particularly, when* (A, B) *is controllable and* (C, A) *is observable,* (A, B, C, D) *is called a* minimal realization *of* $G(s)$.

The state-space realization used in the hard disk drive example described earlier is both controllable and observable, so it is a minimal realization.

Note that realization of $G(s)$ is not unique. For example, using a nonsingular matrix T to transform the state $x(t)$ into a new one

$$z(t) = T^{-1}x(t), \tag{4.8}$$

we get the state equation about $z(t)$

$$\dot{z} = T^{-1}ATz + T^{-1}Bu \tag{4.9}$$

$$y = CTz + Du. \tag{4.10}$$

From the equation

$$CT(sI - T^{-1}AT)^{-1}T^{-1}B + D = C(sI - A)^{-1}B + D = G(s), \tag{4.11}$$

we see that $(T^{-1}AT, T^{-1}B, CT, D)$ is also a realization of $G(s)$. The coordinate transformation of (4.8) is called *similarity transformation*, and T called the transformation matrix. Since a state transformation does not change the input and output, it is natural that the transfer matrix is the same.

4.1.5 Pole

For the transfer matrix $G(s) = (A, B, C, D)$, its poles are defined as follows:

Definition 4.4 *The eigenvalues of matrix* A *are called the* poles *of* (A, B, C, D)*, the realization of* $G(s)$. *When this realization is minimal, the eigenvalues of* A *are called the poles of transfer matrix* $G(s)$.

The denominator polynomial of $C(sI - A)^{-1}B + D$ is $\det(sI - A)$. It is clear that the roots of this polynomial are the poles of its realization. However, we need to distinguish the poles of a transfer matrix from those of its realization. The reason why the eigenvalues of matrix A are called the poles of the realization is that the initial state response of $x(0)$ is equal to

$$\hat{x}(s) = (sI - A)^{-1}x(0) = \frac{\text{adj}(sI - A)}{\det(sI - A)}x(0). \tag{4.12}$$

When the realization of a transfer matrix is not minimal, a part of eigenvalues of A are either uncontrollable (called *uncontrollable modes*) or unobservable (called *unobservable modes*). These uncontrollable or unobservable states will not appear in the transfer matrix. For example, in the system

$$G(s) = \left[\begin{array}{cc|c} 1 & 0 & 1 \\ 0 & 1 & 1 \\ \hline 1 & 1 & 0 \end{array}\right] = \frac{2}{s-1},$$

an eigenvalue of matrix A does not appear in the transfer function. Therefore, the poles of a state-space realization and those of the transfer matrix are not identical.

Poles determine the convergence of the system response. For example, in the linear system

$$G(s) = \left[\begin{array}{cc|c} a & b & 1 \\ -b & a & 0 \\ \hline 1 & 0 & 0 \end{array}\right] = \frac{s-a}{(s-a)^2+b^2},$$

the poles are $a \pm jb$ and its impulse response is $e^{at}\cos bt$. When the real part a of the poles is negative, the impulse response converges; meanwhile, when a is positive, the response diverges (refer to Figure 4.2(a)). Further, the convergence rate is determined by the magnitude of real part. On the other hand, the imaginary part of poles determines the resonant frequency of response. When the imaginary part increases, the oscillation gets intense (refer to Figure 4.2(b)).

4.1.6 Zero

4.1.6.1 Zero of SISO System

For an SISO system, its transfer function is a rational function, and the numerator and denominator are polynomials. The roots of numerator polynomial are called *zeros*. Physically, a zero has the property of blocking some special signal. For example, in the preceding example,

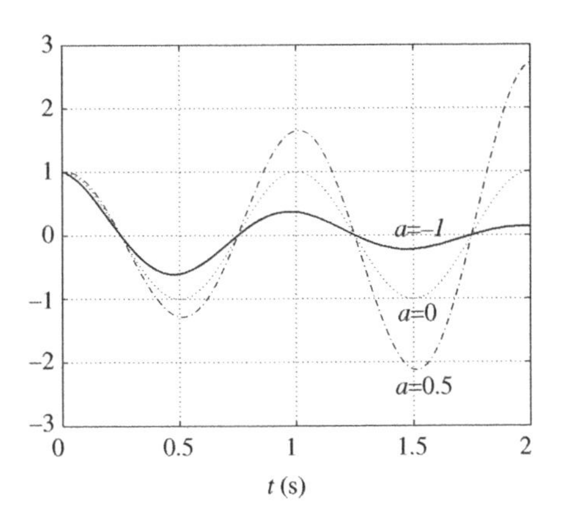

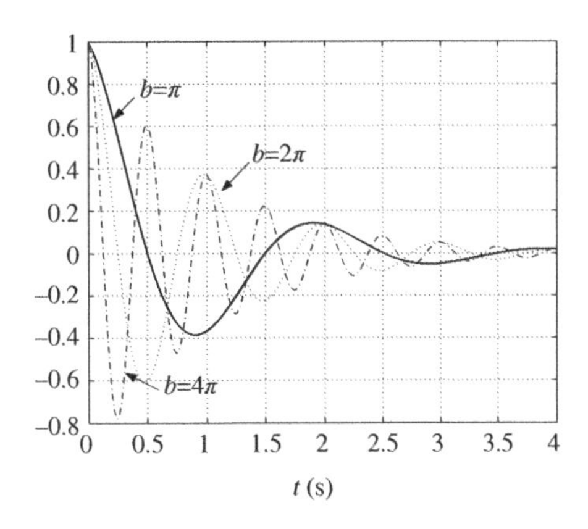

Figure 4.2 Pole position and response (a) $b = 2\pi$, (b) $a = -1$

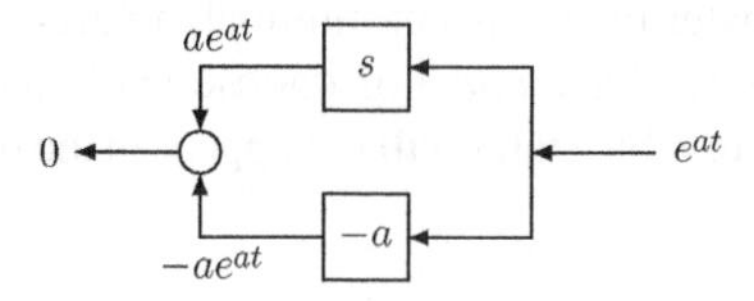

Figure 4.3 Mechanism of zero

when the initial state is $x(0) = [0 \ \ -1/b]^T$ and input is $u(t) = e^{at}$, the output becomes

$$\begin{aligned}\hat{y}(s) &= C(sI - A)^{-1}x(0) + C(sI - A)^{-1}B\hat{u}(s) \\ &= -\frac{1}{(s-a)^2 + b^2} + \frac{1}{(s-a)^2 + b^2} \\ &= 0.\end{aligned}$$

So, the input $u(t) = e^{at}$ is completely blocked and cannot be transmitted to the output.

In this example, the mechanism of zero $z = a$ is as illustrated in Figure 4.3. The numerator polynomial $s - a$ is a parallel connection of a differentiator s and a gain $-a$. When the input e^{at} is applied, the output sum of these two blocks is exactly zero. Therefore, the input e^{at} is blocked from the output. In addition, the input has a jump at $t = 0$, and an impulse signal results in after differentiation. The nonzero initial state in this example is intentionally added in order to offset the effect of this impulse response.

4.1.6.2 Origin of Zero

In general, a physical component seldom has zeros. But when they are interconnected, zeros may appear. Some examples are as follows:

1. Zero resulted from parallel connection
   When systems without any zeros are connected in parallel, zero will be produced. For example,
   $$G_1(s) = \frac{1}{s+a}, \quad G_2(s) = \frac{1}{s+b} \ \Rightarrow G_1 + G_2 = \frac{2s + a + b}{(s+a)(s+b)},$$
   a zero $z = -(a+b)/2$ appears in $G_1 + G_2$.
2. Zero resulted from feedback connection
   In this case, the poles of feedback block become the zeros of closed-loop transfer function. For example,
   $$G_1(s) = \frac{1}{s+a}, \quad G_2(s) = \frac{1}{s+b} \ \Rightarrow \frac{G_1}{1 + G_1G_2} = \frac{s+b}{1 + (s+a)(s+b)},$$
   the pole $-b$ of transfer function G_2 in the feedback path becomes a zero of the closed-loop transfer function.

Example 4.4 *In the two-mass–spring system of Example 4.1, when the frictions are ignored, the transfer functions are, respectively,*

$$P_d(s) = \frac{\frac{k}{J_L^2}}{\left(s^2 + \frac{k}{J_L} + \frac{k}{J_M}\right)s}, \quad P_u(s) = \frac{\frac{1}{J_M}\left(s^2 + \frac{k}{J_L}\right)}{\left(s^2 + \frac{k}{J_L} + \frac{k}{J_M}\right)s}.$$

Clearly, the poles of $P_d(s), P_u(s)$ are located at the origin and on the imaginary axis; $P_d(s)$ has no zeros, while $P_u(s)$ has two zeros on the imaginary axis. It is more difficult to control systems having zeros on the imaginary axis or the right half-plane (refer to Chapter 5 for the specific reason). Particularly, when the load inertia J_L is much smaller than the motor inertia J_M (i.e., the inertia ratio *is far less than 1), the purely imaginary zeros and poles of $P_u(s)$ are very close to each other. In this case, the control of load is extremely difficult. From the physical intuition, when the load is too light, its speed is not easy to detect from the motor speed so that it is difficult to control.*

4.1.6.3 Zero of MIMO System

For a multiple-input multiple-output (MIMO) system, its zero cannot be viewed from the zero of each element of the transfer matrix. In extending the concept of zero to MIMO systems, it is necessary to start from zero's physical property, that is, input blocking. To this end, we need a notion on the rank of function matrix.

Definition 4.5 *Let $Q(s)$ be a $p \times m$ rational function (or polynomial) matrix. The so-called* normal rank *of $Q(s)$ is the maximum possible rank of $Q(s)$ for all $s \in \mathbb{C}$, denoted by* $\text{normalrank}\,(Q(s))$.

Example 4.5 *In the polynomial matrix,*

$$Q(s) = \begin{bmatrix} 1 & s+1 \\ s & s(s+1) \end{bmatrix}$$

the second row is an s multiple of the first row, so they are linearly dependent. Therefore, the normal rank of $Q(s)$ is 1. In contrast, the polynomial matrix

$$Q(s) = \begin{bmatrix} 1 & s+1 \\ s & s+1 \end{bmatrix}$$

has a normal rank 2.

The notion of normal rank is applied to define zeros for MIMO systems.

Definition 4.6 *For a transfer matrix $G(s)$, a complex number z_0 satisfying*

$$\text{rank}(G(z_0)) < \text{normalrank}(G(s))$$

is called a transmission zero. *If $G(z_0) = 0$, then z_0 is called a* blocking zero.

Obviously, a blocking zero is also a transmission zero. For the transfer function of an SISO system, the blocking zeros and transmission zeros are the same. It is worth noting that transmission zero can be defined for nonsquare transfer matrices. When the transfer matrix is square and its normal rank equals its dimension, the transmission zeros are given by all z_0 meeting $\det(G(z_0)) = 0$ (see Exercise 4.3).

Example 4.6 *In the following transfer matrix*

$$G(s) = \begin{bmatrix} \frac{s-1}{s+1} & \frac{1}{s+1} \\ 1 & \frac{s-1}{s+2} \end{bmatrix},$$

elements $(s-1)/(s+1)$ *and* $(s-1)/(s+2)$ *both have a zero* $z = 1$*. But at this point,*

$$G(1) = \begin{bmatrix} 0 & 1/2 \\ 1 & 0 \end{bmatrix}$$

does not lose rank. So $z = 1$ *is not a transmission zero.*

On the other hand, a complex number that originally is not a zero of any element of the transfer matrix may become a transmission zero of the transfer matrix. For example, in

$$G(s) = \begin{bmatrix} \frac{1}{s+1} & \frac{1}{s+3} \\ \frac{1}{s+3} & \frac{1}{s+1} \end{bmatrix}$$

no element has a finite zero. However, at $s = -2$ *the rank of G drops from the normal rank 2 to 1, so it has a transmission zero* -2*.*

In order to know the number of transmission zeros, we need the knowledge on the McMillan canonical form of rational matrices. The details can be found in Refs [49, 100].

The following lemma shows that the transmission zero defined earlier does have the property of input blocking.

Lemma 4.1 *Assume that* $G(s)$ *is a* $p \times m$ *transfer matrix and its minimal realization is* (A, B, C, D)*. If* $z_0 \in \mathbb{C}$ *is a transmission zero of* $G(s)$*, but not a pole of* $G(s)$*, for the nonzero vector* u_0 *satisfying* $G(z_0)u_0 = 0$*, the output w.r.t. the initial state* $x(0) = (z_0 I - A)^{-1}Bu_0$ *and input* $u(t) = u_0 e^{z_0 t}$ *is* $y(t) = 0$*.*

Proof. When the input is $u(t) = u_0 e^{z_0 t}$ ($\hat{u}(s) = u_0/(s - z_0)$) and the initial state is $x(0) = (z_0 I - A)^{-1}Bu_0$, Laplace transform of the corresponding output is

$$\begin{aligned}
\hat{y}(s) &= C(sI - A)^{-1}x(0) + [C(sI - A)^{-1}B + D]\hat{u}(s) \\
&= C(sI - A)^{-1}x(0) + C(sI - A)^{-1}Bu_0(s - z_0)^{-1} + Du_0(s - z_0)^{-1} \\
&= C(sI - A)^{-1}x(0) + C[(sI - A)^{-1} - (z_0 I - A)^{-1}]Bu_0(s - z_0)^{-1} \\
&\quad + G(z_0)u_0(s - z_0)^{-1}
\end{aligned}$$

$$
\begin{aligned}
&= C(sI - A)^{-1}(x(0) - (z_0 I - A)^{-1}Bu_0) + G(z_0)u_0(s - z_0)^{-1} \\
&= G(z_0)u_0(s - z_0)^{-1}
\end{aligned}
$$

in which $(sI - A)^{-1} - (z_0 I - A)^{-1} = -(s - z_0)(sI - A)^{-1}(z_0 I - A)^{-1}$ has been used. Therefore, $y(t) = G(z_0)u_0 e^{z_0 t} = 0$ holds. •

When the normal rank of $G(s)$ is equal to the number of columns, the nonzero vector u_0 meeting $G(z_0)u_0 = 0$ is called a *zero vector*. This lemma shows that a transmission zero z_0 can block the exponential signal with z_0 as the exponent and in the direction of zero vector u_0. Incidentally, in Example 4.6 the zero vector corresponding to zero $z = -2$ is $[1 \quad 1]^T$.

To discuss zeros in the state space, we have the notion of invariant zero.

Definition 4.7 *When a complex number z_0 satisfies*

$$
\text{rank}\begin{bmatrix} A - z_0 I & B \\ C & D \end{bmatrix} < \text{normalrank}\begin{bmatrix} A - sI & B \\ C & D \end{bmatrix},
$$

it is called an invariant zero *of system (A, B, C, D). Further, the matrix function $\begin{bmatrix} A-sI & B \\ C & D \end{bmatrix}$ is called the* system matrix *of system (A, B, C, D).*

In addition to the transmission zeros, the invariant zeros may also contain a part of uncontrollable poles (uncontrollable modes) and unobservable poles (unobservable modes). When the realization of transfer matrix is minimal, the invariant zeros coincide with the transmission zeros.

The reason why a zero defined like this is named as invariant zero is that it is invariant w.r.t. state feedback $u = Fx + v$. After the state feedback, the closed-loop system becomes $(A + BF, B, C + DF, D)$. Due to the relation

$$
\text{rank}\begin{bmatrix} A + BF - sI & B \\ C + DF & D \end{bmatrix} = \text{rank}\begin{bmatrix} A - sI & B \\ C & D \end{bmatrix},
$$

both have the same zero.

Example 4.7 *The normal rank of transfer matrix*

$$
\begin{bmatrix} \frac{1}{s+1} & \frac{1}{s+3} \\ \frac{1}{s+3} & \frac{1}{s+1} \end{bmatrix} = \left[\begin{array}{cccc|cc} -1 & & & & 1 & 0 \\ & -1 & & & 0 & 1 \\ & & -3 & & 1 & 0 \\ & & & -3 & 0 & 1 \\ \hline 1 & 0 & 0 & 1 & 0 & 0 \\ 0 & 1 & 1 & 0 & 0 & 0 \end{array}\right]
$$

is 6. The rank of system matrix drops to 5 at $z = -2$, so $z = -2$ is an invariant zero. It has been shown in the previous example that this point is also a transmission zero.

4.1.7 Relative Degree and Infinity Zero

In the transfer function

$$G(s) = \frac{b_{m+1}s^m + \cdots + b_2 s + b_1}{s^n + a_n s^{n-1} + \cdots + a_2 s + a_1}, \quad b_{m+1} \neq 0, \tag{4.13}$$

the difference between the degree n of denominator polynomial and the degree m of numerator polynomial

$$r = n - m \tag{4.14}$$

is called *relative degree* of the transfer function. For example, the relative degree of

$$G(s) = \frac{5s + 2}{s^3 + 2s^2 + 3s + 4}$$

is $r = 3 - 1 = 2$.

When the relative degree is $r \geq 0$, that is, the denominator degree is no less than the numerator degree, the transfer function is said to be *proper*. When $r > 0$, that is, the denominator degree is higher than the numerator degree, the transfer function is said to be *strictly proper*.

Multiplying this transfer function by s^2, we get a new transfer function

$$s^2 G(s) = s^2 \frac{5s + 2}{s^3 + 2s^2 + 3s + 4}$$

which contains a nonzero direct through term 5. Since s is a differentiator, this means that when the output of system is differentiated twice, the input appears directly for the first time. This point may be seen more clearly in the state space. The transfer function $G(s)$ has the following state-space realization:

$$\dot{x} = Ax + bu, \quad y = cx \tag{4.15}$$

$$A = \begin{bmatrix} 0 & 1 & 0 \\ 0 & 0 & 1 \\ -4 & -3 & -2 \end{bmatrix}, \quad b = \begin{bmatrix} 0 \\ 0 \\ 1 \end{bmatrix}, \quad c = [2 \quad 5 \quad 0].$$

The first- and second-order derivatives of output are, respectively,

$$\dot{y} = c\dot{x} = cAx + cbu \tag{4.16}$$

$$\ddot{y} = cA\dot{x} + cb\dot{u} = cA^2x + cAbu + cb\dot{u}. \tag{4.17}$$

As $cb = 0$, $cAb = 5$, it is only in the second-order derivative or higher that the input u appears in the output y.

Therefore, for an SISO system given by the state-space realization

$$\dot{x} = Ax + bu, \quad y = cx, \tag{4.18}$$

its relative degree can be defined as the positive integer r satisfying

$$cb = cAb = \cdots = cA^{r-2}b = 0, \quad cA^{r-1}b \neq 0. \tag{4.19}$$

Moreover, $G(s) \to 0$ as $s \to \infty$ in a transfer function with a relative degree $r > 0$. In this sense, $s = \infty$ can be regarded as a zero of $G(s)$ at the infinity. So, we say that a transfer function of relative degree $r(> 0)$ has r *infinity zeros*. Since the number of finite zeros is $n - r$, the sum of both is exactly equal to the number of poles. In the preceding example, there are a finite zero and two infinity zeros.

For MIMO systems, we may consider the relative degree in accordance with different outputs. That is, when the input appears for the first time in the r_ith order derivative of the ith output y_i, we say that the relative degree of this output is r_i. Here, assume that the state equations of MIMO system is

$$\dot{x} = Ax + Bu, \quad y = Cx, \tag{4.20}$$

and partition the output matrix C into row vectors:

$$C = \begin{bmatrix} c_1 \\ \vdots \\ c_p \end{bmatrix}. \tag{4.21}$$

Then, the relative degree of the ith output y_i is the positive integer r_i satisfying

$$c_i B = c_i AB = \cdots = c_i A^{r_i - 2} B = 0, \quad c_i A^{r_i - 1} B \neq 0. \tag{4.22}$$

Further, we say that the system as a whole has a (vector) relative degree of $(r_1, \ldots, r_p)$.

4.1.8 Inverse System

When an input $u(t)$ is applied to system $G(s)$, we get the output

$$\hat{y}(s) = G(s)\hat{u}(s).$$

Conversely, in calculating the input $u(t)$ from the output $y(t)$, we may use

$$\hat{u}(s) = G^{-1}(s)\hat{y}(s).$$

$G^{-1}(s)$ is called the *inverse system* of $G(s)$. It should be noticed that even if $G(s)$ is proper, $G^{-1}(s)$ is not necessarily proper in general, that is, its numerator degree may be higher than the denominator degree. For a square matrix $G(s) = (A, B, C, D)$, when D is nonsingular, its inverse system $G^{-1}(s)$ can be calculated by

$$G^{-1}(s) = \left[\begin{array}{c|c} A - BD^{-1}C & BD^{-1} \\ \hline -D^{-1}C & D^{-1} \end{array}\right]. \tag{4.23}$$

4.1.9 System Connections

Interconnections between systems have the following forms: *cascade connection*, *parallel connection*, and *feedback connection*. Consider the case when two linear systems

$$G_1 = \left[\begin{array}{c|c} A_1 & B_1 \\ \hline C_1 & D_1 \end{array}\right], \quad G_2 = \left[\begin{array}{c|c} A_2 & B_2 \\ \hline C_2 & D_2 \end{array}\right]$$

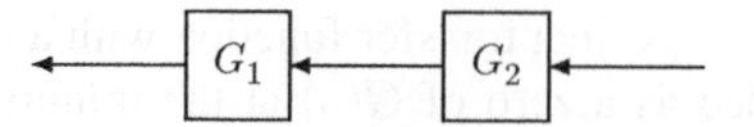

Figure 4.4 Cascade connection

are interconnected. Realization of the resulted system can be computed by transforming the state equations and eliminating the intermediate variables (i.e., the input and output at the connection point).

First, the cascade connection is shown in Figure 4.4, which corresponds to the multiplication of transfer matrices. The computation formula is

$$\begin{aligned} G_1(s)G_2(s) &= \left[\begin{array}{c|c} A_1 & B_1 \\ \hline C_1 & D_1 \end{array}\right] \left[\begin{array}{c|c} A_2 & B_2 \\ \hline C_2 & D_2 \end{array}\right] \\ &= \left[\begin{array}{cc|c} A_1 & B_1C_2 & B_1D_2 \\ 0 & A_2 & B_2 \\ \hline C_1 & D_1C_2 & D_1D_2 \end{array}\right] \qquad (4.24) \\ &= \left[\begin{array}{cc|c} A_2 & 0 & B_2 \\ B_1C_2 & A_1 & B_1D_2 \\ \hline D_1C_2 & C_1 & D_1D_2 \end{array}\right]. \qquad (4.25) \end{aligned}$$

On the other hand, the parallel connection (Figure 4.5) of systems corresponds to the summation of transfer matrices. The computation formula is

$$\begin{aligned} G_1(s) + G_2(s) &= \left[\begin{array}{c|c} A_1 & B_1 \\ \hline C_1 & D_1 \end{array}\right] + \left[\begin{array}{c|c} A_2 & B_2 \\ \hline C_2 & D_2 \end{array}\right] \\ &= \left[\begin{array}{cc|c} A_1 & 0 & B_1 \\ 0 & A_2 & B_2 \\ \hline C_1 & C_2 & D_1 + D_2 \end{array}\right]. \qquad (4.26) \end{aligned}$$

Finally, the feedback connection is shown in Figure 4.6. Here, the transfer matrix between the input and output is given by

$$H_{yr}(s) = \left[\begin{array}{cc|c} A_1 - B_1D_2R_{12}^{-1}C_1 & -B_1R_{21}^{-1}C_2 & B_1R_{21}^{-1} \\ B_2R_{12}^{-1}C_1 & A_2 - B_2D_1R_{21}^{-1}C_2 & B_2D_1R_{21}^{-1} \\ \hline R_{12}^{-1}C_1 & -R_{12}^{-1}D_1C_2 & D_1R_{21}^{-1} \end{array}\right] \qquad (4.27)$$

in which $R_{12} = I + D_1D_2$, $R_{21} = I + D_2D_1$.

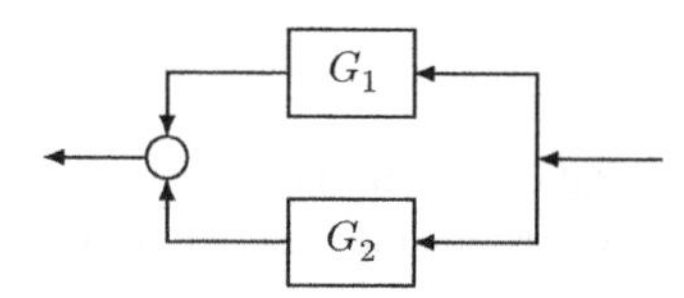

Figure 4.5 Parallel connection

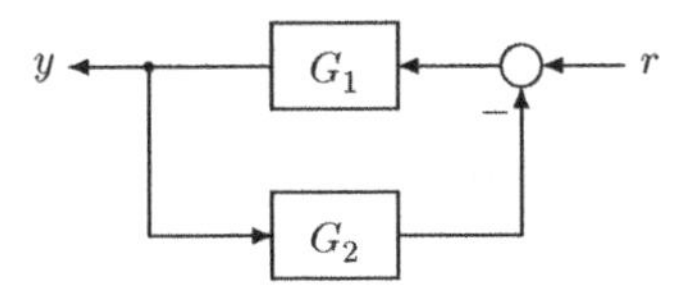

Figure 4.6 Feedback connection

In the feedback connection, it is most often the case that $G_2(s) = D_2$ (i.e., a constant matrix). In this case, the closed-loop transfer matrix reduces to

$$H = \left[\begin{array}{c|c} A_1 - B_1 D_2 R_{12}^{-1} C_1 & B_1 R_{21}^{-1} \\ \hline R_{12}^{-1} C_1 & D_1 R_{21}^{-1} \end{array}\right]. \tag{4.28}$$

It should be noted that even if the realizations of the original systems $G_1(s), G_2(s)$ are controllable and observable, the connected system may not be controllable or observable. Depending on the type of connection, either uncontrollable or unobservable modes may result in. For instance, pole–zero cancellation may occur in cascade connection; the states of two identical systems cannot be controlled independently in parallel connection. Some examples are illustrated in the following.

Example 4.8 *Pole–zero cancellation in cascade connection:*

$$G_1(s) = \frac{s-1}{s+1} = \left[\begin{array}{c|c} -1 & -2 \\ \hline 1 & 1 \end{array}\right], \quad G_2(s) = \frac{1}{s-1} = \left[\begin{array}{c|c} 1 & 1 \\ \hline 1 & 0 \end{array}\right]$$

$$\Rightarrow G_1(s)G_2(s) = \left[\begin{array}{cc|c} -1 & -2 & 0 \\ 0 & 1 & 1 \\ \hline 1 & 1 & 0 \end{array}\right].$$

Its controllability matrix and observability matrix are

$$\mathcal{C} = \begin{bmatrix} 0 & -2 \\ 1 & 1 \end{bmatrix}, \quad \mathcal{O} = \begin{bmatrix} 1 & 1 \\ -1 & -1 \end{bmatrix}.$$

Although the original systems are controllable and observable, after cascade connection, the state is controllable but not observable. The reason is that cancellation between the pole 1 and zero 1 happened.

Parallel connection of identical systems:

$$G_1(s) = G_2(s) = \frac{1}{s-1} = \left[\begin{array}{c|c} 1 & 1 \\ \hline 1 & 0 \end{array}\right]$$

$$\Rightarrow G_1(s) + G_2(s) = \left[\begin{array}{cc|c} 1 & & 1 \\ 1 & 1 & 1 \\ \hline 1 & 1 & 0 \end{array}\right]$$

$$\Rightarrow \mathcal{C} = \mathcal{O} = \begin{bmatrix} 1 & 1 \\ 1 & 1 \end{bmatrix}.$$

After parallel connection, the state vector is neither controllable nor observable. The reason is that in parallel connection the states of two systems with the same dynamics cannot be controlled independently by the same input and the states of systems cannot be distinguished from each other.

4.2 Stability

In this section, we first review the concept of bounded-input bounded-output (BIBO) stability and then introduce a new concept of internal stability which is not treated in the classical control theory.

4.2.1 Bounded-Input Bounded-Output Stability

The most basic requirement on control systems is that when a bounded input is applied to a system, its output must not diverge. This property is called *bounded-input bounded-output stability*, or BIBO stability in short. For example, in Figure 4.7 when a bounded input $u(t)$ is imposed under zero initial state, the output $y(t)$ must be bounded. Mathematically, this can be expressed as

$$|u(t)| \leq c < \infty \ \forall t \Rightarrow |y(t)| \leq M < \infty \ \forall t. \tag{4.29}$$

For MIMO systems, we need to use a vector norm to replace the absolute value. That is, the BIBO stability condition is replaced by

$$\|u(t)\| \leq c < \infty \ \forall t \Rightarrow \|y(t)\| \leq M < \infty \ \forall t. \tag{4.30}$$

Any vector norm may be used.

Testing BIBO stability according to the definition directly needs to consider all bounded inputs, which is certainly impossible. So, let us seek a verifiable BIBO stability criterion. For simplicity, we consider SISO systems. In Figure 4.7, $x(0)$ is the initial state and $u(t)$ is the input; both are bounded. The output is $y(t)$. Since we only consider the input and output, the initial state can be set as $x(0) = 0$. Let the unit impulse response of $G(s)$ be denoted by $g(t)$; then the output $y(t)$ can be calculated by the convolution integral

$$y(t) = \int_0^t g(\tau)u(t-\tau)d\tau. \tag{4.31}$$

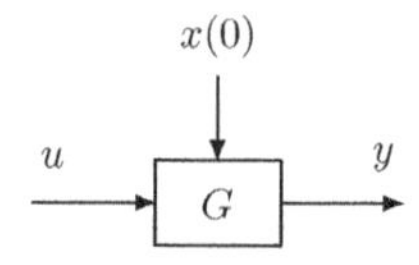

Figure 4.7 bounded-input bounded-output stability

When the input satisfies $|u(t)| \leq c$ for all time t,

$$|y(t)| = \left| \int_0^t g(\tau)u(t-\tau)d\tau \right| \leq \int_0^t |g(\tau)| \cdot |u(t-\tau)|d\tau$$
$$\leq c \int_0^t |g(\tau)|d\tau \leq c \int_0^\infty |g(\tau)|d\tau \tag{4.32}$$

holds. So, whenever $\int_0^\infty |g(t)|dt$ is bounded, $y(t)$ is also bounded. Conversely, for the following special input

$$u(\tau) = \frac{g(t-\tau)}{|g(t-\tau)|} \Rightarrow |u(\tau)| = \frac{|g(t-\tau)|}{|g(t-\tau)|} = 1, \tag{4.33}$$

its output response is

$$y(t) = \int_0^t g(\tau)u(t-\tau)d\tau = \int_0^t g(\tau)\frac{g(\tau)}{|g(\tau)|}d\tau = \int_0^t |g(\tau)|d\tau$$
$$\Rightarrow |y(t)| = \int_0^t |g(\tau)|d\tau. \tag{4.34}$$

$|y(\infty)| < \infty$ holds when the system is stable. Therefore, $\int_0^\infty |g(t)|dt < \infty$ must be true. These arguments lead to the following theorem.

Theorem 4.3 *An SISO system $G(s)$ is BIBO stable iff*

$$\int_0^\infty |g(t)|dt < \infty. \tag{4.35}$$

For an MIMO system, let the unit impulse response matrix be $g(t) = \mathcal{L}^{-1}[G(s)]$. Then, the system is BIBO stable iff

$$\int_0^\infty \|g(t)\|dt < \infty \tag{4.36}$$

in which $\|g(t)\|$ can be any induced norm.

However, this condition requires calculation of the absolute integral of impulse response, which is not so easy. In the sequel, we derive some simpler criteria. To this end, we define the stability of transfer matrix (function) first.

Definition 4.8 *When* all *poles of the transfer matrix (function) $G(s)$ have negative real parts, $G(s)$ is called* stable. *On the contrary, when* all *poles of $G(s)$ have nonnegative real parts, $G(s)$ is called* antistable. *Further, when one or more poles of $G(s)$ have nonnegative real parts, it is called* unstable.

Note that an antistable transfer function is unstable, but not all unstable transfer functions are antistable.

Example 4.9 *The poles of transfer function*

$$G_1(s) = \frac{1}{(s+1)(s+2)}$$

are $p = -1, -2$, *and they are all negative. So,* $G_1(s)$ *is stable. In contrast, poles of the transfer function*

$$G_2(s) = \frac{1}{(s-1)(s-2)}$$

are $p = 1, 2$, *and they are all positive. Thus,* $G_2(s)$ *is antistable. Meanwhile, poles of the transfer function*

$$G_3(s) = \frac{1}{(s+1)(s-2)}$$

are $p = -1, 2$, *one positive and one negative. Hence,* $G_3(s)$ *is unstable, but not antistable.*

The boundedness of $\int_0^\infty |g(t)|dt$ (for MIMO systems, $\int_0^\infty \|g(t)\| dt$) is ensured by the stability of transfer function $G(s)$. That is, the stability of the transfer function ensures the BIBO stability of a system.

Theorem 4.4 *The BIBO stability of a linear system is equivalent to the stability of its transfer matrix.*

Proof. Only the proof for SISO systems is shown. First, consider the case where $G(s)$ has no multiple poles. A partial fraction expansion yields

$$G(s) = \sum \frac{c_i}{s - p_i} + d \Rightarrow g(t) = \sum c_i e^{p_i t} + d\delta(t), t \geq 0.$$

In general, poles are complex numbers and can be written as $p_i = a_i + jb_i$. Since $e^{p_i t} = e^{a_i t} e^{jb_i t}$ and $|e^{jb_i t}| \equiv 1$, we have

$$\int_0^\infty |g(t)| dt \leq \int_0^\infty \left[\sum |c_i| e^{a_i t} + |d|\delta(t)\right] dt.$$

Here, we have used the triangle inequality $|a + b| \leq |a| + |b|$. When all a_i are negative, the right-hand side satisfies

$$|d| + \sum \frac{|c_i|}{-a_i} < \infty.$$

In contrast, when $a_1 \geq 0$, for example, $e^{a_1 t}$ either diverges or converges to a constant, its integral $\int_0^\infty e^{a_1 t} dt$ always diverges.

When there exist multiple poles, for example, when $p = a + jb$ is a pole with a multiplicity of 2, we have

$$G(s) = \frac{c_1}{(s-p)^2} + \frac{c_2}{s-p} + \text{rest terms} \Rightarrow g(t) = c_1 t e^{pt} + c_2 e^{pt} + \text{rest terms}.$$

We need to just check the first integral term. Let $a < 0$; a simple calculation shows that

$$\int_0^\infty |te^{at}| dt = \int_0^\infty te^{at} dt = \frac{1}{a^2} < \infty.$$

Therefore, even though multiple poles exist, the absolute integral of impulse response is still bounded. So, the system is BIBO stable. •

Example 4.10 *The unit impulse response of stable system $G(s) = 1/(s+1)$ is*

$$g(t) = e^{-t}, \quad t \geq 0.$$

Its absolute integral is

$$\int_0^{\infty} |g(t)|\,dt = \int_0^{\infty} e^{-t}\,dt = 1$$

and bounded. Conversely, the unit impulse response of unstable transfer function $G(s) = 1/(s-1)$ is

$$g(t) = e^{t}, \quad t \geq 0.$$

In this case, it is obvious that the integral

$$\int_0^{T} |g(t)|\,dt = e^{T} - 1$$

diverges as $T \to \infty$.

4.2.2 *Internal Stability*

In classical control, the stability of a feedback system is treated from the angle of a single closed-loop transfer function. For example, often considered is the stability of the closed-loop transfer function

$$H_{yr}(s) = \frac{P(s)K(s)}{1 + P(s)K(s)}$$

from reference input r to output y. However, this transfer function alone does not necessarily guarantee the stability of the closed-loop system. Let us look at an example.

Example 4.11 *In the feedback system of Figure 4.8, let the plant and controller be*

$$P(s) = \frac{1}{s-1}, \quad K(s) = \frac{s-1}{s}$$

respectively. Since $P(s)K(s) = 1/s$, the transfer function $H_{yr}(s)$ from reference input r to output y becomes

$$H_{yr}(s) = \frac{P(s)K(s)}{1 + P(s)K(s)} = \frac{1}{s+1}.$$

This transfer function is stable. However, when we check the transfer function $H_{yd}(s)$ from disturbance d to output y, we find that

$$H_{yd}(s) = \frac{P(s)}{1 + P(s)K(s)} = \frac{\frac{1}{s-1}}{1 + \frac{1}{s}} = \frac{s}{(s-1)(s+1)}$$

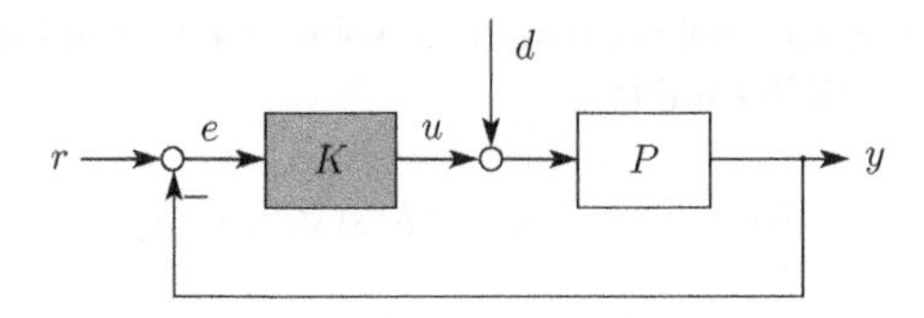

Figure 4.8 Internal stability

which apparently is unstable. The secret lies in that cancellation between unstable pole $p = 1$ and unstable zero $z = 1$ happened between $P(s)$ and $K(s)$. It can also be verified indirectly that pole–zero cancellation has occurred from the fact that the second-order transfer function $H_{yr}(s)$ reduces to first order.

Real-world systems are always subject to disturbance and noise from the surrounding environment. When unstable zero–pole cancellation occurs, it will cause the divergence of system and result in internal damage to the physical system. In order to avoid this from happening, it is necessary to ensure the so-called internal stability below.

Definition 4.9 *In the closed-loop system of Figure 4.8, when all four transfer matrices*

$$\begin{bmatrix} PK(I+PK)^{-1} & P(I+KP)^{-1} \\ K(I+PK)^{-1} & -KP(I+KP)^{-1} \end{bmatrix}$$

from external signals (r, d) to internal signals (y, u) of the system are stable, the closed-loop system in Figure 4.8 is called internally stable.

The internal stability may be tested by the following two methods:

- Characteristic polynomial method (SISO system)
  In this method, we first write the transfer function $P(s), K(s)$ as the fraction of denominator polynomials and numerator polynomials

$$P(s) = \frac{N_P(s)}{M_P(s)}, \quad K(s) = \frac{N_K(s)}{M_K(s)}.$$

  The numerator and denominator must be coprime, that is, they do not contain any common polynomial factor. Then, we compute the characteristic polynomial of the closed-loop system

$$p(s) = N_P N_K + M_P M_K. \tag{4.37}$$

  The stability of characteristic roots can be judged either by solving for the roots of characteristic polynomial directly or by application of Routh–Hurwitz stability criterion. The internal stability is equivalent to that the real parts of all characteristic roots are negative.

Proof. Since $\frac{PK}{1+PK} + \frac{1}{1+PK} = 1$ holds, the stability of $\frac{PK}{1+PK}$ is equivalent to that of $\frac{1}{1+PK}$. Therefore, the internal stability and the stability of transfer matrix

$$H(s) = \begin{bmatrix} \dfrac{PK}{1+PK} & \dfrac{P}{1+PK} \\ \dfrac{K}{1+PK} & \dfrac{1}{1+PK} \end{bmatrix} \tag{4.38}$$

are equivalent. After the substitution of the coprime factorization of $P(s)$ and $K(s)$, $H(s)$ becomes

$$H(s) = \frac{1}{p}\begin{bmatrix} N_P N_K & N_P M_K \\ M_P N_K & M_P M_K \end{bmatrix} = \frac{1}{p}\begin{bmatrix} N_P \\ M_P \end{bmatrix} [N_K \quad M_K].$$

Obviously, if all roots of the characteristic polynomial $p(s)$ are stable, the closed-loop system is internally stable. Conversely, when the closed-loop system is internally stable, if a root of the characteristic polynomial is unstable, this root must be cancelled by the blocking zero of the numerator matrix of $H(s)$. However, due to the assumption of coprimeness, neither $[N_P \quad M_P]^T$ nor $[N_K \quad M_K]$ has zeros, which leads to a contradiction. Therefore, when the closed-loop system is internally stable, the characteristic polynomial cannot have any unstable root. •

- State-space method (MIMO system)
  In this method, we first calculate the state equation of the closed-loop transfer matrix based on those of the subsystems and then determine the internal stability using the eigenvalues of its coefficient matrix A. If the real parts of all eigenvalues are negative, then the system is internally stable. However, it should be pointed out that we must use the state equation of closed-loop system obtained from subsystems directly. If the eigenvalue analysis is done after minimal realization of the state equation, poles cancelled by zeros do not appear in the eigenvalues of matrix A, and we cannot find these unstable poles.

Example 4.12 *For the feedback system of Example 4.11, apply the methods above to validate the internal stability once again.*

Solution

1. Characteristic polynomial method

   Coprimely factorizing the plant and controller as

$$P(s) = \frac{N_P(s)}{M_P(s)}, \quad N_P(s) = 1, \quad M_P(s) = s - 1$$

$$K(s) = \frac{N_K(s)}{M_K(s)}, \quad N_K(s) = s - 1, \quad M_K(s) = s,$$

   we obtain the characteristic polynomial

$$p(s) = (s-1)s + s - 1 = (s-1)(s+1).$$

   This characteristic polynomial has an unstable root 1, so the closed-loop system is not internally stable.

2. State-space method

   A set of realizations of $P(s)$ and $K(s)$ is

$$P(s) = \left[\begin{array}{c|c} 1 & 1 \\ \hline 1 & 0 \end{array}\right], \quad K(s) = \left[\begin{array}{c|c} 0 & -1 \\ \hline 1 & 1 \end{array}\right].$$

By the formula (4.24) of cascade connection, the transfer function of the forward channel is

$$G_1(s) = P(s)K(s) = \left[\begin{array}{cc|c} 1 & 1 & 1 \\ 0 & 0 & -1 \\ \hline 1 & 0 & 0 \end{array}\right].$$

Further, noting that the transfer function of the feedback channel is $G_2(s) = 1$, by the formula (4.28) of feedback connection, matrix A of the closed-loop system and its eigenvalues are computed as

$$A = \begin{bmatrix} 0 & 1 \\ 1 & 0 \end{bmatrix} \Rightarrow \sigma(A) = \{1, -1\}.$$

Thus, we get the same conclusion as the first method, that is, the closed-loop system has an unstable pole 1.

However, if the controller is replaced by a PI compensator

$$K = \frac{3s+1}{s},$$

then the characteristic polynomial turns into

$$p(s) = 3s + 1 + (s-1)s = (s+1)^2$$

whose roots are stable. So the closed-loop system is internally stable. ▽

4.2.3 Pole–Zero Cancellation

In the parallel connected system of Example 4.8, cancellation between pole and zero occurs at the point $s = 1$, which thereby makes the whole system unobservable. This is because the pole–zero cancellation deters the response of the pole of $G_2(s)$ from being transmitted to the output, so the output does not contain the state information of this pole (refer to Figure 4.9). In contrast, in the case of (Figure 4.9)

$$G_1(s) = \frac{1}{s-1}, \quad G_2(s) = \frac{s-1}{s+1},$$

the overall state-space realization is

$$G_1(s)G_2(s) = \left[\begin{array}{cc|c} 1 & 1 & 1 \\ 0 & -1 & -2 \\ \hline 1 & 0 & 0 \end{array}\right].$$

$y \leftarrow G_1 \leftarrow G_2 \leftarrow u$

Figure 4.9 Uncontrollable/unobservable mode stemmed from pole–zero cancellation

Table 4.1 Relations between pole–zero cancellation and controllability/observability (cascade system of Figure 4.9)

Cancellation between zero of G_1 and pole of G_2	Unobservable
Cancellation between pole of G_1 and zero of G_2	Uncontrollable

Its controllability matrix and observability matrix are

$$\mathcal{C} = \begin{bmatrix} 1 & -1 \\ -2 & 2 \end{bmatrix}, \quad \mathcal{O} = \begin{bmatrix} 1 & 0 \\ 1 & 1 \end{bmatrix}.$$

Hence, the realization of the whole system is not controllable. This is because the input is not transmitted to the pole of $G_1(s)$. Conclusions stated here hold true in the general case and are summarized in Table 4.1.

Unstable pole–zero cancellation and pole–zero cancellation near the imaginary axis must be avoided. For example, in the pole–zero cancellation of Example 4.11, since the input cannot reach the cancelled unstable pole of $P(s)$, the disturbance response of $P(s)$ diverges, and the system is not stable.

4.2.4 *Stabilizability and Detectability*

Even if the state-space realizations of subsystems are controllable and observable, pole–zero cancellation may occur when they are connected, resulting in uncontrollable or unobservable phenomenon. In such cases, even though the feedback control cannot guarantee that the state of system converges at a specified rate, at least we hope to ensure its stability. This engineering requirement is explicitly formulated as the notions of stabilizability and detectability.

Definition 4.10 *If there exists a state feedback $u = Fx$ rendering $A + BF$ stable, the dynamic system of (4.1) or (A, B) is called* stabilizable. *Similarly, when there exists L making $A + LC$ stable, (C, A) is called* detectable.

Through generalizations of Theorems 4.1 and 4.2, we obtain the following two theorems.

Theorem 4.5 *The following statements are equivalent:*

1. *(A, B) is stabilizable.*
2. *For all λ satisfying $\Re(\lambda) \geq 0$, the matrix $[A - \lambda I \quad B]$ has full row rank.*
3. *There exists a matrix F such that $A + BF$ is stable.*

Theorem 4.6 *The following statements are equivalent:*

1. *(C, A) is detectable.*
2. *For all λ satisfying $\Re(\lambda) \geq 0$, the matrix $\begin{bmatrix} A-\lambda I \\ C \end{bmatrix}$ has full column rank.*

3. *There exists a matrix* L *such that* $A + LC$ *is stable.*
4. (A^T, C^T) *is stabilizable.*

Example 4.13 *Test the stabilizability and detectability of system*

$$\left[\begin{array}{c|c} A & B \\ \hline C & 0 \end{array}\right] = \left[\begin{array}{cc|c} 1 & 0 & 0 \\ 0 & -1 & 1 \\ \hline 1 & 0 & 0 \end{array}\right].$$

Solution Since

$$\text{rank}[A - I \ \ B] = \text{rank}\begin{bmatrix} 0 & 0 & 0 \\ 0 & -2 & 1 \end{bmatrix} = 1 \neq 2,$$

(A, B) is not stabilizable. In fact, the unstable pole $p_1 = 1$ is uncontrollable. On the other hand,

$$\text{rank}\begin{bmatrix} A - \lambda I \\ C \end{bmatrix} = \text{rank}\begin{bmatrix} 1-\lambda & 0 \\ 0 & -1-\lambda \\ 1 & 0 \end{bmatrix} = 2 \ \forall \Re(\lambda) \geq 0$$

holds, so it is detectable. In fact, the unobservable pole $p_2 = -1$ is stable. ▽

4.3 Lyapunov Equation

In system theory, it is very important how to determine the stability, controllability, and observability accurately. In testing these properties, *Lyapunov equation* is extremely useful. The so-called Lyapunov equation is a linear matrix equation in the form of

$$A^T P + PA + Q = 0 \tag{4.39}$$

in which A and $Q = Q^T$ are given real matrices. The necessary and sufficient condition for this equation to have a unique solution is $\lambda_i(A) + \overline{\lambda}_j(A) \neq 0$ $(\forall i, j)$. So, the equation has a unique solution when A is stable.

Lemma 4.2 *Lyapunov equation (4.39) has a unique solution iff*

$$\lambda_i(A) + \overline{\lambda}_j(A) \neq 0 \quad \forall i, j.$$

Proof. We rewrite the matrix equation (4.39) as a vector equation by using the properties of Kronecker product. First of all,

$$\text{vec}(PA) = \text{vec}(I \cdot P \cdot A) = (A^T \otimes I)\text{vec}(P)$$
$$\text{vec}(A^T P) = \text{vec}(A^T \cdot P \cdot I) = (I \otimes A)\text{vec}(P)$$

holds. So (4.39) is transformed equivalently as

$$(A^T \otimes I + I \otimes A)\text{vec}(P) + \text{vec}(Q) = 0 \Rightarrow (A^T \oplus A)\text{vec}(P) = -\text{vec}(Q).$$

This linear equation has a unique solution iff the coefficient matrix is nonsingular, that is, without any zero eigenvalues. This lemma is true since the eigenvalues of $A^T \oplus A$ are given by $\lambda_i(A) + \lambda_j(A^T) = \lambda_i(A) + \overline{\lambda}_j(A)$, etc. •

The relationship between the stability of A and the solution P and relationship between (Q, A) and P are shown in the following theorem.

Theorem 4.7 *Let P be the solution of Lyapunov equation (4.39). Then, the following statements hold:*

1. *When A is stable, the Lyapunov equation has a unique solution*

$$P = \int_0^{\infty} e^{A^T t} Q e^{At}\, dt. \tag{4.40}$$

2. *When $Q \geq 0$ and (Q, A) is observable, $P > 0$ iff A is stable.*
3. *When $Q > 0$, then $P > 0$ iff A is stable.*
4. *When A is stable and $Q \geq 0$, then $P > 0$ iff (Q, A) is observable.*

Proof.

1. As A is stable, solution of the Lyapunov equation is unique. Substituting the given P into the Lyapunov equation and using $e^{A\infty} = 0$, it can be confirmed that the Lyapunov equation is satisfied:

$$A^T P + PA = \int_0^{\infty} \frac{de^{A^T t} Q e^{At}}{dt} dt = e^{A^T t} Q e^{At}\Big|_0^{\infty} = -Q.$$

2. Let λ be an eigenvalue of A, $u \neq 0$ the corresponding eigenvector. Premultiplying u^* and postmultiplying u to (4.39), we get

$$2\Re(\lambda) u^* P u + u^* Q u = 0. \tag{4.41}$$

When $P > 0$, since $u^* P u > 0$ and $u^* Q u \geq 0$, we have $\Re(\lambda) \leq 0$. If $\Re(\lambda) = 0$, $Qu = 0$ follows from this equation. But

$$(A - \lambda I)u = 0, \quad Qu = 0 \tag{4.42}$$

holds in this case, which contradicts the observability of (Q, A). Thus, $\Re(\lambda) = 0$ is not true and A must be stable.

Conversely, when A is stable, P is positive semidefinite according to (4.40). Noting that the solution of $\dot{x} = Ax$ is $x(t) = e^{At} x(0)$, we multiply $x(0)$ and its transpose to (4.40) and get

$$x^T(0) P x(0) = \int_0^{\infty} x^T(t) Q x(t)\, dt.$$

Since the integral on the right is nonnegative, the quadratic term on the left is positive semidefinite. If P is not positive definite, there exists $x(0) \neq 0$ such that $Px(0) = 0$. Then, the preceding equation shows that $x(t)$ must satisfy

$$Qx(t) \equiv 0, \quad \forall t \geq 0.$$

Repeated differentiation of this equation by the time t yields

$$QA^i x(t) = 0, \quad i = 0, \ldots, n-1, \quad \forall t \geq 0 \tag{4.43}$$

which contradicts the observability of (Q, A). So $P > 0$ is true.

3. Since $Q > 0$, if the solution of Lyapunov equation is positive definite, we see from (4.41) that all eigenvalues λ of A satisfy $\Re(\lambda) < 0$. Accordingly, A is stable. On the contrary, when A is stable, $P > 0$ is obvious from (4.40).
4. In the statement (2), we have proved that $P > 0$ if (Q, A) is observable. Now we prove the converse via reduction to absurdity. Assume that $P > 0$, but (Q, A) is not observable. Then, there are a $\lambda \in \sigma(A)$ and a vector $u \neq 0$ satisfying (4.42). From (4.41) we get $\Re(\lambda) = 0$, which contradicts the stability of A. •

From this lemma, we see that Lyapunov equation not only plays an important role in testing system's stability but also is very effective in testing the controllability and observability of systems. More importantly, this lemma can be applied to not only the stability test of a given system but also the control design, especially the stability proof of closed-loop system in optimal control theories.

However, it should be noticed that even if A is stable, it never means that $Q = -(A^TP + PA)$ is positive definite for any matrix $P > 0$. For example, in

$$A = \begin{bmatrix} -1 & \\ & -2 \end{bmatrix}, \quad P = \begin{bmatrix} 1.1 & 1 \\ 1 & 1 \end{bmatrix} > 0$$

$$\Rightarrow Q = -(A^TP + PA) = \begin{bmatrix} 2.2 & 3 \\ 3 & 4 \end{bmatrix},$$

as $\det(Q) = -0.2$, Q is not positive definite.

Example 4.14 *Let us use a two-dimensional system to verify statement (2) of the theorem. Here, the coefficient matrix of the given system is*

$$A = \begin{bmatrix} 0 & 1 \\ -a_1 & -a_2 \end{bmatrix} \Rightarrow |sI - A| = s^2 + a_2 s + a_1.$$

According to Routh–Hurwitz criterion, it is easy to know that $a_1, a_2 > 0$ is the stability condition. Set the output matrix c as $c = [1 \ \ 0]$; then (c, A) is observable. So (c^Tc, A) is also observable. Let the unknown matrix be $P = \begin{bmatrix} p_1 & p_2 \\ p_2 & p_3 \end{bmatrix}$ in Lyapunov equation

$$PA + A^TP + c^Tc = 0.$$

Expansion of this equation leads to

$$\begin{cases} p_1 - a_2 p_2 - a_1 p_3 = 0 \\ p_2 - a_2 p_3 = 0 \\ 2a_1 p_2 = 1 \end{cases} \Rightarrow \begin{bmatrix} 1 & -a_2 & -a_1 \\ 0 & 1 & -a_2 \\ 0 & 2a_1 & 0 \end{bmatrix} \begin{bmatrix} p_1 \\ p_2 \\ p_3 \end{bmatrix} = \begin{bmatrix} 0 \\ 0 \\ 1 \end{bmatrix}. \tag{4.44}$$

This equation has a unique solution iff the coefficient matrix on the left-hand side is nonsingular. Due to the block triangular structure of this matrix, it is nonsingular iff so is the 2×2 *block in the lower right corner. That is,*

$$\begin{vmatrix} 1 & -a_2 \\ 2a_1 & 0 \end{vmatrix} \neq 0 \iff a_1 a_2 \neq 0 \iff a_1 \neq 0, a_2 \neq 0 \tag{4.45}$$

is the condition for the existence of unique solution.

Further, solving the expanded Lyapunov equation successively, we get

$$p_2 = \frac{1}{2a_1}, \quad p_3 = \frac{1}{a_2} p_2 = \frac{1}{2a_1 a_2}, \quad p_1 = a_2 p_2 + a_1 p_3 = \frac{a_1 + a_2^2}{2a_1 a_2}.$$

Therefore, the solution of Lyapunov equation is

$$P = \frac{1}{2a_1 a_2} \begin{bmatrix} a_1 + a_2^2 & a_2 \\ a_2 & 1 \end{bmatrix} = \frac{1}{2a_1 a_2} \begin{bmatrix} 1 & a_2 \\ 0 & 1 \end{bmatrix} \begin{bmatrix} a_1 & 0 \\ 0 & 1 \end{bmatrix} \begin{bmatrix} 1 & 0 \\ a_2 & 1 \end{bmatrix}.$$

Moreover, the following conditions are equivalent:

$$P > 0 \iff 2a_1 a_2 > 0, \quad a_1 > 0 \iff a_1 > 0, \quad a_2 > 0. \tag{4.46}$$

This condition coincides with the stability condition obtained from Routh–Hurwitz criterion.

4.3.1 Controllability Gramian and Observability Gramian

Solution L_o of a Lyapunov equation about (C, A)

$$A^T L_o + L_o A + C^T C = 0 \tag{4.47}$$

is called the *observability Gramian*. As $C^T C \geq 0$, according to Theorem 4.7(1), we have $L_o \geq 0$ when A is stable. Further, $L_o > 0$ iff $(C^T C, A)$ is observable. As the observability of $(C^T C, A)$ is equivalent to that of (C, A)[2], $L_o > 0$ is equivalent to the observability of (C, A).

Similarly, solution L_c of the Lyapunov equation

$$AL_c + L_c A^T + BB^T = 0 \tag{4.48}$$

is called the *controllability Gramian*. From the duality of Theorem 4.7, we know that $L_c \geq 0$ holds when A is stable. Further, the controllability of (A, B) is equivalent to the positive definiteness of the controllability Gramian L_c. Furthermore, by the semipositive definiteness of L_c and L_o, their eigenvalues are all nonnegative real numbers. From the viewpoint of system engineering, eigenvalue magnitudes of the controllability Gramian and observability Gramian indicate the degree of difficulty on the controllability and observability of system states. The following example illustrates this.

[2] Since $Cu = 0$ is equivalent to $C^T Cu = 0$, $(A - \lambda I)u = 0$ and $Cu = 0$ is equivalent to $(A - \lambda I)u = 0$ and $C^T Cu = 0$. That is, when a pair of $(C^T C, A)$ and (C, A) is not observable, another pair is not observable too. So the observability of both is equivalent.

Example 4.15 *Assume that $\lambda_1, \lambda_2, \epsilon > 0$. Consider a stable system*

$$\dot{x} = \begin{bmatrix} -\lambda_1 & \\ & -\lambda_2 \end{bmatrix} x + \begin{bmatrix} 1 & \\ & \epsilon \end{bmatrix} u, \quad y = [1 \quad 1]x.$$

Computing its controllability Gramian, we get

$$L_c = \begin{bmatrix} \frac{1}{2\lambda_1} & \\ & \frac{\epsilon}{2\lambda_2} \end{bmatrix}.$$

It is observed from $\dot{x}_2 = -\lambda_2 x_2 + \epsilon u_2$ that the state x_2 is not easy to control as $\epsilon \to 0$. The eigenvalue of controllability Gramian corresponding to state x_2 is $\frac{\epsilon}{2\lambda_2}$, which gets very small as $\epsilon \to 0$. That is, when an eigenvalue of the controllability Gramian gets smaller, the corresponding state is more difficult to control. The same conclusion is true for the observability Gramian. That is, when an eigenvalue of the observability Gramian gets smaller, the corresponding state is more difficult to observe.

In this example, the transfer function is

$$G(s) = \frac{1}{s+\lambda_1} + \frac{\epsilon}{s+\lambda_2}.$$

So when ϵ is small enough, even if we ignore the state x_2 it has little impact on the input–output relation. It seems that even if we ignore the states corresponding to small eigenvalues of controllability Gramian or observability Gramian, the resulted discrepancy in transfer function will be small too. However, this hypothesis is not correct. The following example given in Ref. [100] clearly illustrates where the pitfall lies in. Here, we examine a stable transfer function

$$G(s) = \frac{3s+18}{s^2+3s+18}.$$

This transfer function has the following special state-space realization:

$$G(s) = \left[\begin{array}{cc|c} -1 & -4/\alpha & 1 \\ 4\alpha & -2 & 2\alpha \\ \hline -1 & 2/\alpha & 0 \end{array}\right]$$

in which $\alpha \neq 0$ is an arbitrary real number. The controllability Gramian of this realization is

$$L_c = \text{diag}(0.5, \ \alpha^2).$$

When we select small α, the eigenvalue α^2 of L_c gets so small that the controllability of its corresponding state is weakened. Eliminating the state x_2 corresponding to this eigenvalue, we get a first-order transfer function

$$\hat{G}(s) = \left[\begin{array}{c|c} -1 & 1 \\ \hline -1 & 0 \end{array}\right] = \frac{-1}{s+1}.$$

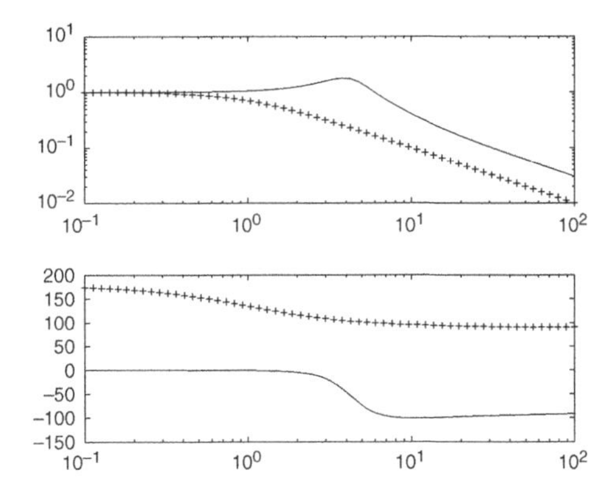

Figure 4.10 Bode plots of G (solid line) and $\hat{G}$ (+)

However, from the Bode diagram of Figure 4.10, it is obvious that this transfer function is by no means close to the original one. Analyzing the observability Gramian

$$L_o = \text{diag}\left(0.5, \frac{1}{\alpha^2}\right)$$

reveals where the problem is. When α is small, the eigenvalue $1/\alpha^2$ of L_o gets large. Therefore, the observability of the corresponding state x_2 is very strong. In other words, although the controllability of x_2 is weak, its observability is so strong that the effect of x_2 on the input transmission cannot be ignored. In summary, we cannot correctly determine which state dominates the input transmission by using only controllability Gramian or observability Gramian.

4.3.2 Balanced Realization

As shown in the previous subsection, controllability Gramian or observability Gramian alone cannot correctly determine which state has little effect on the input–output relationship. To resolve this problem, this subsection introduces a realization with special controllability and observability Gramians.

Assume that $G(s) = (A, B, C, D)$ is stable, that is, A is stable. Its controllability Gramian L_c and observability Gramian L_o satisfy (4.48) and (4.47), respectively, and $L_c \geq 0$, $L_o \geq 0$. Further, (A, B) is controllable iff $L_c > 0$; (C, A) is observable iff $L_o > 0$.

Doing a state transformation $z = T^{-1}x$ using a nonsingular matrix T, we get a new realization

$$G(s) = \left[\begin{array}{c|c} \hat{A} & \hat{B} \\ \hline \hat{C} & \hat{D} \end{array}\right] = \left[\begin{array}{c|c} T^{-1}AT & T^{-1}B \\ \hline CT & D \end{array}\right].$$

Its controllability Gramian and observability Gramian are $\hat{L}_c = T^{-1}L_cT^{-T}$ and $\hat{L}_o = T^TL_oT$ (Exercise 4.13). By $\hat{L}_c\hat{L}_o = T^{-1}L_cL_oT$, it is seen that the eigenvalues of the product of these two Gramians are invariant to state transformation. Thus, if we split the eigenvalue of L_cL_o equally to L_c and L_o, we can clearly see the impact of each state on the transfer function.

In fact, such transformation always exists. Particularly, for a minimal realization, we have that

$$\hat{L}_c = T^{-1} L_c T^{-T} = \Sigma, \quad \hat{L}_o = T^T L_o T = \Sigma \tag{4.49}$$

$$\Sigma = \text{diag}(\sigma_1, \sigma_2, \ldots, \sigma_n). \tag{4.50}$$

This new realization is called the *balanced realization.* σ_i aligned in the order of $\sigma_1 \geq \sigma_2 \geq \ldots \geq \sigma_n \geq 0$ are known as *Hankel singular values* of the system.

More generally, for any nonminimal realization of system, there exists a state transformation such that the controllability Gramian and observability Gramian of the realization are diagonal matrices after transformation. Moreover, the controllable and observable subsystem is balanced. The specific conclusions are given by the following theorem.

Theorem 4.8 *For a stable transfer matrix $G(s) = (A, B, C, D)$, there exists a transformation matrix T such that $G(s) = (T^{-1}AT, T^{-1}B, CT, D)$ has the following controllability Gramian L_c and observability Gramian L_o*

$$L_c = \begin{bmatrix} \Sigma_1 & & & \\ & \Sigma_2 & & \\ & & 0 & \\ & & & 0 \end{bmatrix}, \quad L_o = \begin{bmatrix} \Sigma_1 & & & \\ & 0 & & \\ & & \Sigma_3 & \\ & & & 0 \end{bmatrix}$$

in which the matrix $\Sigma_1, \Sigma_2, \Sigma_3$ are positive definite diagonal matrix.

The states corresponding to zero eigenvalues of L_c, L_o can be eliminated from the transfer matrix. That is, when we decompose the new realization of $G(s)$ as

$$G(s) = \left[\begin{array}{c|c} T^{-1}AT & T^{-1}B \\ \hline CT & D \end{array}\right] = \left[\begin{array}{ccc|c} A_{11} & \cdots & A_{14} & B_1 \\ \vdots & & \vdots & \vdots \\ A_{41} & \cdots & A_{44} & B_4 \\ \hline C_1 & \cdots & C_4 & D \end{array}\right] \tag{4.51}$$

in accordance with the partition of L_c, L_o (A_{ii} has the same dimension as Σ_i),

$$G(s) = \left[\begin{array}{c|c} A_{11} & B_1 \\ \hline C_1 & D \end{array}\right] \tag{4.52}$$

holds. Of course, if there exist some very small eigenvalues in A_{11}, the corresponding states can also be approximately eliminated which is called *truncation.* In system theory, approximation of a high-order transfer matrix by a low-order one is called *model reduction.* The truncation of balanced realization is actually a model reduction method.

4.4 Linear Fractional Transformation

In mathematics, a function in the form of

$$\frac{cx + d}{ax + b}$$

is called a linear fractional transformation of variable x. This equation can also be written as

$$\frac{cx+d}{ax+b}=\frac{d}{b}+\frac{d}{b}\frac{x}{1+\frac{a}{b}x}\left(\frac{c}{d}-\frac{a}{b}\right).$$

Its extension to matrix is the *linear fractional transformation*, or LFT for short, to be touched in this section.

Now, partition the matrix M as

$$M=\begin{bmatrix}M_{11} & M_{12}\\ M_{21} & M_{22}\end{bmatrix}$$
$$M_{11}\in\mathbb{F}^{p_1\times q_1},\quad M_{12}\in\mathbb{F}^{p_1\times q_2},\quad M_{21}\in\mathbb{F}^{p_2\times q_1},\quad M_{22}\in\mathbb{F}^{p_2\times q_2}.$$

Assume that $X_\ell\in\mathbb{F}^{q_2\times p_2}$, $X_u\in\mathbb{F}^{q_1\times p_1}$. When $(I-M_{22}X_\ell)$ is nonsingular, the *lower LFT* about X_ℓ is defined as

$$\mathcal{F}_\ell(M,X_\ell):=M_{11}+M_{12}X_\ell(I-M_{22}X_\ell)^{-1}M_{21}. \tag{4.53}$$

Similarly, when $(I-M_{11}X_u)$ is nonsingular, the *upper LFT* about X_u is defined as

$$\mathcal{F}_u(M,X_u)=M_{22}+M_{21}X_u(I-M_{11}X_u)^{-1}M_{12}. \tag{4.54}$$

They all are referred to as LFT.

The matrix M in LFT is called *coefficient matrix*. From a system viewpoint, these LFT correspond to the feedback systems shown in the block diagrams of Figure 4.11, respectively. In Figure 4.11, the left corresponds to equations

$$\begin{bmatrix}z_1\\ y_1\end{bmatrix}=M\begin{bmatrix}w_1\\ u_1\end{bmatrix}=\begin{bmatrix}M_{11} & M_{12}\\ M_{21} & M_{22}\end{bmatrix}\begin{bmatrix}w_1\\ u_1\end{bmatrix} \tag{4.55}$$
$$u_1=X_\ell\, y_1 \tag{4.56}$$

while the right corresponds to equations

$$\begin{bmatrix}y_2\\ z_2\end{bmatrix}=M\begin{bmatrix}u_2\\ w_2\end{bmatrix}=\begin{bmatrix}M_{11} & M_{12}\\ M_{21} & M_{22}\end{bmatrix}\begin{bmatrix}u_2\\ w_2\end{bmatrix} \tag{4.57}$$
$$u_2=X_u\, y_2. \tag{4.58}$$

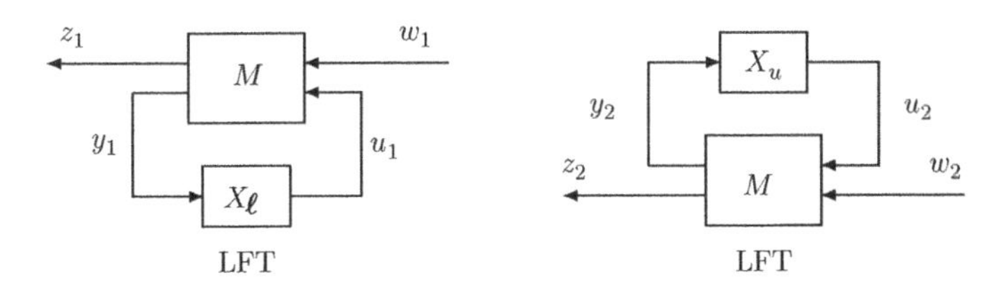

Figure 4.11 Graphical representation of LFT

It is easy to know that the transfer matrix of $w_1 \mapsto z_1$ in the left figure is $\mathcal{F}_\ell(M, X_\ell)$, that is,

$$z_1 = \mathcal{F}_\ell(M, X_\ell)w_1. \tag{4.59}$$

So, $\mathcal{F}_\ell(M, X_\ell)$ is obtained from closing the loop *from below* in the left of Figure 4.11. This is why it is called the *lower LFT*. On the other hand, the transfer matrix of $w_2 \mapsto z_2$ in the right figure is $\mathcal{F}_u(M, X_u)$, namely,

$$z_2 = \mathcal{F}_u(M, X_u)w_2. \tag{4.60}$$

Therefore, $\mathcal{F}_u(M, X_u)$ is obtained from closing the loop *from above* in the right of Figure 4.11 and called the *upper LFT*. The feedback, cascade, and parallel connections of transfer functions are special cases of LFT. For example, when $M_{22} = 0$, $\mathcal{F}_\ell(M, I)$ becomes the parallel connection of M_{11} and $M_{12}M_{21}$, while when $M_{11} = 0$, $\mathcal{F}_\ell(M, I)$ is the cascade connection of M_{12} and M_{21}.

Now we compute the state-space realization of closed-loop transfer matrix $\mathcal{F}_\ell(P, K)$ when transfer matrices

$$P(s) = \left[\begin{array}{c|cc} A & B_1 & B_2 \\ \hline C_1 & D_{11} & D_{12} \\ C_2 & D_{21} & 0 \end{array}\right], \quad K(s) = \left[\begin{array}{c|c} A_K & B_K \\ \hline C_K & D_K \end{array}\right] \tag{4.61}$$

are LFT connected. To this end, let the state of $P(s)$ be x and the state of $K(s)$ be x_K. Further, assume that their input–output relations are, respectively,

$$\begin{bmatrix} \hat{z} \\ \hat{y} \end{bmatrix} = P \begin{bmatrix} \hat{w} \\ \hat{u} \end{bmatrix}, \quad \hat{u} = K\hat{y}. \tag{4.62}$$

We need just derive the relation between $\hat{w}(s)$ and $\hat{z}(s)$. First, substituting $y = C_2x + D_{21}w$ into u and $\dot{x}_K$, we get

$$\begin{aligned} u &= C_Kx_K + D_K(C_2x + D_{21}w) = D_KC_2x + C_Kx_K + D_KD_{21}w \\ \dot{x}_K &= A_Kx_K + B_K(C_2x + D_{21}w) = B_KC_2x + A_Kx_K + B_KD_{21}w. \end{aligned}$$

Then, substitution of this u into $\dot{x}$ and z yields

$$\begin{aligned} \dot{x} &= Ax + B_1w + B_2(D_KC_2x + C_Kx_K + D_KD_{21}w) \\ &= (A + B_2D_KC_2)x + B_2C_Kx_K + (B_1 + B_2D_KD_{21})w \\ z &= C_1x + D_{11}w + D_{12}(D_KC_2x + C_Kx_K + D_KD_{21}w) \\ &= (C_1 + D_{12}D_KC_2)x + D_{12}C_Kx_K + (D_{11} + D_{12}D_KD_{21})w. \end{aligned}$$

Arranging these equations in vector form, we get the desired result:

$$\mathcal{F}_\ell(P, K) = \left[\begin{array}{cc|c} A + B_2D_KC_2 & B_2C_K & B_1 + B_2D_KD_{21} \\ B_KC_2 & A_K & B_KD_{21} \\ \hline C_1 + D_{12}D_KC_2 & D_{12}C_K & D_{11} + D_{12}D_KD_{21} \end{array}\right]. \tag{4.63}$$

Exercises

4.1 Test the controllability and observability of the state equation of two-mass–spring system in Example 4.1.

4.2 Consider the linear system

$$\dot{x} = \begin{bmatrix} 1 & 0 \\ 0 & 1 \end{bmatrix} x + \begin{bmatrix} 1 \\ 1 \end{bmatrix} u, \quad y = [1, 0]x.$$

(a) Show that this system is neither controllable nor observable.
(b) Use block diagram to explain the reason.

4.3 Prove that the transmission zeros of a square transfer matrix $G(s)$ are given by the set $\{z | \det(G(z)) = 0\}$ when its normal rank is equal to its dimension.

4.4 Use the result of the preceding exercise to calculate the transmission zeros of transfer matrix

$$G(s) = \begin{bmatrix} \dfrac{s-1}{s+1} & \dfrac{1}{s+1} \\ 1 & \dfrac{s-1}{s+2} \end{bmatrix}.$$

4.5 For the two-mass–spring system shown in Example 4.1, analyze the pole and zero properties of transfer functions $P_u(s)$, $P_d(d)$ and their relations with the inertia ratio when the output y is selected as the load speed ω_L. To simplify the analysis, all damping ratios may be set as zeros.

4.6 Prove the following equation:

$$\text{rank} \begin{bmatrix} A + BF - sI & B \\ C + DF & D \end{bmatrix} = \text{rank} \begin{bmatrix} A - sI & B \\ C & D \end{bmatrix}.$$

4.7 Prove the formula for the realization of inverse system by using state equation.

4.8 Prove the formulae for the realizations of cascade and feedback connections of systems.

4.9 Prove the stability condition for MIMO systems given in Theorem 4.3 following the steps below and using the two-norms of vector and matrix.
(a) Starting from $\|g(\tau)u(t-\tau)\|_2 \leq \|g(\tau)\|_2 \|u(t-\tau)\|_2$ and the integral inequality $\left\| \int g(\tau)u(t-\tau)d\tau \right\|_2 \leq \int \|g(\tau)u(t-\tau)\|_2 d\tau$, prove the sufficiency.
(b) At any time t, $\|g(t)\|_2 = \sigma_{\max}(g(t))$ has a unit right singular vector $v_1(t)$ and a unit left singular vector $u_1(t)$ satisfying $g(t)v_1(t) = \|g(t)\|_2 u_1(t)$. Use the special input vector $u(t-\tau) = v_1(\tau)$ to prove the necessity.

4.10 Prove that all poles of the autonomous system $\dot{x} = Ax$ satisfy $\Re(\lambda_i(A)) < -\sigma(< 0)$ iff the LMI

$$PA + A^T P + 2\sigma P < 0$$

has a positive definite solution P.

4.11 We want to use a state feedback $u = Fx$ to place all closed-loop poles of the linear system $\dot{x} = Ax + Bu$ in the half-plane $\Re(s) < -\sigma(< 0)$. Derive the design condition by using the methods of variable elimination and variable change, respectively, based on the result of previous exercise.

4.12 Prove that the output two-norm of the stable autonomous system

$$\dot{x} = Ax, \quad y = Cx, \quad x(0) \neq 0$$

can be calculated as follows:

$$\|y\|_2^2 = \int_0^\infty y(t)^T y(t) dt = x(0)^T P x(0)$$

where P is the solution of Lyapunov equation.

$$PA + A^T P + C^T C = 0.$$

4.13 For a stable transfer matrix $G = (A, B, C, D)$, let the controllability Gramian be L_c and the observability Gramian be L_o. Prove that, after a similarity transformation, the controllability Gramian and observability Gramian corresponding to the realization $G = (T^{-1}AT, T^{-1}B, CT, D)$ are $\hat{L}_c = T^{-1}L_cT^{-T}$, $\hat{L}_o = T^T L_o T$.

Notes and References

Chen [12], Kailath [44] and Anderson and Moore [2] are popular textbooks on the fundamentals of linear systems. LFT transformation and related topics are discussed in References [27, 100].

5

System Performance

The ultimate goal of feedback control is to achieve reference tracking and disturbance rejection. These goals are usually referred to as *performance*. In order to evaluate the pros and cons of system performance objectively, we need performance criteria. In control engineering, the performance of a system is judged from two angles: transient response and steady-state response. This chapter provides a detailed analysis on these performance criteria.

The objective of control is to make the physical variables (i.e., output) of the physical system vary as desired. If the desired dynamic response of output is specified using a signal called *reference input*, then the goal is to keep the output consistent with the reference input. On the other hand, any system works in the physical world and is always subject to the influence of the environment. For example, an aircraft in flight is disturbed by turbulences, and a sailing ship is affected by waves. The turbulence and wave are the influence from the environment when the aircraft and ship are regarded as systems. In control engineering, such influence coming from the environment is called *disturbance*[1]. We need to ensure that the system output does not deviate from the reference input even under the influence of disturbance. Shown in Figure 5.1 is a typical operation situation of system in which r represents the reference input and d the disturbance. The control to make the output consistent with the reference input is called *reference tracking*. Meanwhile, the control to suppress the influence of disturbance on the system output is called *disturbance suppression*. Usually, we need to achieve these two requirements simultaneously in practice.

For a designed system, we need to evaluate its performance by using time response. The time response can be further divided into transient response and steady-state response. The following sections describe the evaluation methods for these responses. This chapter mainly treats single-input single-output (SISO) systems.

[1] In mechanical systems, the disturbance affects a system always in the form of force or torque. It brings about an adverse impact on the output such as position, velocity, and so on via the dynamics of the system. In this sense, the disturbance does not directly affect the system output. Therefore, the notion of output disturbance which acts directly on the output is not correct physically.

Robust Control: Theory and Applications, First Edition. Kang-Zhi Liu and Yu Yao.

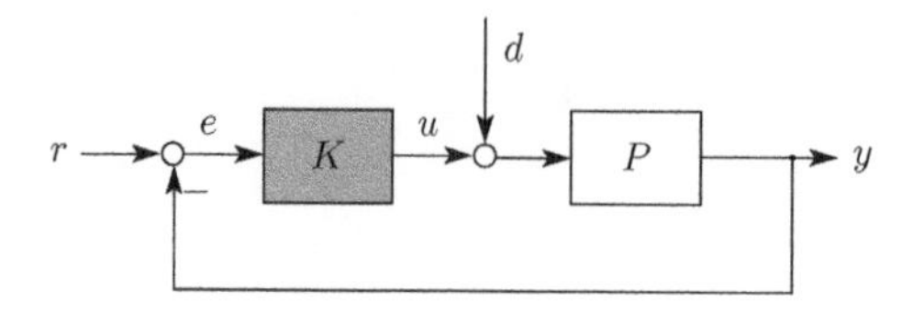

Figure 5.1 Reference tracking and disturbance control

5.1 Test Signal

5.1.1 Reference Input

Consider the tracking control of reference input $r(t)$ in the system shown in Figure 5.1. In practice, there are so many kinds of reference inputs to be tracked, so it is impractical to consider all of them in the design of control systems. Therefore, some special signals called *test signals* are usually used to judge the tracking performance of a system in system engineering. Typical test signals include *step signal*, *ramp signal*, and *sinusoidal signal*. The background behind dealing with signal tracking in this way is that the target trajectory is always a slowly varying smooth signal of time in most cases. If we partition a smooth signal into many segments and treat it piecewise, we may approximate the target trajectory quite accurately by using these test signals in each time segment. An example is given in Figure 5.2. It is obvious that when a system is able to track the test signals fast enough, it certainly can track the actual target trajectory with a high accuracy. Therefore, we need only to consider the tracking of test signals, on the premise that the system response is sufficiently fast.

Typical test signals are shown in Figure 5.3. Their mathematical descriptions are as follows:

Step signal

$$r(t) = \begin{cases} k, & t \geq 0 \\ 0, & t < 0 \end{cases}$$

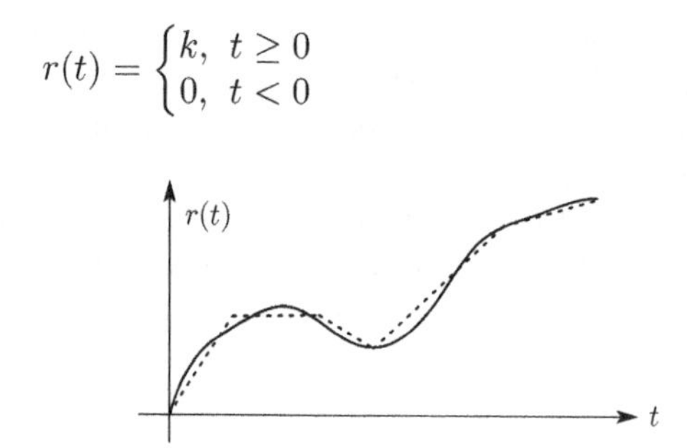

Figure 5.2 Approximation of target trajectory

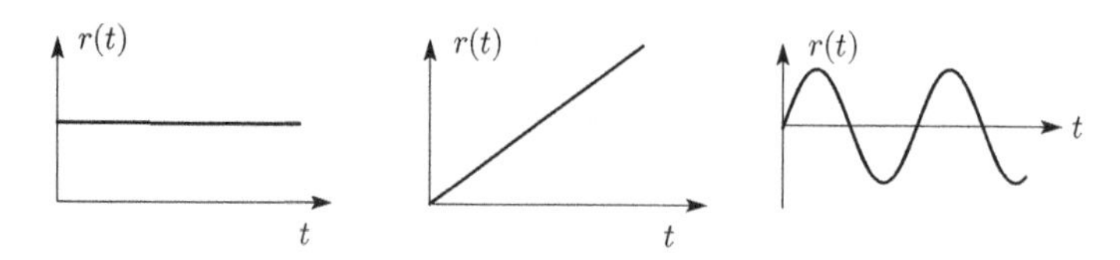

Figure 5.3 Typical test signals

Ramp signal

$$r(t) = \begin{cases} kt, & t \geq 0 \\ 0, & t < 0 \end{cases}$$

Sinusoidal signal

$$r(t) = \begin{cases} k \sin \omega t, & t \geq 0 \\ 0, & t < 0 \end{cases}$$

Particularly, the step signal with an amplitude $k = 1$ is called the *unit step signal* and denoted by $1(t)$. With the unit step signal, we can simply write a ramp signal as $kt \cdot 1(t)$ and sinusoidal signal as $k \sin \omega t \cdot 1(t)$.

Physically, a step signal corresponds to the switching of the working states of system, and a ramp signal corresponds to the acceleration and deceleration. For example, when a train brakes suddenly, the target speed is a step signal, while in the normal departure and arrival the target speed is a ramp signal. Typical target speed in the operation of trains is depicted in Figure 5.4. Shown in the figure is the entire procedure of a train's operation: the train leaves a station and accelerates until reaching the prescribed speed, then drives at this constant speed, and slows down when approaching the next stop. Compared with this, a sinusoidal signal appears in cases such as a circular orbit in the two-dimensional space. In system design, it depends on the practical problem which kind of test signal should be used. Usually, the step signal is selected.

5.1.2 Persistent Disturbance

Similarly, the suppression of persistent disturbance is also a major control problem. For instance, the gravity disturbance acting on a climbing car may be regarded as a step signal so long as the slope is roughly the same; meanwhile, the friction or air resistance resulting from a motion with a constant acceleration may be treated as a ramp signal. Also, a signal changing periodically with certain amplitude, such as the motion of a wave, can be regarded as a sinusoidal signal. The common feature of these disturbances is that they do not decay with time. So, they are called *persistent disturbances*.

5.1.3 Characteristic of Test Signal

The character of a test signal is that it never converges to zero. For this reason, it is called a *persistent signal*. Moreover, focusing on their Laplace transforms

$$\mathcal{L}[1(t)] = \frac{1}{s}, \quad \mathcal{L}[t \cdot 1(t)] = \frac{1}{s^2}, \quad \mathcal{L}[\sin \omega t \cdot 1(t)] = \frac{\omega}{s^2 + \omega^2}, \tag{5.1}$$

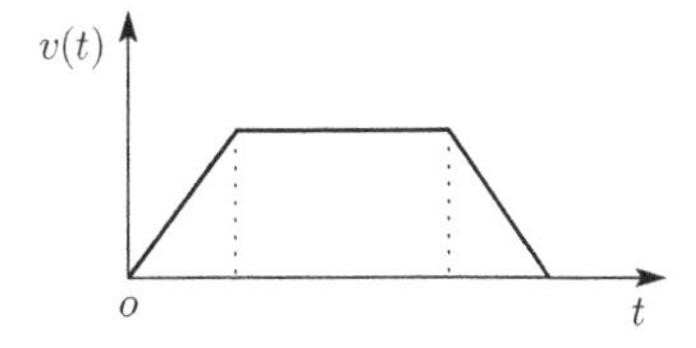

Figure 5.4 Target speed

we find that these functions have singularities on the imaginary axis. Any signal is the unit impulse response of its Laplace transform, so we can regard the Laplace transform as its model. In this sense, the transfer functions on the right-hand side of the equations in (5.1) are models of test signals. These transfer functions have poles on the imaginary axis, so they are *unstable* models.

5.2 Steady-State Response

The output response $\lim_{t\to\infty} y(t)$ of a system after a sufficiently long time is called the *steady-state response*. This section discusses what is a good steady-state response and what condition is needed to achieve a good steady-state response.

5.2.1 Analysis on Closed-Loop Transfer Function

First, we analyze the conditions required to achieve a good steady-state response for a given transfer matrix $G(s)$. Let the input of $G(s)$ be $u(t)$ and the output be $y(t)$. Since the closed-loop system must be stable, the stability of $G(s)$ is assumed.

5.2.1.1 Relation between Steady-State Output and Frequency Response

The following theorem shows that if the input of a stable system is a step or sinusoid, the output also has the same feature in the steady state.

Theorem 5.1 *When the input of stable system $G(s)$ is $u(t) = \cos\omega t \cdot 1(t)$, the steady-state output $y(t)$ is given by*

$$\lim_{t\to\infty} y(t) = |G(j\omega)| \cos(\omega t + \arg G(j\omega)). \tag{5.2}$$

Proof. Since $G(s)$ is stable, the Fourier transform of its impulse response $g(t)$ equals $\mathcal{F}[g(t)] = \int_0^\infty g(\tau)e^{-j\omega\tau}d\tau = G(j\omega)$. In the polar coordinate, $G(j\omega)$ can be written as $|G(j\omega)|e^{j\,\arg G(j\omega)}$. Then, applying $\cos\omega t = \Re(e^{j\omega t})$ and $\lim_{t\to\infty} e^{j\omega t} = e^{j\omega t}$, we see that

$$\begin{aligned}\lim_{t\to\infty} y(t) &= \lim_{t\to\infty}\int_0^t g(\tau)u(t-\tau)d\tau = \lim_{t\to\infty}\int_0^t g(\tau)\Re\left(e^{j\omega(t-\tau)}\right)d\tau \\ &= \Re\left(e^{j\omega t}\lim_{t\to\infty}\int_0^t g(\tau)e^{-j\omega\tau}d\tau\right) = \Re\left(e^{j\omega t}G(j\omega)\right) \\ &= |G(j\omega)|\Re\left(e^{j(\omega t+\arg G(j\omega))}\right) = |G(j\omega)|\cos(\omega t + \arg G(j\omega))\end{aligned}$$

holds. •

As $1 = \cos(0\cdot t)$, the unit step signal can be regarded as a cosine function with a frequency $\omega = 0$. So, its steady-state response becomes

$$\lim_{t\to\infty} y(t) = |G(j0)|\cos\arg G(j0).$$

$G(s)$ is a rational function with real coefficients, hence so is $G(j0)$. Its phase angle $\arg G(j0)$ is either 0 rad or π rad. Therefore, we obtain the following relation from the previous equation:

$$\lim_{t\to\infty} |y(t)| = |G(j0)|. \tag{5.3}$$

From this theorem, we see that *in order to reduce the negative effect of persistent disturbance $u(t) = \cos(\omega t)1(t)$ on the steady-state performance of system $G(s)$, it is necessary and sufficient to lower the gain $|G(j\omega)|$* at the same frequency ω. For a step disturbance, this means reducing the low-frequency gain $|G(j0)|$.

5.2.1.2 Condition for Asymptotic Convergence

The so-called asymptotic convergence of output means that

$$\lim_{t\to\infty} y(t) = 0. \tag{5.4}$$

In the analysis of the required conditions, the following lemma is fundamental.

Lemma 5.1 *Suppose that signal $f(t)$ is defined over $t \geq 0$ and its Laplace transform $\hat{f}(s)$ is a rational function. Then, $\lim_{t\to\infty} f(t) = 0$ iff $\hat{f}(s)$ is stable.*

Proof. Sufficiency follows immediately from the final value theorem.

Necessity: $\hat{f}(s)$ is a rational function and can be expanded into partial fractions. If $\hat{f}(s)$ contains a pole on the imaginary axis or in the right half-plane, inverse Laplace transform of the corresponding partial fraction will be either an exponential function whose exponent has a nonnegative real part or a polynomial of time t, or a sine wave. These functions never converge to zero. Therefore, $\hat{f}(s)$ must be stable. •

Therefore, for a linear system $G(s)$, its output $y(t)$ converges asymptotically iff $\hat{y}(s)$ is stable. In the sequel, we analyze the required conditions from the angle of transfer function.

As a simple example, consider the step input. In this case, the model of input $u(t) = 1(t)$ is given by

$$\hat{u}(s) = \frac{1}{s}$$

which has a pole at the origin. For the stability of $\hat{y}(s) = G(s)\hat{u}(s)$, the unstable pole of $\hat{u}(s)$ must be cancelled by the zero of $G(s)$. So, $G(s)$ must satisfy

$$G(j0) = 0. \tag{5.5}$$

Similarly, for the sinusoidal input $u(t) = \sin \omega t \cdot 1(t)$, to make its steady-state output zero, there must hold

$$G(j\omega) = 0. \tag{5.6}$$

That is, in order to ensure that the steady-state output of a test signal is zero, the transfer function $G(s)$ *must have zeros at the locations of the poles of test signal model.*

In the following subsections, we will derive conditions on the open-loop system in order to achieve reference tracking and suppression of persistent disturbance, based on the analysis in this subsection.

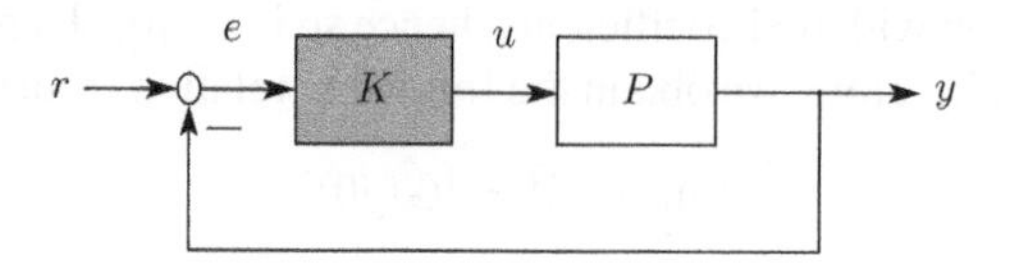

Figure 5.5 Reference tracking problem

5.2.2 *Reference Tracking*

Consider the reference tracking problem shown in Figure 5.5 in which $r(t)$ is the reference input. The control purpose is to let the plant output $y(t)$ follow the reference input $r(t)$. In order to evaluate the reference tracking, we look at the tracking error

$$e(t) = r(t) - y(t). \tag{5.7}$$

Let the transfer function from reference input to tracking error be $H_{er}(s)$ and the model of reference input be $R(s)$. Then, the tracking error becomes

$$\hat{e}(s) = H_{er}(s)R(s). \tag{5.8}$$

Since physical systems have inertia, their output can only vary smoothly. Therefore, the output of a system cannot track the test signal instantaneously; only after the transient process can it track the test signal. However, at least we hope to reduce the error $\lim_{t\to\infty} e(t)$ between the output and the reference input after a sufficiently long time, that is, in the steady state $(t \to \infty)$. Ideally, we hope that the output matches the reference input perfectly, that is, $e(\infty) = 0$. This is called *asymptotic tracking.*

5.2.2.1 Relation between Tracking Error and Loop Gain

In order to derive conditions on the open-loop transfer function, we define the loop gain as

$$L(s) = P(s)K(s) \tag{5.9}$$

for simplicity. The closed-loop transfer function from reference input to tracking error is given by

$$H_{er}(s) = \frac{1}{1 + L(s)}. \tag{5.10}$$

To reduce the steady-state tracking error w.r.t. the step reference input, we need

$$|H_{er}(j0)| = \left| \frac{1}{1 + L(j0)} \right| \ll 1. \tag{5.11}$$

So, there must be

$$|L(j0)| \gg 1, \tag{5.12}$$

that is, the loop gain $L(s)$ must have a sufficiently high gain at low frequency. Similarly, in order to reduce the steady-state tracking error of the sinusoidal reference input with an angular frequency ω_0, there should be

$$|L(j\omega_0)| \gg 1. \tag{5.13}$$

5.2.2.2 Asymptotic Tracking Conditions

To guarantee the asymptotic tracking of step signal, we need $H_{er}(j0) = 0$. From it, we have

$$H_{er}(j0) = \frac{1}{1 + L(j0)} = 0 \quad \Leftrightarrow \quad |L(j0)| = \infty$$
$$\Leftrightarrow \quad L(s) \text{ has a pole at } s = 0. \tag{5.14}$$

In other words, the loop gain $L(s)$ should contain the model $R(s) = 1/s$ of the reference input. As shown in Figure 5.5, the output $y(t)$ of plant $P(s)$ is also the output of loop gain of $L(s)$. $L(s)$ is driven by the tracking error $e(t)$ which should converge to zero. In order to ensure that the output $y(t)$ is able to maintain a constant value, the loop gain $L(s)$ must have some component that can keep the past data (energy), that is, an integrator (see Figure 5.6). However, the integrator is also the model of step signal. This indicates that the loop gain $L(s)$ should contain the model of step signal in order to guarantee the asymptotic tracking of step signal.

Furthermore, when the plant $P(s)$ has a zero at the origin, this unstable zero cannot be cancelled by the pole of controller $K(s)$ in order to ensure the internal stability. That is, there cannot be any integrator in the loop gain of $L(s) = P(s)K(s)$ so that the output is unable to track the step signal.

The same conclusion applies to other test signals. Its generalization is the famous *internal model principle*.

Theorem 5.2 (Internal model principle 1) *Suppose that the SISO system in Figure 5.5 is internally stable and the model of reference input is $R(s) = N_R(s)/M_R(s)$ and antistable. Then, the following statements are true:*

1. *To achieve asymptotic tracking, the zeros of plant $P(s)$ cannot coincide with the poles of $R(s)$.*
2. *When condition (1) is true, the necessary and sufficient condition for the asymptotic tracking is that the loop gain $L(s)$ contains $1/M_R(s)$.*

Proof. (1) Coprimely factorized the plant and controller as $P = N_P/M_P$ and $K = N_K/M_K$. Then, the numerator and denominator polynomials of the loop gain $L(s) = N(s)/M(s)$ become $N = N_P N_K$, $M = M_P M_K$, respectively. If N_P and M_R have a common factor $(s - \lambda)(\Re(\lambda) \geq 0)$, this factor cannot be cancelled by the factor of $M_K(s)$ (corresponding to controller poles) in order to ensure the internal stability. Therefore, the

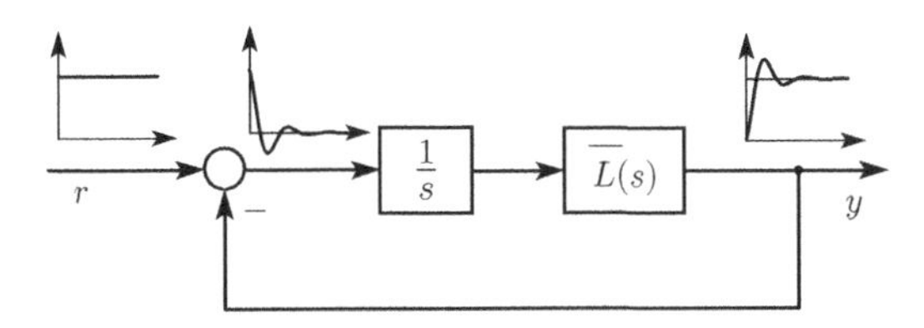

Figure 5.6 Signals in reference tracking

factor $(s - \lambda)$ appears in $N(s)$ such that $L(\lambda) = 0$. So, at $s = \lambda$, the transfer function $H_{er}(s)$ from reference input to tracking error becomes

$$H_{er}(\lambda) = \frac{1}{1 + L(\lambda)} = 1.$$

Hence, it has no zeros at $s = \lambda$. Therefore, $\hat{e}(s) = H_{er}(s)R(s)$ is unstable, and the tracking error cannot converge to zero.

Further, the Laplace transform of tracking error can be written as

$$\hat{e}(s) = \frac{1}{1 + L(s)} R(s) = \frac{M(s)}{M(s) + N(s)} \frac{N_R(s)}{M_R(s)}.$$

Since all roots of $M_R(s)$ are located in the right half-plane, we must cancel $M_R(s)$ using a factor of $M(s)$ in order to stabilize $\hat{e}(s)$. So it is necessary and sufficient for $M(s)$ to contain $M_R(s)$. •

Since the loop gain $L(s)$ contains the model of reference input, namely, there is a model of reference input inside the closed loop, $1/M_R(s)$ in $L(s)$ is called the *internal model*. The internal model principle is also true for multiple-input multiple-output (MIMO) systems. The difference lies in that the internal model is no longer simply a copy of the model of reference input. The internal model gets rather complicated due to the coupling between inputs and outputs. An example is illustrated in the following .

Example 5.1 *Consider the system shown in Figure 5.7. The control purpose is to ensure that the output $y_1(t)$ tracks the sinusoidal signal $r_1(t) = \sin\omega t \cdot 1(t)$ and the output $y_2(t)$ tracks the step signal $r_2(t) = 1(t)$. The transfer matrices of the two inputs, two outputs plant and controller are*

$$P(s) = \begin{bmatrix} P_1(s) & 0 \\ \varepsilon P_1(s) & P_2(s) \end{bmatrix}, \quad K(s) = \begin{bmatrix} K_1(s) & 0 \\ 0 & K_2(s) \end{bmatrix},$$

respectively. Here, set the loop gain of these two subsystems as $L_1(s) = P_1(s)K_1(s)$ and $L_2(s) = P_2(s)K_2(s)$, respectively. It follows from the block diagram that

$$\hat{y}_1(s) = \frac{L_1}{1 + L_1}\hat{r}_1(s), \quad \hat{y}_2(s) = \frac{L_2}{1 + L_2}\hat{r}_2(s) + \frac{\varepsilon}{1 + L_2}\hat{y}_1(s).$$

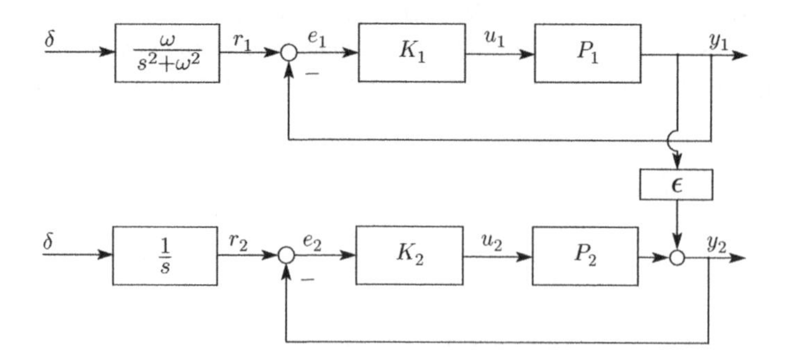

Figure 5.7 MIMO system with a coupling term

This means that the output y_1 acts as a disturbance for the output y_2. Therefore, the tracking errors become

$$\hat{e}_1(s) = \frac{1}{1+L_1}\frac{\omega}{s^2+\omega^2}, \quad \hat{e}_2(s) = \frac{1}{1+L_2}\frac{1}{s} - \frac{\varepsilon}{1+L_2}\hat{y}_1(s), \tag{5.15}$$

respectively. Obviously, $e_1(\infty) = 0$ iff $L_1(s)$ contains $1/(s^2+\omega^2)$. Meanwhile, as $\hat{y}_1(s)$ has $1/(s^2+\omega^2)$, in order to ensure that $e_2(\infty) = 0$, $L_2(s)$ must contain $1/(s^2+\omega^2)$ in addition to $1/s$. Thus, the overall loop gain of the system

$$L(s) = P(s)K(s) = \begin{bmatrix} L_1 & 0 \\ \varepsilon L_1 & L_2 \end{bmatrix}$$

should contain the following internal model:

$$\begin{bmatrix} \dfrac{1}{(s^2+\omega^2)} & 0 \\ 0 & \dfrac{1}{s(s^2+\omega^2)} \end{bmatrix}. \tag{5.16}$$

Moreover, it is observed from the previous description that the asymptotic tracking of reference input is achieved by producing, at the output port of open-loop transfer function $L(s)$, a signal identical to the reference input by using the internal model. This does not depend on the parameters of the system. It implies that the asymptotic tracking is guaranteed as long as the closed-loop system is stable, even if the system parameters change. In this sense, the internal model offers a guarantee for the *robust asymptotic tracking*.

5.2.3 Disturbance Suppression

Figure 5.8 shows a plant subject to the influence of a disturbance. In practical systems, examples of disturbance include mechanical disturbance caused by turbulence, wave, and so on; friction; gravity disturbance; as well as temperature variation caused by the ambient air, and so on. The control objective is to minimize the negative effect on the system output brought by persistent disturbances.

5.2.3.1 Conditions for Steady-State Disturbance Suppression

The transfer function between the disturbance $d(t)$ and output $y(t)$ in the system shown in Figure 5.8 is

$$H_{yd}(s) = \frac{P(s)}{1+P(s)K(s)}. \tag{5.17}$$

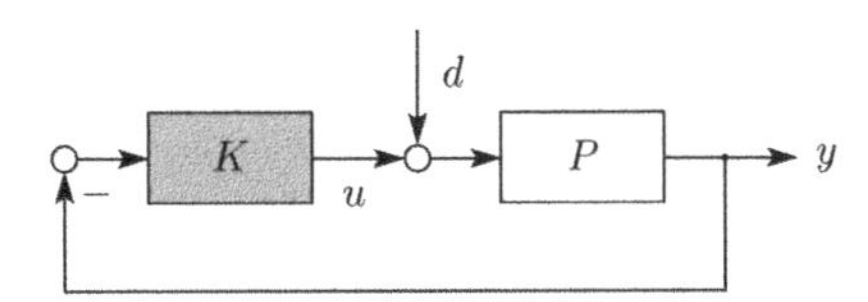

Figure 5.8 Asymptotic rejection of persistent disturbance

To suppress the step disturbance, $|H_{yd}(j0)| \ll 1$ is needed. There are two ways to accomplish this:

$$(1)|P(j0)| \ll 1 \quad \text{or} \quad (2)|K(j0)| \gg 1.$$

Plant satisfying (1) seldom exists in reality. This is because the main objective of control system is reference tracking. In order to track low-frequency sinusoidal signals well, it is necessary to raise the loop gain $|L| = |PK|$ sufficiently high in the same frequency band. So, plants are usually designed as systems with high low-frequency gain.

On the other hand, (2) is a requirement on the controller and not contradictory with the objective of raising the loop gain. So it can be realized. Similarly, in order to suppress a sinusoidal disturbance with an angular frequency ω, it is sufficient for the controller to satisfy $|K(j\omega)| \gg 1$.

5.2.3.2 Asymptotic Rejection of Persistent Disturbance

The ideal case is to completely eliminate the influence of disturbance on the steady-state output when the model of disturbance is known precisely. According to the system structure in Figure 5.8, it can be envisioned that the disturbance can be rejected asymptotically only if the controller $K(s)$ can produce the same signal to cancel the disturbance (Figure 5.9). Therefore, the controller should contain the model $D(s)$ of disturbance (internal model).

For example, to eliminate a step disturbance asymptotically, the controller $K(s)$ must have the disturbance model $D(s) = 1/s$. Mathematically, this is proved as follows. Here, factorize the plant and controller as

$$P(s) = \frac{N_P(s)}{M_P(s)}, \quad K(s) = \frac{N_K(s)}{M_K(s)} \tag{5.18}$$

by using coprime numerator and denominator polynomials. Then, the output can be written as

$$\hat{y}(s) = \frac{P(s)}{1 + P(s)K(s)} D(s) = \frac{N_P(s)M_K(s)}{M_P(s)M_K(s) + N_P(s)N_K(s)} \times \frac{1}{s}. \tag{5.19}$$

Since the asymptotic convergence of $y(t)$ to zero is equivalent to the stability of $\hat{y}(s)$, we must cancel the unstable pole $p = 0$ by the zero of $N_P(s)M_K(s)$. Therefore,

$$N_P(0)M_K(0) = 0 \quad \Leftrightarrow \quad N_P(0) = 0 \text{ or } M_K(0) = 0 \tag{5.20}$$

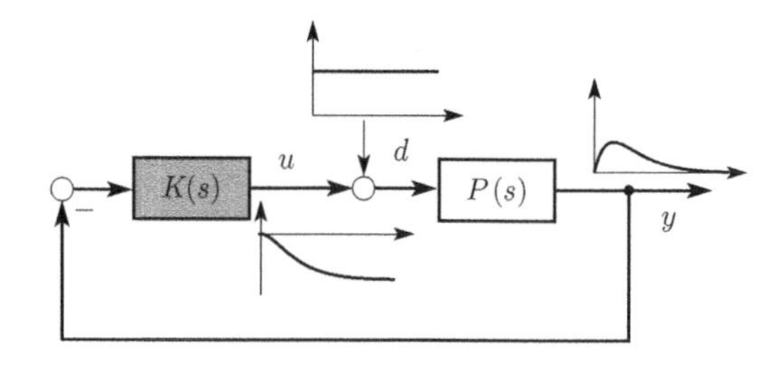

Figure 5.9 Signals in the control of persistent disturbance

is necessary. That is, either $P(s)$ has a zero $z = 0$ or $K(s)$ has a pole $p = 0$. However, it is very rare that the plant possesses a zero at the origin. So the controller should have an integrator.

Generalizing the previous analysis, we obtain the following theorem.

Theorem 5.3 (Internal model principle 2) *Suppose that the SISO system in Figure 5.8 is internally stable and the disturbance model is $D(s) = N_D(s)/M_D(s)$ and antistable. Furthermore, assume that the zeros of plant $P(s)$ are different from the poles of $D(s)$. Then, the necessary and sufficient condition for the asymptotic rejection of disturbance ($y(t) \to 0$) is that the controller $K(s)$ contains $1/M_D(s)$.*

Proof. Coprimely factorizing the plant $P(s)$ and controller $K(s)$ as (5.18), Laplace transform of the output can be written as

$$\hat{y}(s) = \frac{P(s)}{1 + P(s)K(s)} D(s) = \frac{N_P(s)M_K(s)}{M_P(s)M_K(s) + N_P(s)N_K(s)} \frac{N_D(s)}{M_D(s)}.$$

Since all roots of $M_D(s)$ are unstable and N_P, M_D are coprime, to stabilize $\hat{y}(s)$, we must cancel $M_D(s)$ by a factor of $M_K(s)$. That is, it is necessary and sufficient for $M_K(s)$ to have $M_D(s)$. •

Further, when the plant has zeros the same as some of the poles of disturbance model, it is sufficient for the controller to contain the remaining poles of $D(s)$. In this sense, the zeros of plant are not a barrier for the asymptotical rejection of disturbance.

In the case of asymptotic tracking, it is not necessary for the controller to contain an internal model so long as the plant has an internal model. In contrast, the controller $K(s)$ must contain an internal model in the asymptotic rejection of disturbance. This distinction is worth noting.

In the case where reference input and persistent disturbance exist simultaneously, the condition for the asymptotic tracking of reference input is as follows.

Corollary 5.1 *In the system of Figure 5.1, assume that the reference input model is $R(s) = N_R(s)/M_R(s)$ and disturbance model is $D(s) = N_D(s)/M_D(s)$. Both are coprime factorizations and antistable. The least common multiple of $M_R(s)$ and $M_D(s)$ is $\phi(s)$. When the zeros of plant are different from those of $\phi(s)$, the output can track the reference input $r(t)$ asymptotically as long as the controller contains $1/\phi(s)$.*

Proof. As $M_K(s)$ contains $\phi(s)$, it also contains $M_D(s)$. According to Theorem 5.3, the steady-state response of disturbance is zero. Moreover, the loop gain $L(s) = P(s)K(s)$ contains $1/M_R(s)$ owing to the given condition. Therefore, the steady-state output is the same as the reference input according to Theorem 5.2. The output of plant is the sum of these two responses. So, the output asymptotically tracks the reference input. •

Further, when the poles of $P(s)$ and $R(s)$ are distinct from each other and the zeros of $P(s)$ are distinct from the roots of $\phi(s)$, this condition becomes necessary.

5.3 Transient Response

The so-called transient response indicates the response before the system output converges to the steady state. In reference tracking and disturbance suppression problems, the output is required not only to converge quickly to the steady state but also have small overshoot.

5.3.1 Performance Criteria

In the evaluation of the transient response of a system, the most typical method is to use the step response as an indicator to judge the quality of transient response. Typical transient response of a stable system is shown in Figure 5.10, in which the output $y(t)$ has been normalized by the steady-state value $y(\infty)$. We use the following performance indicators to measure the quality of response:

Rise time t_r: The time needed for the output $y(t)$ to rise from 10% to 90% of the steady-state value

Settling time t_s: The time needed for the output to converge to a prescribed range ε around its steady-state value (usually $\pm 1\%$ or $\pm 5\%$)

Overshoot M_p: The relative error $(y_{\max} - y(\infty))/y(\infty)$ between the maximum output $y_{\max}$ and the steady-state value

These quantities are called *specifications*; their meanings are illustrated in Figure 5.10. The tracking control aims at following the reference input as quickly as possible while suppressing the overshoot as more as possible. Therefore, *the smaller these specifications are, the better the system performance is*. The responses of stable linear systems all possess a form of exponential function of time. Mathematically, it takes infinitely long time to converge to the

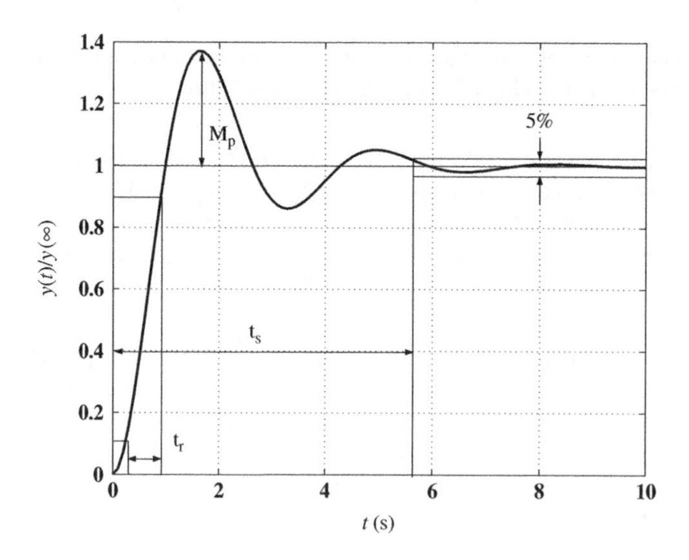

Figure 5.10 Transient response

steady state completely. In this sense, it may be said that the convergence time of response is the same for any stable system. However, from our experience we know that the convergence rates of systems are not the same. One of the indicators specifying this difference is the settling time. That is to say, *the settling time is regarded as the convergence time* in engineering practice.

5.3.2 Prototype Second-Order System

Let us look at the relation between these specifications and system parameters, particularly the relation with pole position. For general systems, this relationship is too complicated to provide a universal guide. However, in most cases, the poles of a system include a pairs of complex poles relatively closed to the imaginary axis, and the other poles are far away from the imaginary axis. From the partial fraction expansion of transfer function, it is seen that the output response is composed of terms corresponding to all poles. Each term is called a *mode*. The pair of complex poles that are the closest to imaginary axis are called the *dominant poles*. Compared to the dominant poles, the other modes converge to zero much faster. That is, the output response can be approximated by that of the dominant poles, excluding the initial period of time. Therefore, as long as we know the relationship between the parameters of low-order system and the response, it may also be instructive for high-order systems.

Example 5.2 *Compare the step responses of the following two systems:*

$$G_1(s) = \frac{8}{s^2 + 4s + 8}, \quad G_2(s) = \frac{80}{(s^2 + 4s + 8)(s + 10)}.$$

From the responses in Figure 5.11, it is clear that there is no essential difference except the rise times. In fact, although G_2 has a pole -10, it is far away from the imaginary axis compared to the dominant poles $-2 \pm j2$ (which are also the poles of G_1). The response of this pole converges five times faster than that of the dominant poles and disappears soon. Mathematically,

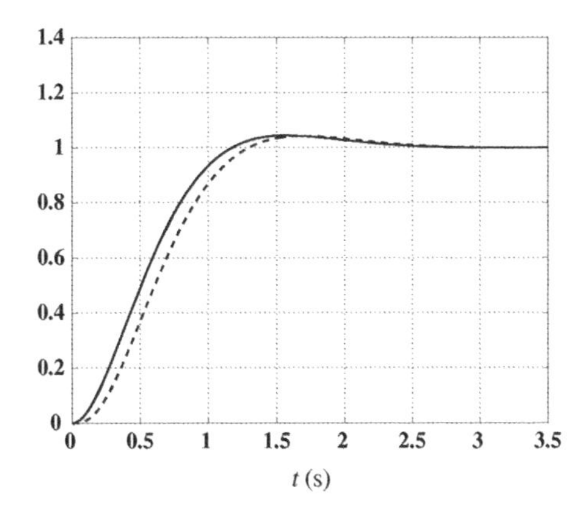

Figure 5.11 Comparison of step responses (solid: G_1, dotted: G_2)

G_2 can be approximated by G_1:

$$G_2(s) = \frac{8}{s^2+4s+8} \times \frac{1}{s/10+1} \approx \frac{8}{s^2+4s+8} = G_1(s).$$

This approximation is precise enough in the low frequency. Since the low-frequency components are the major part in a step signal, their transmission is mainly determined by the low-frequency characteristics of transfer function. This is why the step responses between G_2 and G_1 make no big difference.

For this reason, we usually consider the following *prototype second-order system*

$$H(s) = \frac{\omega_n^2}{s^2 + 2\zeta\omega_n s + \omega_n^2}. \tag{5.21}$$

Here, ζ is the *damping ratio* and ω_n the *natural frequency*. The poles of $H(s)$ are

$$p_{1,2} = -\zeta\omega_n \pm j\omega_n\sqrt{1-\zeta^2}. \tag{5.22}$$

The feature of this system is that it has no finite zeros and the low-frequency gain $H(j0)$ is equal to 1. By the final value theorem, when the step input is applied to the $H(s)$, the steady-state response is given by

$$y(\infty) = \lim_{s\to 0} s\hat{y}(s) = \lim_{s\to 0} s \times H(s)\frac{1}{s} = H(0) = 1. \tag{5.23}$$

So, the output can asymptotically track step reference input. Next, we discuss the relationship between the specifications on the transient response and the parameters of $H(s)$.

5.3.2.1 Specification and Parameter

It can be proved that when the damping ratio is in the range of $0 < \zeta < 1$, the following relations [30] hold between the parameters of prototype second-order system and the specifications (refer to Exercise 5.6).

$$t_r \approx \frac{1.8}{\omega_n} \tag{5.24}$$

$$M_p = e^{-\frac{\pi\zeta}{\sqrt{1-\zeta^2}}} \tag{5.25}$$

$$t_s \approx \frac{4.6}{\zeta\omega_n} \quad (\varepsilon = \pm 1\%), \quad \frac{3}{\zeta\omega_n} \quad (\varepsilon = \pm 5\%). \tag{5.26}$$

We look into the case of $\varepsilon = \pm 5\%$. The following conclusions can be drawn from the previous relations:

1. The overshoot M_p only depends on the damping ratio ζ. M_p drops when ζ increases (see Figure 5.12).
2. The product of ζ and ω_n equals the magnitude of the pole's real part. The settling time t_s decreases when $\zeta\omega_n$ increases.

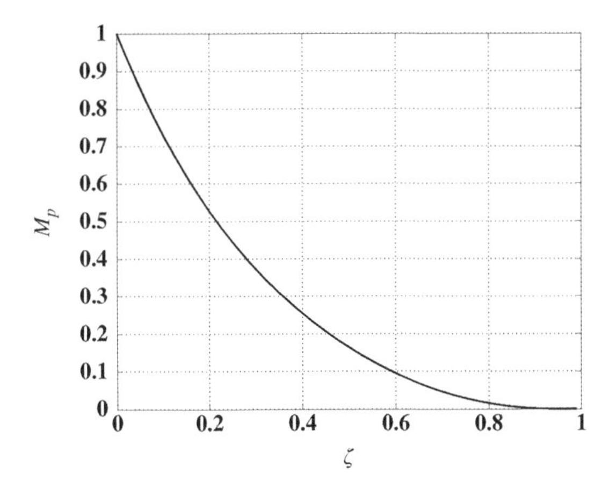

Figure 5.12 Relation between overshoot M_p and damping ratio ζ

Hence, given the overshoot M_p and settling time t_s, the parameter ranges satisfying the specifications are as follows:

$$\zeta \geq \zeta(M_p) = \frac{|\ln M_p|}{\sqrt{\pi^2 + |\ln M_p|^2}} \tag{5.27}$$

$$\zeta\omega_n \geq \frac{3}{t_s}. \tag{5.28}$$

5.3.2.2 Specification and Pole Position

Now, we investigate the relation between the overshoot, settling time, and the poles of prototype second-order system. First, according to (5.28), the real parts of these two complex poles must satisfy

$$\Re(p_{1,2}) = -\zeta\omega_n \leq -\frac{3}{t_s}. \tag{5.29}$$

Second, the poles of prototype second-order system can be expressed as

$$p_{1,2} = -\zeta\omega_n \pm j\omega_n\sqrt{1-\zeta^2} = \omega_n e^{j(\pi\pm\theta)}, \ \theta = \arccos\zeta$$

in the polar coordinate, in which $\arccos\zeta$ is a decreasing function of ζ. So, owing to (5.27), the angle θ should meet

$$\theta \leq \theta_p := \arccos\zeta(M_p). \tag{5.30}$$

Drawing the allowable region of poles in the complex plane, we get the shaded part in Figure 5.13. For example, the angle $\theta_p = 45^o$ corresponds to $\zeta(M_p) = 0.707$, that is, $M_p = 5\%$. This relationship will play an important role in the pole placement design in Chapters 6 and 19.

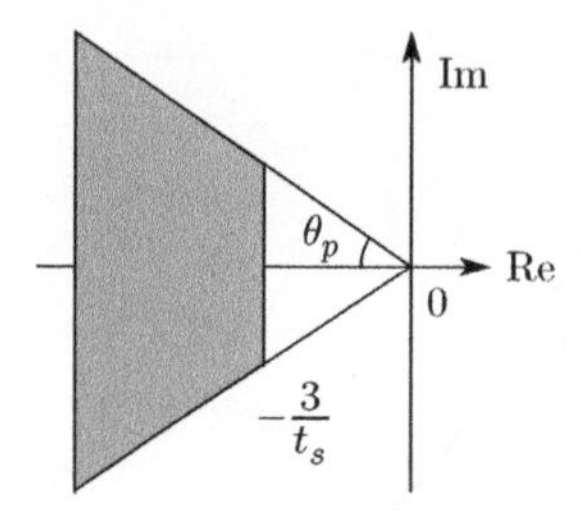

Figure 5.13 Allowable region of poles

Example 5.3 *Determine the ranges of parameters* (ζ, ω_n) *such that* $t_r \leq 0.6[s]$*,* $M_p \leq 10\%$*, and* $t_s \leq 3[s]$*.*

First, the damping ratio corresponding to $M_p = 10\%$ *is*

$$\zeta(M_p) = \frac{|\ln 0.1|}{\sqrt{\pi^2 + |\ln 0.1|^2}} \approx 0.59.$$

So, $\zeta \geq 0.59$ *must hold. Second, from the specifications on the rise and settling times, we obtain*

$$\omega_n \approx \frac{1.8}{t_r} \geq 3, \quad \zeta\omega_n \geq \frac{3}{t_s} \geq 1.$$

When $\zeta \geq 0.59$*,* $\omega_n \geq 3$*, the condition* $\zeta\omega_n \geq 1$ *is satisfied automatically. So, finally the ranges of allowable parameters become*

$$\zeta \geq 0.59, \quad \omega_n \geq 3.$$

5.3.3 Impact of Additional Pole and Zero

5.3.3.1 Impact of Additional Zero

Let us analyze the change of response when a zero is added to the prototype second-order system (5.21). The new transfer function is

$$H(s) = \frac{\omega_n^2\left(1 + \frac{1}{a\zeta\omega_n}s\right)}{s^2 + 2\zeta\omega_n s + \omega_n^2}, \quad a \in \mathbb{R}. \tag{5.31}$$

Note that its low-frequency gain $H(j0)$ has been adjusted to that of the prototype second-order system. The additional zero is

$$z = -a\zeta\omega_n. \tag{5.32}$$

Especially, this zero approaches the imaginary axis as $a \to 0$. Calculating the step responses for different a, we get the results of Figure 5.14. Compared to the prototype second-order system, it has the following features:

- The settling times are almost the same.

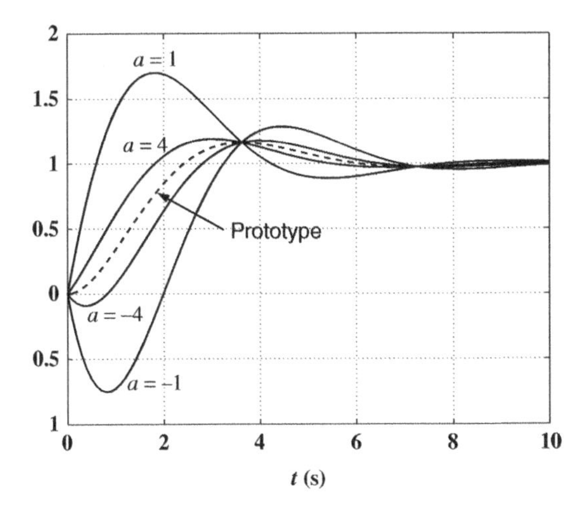

Figure 5.14 Step response with an additional zero

- When $a > 0$, the smaller a is, the bigger the overshoot is. Meanwhile, when $a \gg 1$, the response approaches that of the prototype second-order system.
- When $a < 0$, *undershoot* (the response with a sign opposite to the input) appears at the initial stage. The smaller $|a|$ is, the bigger the undershoot is. Note that the zero is unstable in this case.

That is to say, a zero close to the imaginary axis either increases the overshoot of the step response or causes a big undershoot. This phenomenon is explained qualitatively as follows. The transfer function (5.31) can be decomposed as

$$H(s) = \frac{\omega_n^2}{s^2 + 2\zeta\omega_n s + \omega_n^2} + \frac{1}{a\zeta\omega_n} \times s \times \frac{\omega_n^2}{s^2 + 2\zeta\omega_n s + \omega_n^2}. \tag{5.33}$$

Let the step response of prototype second-order system be $y_0(t)$. Then, since the variable s represents a differentiator, the step response of transfer function $H(s)$ becomes

$$y(t) = y_0(t) + \frac{1}{a\zeta\omega_n}\dot{y}_0(t). \tag{5.34}$$

$y_0(t)$ rapidly rises at the initial stage, so the rate of change (value of derivative) is large. Therefore, when the absolute value of a is small, positive a will cause a large overshoot, while negative a will cause a large undershoot. Moreover, the smaller $|a|$ is, the stronger the effect of the second term is. When $y_0(t)$ gradually gets close to the steady state, its derivative also gradually converges to zero such that its impact on the settling time diminishes. In contrast, when $|a|$ is big, the differentiation impact decreases greatly. Then, the response also gets close to the prototype. In summary, *the response of a system depends not only on the poles but also on the zeros.*

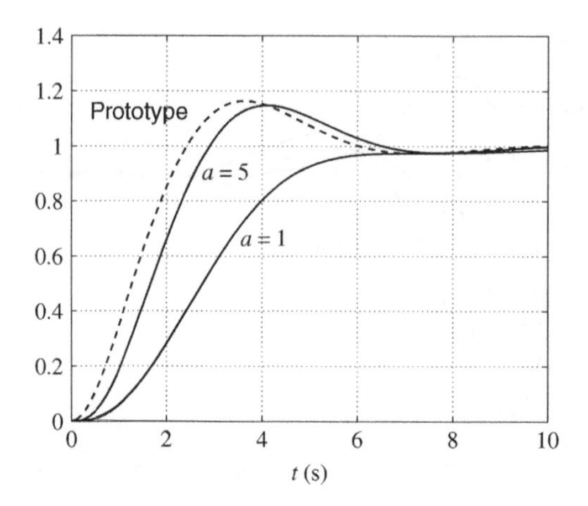

Figure 5.15 Step response with additional pole

5.3.3.2 Influence of Additional Pole

Adding to the prototype second-order system of (5.21) a new pole which is less than the real part of the dominant poles, the transfer function becomes

$$H(s) = \frac{\omega_n^2}{\left(1 + \frac{1}{a\zeta\omega_n}s\right)(s^2 + 2\zeta\omega_n s + \omega_n^2)}, \quad a \geq 1. \tag{5.35}$$

Here, we have adjusted the low-frequency gain of the first-order system $1/(1 + s/a\zeta\omega_n)$ such that the low-frequency gain $H(j0)$ is 1. The additional pole is

$$p = -a\zeta\omega_n < 0. \tag{5.36}$$

When $a \to 1$, this pole gets close to the real part of the poles of prototype system. Simulations w.r.t. various values of a yield the step responses shown in Figure 5.15. In this figure, we see that the rise time gets longer. This impact is very strong especially when $a \to 1$. The reason is that after adding a first-order system, output of the prototype system has to pass through this first-order system before reaching the output port.

Here, we take the poles of prototype system as the dominant poles. So the case of $a < 1$ is not investigated.

5.3.4 *Overshoot and Undershoot*

In reference tracking problems, the response of some system may take a sign opposite to the final value in certain period of time. The part of response with a sign opposite to the final value is called *undershoot*. Let us examine the following example.

Example 5.4 *Calculating the step responses for two transfer functions*

$$G_1(s) = \frac{8(1-s)}{s^2 + 4s + 8}, \quad G_2(s) = \frac{8(2-s)(1-s)}{(s+2)(s^2 + 4s + 8)},$$

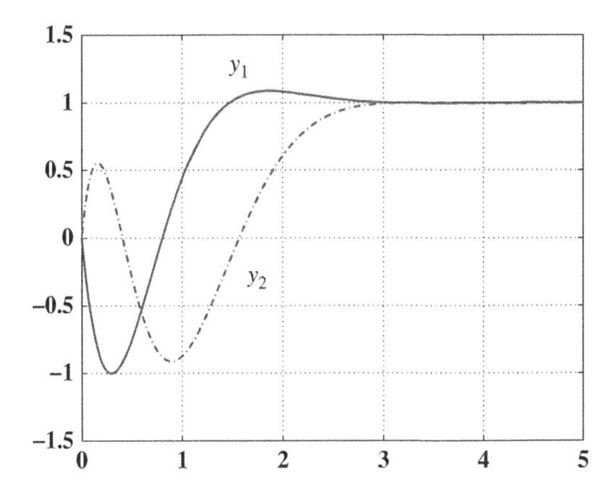

Figure 5.16 Various undershoot phenomena

we get the result shown in Figure 5.16. The step response $y_1(t)$ of $G_1(s)$ has a sign opposite to the final value around $t = 0$ s, while the step response $y_2(t)$ of $G_2(s)$ has a opposite sign around $t = 1$ s.

In order to distinguish the undershoot of $y_1(t)$ from that of $y_2(t)$, we call the undershoot occurring around $t = 0$ like $y_1(t)$ as *type A undershoot* and call the undershoot occurring around $t > 0$ like $y_2(t)$ as *type B undershoot.*

In engineering practice, overshoot or undershoot phenomenon is not allowed in most cases. For example, in the gantry crane used for loading and unloading cargos in port, type A undershoot in cargo position response means that the cargo will collide with the crane at the starting, and overshoot implies that the cargo will collide with the crane at the stopping. These two cases are not allowed. Then, what is the cause for the occurrence of these phenomena? This section will give a detailed discussion on this query and reveal the mechanism why they happen [84].

Theorem 5.4 *Assume that the transfer function $H(s)$ satisfies $H(0) \neq 0$. The step response undershoots if $H(s)$ has positive real zeros.*

Proof. Let this zero be $z > 0$; then $H(z) = 0$. Noting that z is located in the convergence domain of the Laplace transform $\hat{y}(s)$ of step response $y(t)$, we have

$$0 = H(z)\frac{1}{z} = \hat{y}(z) = \int_0^{\infty} y(t)e^{-zt}\,dt.$$

Consider the case of $H(0) > 0$ first. Since $y(\infty) = H(0) > 0$, $y(t) > 0 \Rightarrow y(t)e^{-zt} > 0$ for sufficiently large t. Thus, to satisfy the previous equation, there must be a time period in which $y(t) < 0$. The conclusions for the $H(0) < 0$ case are proved similarly. •

The following theorem gives the condition for type A undershoot [69].

Theorem 5.5 *The step response of stable transfer function*

$$H(s) = K\frac{(s-z_1)\cdots(s-z_m)}{(s-p_1)\cdots(s-p_n)} = \frac{N(s)}{M(s)}, \quad K>0,\ n \geq m$$

has type A undershoot iff the number of positive real zeros is odd.

Proof. When type A undershoot occurs, the sign of initial output response is opposite to the final value. For any complex pole p_i, its conjugate $\overline{p_i}$ is also a pole. Then, $(-p_i)(-\overline{p_i}) = |p_i|^2 > 0$ holds. For a real pole p_j, we know that $-p_j > 0$ from the stability of system. Hence, $M(0) = (-p_1)\cdots(-p_n) > 0$. On the other hand, when positive real zeros exist, $N(0) = (-z_1)\cdots(-z_m) < 0$ when its number is odd and $N(0) > 0$ when its number is even. Further, $N(0) > 0$ in the absence of positive real zeros. From $\hat{y}(s) = H(s)/s$ and the final value theorem, we obtain

$$y(\infty) = \lim_{s\to 0} s \cdot H(s)\frac{1}{s} = H(0) = \frac{N(0)}{M(0)}. \tag{5.37}$$

Therefore, $y(\infty) < 0$ when the number of positive real zeros is odd and $y(\infty) > 0$ otherwise.

Next, we note that the relative degree of $H(s)$ is $n-m$ and use it to examine the sign of the initial response. Based on the Laplace transform of derivative, initial value theorem, and relative degree, we see that

$$\begin{aligned}
y(0_+) &= \lim_{s\to\infty} s\hat{y}(s) = \lim_{s\to\infty} s \cdot H(s)\frac{1}{s} = \lim_{s\to\infty} H(s) = 0\\
\dot{y}(0_+) &= \lim_{s\to\infty} s \cdot (s\hat{y}(s) - y(0)) = \lim_{s\to\infty} sH(s) = 0\\
&\ \vdots\\
y^{(n-m-1)}(0_+) &= \lim_{s\to\infty} s^{n-m-1}H(s) = 0\\
y^{(n-m)}(0_+) &= \lim_{s\to\infty} s^{n-m}H(s) = K > 0.
\end{aligned}$$

By the smoothness of response, $y^{(n-m)}(t) \approx y^{(n-m)}(0_+) = K\ \ (0 < t \leq \varepsilon)$ holds for a sufficiently small $\varepsilon > 0$. So

$$y(\varepsilon) = \overbrace{\int_0^{\varepsilon}\cdots\int_0^{t_{n-m+1}}}^{n-m} y^{(n-m)}(t_{n-m})dt_{n-m}\cdots dt_1 \approx \frac{K}{(n-m)!}\varepsilon^{n-m} > 0$$

is proved, and the theorem is concluded. •

Moreover, the following result holds about the overshoot [84]. Its proof is similar to Theorem 5.4 and can be accomplished by analyzing the tracking error $\hat{e}(p)$ at the positive real pole p of $L(s)$ (Exercise 5.7).

Theorem 5.6 *In the unity feedback system of Figure 5.5, assume that the loop gain $L(s) = P(s)K(s)$ contains positive real poles and more than one integrator. Then, the step response undershoots.*

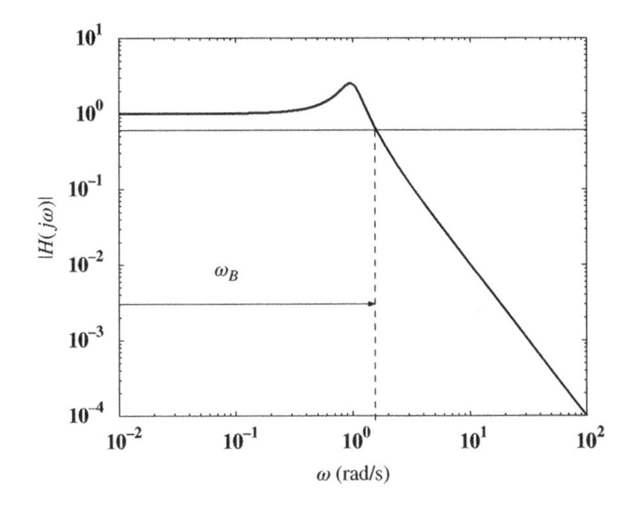

Figure 5.17 Frequency response and bandwidth of prototype second-order system

5.3.5 Bandwidth and Fast Response

Drawing the frequency response of closed-loop transfer function $H(s)$ from reference input to output, we get a shape as illustrated in Figure 5.17 in general. Its low-frequency gain is almost 1 and have a low-pass property. The input–output relationship of $H(s)$ is

$$\hat{y}(j\omega) = H(j\omega)\hat{u}(j\omega). \tag{5.38}$$

The output and input are in good agreement in the frequency band where the gain is 1. Now, we consider how to judge whether the output is consistent with the input. The commonly used rule in control engineering is that as long as the output power $|\hat{y}(j\omega)|^2$ is not lower than half of the input power $|\hat{u}(j\omega)|^2$, both are regarded as consistent[2]. This means that in the band in which

$$|H(j\omega)| \geq \frac{1}{\sqrt{2}}, \tag{5.39}$$

the reference tracking performance is good enough. Here, we call the smallest frequency ω satisfying

$$|H(j\omega)| = \frac{1}{\sqrt{2}} \approx -3[\text{dB}] \tag{5.40}$$

as the *bandwidth*[3], denoted by ω_B. Obviously, (5.39) holds as long as $\omega \leq \omega_B$. Therefore, the bandwidth ω_B represents the frequency range in which the output can track the input.

For the prototype second-order system, the rapidity of response (i.e., rise time) is roughly inversely proportional to the bandwidth:

$$t_r \sim \frac{1}{\omega_B}. \tag{5.41}$$

[2] In fact, this is a concept borrowed from communication engineering.
[3] The bandwidth is defined by $|H(j\omega_B)| = |H(j0)|/\sqrt{2}$ for low-pass transfer functions. But for high-pass transfer functions, the bandwidth should be defined by $|H(j\omega_B)| = |H(j\infty)|/\sqrt{2}$.

To understand this, it suffices to analyze the relationship between the bandwidth ω_n and the characteristic frequency ω_n. For the prototype second-order system, we get by solving (5.40) that

$$\omega_B = \omega_n \sqrt{1 - 2\zeta^2 + \sqrt{1 + (1 - 2\zeta^2)^2}}. \tag{5.42}$$

Obviously, ω_B is proportional to ω_n. Since the rise time is $t_r \approx 1.8/\omega_n$, t_r is inversely proportional to ω_B. As for systems other than the prototype second-order system, so long as their frequency responses have a shape similar to the prototype second-order system in the low-frequency band and its gain rolls-off in other frequency bands, this relationship is approximately true. So, the bandwidth is an important indicator of fast response for this type of systems. It should be noted, however, that this relationship is only true for the prototype second-order system and systems with a similar frequency characteristics. Hence, one should be cautious not to stretch this interpretation.

5.4 Comparison of Open-Loop and Closed-Loop Controls

In this section, we compare the reference tracking and disturbance rejection performances for the open-loop system of Figure 5.18 and the closed-loop system of Figure 5.19, and discuss their advantages and disadvantages.

5.4.1 Reference Tracking

To track any reference input perfectly and instantaneously, *the transfer function from reference input $r(t)$ to output $y(t)$ must be 1*. This may be said to be the ultimate goal of control system design.

Consider the open-loop control system first, where the controller is denoted by $K_o(s)$. Assume that the plant $P(s)$ is stable and its zeros are also stable. This kind of systems is called *minimum phase* systems. Then, the inverse system $P^{-1}(s)$ is proper and stable if the

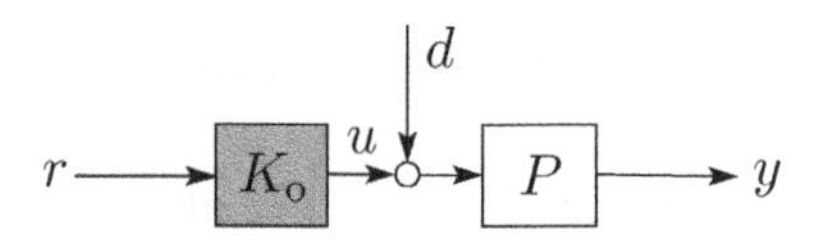

Figure 5.18 Open-loop control system

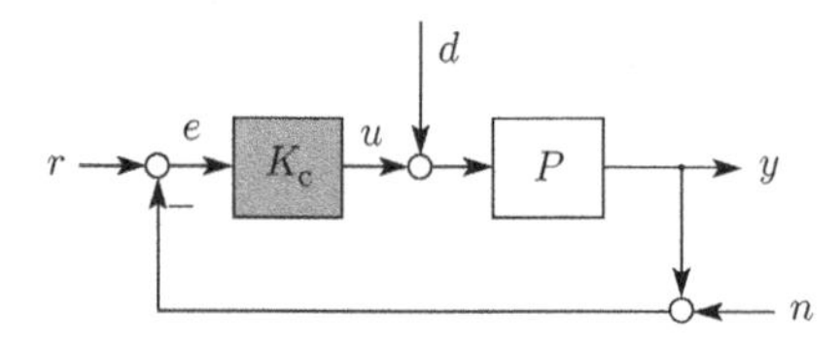

Figure 5.19 Closed-loop control system

relative degree of $P(s)$ is zero. Therefore, from

$$K_o(s) = P^{-1}(s) \quad \Rightarrow \quad L(s) = P(s)K_o(s) = 1 \tag{5.43}$$

we see that the open-loop transfer function becomes 1. However, any physical system has inertia so that its relative degree is greater than 1 (this can be understood by considering the transfer function from force to speed/displacement in mechanical systems). For this reason, the controller designed above is not proper and cannot be realized.

However, for a minimum phase plant with a relative order $\gamma > 0$, the controller

$$K_o(s) = P^{-1}(s)\frac{1}{(\varepsilon s + 1)^\gamma}, \quad \varepsilon > 0 \tag{5.44}$$

is proper and can be realized. In this case, the open-loop transfer function becomes

$$L(s) = P(s)K_o(s) = \frac{1}{(\varepsilon s + 1)^\gamma}. \tag{5.45}$$

Taking a small enough parameter ε, the system is able to track the reference input (Figure 5.20) over a fairly wide band. However, such open-loop control cannot be realized if the plant is not minimum phase.

On the other hand, the closed-loop transfer function in feedback control is

$$H(s) = \frac{P(s)K_c(s)}{1 + P(s)K_c(s)} \tag{5.46}$$

in which the feedback controller is denoted by $K_c(s)$. Under the same conditions of minimum phase and the relative degree, to achieve the same transfer function as the open-loop control, we may compute the controller from $H(s) = 1/(\varepsilon s + 1)^\gamma$. What we obtain is the following proper transfer function:

$$K_c(s) = P^{-1}(s)\frac{1}{(\varepsilon s + 1)^\gamma - 1}, \quad 0 < \varepsilon \ll 1. \tag{5.47}$$

Factorizing the plant coprimely as $P(s) = N(s)/M(s)$ by numerator and denominator polynomials, the characteristic polynomial of closed-loop system is obtained as

$$p(s) = M(s)N(s)(\varepsilon s + 1)^\gamma. \tag{5.48}$$

From the minimum phase assumption on the plant, the roots of $M(s)$ and $N(s)$ are all stable. Therefore, all characteristic roots are stable. But this feedback controller $K_c(s)$ is sensitive

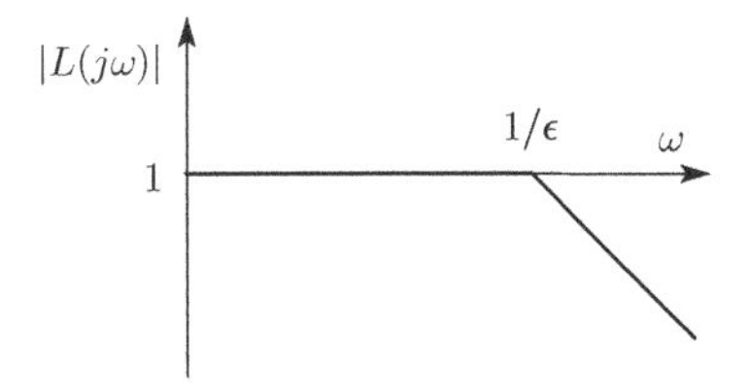

Figure 5.20 Gain plot of open-loop transfer function $L(s)$

to sensor noise $n(t)$. That is, when comparing the open-loop control input $u_o(t)$ with the closed-loop control input $u_c(t)$, we can see clearly that the closed-loop control is affected by noise from

$$\hat{u}_o(s) = K_o(s)\hat{r}(s) = P^{-1}(s)\frac{1}{(\varepsilon s+1)^\gamma}\hat{r}(s) \tag{5.49}$$

$$\hat{u}_c(s) = \frac{K_c(s)}{1+P(s)K_c(s)}[\hat{r}(s)-\hat{n}(s)] = P^{-1}(s)\frac{1}{(\varepsilon s+1)^\gamma}[\hat{r}(s)-\hat{n}(s)]. \tag{5.50}$$

In addition, it is also seen from the equation

$$\hat{y}_c(s) = H(s)[\hat{r}(s)-\hat{n}(s)] \tag{5.51}$$

about the output $y_c(t)$ of closed-loop control system that a good tracking performance cannot be expected in the frequency band where the noise $\hat{n}(s)$ has a large amplitude. So, for a minimum phase system, we conclude that open-loop control is superior to closed-loop control when its model is accurately known.

5.4.2 *Impact of Model Uncertainty*

In the previous section, we have compared the open-loop control and closed-loop control under the condition that the plant model $P(s)$ is known precisely. However, it is unrealistic to obtain an accurate system model in engineering practice. There is a variety of uncertainties in plant model. As an example, we make a comparison using an operational amplifier with parameter uncertainty.

Consider the operational amplifier shown in Figure 5.21(a). There is a strong nonlinearity in the operational amplifier, that is, the gain varies with the input amplitude. Assume that the gain of operational amplifier A is in the range of $10^3 \sim 10^4$. Then, the relative error of the output is equal to

$$\frac{|y_{\max}|-|y_{\min}|}{|y_{\min}|} = \frac{10^4u-10^3u}{10^3u} = 900\%. \tag{5.52}$$

This means that the maximum error of the output in the open-loop control is up to nine times.

Compared to this, when two resistors R and r are connected as in Figure 5.21(b), the current flowing into the operational amplifier is almost zero since the input impedance of the

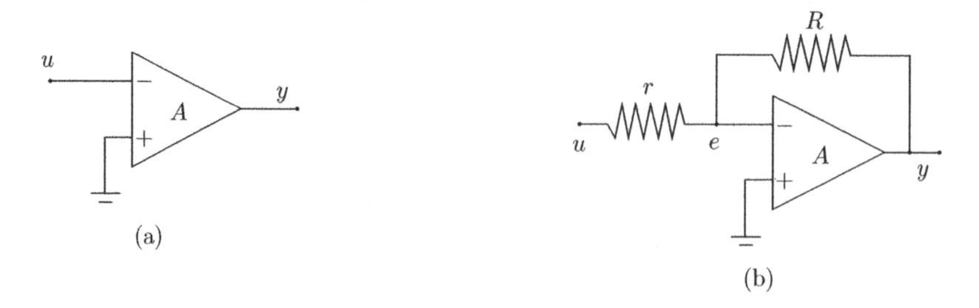

Figure 5.21 Operational amplifier (a) Single body (b) With feedback

operational amplifier is very large. So, the same current flows through the resistances R and r so that

$$\frac{u-e}{r}=\frac{e-y}{R}$$

holds. Solving from this equation for the input voltage e of the operational amplifier, we get

$$e=\frac{rR}{r+R}\left(\frac{u}{r}+\frac{y}{R}\right). \tag{5.53}$$

The relation between the input and output voltages of the operational amplifier is

$$y=-Ae. \tag{5.54}$$

Plotting these equations in a block diagram, we obtain the feedback system of Figure 5.22. From this figure, we see that the gain between the input and output is

$$\frac{y}{u}=-\frac{R}{r+\frac{r+R}{A}}. \tag{5.55}$$

Setting $R=10r$, then

$$\frac{y}{u}=-\frac{10}{1+\frac{11}{A}}=-9.881\sim-9.989$$

and the relative error becomes

$$\frac{9.989-9.881}{9.881}\approx 1.09\%.$$

That is, the error is reduced drastically. To achieve a gain of 10^3, it suffices to simply connect three operational amplifiers in series. In this case, the relative error

$$\frac{9.989^3-9.881^3}{9.881^3}\approx 2.9\% \tag{5.56}$$

is still very small. Therefore, via feedback control we have obtained a stable gain. This shows that feedback control has a strong resistance ability to model uncertainty, which is known as *robustness*. However, the price paid is that three operational amplifiers have to be used in the closed-loop control in order to achieve the same gain. Further, feedback control requires sensors in general, which is a major cause for pushing up the product price. Therefore, if more emphasis is placed on the product price rather than the control precision, we should use the open-loop control. Typical examples are microwaves, washing machines, and so on.

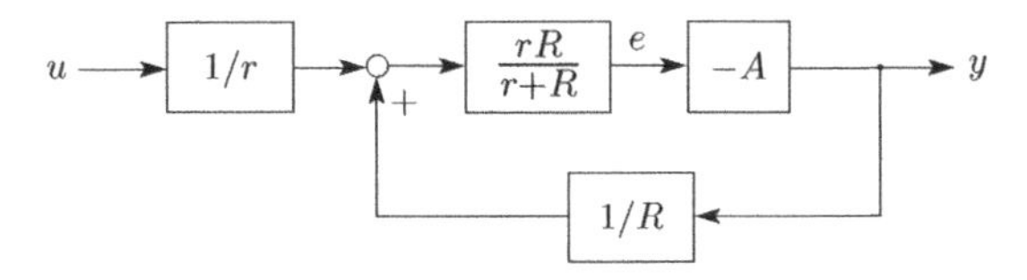

Figure 5.22 Equivalent block diagram for the operational amplifier with feedback

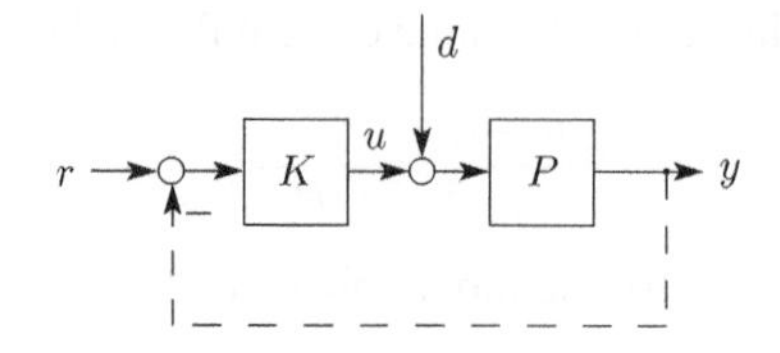

Figure 5.23 The case where the disturbance is not measured

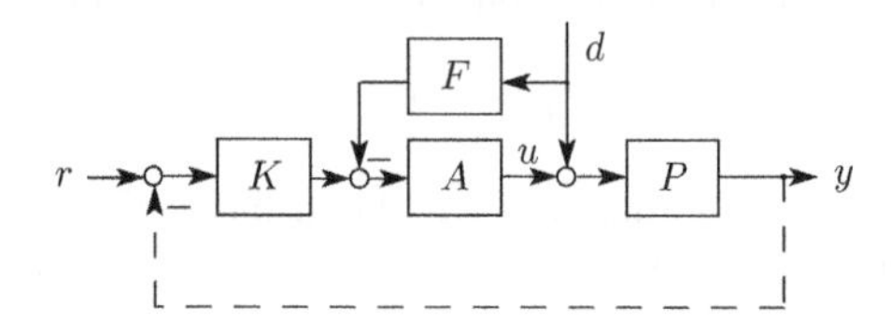

Figure 5.24 The case where the disturbance is measured

5.4.3 Disturbance Suppression

Next, we consider the influence on the output by a disturbance $d(t)$ entering the closed loop from the input port. Here, we will investigate two cases: (1) the disturbance is not measured, and (2) the disturbance is measured. The part excluding the dashed line in Figures 5.23 and 5.24 represent the open-loop system.

5.4.3.1 Case 1: Unmeasured Disturbance

In the open-loop system, we get from Figure 5.23 that

$$\hat{y}_{\mathrm{o}}(s) = P(s)K(s)\hat{r}(s) + P(s)\hat{d}(s). \tag{5.57}$$

The deviation of output caused by the disturbance is $\hat{y}_{\mathrm{od}}(s) = P(s)\hat{d}(s)$. On the other hand, the output of closed-loop system is

$$\hat{y}_{\mathrm{c}}(s) = \frac{P(s)K(s)}{1 + P(s)K(s)}\hat{r}(s) + \frac{P(s)}{1 + P(s)K(s)}\hat{d}(s) \tag{5.58}$$

in which the deviation of output caused by the disturbance is $\hat{y}_{\mathrm{cd}}(s) = P\hat{d}/(1 + PK)$. They have the relationship

$$\hat{y}_{\mathrm{cd}}(s) = \frac{1}{1 + P(s)K(s)}\hat{y}_{\mathrm{od}}(s). \tag{5.59}$$

Raising the gain of controller $K(s)$ can reduce the gain of $1/(1 + PK)$ so that $|\hat{y}_{\mathrm{cd}}(j\omega)| \ll |\hat{y}_{\mathrm{od}}(j\omega)|$. Therefore, the closed-loop control is insensitive to disturbance.

Then, why there is such a huge difference between the open-loop and closed-loop controls? From Figure 5.23, we see clearly that the input of controller does not contain any information about the disturbance in the open-loop control, so the open-loop control is powerless to disturbance. Meanwhile, in closed-loop control the information of disturbance is reflected indirectly by its influence on the output. The key of closed-loop control is utilizing this information. Therefore, it is possible to suppress the disturbance effectively if the controller is designed appropriately.

5.4.3.2 Case 2: Measured Disturbance

When the disturbance is measured online, we can offset the disturbance directly as shown in Figure 5.24, in which $A(s)$ is the transfer function of the *actuator*. In the open-loop control, the deviation of output caused by disturbance is

$$\hat{y}_{\rm od}(s) = P(s)[1 - A(s)F(s)]\hat{d}(s). \tag{5.60}$$

When $A(s)$ is minimum phase and the relative degree is γ, if we apply a *prefilter*

$$F(s) = A^{-1}(s)\frac{1}{(\varepsilon s + 1)^{\gamma}}, \tag{5.61}$$

then the output error caused by the disturbance becomes

$$\hat{y}_{\rm od}(s) = P(s)\left[1 - \frac{1}{(\varepsilon s + 1)^{\gamma}}\right]\hat{d}(s). \tag{5.62}$$

Letting $\varepsilon \to 0$, it is possible to eliminate the influence of disturbance over a sufficiently wide frequency band.

On the other hand, when we offset the disturbance in the same way in the closed-loop control system, the deviation of output caused by the disturbance is

$$\begin{aligned}\hat{y}_{\rm cd}(s) &= \frac{P(s)}{1 + P(s)A(s)K(s)}[1 - A(s)F(s)]\hat{d}(s)\\ &= \frac{1}{1 + P(s)A(s)K(s)}\hat{y}_{od}(s).\end{aligned} \tag{5.63}$$

In this case, the closed-loop control is still superior to the open-loop control. However, if we only suppress the disturbance with feedback, instead of direct cancellation, the output deviation becomes

$$\hat{y}_{\rm cd}(s) = \frac{P(s)}{1 + P(s)A(s)K(s)}\hat{d}(s). \tag{5.64}$$

Since we cannot lower the gain of closed-loop transfer function over a wide band, its disturbance suppression performance is degraded.

Summarizing these discussions, we have the following conclusions:

1. The cost of open-loop control is low, but the control accuracy is also low, and it is not suitable for unstable systems as well as nonminimum phase systems.
2. The closed-loop control can achieve high-precision control, but the cost is high.
3. If we can obtain the information about disturbance, no matter directly or indirectly, it is extremely effective to apply feedforward compensation of disturbance.

Particularly, persistent disturbances can be computed via the so-called *disturbance observer* [81]. Moreover, the advantages of feedback control can be summarized as follows:

1. It can suppress the influence of model uncertainty.
2. Its capability of disturbance suppression is superior.
3. It can precisely control unstable/nonminimum phase systems.

Exercises

5.1 In the closed-loop system of Figure 5.1, let the transfer functions be

$$P(s) = \frac{1}{s+1}, \quad K(s) = \frac{k}{s^2+4}.$$

(a) Show that there is a k that can stabilize the closed-loop system and find its range.
(b) When $r(t) = 0, d(t) = \sin 2t \ (t \geq 0)$, prove that $y(\infty) = 0$ holds.
(c) When $r(t) = \sin 2t, d(t) = 0 \ (t \geq 0)$, find the steady-state output $\lim_{t\to\infty} y(t)$ via the computation of $\lim_{t\to\infty}(r(t) - y(t))$.

5.2 In the closed-loop system of Figure 5.1, the transfer functions are as follows:

$$P(s) = \frac{1}{s(s+1)}, \quad K(s) = k.$$

Solve the following problems and discuss the results of (b) and (c).
(a) Find the range of gain k such that the closed-loop system is stable.
(b) Compute the steady-state tracking error $e(\infty)$ when $d(t) = 0, \ r(t) = 1(t)$.
(c) Compute the steady-state tracking error $y(\infty)$ when $r(t) = 0, \ d(t) = 1(t)$.

5.3 Replace the plant and controller in Exercise 5.2 with

$$P(s) = \frac{1}{s+1}, \quad K(s) = \frac{k}{s}$$

and solve the same problem. Then, discuss the difference from the previous problem.

5.4 In Figure 5.1, the plant and the controller are

$$P(s) = \frac{1}{s-1}, \quad K(s) = 3 + \frac{k}{s}.$$

(a) Seek the range of k ensuring the stability of closed-loop system.
(b) For a ramp input $r(t) = t \quad (t \geq 0)$, find the range of k such that the tracking error $e(t) = r(t) - y(t)$ satisfies $|e(\infty)| < 0.05$.

5.5 For a stable plant $G(s)$, propose a method for the identification of $G(j\omega)$ via experiments by using a sinusoidal signal generator and an oscilloscope, based on the result of Theorem 5.1.

5.6 Solve the following problems for the prototype second-order system (5.21).
(a) Compute its unit step response.
(b) Derive the formulae of overshoot and settling time.

5.7 Prove Theorem 5.6. (Hint: The tracking error $e(t)$ is negative when the overshoot occurs.)

5.8 What kind of undershoot occurs when the number of positive real zeros is even in a transfer function $H(s)$?

5.9 Follow the following steps to investigate numerically the relationship between the response speed and bandwidth of the closed-loop system in Exercise 5.2.
(a) Calculate and plot the output $y(t)$ for $k = 0.25, 1, 5$ when $d(t) = 0,\ \ r(t) = 1(t)$.
(b) Draw the Bode plot for the closed-loop transfer function $H_{yr}(s)$ from r to y for $k = 0.25, 1, 5$, and measure the frequency bandwidth ω_B.
(c) Analyze the relationship between the response speed and bandwidth based on these results.

Notes and References

Frankline *et al.* [30] is a widely adopted textbook on the classical control and contains a detailed description on system performance. Mita [69] exposed theoretically the influence of system zeros on the response. A detailed description on the overshoot and undershoot can be found in Ref. [84].

6

Stabilization of Linear Systems

In the design of control systems, the stability of closed-loop system must be guaranteed first, which is called the *stabilization* problem. In this chapter, we show the most basic methods of system stabilization. Specifically, we will introduce the state feedback control method and the dynamic output feedback control method based on state feedback and observer.

The stabilization of a general plant described by

$$\dot{x}(t) = Ax(t) + Bu(t) \tag{6.1}$$

$$y(t) = Cx(t) \tag{6.2}$$

is considered. In this state equation, $x \in \mathbb{R}^n, u \in \mathbb{R}^m, y \in \mathbb{R}^p$.

6.1 State Feedback

We examine the case where all states are measured by sensors. Since all information of the system are contained in the states, states multiplied by appropriate coefficients must be able to stabilize the closed-loop system. That is, the input is constructed as

$$u = Fx, \quad F = \begin{bmatrix} f_{11} & \cdots & f_{1n} \\ \vdots & \ddots & \vdots \\ f_{m1} & \cdots & f_{mn} \end{bmatrix}. \tag{6.3}$$

This stabilization approach is called the *state feedback*, and the coefficient matrix F called the *state feedback gain* . Substituting (6.3) into (6.1), we obtain a closed-loop system called *state feedback system* whose state equation is

$$\dot{x} = (A + BF)x. \tag{6.4}$$

Obviously, the stability of state feedback system depends on the stability of its coefficient matrix $A + BF$. In this section, we investigate the condition required for the stabilization of closed-loop system, as well as concrete design method for the feedback gain matrix F.

Robust Control: Theory and Applications, First Edition. Kang-Zhi Liu and Yu Yao.

There have been numerous methods proposed for the design of state feedback. Among them, the simplest one is the *pole placement* method. Its idea is to select the desired closed-loop poles first and then determine an appropriate state feedback gain to ensure that the poles of the actual closed-loop system are equal to these specified poles.

The poles are roots of the characteristic polynomial of system matrix A. Once the closed-loop poles are specified, so is the characteristic polynomial. For example, if the poles of an nth-order system are specified as $\{p_1, \ldots, p_n\}$, then the corresponding characteristic polynomial is $(s - p_1) \cdots (s - p_n)$. On the other hand, as the coefficient matrix of the closed-loop system is $A + BF$, its characteristic polynomial is $\det\,(sI - (A + BF))$. They must be the same, that is, satisfy the following identity about variable s:

$$\det\,(sI - (A + BF)) \equiv (s - p_1) \cdots (s - p_n). \tag{6.5}$$

Both sides of the equation are polynomials. For this identity to hold, the coefficients of the terms having the same degree must be equal. So, by comparing the coefficients on both sides of the equation, we get n simultaneous equations on the elements of matrix F. In a single-input system, F is a $1 \times n$ row vector, and the number of elements is n. So, the solution is unique if it exists. For an m-input system, F is an $m \times n$ matrix, and the number of elements is $m \times n$. Then, even if the solution exists, it is not unique.

It should be noticed that the coefficients of this characteristic polynomial are real since $A + BF$ is a real matrix. Thus, when λ is a complex characteristic root, so will be its conjugate $\overline{\lambda}$. Further, λ and $\overline{\lambda}$ have the same multiplicity. So, the *set of specified eigenvalues must be symmetrical w.r.t. the real axis*. In the sequel, we limit the specifiable eigenvalues to such a set.

Next, let us see what condition is needed for pole placement through an example.

Example 6.1 *Examine the pole placement of a system with coefficient matrices*

$$A = \begin{bmatrix} 1 & 1 \\ 0 & -2 \end{bmatrix}, \quad b = \begin{bmatrix} 1 \\ 0 \end{bmatrix}.$$

Set the feedback gain as $f = [f_1 \;\; f_2]$. *Then, the characteristic polynomial of* $A + bf$ *becomes*

$$\det \begin{bmatrix} s - (1 + f_1) & -(1 + f_2) \\ 0 & s + 2 \end{bmatrix} = (s - 1 - f_1)(s + 2).$$

Its characteristic roots are $(1 + f_1, -2)$. *That is, the open-loop pole* -2 *becomes a closed-loop pole and cannot be changed by state feedback. So, for this system we cannot use state feedback to achieve arbitrary pole placement of closed-loop poles. In fact, this system is not controllable.*

From this example, we may speculate that in order to guarantee arbitrary pole placement of the closed-loop system, the open-loop system must be controllable. The following theorem confirms this speculation.

Theorem 6.1 *In order to place the eigenvalues of* $A + BF$ *arbitrarily,* (A, B) *must be controllable.*

Proof. When (A, B) is not controllable, there is a similarity transformation such that (A, B) is transformed into

$$A = T\begin{bmatrix} A_1 & A_{12} \\ 0 & A_2 \end{bmatrix} T^{-1}, \quad B = T\begin{bmatrix} B_1 \\ 0 \end{bmatrix}$$

(see Exercise 6.1). Then, for any F, by setting $\overline{F} = FT = [\overline{F}_1 \ \ \overline{F}_2]$, we have

$$A + BF = T\begin{bmatrix} A_1 + B_1\overline{F}_1 & A_{12} + B_1\overline{F}_2 \\ 0 & A_2 \end{bmatrix} T^{-1}.$$

Since $\det(sI - A - BF) = \det(sI - A_1 - B_1\overline{F}_1)\det(sI - A_2)$, clearly the eigenvalues of block A_2 cannot be moved. This means that no matter what state feedback gain F is used, it is impossible to achieve arbitrary placement of the eigenvalues of $A + BF$. •

6.1.1 Canonical Forms

For the ease of finding a sufficient condition for pole placement, we will use special state realizations called *canonical forms* hereafter. Two canonical forms are presented. As a preparation, we first show the following lemma.

Lemma 6.1 *The following statements hold for an $n \times n$ matrix:*

$$A = \begin{bmatrix} 0 & 1 & & \\ & 0 & \ddots & \\ & & \ddots & 1 \\ -a_1 & -a_2 & \cdots & -a_n \end{bmatrix}. \tag{6.6}$$

1. *The characteristic polynomial of matrix A is $p(s) = \det(sI - A) = s^n + a_n s^{n-1} + \cdots + a_2 s + a_1$.*
2. *Denote by e_i an n-dimensional vector whose ith entry is 1 and the rest are all 0. Set a function vector $z(s)$ as $z(s) = (sI - A)^{-1}e_i$. Then, its elements are given as follows:*
   (a) When $i = 1$,

$$z_1 = \frac{s^{n-1} + a_n s^{n-2} + \cdots + a_2}{p(s)}$$

$$z_2 = -\frac{a_1}{p(s)}, \quad z_3 = -\frac{a_1 s}{p(s)}, \quad \ldots, \quad z_n = -\frac{a_1 s^{n-2}}{p(s)}.$$

   (b) When $2 \le i \le n - 1$,

$$z_1 = \frac{s^{n-i} + a_n s^{n-(i+1)} + \cdots + a_{i+1}}{p(s)}$$

$$z_2 = sz_1, \quad \ldots, \quad z_i = s^{i-1}z_1$$

$$z_{i+1} = -\frac{a_i s^{i-1} + \cdots + a_2 s + a_1}{p(s)}$$

$$z_{i+2} = s z_{i+1}, \quad \ldots, \quad z_n = s^{n-(i+1)} z_{i+1}.$$

(c) When $i = n$,

$$z = \frac{1}{p(s)}[1 \quad s \quad \cdots \quad s^{n-1}]^T.$$

Here, z_j *denotes the* j*th element of* $z(s)$.

Proof. Statement 1 can be proved by converting the determinant into that of a lower triangle matrix via elementary transformations or by Laplace expansion. The detail is left to the readers.

For statement 2, we only give the proof for the case when $2 \leq i \leq n-1$. Other cases can be proved similarly. $z(s) = (sI - A)^{-1} e_i$ is equivalent to the following linear algebraic equation:

$$\begin{bmatrix} s & -1 & & & \\ & s & -1 & & \\ & & \ddots & \ddots & \\ & & & s & -1 \\ a_1 & a_2 & \cdots & a_{n-1} & s + a_n \end{bmatrix} \begin{bmatrix} z_1 \\ \vdots \\ z_i \\ \vdots \\ z_n \end{bmatrix} = \begin{bmatrix} 0 \\ \vdots \\ 1 \\ \vdots \\ 0 \end{bmatrix}.$$

Expanding this equation, we get the following simultaneous linear equations:

$$\begin{aligned} & s z_1 - z_2 = 0, \quad s z_2 - z_3 = 0, \quad \ldots, \quad s z_{i-1} - z_i = 0 \\ & s z_i - z_{i+1} = 1 \\ & s z_{i+1} - z_{i+2} = 0, \quad \ldots, \quad s z_{n-1} - z_n = 0 \\ & a_1 z_1 + a_2 z_2 + \cdots + a_{n-1} z_{n-1} + (s + a_n) z_n = 0. \end{aligned}$$

Solving the equations successively from the first one yields

$$\begin{aligned} & z_2 = s z_1, \quad \ldots, \quad z_i = s z_{i-1} = s^{i-1} z_1 \\ & z_{i+1} = s z_i - 1 = s^i z_1 - 1, \quad z_{i+2} = s z_{i+1} = s^{i+1} z_1 - s, \quad \ldots \\ & z_n = s z_{n-1} = s^{n-1} z_1 - s^{n-(i+1)}. \end{aligned}$$

Substituting them into the last equation, we obtain

$$z_1 = \frac{s^{n-i} + a_n s^{n-i-1} + \cdots + a_{i+2} s + a_{i+1}}{p(s)}.$$

Therefore,

$$z_{i+1} = s^i z_1 - 1 = -\frac{a_i s^{i-1} + \cdots + a_2 s + a_1}{p(s)}.$$

•

When the system (6.1) is controllable, we can transform it into a special realization called the *controllability canonical form* by similarity transformation. Here, we consider a single-input system.

Lemma 6.2 *Let the realization of a system be $(A, b, c, 0)$ and the characteristics polynomial of matrix A be*

$$\det\,(sI - A) = s^n + a_n s^{n-1} + \cdots + a_2 s + a_1. \tag{6.7}$$

Further, assume that the transfer function corresponding to $(A, b, c, 0)$ is

$$c(sI - A)^{-1}b = \frac{\beta_n s^{n-1} + \beta_{n-1} s^{n-2} + \cdots + \beta_2 s + \beta_1}{s^n + a_n s^{n-1} + \cdots + a_2 s + a_1}. \tag{6.8}$$

When (A, b) is controllable, there exists a similarity transformation matrix T satisfying the following conditions:

$$\overline{A} := T^{-1}AT = \begin{bmatrix} 0 & 1 & 0 & \cdots & 0 \\ 0 & 0 & 1 & \cdots & 0 \\ \vdots & \vdots & \vdots & \ddots & \vdots \\ 0 & 0 & 0 & \cdots & 1 \\ -a_1 & -a_2 & -a_3 & \cdots & -a_n \end{bmatrix}, \quad \overline{b} := T^{-1}b = \begin{bmatrix} 0 \\ 0 \\ \vdots \\ 0 \\ 1 \end{bmatrix} \tag{6.9}$$

$$\overline{c} := cT = [\beta_1 \quad \beta_2 \quad \cdots \quad \beta_{n-1} \quad \beta_n]. \tag{6.10}$$

This new realization $(\overline{A}, \overline{b}, \overline{c}, 0)$ is called the controllability canonical form.

Proof. By the controllability assumption, the controllability matrix $C = [b \;\; Ab \;\; \cdots \;\; A^{n-2}b \;\; A^{n-1}b]$ is nonsingular. Obviously, the matrix

$$U = \begin{bmatrix} a_2 & a_3 & \cdots & a_n & 1 \\ a_3 & a_4 & \cdots & 1 & 0 \\ \vdots & \vdots & \vdots & \vdots & \vdots \\ a_n & 1 & \cdots & 0 & 0 \\ 1 & 0 & \cdots & 0 & 0 \end{bmatrix} \tag{6.11}$$

is also nonsingular. So, the matrix

$$T := CU = [t_1 \;\; t_2 \;\; \cdots \;\; t_{n-1} \;\; t_n] \tag{6.12}$$

is invertible. Expanding T, its columns are

$$\begin{aligned} t_1 &= (A^{n-1} + a_n A^{n-2} + \cdots + a_3 A + a_2 I)b \\ t_2 &= (A^{n-2} + a_n A^{n-3} + \cdots + a_3 I)b \\ &\vdots \\ t_{n-1} &= (A + a_n I)b \\ t_n &= b, \end{aligned}$$

respectively. Based on this and the fact that $b = t_n$, we have

$$At_2 = t_1 - a_2 t_n, \ \ldots, \ At_{n-1} = t_{n-2} - a_{n-1} t_n, \ At_n = t_{n-1} - a_n t_n. \tag{6.13}$$

Furthermore, $A^n + a_n A^{n-1} + \cdots + a_2 A + a_1 I = 0$ is true according to Cayley–Hamilton theorem. Multiplying this equation from the right by $b = t_n$, we get

$$At_1 = -a_1 t_n. \tag{6.14}$$

Summarizing these equations together, we obtain

$$AT = T\overline{A} \quad \Rightarrow \quad \overline{A} = T^{-1}AT. \tag{6.15}$$

In addition, it is easy to know that $T\,\overline{b} = b$, so $T^{-1}b = \overline{b}$ holds. Finally, since similarity transformation does not change the transfer function, we can easily obtain the equation about $\overline{c}$ using $c(sI - A)^{-1}b = \overline{c}(sI - \overline{A})^{-1}\overline{b}$ and Lemma 6.1(2c). •

As a dual of this lemma, we have the following lemma.

Lemma 6.3 *Let the realization of a system be* $(A, b, c, 0)$, *the characteristic polynomial of matrix* A *be*

$$\det(sI - A) = s^n + a_n s^{n-1} + \cdots + a_2 s + a_1, \tag{6.16}$$

and the transfer function of $(A, b, c, 0)$ *be*

$$c(sI - A)^{-1}b = \frac{\beta_n s^{n-1} + \beta_{n-1} s^{n-2} + \cdots + \beta_2 s + \beta_1}{s^n + a_n s^{n-1} + \cdots + a_2 s + a_1}. \tag{6.17}$$

When (c, A) *is observable, there exists a similarity transformation matrix* S *satisfying*

$$SAS^{-1} = \begin{bmatrix} 0 & 0 & \cdots & 0 & -a_1 \\ 1 & 0 & \cdots & 0 & -a_2 \\ 0 & 1 & \ddots & \vdots & -a_{n-1} \\ \vdots & \ddots & \ddots & 0 & \vdots \\ 0 & \cdots & 0 & 1 & -a_n \end{bmatrix}, \quad Sb = \begin{bmatrix} \beta_1 \\ \beta_2 \\ \vdots \\ \beta_{n-1} \\ \beta_n \end{bmatrix} \tag{6.18}$$

$$cS^{-1} = [0 \ \ 0 \ \ \cdots \ \ 0 \ \ 1]. \tag{6.19}$$

This new realization is called the observability canonical form .

Proof. The conclusion is obtained by converting the dual system $(A^T, c^T, b^T, 0)$ into the controllability canonical form and then calculating its dual system again. Note that the transformation matrix is $S = U\mathcal{O}$ where U is the matrix defined by (6.11) and $\mathcal{O}$ is the observability matrix. •

These canonical forms can be extended to multiple-input multiple-output (MIMO) systems, but the descriptions get very complicated and not explored here. Interested readers may consult Ref. [44].

6.1.2 Pole Placement of Single-Input Systems

For a single-input dynamic system

$$\dot{x} = Ax + bu, \tag{6.20}$$

the condition needed for achieving arbitrary pole placement via state feedback is given by the following theorem.

Theorem 6.2 *For the single-input system (6.20), the eigenvalues of $A + bf$ can be arbitrarily placed using state feedback $u = fx(f^T \in \mathbb{R}^n)$ iff (A, b) is controllable.*

Proof. The necessity has been proved in Theorem 6.1, so only the sufficiency is proved here.
First, we convert (A, b) into the controllability canonical form $(\overline{A}, \overline{b})$ using the transformation matrix T in (6.12). Setting $fT = \overline{f} = [\overline{f}_1 \quad \cdots \quad \overline{f}_n]$, its substitution leads to

$$\begin{aligned} T^{-1}(A + bf)T &= \overline{A} + \overline{b}\,\overline{f} \\ &= \begin{bmatrix} 0 & 1 & 0 & \cdots & 0 \\ 0 & 0 & 1 & \cdots & 0 \\ \vdots & \vdots & \vdots & \ddots & \vdots \\ 0 & 0 & 0 & \cdots & 1 \\ -(a_1 - \overline{f}_1) & -(a_2 - \overline{f}_2) & -(a_3 - \overline{f}_3) & \cdots & -(a_n - \overline{f}_n) \end{bmatrix}. \end{aligned}$$

Then, based on the equation $\det(sI - T^{-1}XT) = \det(T^{-1}(sI - X)T) = \det(T^{-1}) \det(sI - X) \det(T) = \det(sI - X)$ and Lemma 6.1(1), we obtain the following equation:

$$\begin{aligned} \det(sI - (A + bf)) &= \det(sI - (\overline{A} + \overline{bf})) \\ &= s^n + (a_n - \overline{f}_n)s^{n-1} + \cdots + (a_1 - \overline{f}_1). \end{aligned} \tag{6.21}$$

When the specified closed-loop characteristic polynomial is $s^n + \gamma_n s^{n-1} + \cdots + \gamma_2 s + \gamma_1$, we obtain $\overline{f} = [a_1 - \gamma_1 \quad \cdots \quad a_n - \gamma_n]$ through the comparison of coefficients. Therefore, the state feedback gain

$$f = \overline{f}T^{-1} = [a_1 - \gamma_1 \quad \cdots \quad a_n - \gamma_n]T^{-1} \tag{6.22}$$

is able to place the poles to the assigned locations. •

In fact, the proof also provides a method for calculating the state feedback gain f.

Example 6.2 *Consider the 1-DOF vibration system consisting of a spring and a mass as shown in Figure 6.1, where M denotes the mass and K the spring constant and the damping is zero. y is the displacement from the balanced position and u is the external force. If the state variables are chosen as the displacement and speed of the mass, the coefficient matrices of state equation become*

$$A = \begin{bmatrix} 0 & 1 \\ -\dfrac{K}{M} & 0 \end{bmatrix}, \quad b = \begin{bmatrix} 0 \\ \dfrac{1}{M} \end{bmatrix},$$

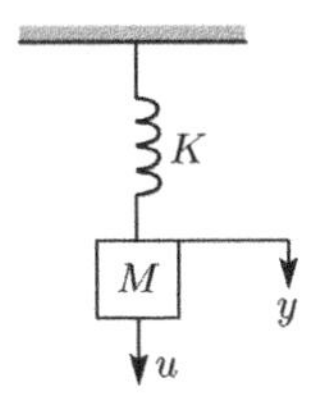

Figure 6.1 1-DOF vibration system

respectively. For this 1-DOF vibration system, we design a state feedback gain $f = [f_1\ f_2]$ *so as to place the eigenvalue of* $A + bf$ *to* $\{p_1, p_2\}$*. According to the definition of eigenvalue, we have the following identity:*

$$\det(sI - (A + bf)) = (s - p_1)(s - p_2)$$
$$\Rightarrow\ s^2 - \frac{f_2}{M}s + \left(\frac{K}{M} - \frac{f_1}{M}\right) = s^2 - (p_1 + p_2)s + p_1 p_2.$$

Comparing the coefficients on both sides of the equation, we get the simultaneous equation

$$\frac{f_2}{M} = p_1 + p_2, \quad \frac{K}{M} - \frac{f_1}{M} = p_1 p_2.$$

Solving this equation, we obtain the unique solution

$$f = [K - Mp_1p_2,\ M(p_1 + p_2)].$$

For systems with a degree lower than 3, it is easy to solve directly the simultaneous linear algebraic equation derived from the characteristic polynomial, as we just did in this example. However, for systems of higher degree, it is more convenient to apply the formula (6.22).

Example 6.3 *For a third-order system*

$$\dot{x} = \begin{bmatrix} 1 & 1 & -2 \\ 0 & 1 & 1 \\ 0 & 0 & 1 \end{bmatrix} x + \begin{bmatrix} 1 \\ 0 \\ 1 \end{bmatrix} u,$$

design a state feedback gain such that the closed-loop poles are placed at $-2, -1 \pm j1$.
Solution The controllability matrix is computed as

$$\mathcal{C} = [b\ \ Ab\ \ A^2b] = \begin{bmatrix} 1 & -1 & -2 \\ 0 & 1 & 2 \\ 1 & 1 & 1 \end{bmatrix}$$

whose rank is 3. So, the system is controllable, and we can use state feedback to place the poles. From the characteristic polynomial of open-loop system

$$\det(sI - A) = s^3 - 3s^2 + 3s - 1,$$

we get the coefficients $a_1 = -1, a_2 = 3, a_3 = -3$. Then, the transformation matrix for the controllability canonical form becomes

$$T = \mathcal{C}\begin{bmatrix} a_2 & a_3 & 1 \\ a_3 & 1 & 0 \\ 1 & 0 & 0 \end{bmatrix} = \begin{bmatrix} 4 & -4 & 1 \\ -1 & 1 & 0 \\ 1 & -2 & 1 \end{bmatrix}.$$

Further, the characteristic polynomial corresponding to the specified close-loop poles is

$$(s+2)(s+1-j)(s+1+j) = s^3 + 4s^2 + 6s + 4$$

whose coefficients are $\gamma_1 = 4, \gamma_2 = 6, \gamma_3 = 4$. So, the state feedback gain is obtained via some calculation as

$$f = [a_1 - \gamma_1 \quad a_2 - \gamma_2 \quad a_3 - \gamma_3]T^{-1} = [-15 \ \ -47 \ \ 8].$$

▽

6.1.3 *Pole Placement of Multiple-Input Systems**

Here, we examine that what is the condition for arbitrarily placing the eigenvalues of $A + BF$ for multiple-input systems. The conclusion is that the eigenvalues of $A + BF$ can be arbitrarily placed iff (A, B) is controllable. This conclusion is proved by converting the pole placement of a multiple-input system equivalently into that of a single-input system.

Lemma 6.4 *When an m-input system (A, B) is controllable, then there must be a matrix $K \in \mathbb{R}^{m\times n}$ such that $(A + BK, b)$ is controllable for any nonzero vector $b \in \text{Im}B$.*

Proof. Define $b_1 = b$ and a subspace $U_1 = \text{span}\{b_1, Ab_1, \ldots, A^{n-1}b_1\}$. If dim $(U_1) = n$, then $K = 0$ meets the requirement. $\{b_1, Ab_1, \ldots, A^{n_1-1}b_1\}$ becomes a basis of U_1 [1] when dim $(U_1) = n_1 < n$. Now, we construct a sequence of n_1 vectors:

$$x_1 = b_1, \quad x_j = Ax_{j-1} + b_1, \quad j = 2, \ldots, n_1. \tag{6.23}$$

The vector sequence given by this recursive formula has the property of $x_j - x_{j-1} = A(x_{j-1} - x_{j-2})$ $(j \geq 3)$. Starting from $x_2 - x_1 = Ab_1$ and using this property successively, we obtain the following relations:

$$b_1 = x_1, \ \ Ab_1 = x_2 - x_1, \ldots, \ \ A^{n_1-1}b_1 = x_{n_1} - x_{n_1-1}.$$

Vector sequences $\{x_i\}$ and $\{A^{i-1}b_1\}$ are in a one-to-one relationship, so $\{x_1, \ldots, x_{n_1}\}$ is also a basis of subspace U_1.

[1] Assume that the vectors $\{b_1, Ab_1, \ldots, A^{k-1}b_1\}$ are linearly independent and $A^k b_1$ can be expressed as their linear combination. Then all $A^{k+i}b_1 (i \geq 1)$ can also be expressed as linear combinations of $\{b_1, Ab_1, \ldots, A^{k-1}b_1\}$. $k = n_1$ follows from $\dim(U_1) = n_1$.

We know from the controllability that there must be a vector $b_2 \in \mathrm{Im}B$ satisfying $b_2 \notin U_1$[2]. Next, let n_2 be the largest integer such that vectors

$$x_1, \ldots, x_{n_1}, b_2, Ab_2, \ldots, A^{n_2-1}b_2$$

are linearly independent. We construct a new vector sequence as follows:

$$x_{n_1+i} = Ax_{n_1+i-1} + b_2, \quad i = 1, \ldots, n_2. \tag{6.24}$$

From $x_{n_1+1} - x_{n_1} = (A - I)x_{n_1} + b_2$, we know that

$$\begin{aligned} x_{n_1+i} - x_{n_1+i-1} &= A(x_{n_1+i-1} - x_{n_1+i-2}) = \cdots \\ &= A^{i-1}(x_{n_1+1} - x_{n_1}) = A^{i-1}b_2 + A^{i-1}(A - I)x_{n_1}. \end{aligned}$$

That is, a linearly independent column vector $A^{i-1}b_2$ is added whenever we construct a new vector. Since U_1 is A-invariant, $A^{i-1}(A - I)x_{n_1} \in U_1$ holds. Therefore, $x_1, \ldots, x_{n_1}, \ldots, x_{n_1+n_2}$ are linearly independent.

The rank of $\mathcal{C}$ is n. Continuing this process, we must be able to construct a linearly independent vector sequence $x_1, \ldots, x_n$ in finite steps. And they satisfy the relation

$$x_{i+1} = Ax_i + \bar{b}_i, \quad i = 1, \ldots, n - 1. \tag{6.25}$$

Here, for simplicity, we have changed b_1, b_2, and so on to new subscripts in correspondence with x_i, denoted by $\bar{b}_i$. Certainly, there holds $\bar{b}_i \in \mathrm{Im}B$, and it can be expressed as $\bar{b}_i = Bg_i (g_i \in \mathbb{R}^m)$. Set $X = [x_1 \ \cdots \ x_n]$. Then X is nonsingular. Let

$$K = [g_1 \ \cdots \ g_n]X^{-1} \tag{6.26}$$

in which $g_n \in \mathbb{R}^m$ is an arbitrary vector. It is easy to see that

$$BKx_i = \bar{b}_i, \quad i = 1, \ldots, n - 1.$$

Substituting it into the recursive formula (6.25) about x_i, we get

$$x_{i+1} = (A + BK)x_i \ \Rightarrow \ x_i = (A + BK)^{i-1}x_1 = (A + BK)^{i-1}b. \tag{6.27}$$

Therefore, the rank of $[b \ (A + BK)b \ \cdots \ (A + BK)^{n-1}b]$ equals n and $(A + BK, b)$ is controllable. •

Even if we do not construct the feedback gain K like what we did in the proof, but choose K at random, $(A + BK, Bg)$ is controllable in almost all cases [12]. Here, g is an arbitrary nonzero vector.

Based on this lemma and Theorem 6.2, it is easy to deduce the following theorem on the pole placement of multiple-input systems.

[2] Noting that $AU_1 \subset U_1$, if there is not such a b_2, then any $b' \in \mathrm{Im}B$ is contained in U_1 so that $A^i b' \in U_1 \quad (\forall i \geq 0)$ holds. That is, all columns of the controllability matrix $\mathcal{C}$ are contained in U_1. Since $n_1 < n$, this contradicts the controllability of (A, B).

Theorem 6.3 *For an m- input system*

$$\dot{x} = Ax + Bu,$$

there exists a state feedback gain $F \in \mathbb{R}^{m\times n}$ such that the eigenvalues of $A + BF$ are placed arbitrarily iff (A, B) is controllable.

Proof. Note that in the previous lemma, b can be written as $Bg(g \in \mathbb{R}^m)$. After the input transformation $u = Kx + gv$, this system turns into a controllable single-input system $\dot{x} = (A + BK)x + bv$. According to Theorem 6.2, the poles of this single-input system can be arbitrarily placed with a state feedback $v = fx$. Therefore,

$$F = K + gf \tag{6.28}$$

gives the feedback gain which we are searching for. •

Example 6.4 *Design a state feedback gain F to place the poles of two-input system*

$$\dot{x} = Ax + Bu = \begin{bmatrix} 1 & 0 & 0 & 0 \\ 0 & 0 & 1 & 0 \\ 1 & 0 & 0 & 0 \\ 0 & 0 & 0 & 0 \end{bmatrix} x + \begin{bmatrix} 1 & 0 \\ 1 & 0 \\ 0 & 0 \\ 0 & 1 \end{bmatrix} u$$

to $\{-1, -1, -1 + j, -1 - j\}$.

First of all, set $b = b_1 = [0 \quad 0 \quad 0 \quad 1]^T$, $b_2 = [1 \quad 1 \quad 0 \quad 0]^T$ (note that the subscript of vector is different from the column number of matrix B). It may be verified that $n_1 = 1, n_2 = 3$. Then, computing according to the algorithm given in Lemma 6.4, we get

$$X = [x_1 \quad x_2 \quad x_3 \quad x_4] = \begin{bmatrix} 0 & 1 & 2 & 3 \\ 0 & 1 & 1 & 2 \\ 0 & 0 & 1 & 2 \\ 1 & 0 & 0 & 0 \end{bmatrix}$$

where x_2, x_3, x_4 are obtained based on the calculation of $x_2 = Ax_1 + b_2 = Ax_1 + \bar{b}_1, \ldots, x_4 = Ax_3 + b_2 = Ax_3 + \bar{b}_3$, so $\bar{b}_1 = \bar{b}_2 = \bar{b}_3 = b_2 = B[1 \quad 0]^T$. Therefore,

$$g_1 = g_2 = g_3 = \begin{bmatrix} 1 \\ 0 \end{bmatrix}$$

By setting $g_4 = [0 \ \ 0]^T$, we get

$$K = [g_1 \quad g_2 \quad g_3 \quad g_4]X^{-1} = \begin{bmatrix} 2 & -1 & -2 & 1 \\ 0 & 0 & 0 & 0 \end{bmatrix}$$

Calculating the eigenvalues of $A + BK$, we obtain $\det(sI - (A + BK)) = s^4 - 2s^3 + s^2 + s$. *So $a_1 = 0, a_2 = a_3 = 1, a_4 = -2$. On the other hand, the specified characteristic polynomial of closed-loop system is*

$$(s + 1)(s + 1)(s + 1 - j)(s + 1 + j) = s^4 + 4s^3 + 7s^2 + 6s + 2$$

so that $\gamma_1 = 2, \gamma_2 = 6, \gamma_3 = 7, \gamma_4 = 4$*. Next, the matrix that transforms* $(A + BK, b)$ *into the controllable canonical form is computed according to (6.12):*

$$T = \begin{bmatrix} 0 & 0 & 1 & 0 \\ 1 & -1 & 1 & 0 \\ 0 & 1 & 0 & 0 \\ 1 & 1 & -2 & 1 \end{bmatrix}, \quad T^{-1} = \begin{bmatrix} -1 & 1 & 1 & 0 \\ 0 & 0 & 1 & 0 \\ 1 & 0 & 0 & 0 \\ 3 & -1 & -2 & 1 \end{bmatrix}.$$

In this way, the state feedback gain f *is obtained according to (6.22):*

$$f = [a_1 - \gamma_1 \quad a_2 - \gamma_2 \quad a_3 - \gamma_3 \quad a_4 - \gamma_4]T^{-1} = [-22 \quad 4 \quad 5 \quad -6].$$

Meanwhile, from $b = B[0 \ 1]^T$ *we have* $g = [0 \ 1]^T$*. In the end, the state feedback gain becomes*

$$F = K + gf = \begin{bmatrix} 2 & -1 & -2 & 1 \\ -22 & 4 & 5 & -6 \end{bmatrix}.$$

6.1.4 Principle of Pole Selection

In using the pole placement method to design a state feedback, the key is how to set the closed-loop poles. In selecting the closed-loop poles, we should not only consider the stability of system but also pay attention to its performance (viz., transient response). This is because the ultimate goal of control is to achieve performance improvement. In addition, practical restriction also needs to be considered, which mainly stems from actuator constraint and is reflected in the limitation of control input. Particularly, there is only a qualitative relationship between the amplitude of control input and the location of poles. Therefore, trial and error is inevitable in finding appropriate poles. The principle is summarized as follows:

1. To ensure the required convergence rate, the poles should keep a certain distance from the imaginary axis. Namely, they need to satisfy

$$\Re(p_i) \le -\sigma \ \forall\, i.$$

   Here, $\sigma > 0$ is the parameter specifying the response speed. Generally speaking, the amplitude of transient response roughly depends on the exponential function $e^{-\sigma t}$. Since $e^{-3} \approx 5\%$, the settling time t_s can be estimated by $\sigma t_s \approx 3$. According to this relation, given the settling time t_s, the parameter σ can be determined based on $\sigma \approx 3/t_s$.
2. The imaginary parts of poles correspond to the resonant frequencies of the response. To reduce the number of vibrations per second, we should reduce the imaginary parts. Usually the imaginary part is taken to be less than the magnitude of real part. This is because the period corresponding to the resonant frequency $\Im(p)$ is $2\pi/\Im(p)$, when $|\Re(p)| \ge \Im(p)$ the amplitude of response reduces to $e^{-|\Re(p)|\cdot 2\pi/\Im(p)} \le e^{-2\pi} = 0.19\%$ in a period. In addition, for second-order systems, when the magnitude of real part is equal to the imaginary part, the characteristic polynomial is

$$\left(s + \frac{\omega_n}{\sqrt{2}}\right)^2 + \left(\frac{\omega_n}{\sqrt{2}}\right)^2 = s^2 + 2 \cdot \frac{1}{\sqrt{2}} \cdot \omega_n s + \omega_n^2$$

   whose damping ratio is $1/\sqrt{2}$, which is usually regarded as the minimum of allowable damping (according to Figure 5.12, when $\zeta = 1/\sqrt{2}$ the overshoot is about 5%).

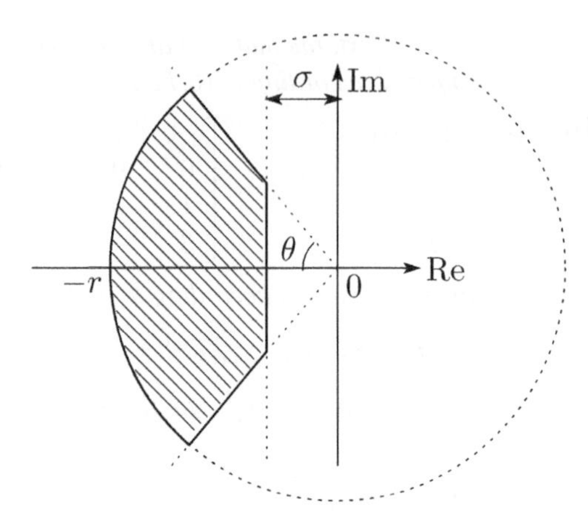

Figure 6.2 Allowable region of poles

3. Basically, the magnitude of input increases with the magnitudes of poles. In order to limit the input, the poles cannot be too big. That is, they should satisfy

$$|p_i| < r \ \forall \, i$$

for certain specified value $r > 0$. Besides, when a pole approaches the negative infinity, the maximum amplitude of response will diverge gradually, which brings a bad impact to the response [69].

In summary, the poles should be selected in the shaded region shown in Figure 6.2.

However, it should be noted that the damping ratio will get too big if the ratio $\Im(p)/|\Re(p)|$ of imaginary part to real part is too small, which will lower the bandwidth and consequently prolong the rise time. Therefore, when a short rise time is desirable, this ratio should be in the interval $[1, 2]$.

A design example is illustrated in the following.

Example 6.5 *Consider the two-mass–spring system in Example 1.2. The control purpose is to suppress the variation of load speed ω_L when the load rotating at a constant speed is affected by a torque disturbance d. As a numerical example, we set $J_M = J_L = 1$, $D_M = D_L = 0$, and $k = 100$ and place the closed-loop poles at $\{-4 \pm j4, -8\}$ by a state feedback. The feedback gain obtained is*

$$F = [13.4 \quad 104 \quad -16].$$

The responses of ω_M, ω_L to a unit impulse disturbance is shown in Figure 6.3.

6.2 Observer

In order to implement the state feedback, all states have to be measured by sensors, which usually are rather expensive. Thus, state feedback is an extravagant control method. Moreover,

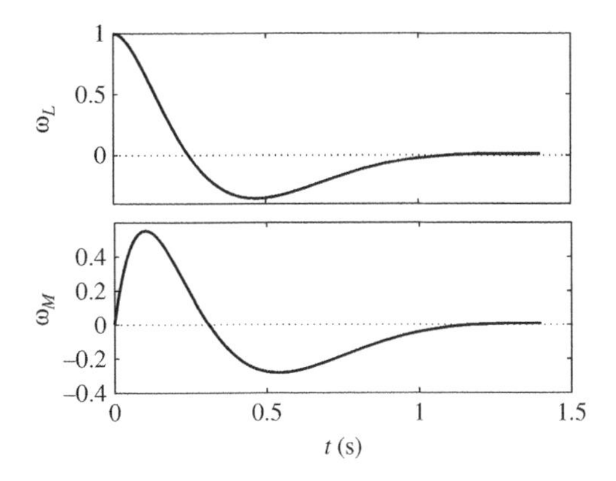

Figure 6.3 Impulse response of state feedback system

in engineering practice there exist many cases in which not all states can be measured. For example, in a flexible system such as the solar panel of space station, there are countless states because of the flexibility. Therefore, we usually need to estimate the states using the measured output and control input through software computing. The estimation algorithm of states is called an *observer*.

6.2.1 Full-Order Observer

A typical observer is given by

$$\dot{\overline{x}} = A\overline{x} + Bu + L(C\overline{x} - y). \tag{6.29}$$

The order of this observer is equal to that of the plant, so it is called a *full-order observer*, also known as *Luenberger observer*.

The mechanism of full-order observer is to calculate the state estimate $\overline{x}$ by using the plant dynamics (the first two terms on the right) with software and then compensate the state estimate $\overline{x}$ based on the difference between the output $C\overline{x}$ of observer and the measured output y of plant (the third term on the right). Matrix L is called the *observer gain* . Generally, the initial value of observer state is chosen as $\overline{x}(0) = 0$.

Now, we examine the condition needed for state estimation. Here, we set the estimation error as

$$e = \overline{x} - x. \tag{6.30}$$

Since $\overline{x} \to x$ is equivalent to $e \to 0$, we only need consider the convergence of estimation error. Therefore, let us look at the dynamics of $e(t)$. Subtracting (6.1) from (6.29), we obtain the model of estimate error:

$$\begin{aligned}\dot{e} &= A\overline{x} + Bu + L(C\overline{x} - Cx) - (Ax + Bu)\\ &= (A + LC)e.\end{aligned} \tag{6.31}$$

In order to ensure $e(t) \to 0$ for any initial error $e(0)$, $(A + LC)$ must be stable. Hereafter, we call the eigenvalues of $(A + LC)$ as the *observer poles* .

Noting that the eigenvalues of $(A + LC)$ and $(A + LC)^T = A^T + C^T L^T$ are the same and the observability of (C, A) is equivalent to the controllability of (A^T, C^T), it is easy to see that the following theorem holds.

Theorem 6.4 *For the full-order observer (6.29), there exists an observer gain $L \in \mathbb{R}^{n\times p}$ such that the eigenvalues of $A + LC$ can be placed arbitrarily iff (C, A) is observable.*

Similar to state feedback, the pole placement method is the simplest in the design of observer. It is explained in the following. Let the assigned observer poles be $\{r_1, \dots, r_n\}$. Then, the characteristic polynomial of observer is

$$\det(sI - (A + LC)) \equiv (s - r_1)\cdots(s - r_n). \tag{6.32}$$

Comparing the coefficients on both sides, we obtain n simultaneous linear equations about the elements of L. For single-output systems, the solution can be obtained by solving the simultaneous equations. Meanwhile, the design of multiple-output systems is based on the following procedure. Since

$$\det(sI - (A + LC)) = \det(sI - (A^T + C^T L^T)), \tag{6.33}$$

we first set $\overline{A} = A^T, \overline{B} = C^T, \overline{F} = L^T$ and calculate $\overline{F}$ based on the algorithm for state feedback gain, then the observer gain is obtained as $L = \overline{F}^T$.

6.2.1.1 Selection of Observer Poles

The function of observer is to estimate the state of plant so as to realize the state feedback. To achieve this goal, the state estimate must converge to the true state fast enough. This means that the *observer poles should be sufficiently far from the imaginary axis compared with the poles of state feedback system.* We usually choose the distance between observer poles and the imaginary axis $2 \sim 5$ times greater than that of state feedback poles. This is a rule of thumb for observer design.

Next, the two-mass–spring system is used as example to demonstrate the design.

Example 6.6 *In the two-mass–spring system of Example 1.2, it is very difficult to mount the rotational speed sensor on the load, so generally we only measure the motor speed ω_M. Here, we design an observer that can estimate the states based on this output. The designated observer poles are $\{-12 \pm j19, -24\}$. The observer gain designed is*

$$L = [-73.2 \quad 8.8 \quad -48].$$

The response of estimation error w.r.t. initial condition $x(0) = [1 \quad 0 \quad 0]^T$ is shown in Figure 6.4. Illustrated successively from top to bottom are the errors of ω_L, ϕ, and ω_M. Obviously, all state estimates converge rapidly to the true states.

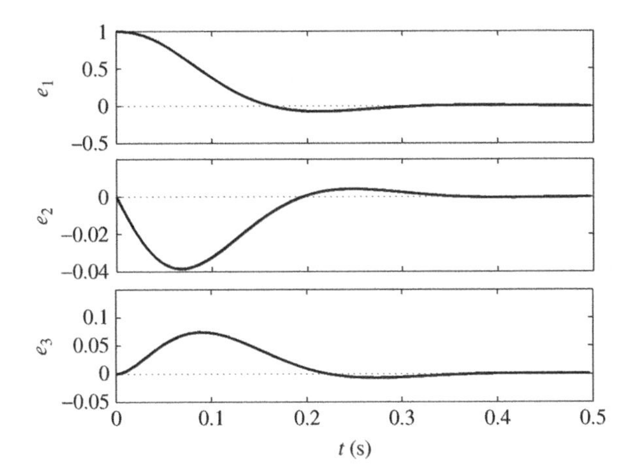

Figure 6.4 Estimation error responses of observer

6.2.2 *Minimal Order Observer*

Usually, the matrix C in the output equation $y = Cx$ has full row rank because redundant sensors exist if not. Since the measured output $y = Cx$ itself already contains a part of states, we do not need to estimate all states. What is needed is to estimate the unknown part. Therefore, we consider how to use the output information to estimate a signal

$$z = Vx. \tag{6.34}$$

The reason why we estimate z is that since

$$\begin{bmatrix} y \\ z \end{bmatrix} = \begin{bmatrix} C \\ V \end{bmatrix} x, \tag{6.35}$$

the state x can be calculated from (y, z) uniquely so long as $\left[\begin{smallmatrix} C \\ V \end{smallmatrix}\right]$ is nonsingular. In this case, $z \in \mathbb{R}^{(n-p)}$. Conversely, x cannot be computed uniquely when $\left[\begin{smallmatrix} C \\ V \end{smallmatrix}\right]$ is a wide matrix; and there will be redundant elements in z when $\left[\begin{smallmatrix} C \\ V \end{smallmatrix}\right]$ is a tall matrix. Therefore, in order to compute the state x uniquely, the minimal dimension of signal z must be $n - p$. The corresponding observer is called the *minimal order observer*, whose dimension is $n - p$.

Now, we consider the estimation of z. Let the estimate of z be $\overline{z}$ and the estimation error be

$$e := \overline{z} - z. \tag{6.36}$$

First of all, when the initial estimation error is $e(0) = 0$, the identity $e(t) \equiv 0$ must be true. So, the estimation error needs having a dynamics of

$$\dot{e} = Te. \tag{6.37}$$

Here, the sufficient and necessary condition for $e(\infty) = 0$ is that $T \in \mathbb{R}^{(n-p)\times(n-p)}$ is a stable matrix. Only in this case it is possible to estimate the signal z. Next, we start from this condition to investigate the structure necessary for the minimal order observer. $\dot{z} = V\dot{x} = VAx + VBu$

combined with the previous two equations gives

$$\dot{\overline{z}} = \dot{e} + \dot{z} = T\overline{z} - Tz + \dot{z} = T\overline{z} + VAx + VBu - Tz.$$

Setting

$$A\begin{bmatrix} C \\ V \end{bmatrix}^{-1} = [K \quad M] \Rightarrow A = KC + MV \tag{6.38}$$

and using (y, z) to express x, we have

$$\dot{\overline{z}} = T\overline{z} + VKy + VBu + (VM - T)z.$$

Since we can only use the known signals (y, u) in state estimation, there must be

$$VM - T = 0. \tag{6.39}$$

Further, this condition is equivalent to

$$V(A - KC) = TV. \tag{6.40}$$

Why? According to $A - KC = MV$, when $VM = T$, we have $V(A - KC) = VMV = TV$. Conversely, when $V(A - KC) = TV$, there holds $VMV = TV$, that is, $(VM - T)V = 0$. Since V has full row rank, there must be $VM - T = 0$.

Finally, substitution of $\overline{z}$ into z in (6.35) yields the state estimate $\overline{x}$. In summary, we have the following theorem.

Theorem 6.5 *Let $T \in \mathbb{R}^{(n-p)\times(n-p)}$ be any given stable matrix. When there exist matrices $V \in \mathbb{R}^{(n-p)\times n}$, $K \in \mathbb{R}^{n\times p}$ satisfying*

$$\text{rank}\begin{bmatrix} C \\ V \end{bmatrix} = n, \quad V(A - KC) = TV,$$

the state x can be estimated using the observer

$$\dot{\overline{z}} = T\overline{z} + VKy + VBu, \tag{6.41}$$

and the estimate $\overline{x}$ is

$$\overline{x} = \begin{bmatrix} C \\ V \end{bmatrix}^{-1} \begin{bmatrix} y \\ \overline{z} \end{bmatrix}. \tag{6.42}$$

Now, we prove that the observability of (C, A) is the condition for the existence of matrices satisfying this theorem.

Theorem 6.6 *For any given eigenvalue set $\Lambda = \{\lambda_1, \ldots, \lambda_{n-p}\}$, there are matrices $V \in \mathbb{R}^{(n-p)\times n}$, $K \in \mathbb{R}^{n\times p}$, and $T \in \mathbb{R}^{(n-p)\times(n-p)}$ satisfying*

$$\text{rank}\begin{bmatrix} C \\ V \end{bmatrix} = n, \quad V(A - KC) = TV, \quad \sigma(T) = \Lambda$$

iff (C, A) is observable.

Proof. The sufficiency can be proved by concretely constructing such matrices based on the Gopinath algorithm to be illustrated later. So, here we only prove its necessity.

When (C, A) is not observable, we prove that even if the first two conditions are true, the third one is not satisfied. To this end, let λ be an unobservable eigenvalue of (C, A) and $u \neq 0$ the corresponding eigenvector. Then

$$Au = \lambda u, \quad Cu = 0. \tag{6.43}$$

Further, $Vu \neq 0$ follows from the nonsingularity of $\left[\begin{smallmatrix} C \\ V \end{smallmatrix}\right]$. Hence,

$$V(A - KC)u = TVu \ \Rightarrow \ \lambda \cdot Vu = T \cdot Vu \tag{6.44}$$

holds, that is, $\lambda \in \sigma(T)$. This means that the eigenvalues of matrix T cannot be chosen arbitrarily. •

6.2.2.1 Gopinath Algorithm

The algorithm introduced here was proposed by Gopinath specifically for the design of minimal order observer:

1. Determine a matrix $D \in \mathbb{R}^{(n-p)\times n}$ such that the following matrix is nonsingular:
$$S := \begin{bmatrix} C \\ D \end{bmatrix}.$$
2. Carry out the following similarity transformation:
$$SAS^{-1} = \begin{bmatrix} A_{11} & A_{12} \\ A_{21} & A_{22} \end{bmatrix}.$$
3. Set $T = A_{22} + LA_{12}$ and find a matrix $L \in \mathbb{R}^{(n-p)\times p}$ such that $\sigma(T) = \Lambda$.
4. Set $V = D + LC, K = -AS^{-1}\left[\begin{smallmatrix} -I_p \\ L \end{smallmatrix}\right]$.

Now, we prove that the matrices (V, K, T) determined by this algorithm satisfy the conditions of Theorem 6.6. First of all, since C is a row full rank matrix, there is a matrix D such that S is nonsingular. Since $SS^{-1} = I$, it is easy to know that $CS^{-1} = [I_p \ 0]$. From it, we can prove that (A_{12}, A_{22}) is also observable based on the observability of (C, A) and its invariance to similarity transformation (Exercise 6.10). Hence, by using the pole placement method, we can obtain a gain matrix $L \in \mathbb{R}^{(n-p)\times p}$ satisfying $\sigma(A_{22} + LA_{12}) = \Lambda$. According to Gopinath algorithm, we have $V = [L \ I_{n-p}]S$ and $T = [L \ I_{n-p}]SAS^{-1}\left[\begin{smallmatrix} 0 \\ I_{n-p} \end{smallmatrix}\right]$, so it can be proved that

$$\begin{aligned} V(A - KC) &= [L \ I_{n-p}]SA + [L \ I_{n-p}]SAS^{-1}\begin{bmatrix} -I_p \\ L \end{bmatrix} C \\ &= [L \ I_{n-p}]SAS^{-1}\left(S + \begin{bmatrix} -I_p \\ L \end{bmatrix} C\right) \end{aligned}$$

$$= [L\ I_{n-p}]SAS^{-1}\begin{bmatrix} 0 \\ I_{n-p} \end{bmatrix}(D+LC)$$
$$= TV.$$

Finally

$$\begin{bmatrix} C \\ V \end{bmatrix} = \begin{bmatrix} I_p & 0 \\ L & I_{n-p} \end{bmatrix} S \;\Rightarrow\; \text{rank}\begin{bmatrix} C \\ V \end{bmatrix} = n$$

also holds. This proves the sufficiency of Theorem 6.6.

Example 6.7 *Design a minimal order observer for system*

$$\dot{x} = \begin{bmatrix} 1 & 1 & -2 \\ 0 & 1 & 1 \\ 0 & 0 & 1 \end{bmatrix} x + \begin{bmatrix} 1 \\ 0 \\ 1 \end{bmatrix} u, \quad y = [1 \quad 0 \quad 0]x.$$

Solution First, calculating the observability matrix, we get

$$\mathcal{O} = \begin{bmatrix} 1 & 0 & 0 \\ 1 & 1 & -2 \\ 1 & 2 & -3 \end{bmatrix}.$$

Its rank is 3, so this system is observable. In addition, the dimension of minimal order observer is $n - p = 3 - 1 = 2$. Here, we design an observer with poles $\{-4, -4\}$. Select the matrix D as

$$D = \begin{bmatrix} 0 & 1 & 0 \\ 0 & 0 & 1 \end{bmatrix} \;\Rightarrow\; S = \begin{bmatrix} C \\ D \end{bmatrix} = I_3.$$

Then, SAS^{-1} can be partitioned as

$$SAS^{-1} = \begin{bmatrix} A_{11} & A_{12} \\ A_{21} & A_{22} \end{bmatrix} = \left[\begin{array}{c|cc} 1 & 1 & -2 \\ \hline 0 & 1 & 1 \\ 0 & 0 & 1 \end{array}\right].$$

Let $L = [l_1\ l_2]^T$. We have

$$\det\,(sI - (A_{22} + LA_{12})) = s^2 + (2l_2 - l_1 - 2)s - (3l_2 - l_1 - 1).$$

On the other hand, the characteristic polynomial of T is $s^2 + 8s + 16$. Comparing the coefficients of both, we get

$$2l_2 - l_1 - 2 = 8,\ 3l_2 - l_1 - 1 = -16 \;\Rightarrow\; l_1 = -60,\ l_2 = -25.$$

Finally, computation following the Gopinath algorithm leads to

$$V = D + LC = \begin{bmatrix} -60 & 1 & 0 \\ -25 & 0 & 1 \end{bmatrix}, \quad T = A_{22} + LA_{12} = \begin{bmatrix} -59 & 121 \\ -25 & 51 \end{bmatrix}$$

$$VB = -\begin{bmatrix} 60 \\ 24 \end{bmatrix}, \quad VK = -\begin{bmatrix} 575 \\ 250 \end{bmatrix}.$$

Therefore, the minimal order observer is

$$\dot{z} = \begin{bmatrix} -59 & 121 \\ -25 & 51 \end{bmatrix} z - \begin{bmatrix} 575 \\ 250 \end{bmatrix} y - \begin{bmatrix} 60 \\ 24 \end{bmatrix} u$$

and the estimate of state x is given by

$$\overline{x} = \begin{bmatrix} C \\ V \end{bmatrix}^{-1} \begin{bmatrix} y \\ z \end{bmatrix} = \begin{bmatrix} 0 \\ z \end{bmatrix} + \begin{bmatrix} 1 \\ 60 \\ 25 \end{bmatrix} y.$$

(MATLAB drill: Do simulations with this minimal order observer.) ▽

6.3 Combined System and Separation Principle

When only the output y is measured, we consider how to realize the state feedback by using the state estimate of observer. Concretely, the input is constructed by

$$u = F\overline{x} + v. \tag{6.45}$$

Here, signal v denotes an external input (e.g., signals used for reference tracking). The closed-loop system composed of a state feedback and an observer is called a *combined system* (see Figure 6.5). In the sequel, we examine the property of closed-loop system for the cases of full-order observer and minimal order observer separately.

6.3.1 *Full-Order Observer Case*

In this case, the estimation error is $e = \overline{x} - x$. Select the state of closed-loop system as (x, e). From the control input u given in (6.45) as well as (6.1), we get

$$\dot{x} = Ax + BF\overline{x} + Bv = (A + BF)x + BFe + Bv.$$

Combining it together with (6.31), the state equation of closed-loop system is obtained as

$$\begin{bmatrix} \dot{x} \\ \dot{e} \end{bmatrix} = \begin{bmatrix} A + BF & BF \\ 0 & A + LC \end{bmatrix} \begin{bmatrix} x \\ e \end{bmatrix} + \begin{bmatrix} B \\ 0 \end{bmatrix} v. \tag{6.46}$$

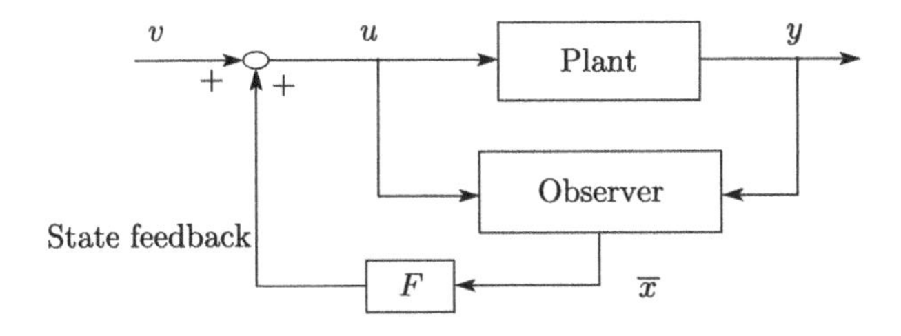

Figure 6.5 Structure of combined system

From this equation, we see that the poles of closed-loop system are the roots of the following characteristic polynomial:

$$\det\begin{bmatrix} sI-(A+BF) & -BF \\ 0 & sI-(A+LC)\end{bmatrix}$$
$$= \det(sI-(A+BF))\det(sI-(A+LC)). \tag{6.47}$$

Obviously, they are in the union of pole sets of the state feedback system and the observer. Therefore, as long as stable state feedback system and observer are designed, respectively, the stability of the whole closed-loop system is automatically guaranteed. That is, from the perspective of closed-loop poles, the designs of state feedback and observer are independent of each other and can be designed separately. This property, called the *separation principle*, is very important in system theory.

It should be remembered that the separation principle never means that good reference tracking and disturbance responses can be taken as granted so long as the responses of state feedback and observer are sufficiently fast. This is because the transient response depends not only on poles but also on zeros. For example, in the case of reference tracking, the transfer zeros of plant are also the zeros of closed-loop system. When there are stable or unstable zeros near the imaginary axis, if the closed-loop poles are too big, the output response will change sharply in the initial stage, which brings about a very big overshoot or undershoot. One may consider cancelling the zeros of plant with the poles of controller. But from the viewpoint of disturbance response, pole–zero cancellation near the imaginary axis is not allowed.

Finally, substitution of $u = F\overline{x}$ into (6.29) yields

$$\dot{\overline{x}} = (A+BF+LC)\overline{x} - Ly.$$

So, the dynamic output feedback controller of $y \mapsto u$ becomes

$$K(s) = \left[\begin{array}{c|c} A+BF+LC & -L \\ \hline F & 0 \end{array}\right]. \tag{6.48}$$

This $K(s)$ is the dynamic output controller formed by the combination of state feedback and observer.

Example 6.8 *In the two-mass–spring system, we combine the state feedback and the observer designed in Examples 6.5 and 6.6 to form the closed-loop system. Imposing a unit impulse torque disturbance on the load, the responses of motor angular velocity ω_M and load angular velocity ω_L obtained via simulation are shown in Figure 6.6.*

6.3.2 *Minimal Order Observer Case*

In this case, from the estimation error $e = \overline{z} - z$, we see that the estimated signal $\overline{z}$ is $\overline{z} = z + e$. For simplicity, let

$$\begin{bmatrix} C \\ V \end{bmatrix}^{-1} = [G \quad H]. \tag{6.49}$$

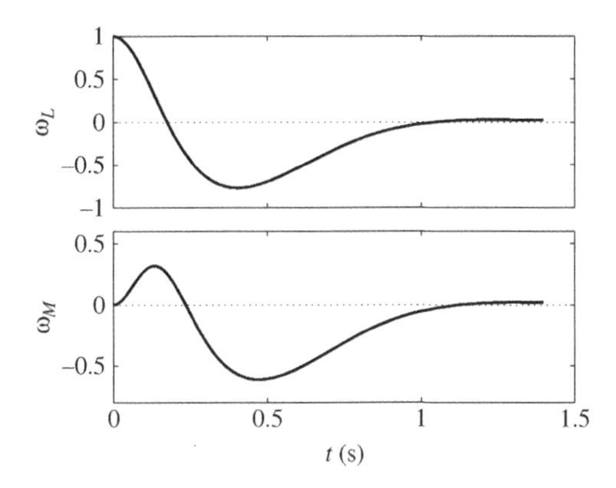

Figure 6.6 Impulse response of combined system

Then, according to (6.42), (6.35), the state estimate $\overline{x}$ can be written as

$$\overline{x} = \begin{bmatrix} C \\ V \end{bmatrix}^{-1} \begin{bmatrix} y \\ \overline{z} \end{bmatrix} = \begin{bmatrix} C \\ V \end{bmatrix}^{-1} \begin{bmatrix} y \\ z + e \end{bmatrix} = x + He. \tag{6.50}$$

Therefore, after the substitution of input $u = F\overline{x} + v$, the state equation of x becomes

$$\dot{x} = Ax + BF\overline{x} + Bv = (A + BF)x + BFHe + Bv.$$

Combined with the state equation $\dot{e} = Te$ of estimation error, state equation of the closed-loop system is obtained as

$$\begin{bmatrix} \dot{x} \\ \dot{e} \end{bmatrix} = \begin{bmatrix} A + BF & BFH \\ 0 & T \end{bmatrix} \begin{bmatrix} x \\ e \end{bmatrix} + \begin{bmatrix} B \\ 0 \end{bmatrix} v. \tag{6.51}$$

Therefore, even in the case of minimal order observer, poles of the closed-loop system are still equal to the union of the sets of state feedback system poles and observer poles. That is, the separation principle is still true.

Further, from $\overline{x} = Gy + H\overline{z}$, we know that the input u can be expressed as

$$u = F\overline{x} = FH\overline{z} + FGy. \tag{6.52}$$

Substituting this equation into the minimal order observer (6.41), its state equation turns into

$$\dot{\overline{z}} = T\overline{z} + VKy + VBu = (T + VBFH)\overline{z} + V(K + BFG)y.$$

As such, we have obtained the dynamic output controller consisting of state feedback and minimal order observer:

$$K(s) = \left[\begin{array}{c|c} T + VBFH & V(K + BFG) \\ \hline FH & FG \end{array}\right] \tag{6.53}$$

whose order is $(n - p)$.

Exercises

6.1 Following the procedure below to prove that, when (A, B) is uncontrollable, there must be a similarity transformation converting (A, B) into

$$A = T\begin{bmatrix} A_1 & A_{12} \\ 0 & A_2 \end{bmatrix} T^{-1}, \quad B = T\begin{bmatrix} B_1 \\ 0 \end{bmatrix}.$$

(a) Assume that $q_1, \ldots, q_k$ are the linearly independent columns of the controllability matrix $\mathcal{C}$. Use Cayley–Hamilton theorem to prove that Im $\mathcal{C}$ is A-invariant.

(b) Prove that there is a matrix A_1 satisfying $A[q_1 \ \cdots \ q_k] = [q_1 \ \cdots \ q_k]A_1$ based on the property of A-invariant space.

(c) Choose real vectors $q_{k+1}, \ldots, q_n$ as such that matrix $T = [q_1 \ \cdots \ q_k \ q_{k+1} \ \cdots \ q_n]$ is nonsingular. Prove that T is the transformation matrix satisfying the property above.

6.2 For the plant

$$\dot{x} = \begin{bmatrix} 0 & 1 \\ 3 & -2 \end{bmatrix} x + \begin{bmatrix} 0 \\ 1 \end{bmatrix} u, \ \ y = [1, 0]x,$$

we hope to design a state feedback $u = fx$ to place the closed-loop system poles at $\{-2, -3\}$. Compute the state feedback gain f.

6.3 Answer the following questions about linear system

$$\dot{x} = \begin{bmatrix} 0 & 1 \\ -2 & 0 \end{bmatrix} x + \begin{bmatrix} 0 \\ 1 \end{bmatrix} u, \ \ y = [0 \ 1]x.$$

(a) Design a state feedback $u(t) = fx(t)$ such that the closed-loop poles are $\{-1, -2\}$.

(b) Discuss whether this pole placement problem can be realized by a static output feedback $u(t) = ky(t)$.

6.4 Consider the following linear system

$$\dot{x} = \begin{bmatrix} 2 & 1 & 0 \\ 0 & 2 & 0 \\ 0 & 0 & -1 \end{bmatrix} x + \begin{bmatrix} 0 \\ 1 \\ 0 \end{bmatrix} u.$$

(a) Can you place the closed-loop poles at $\{-2, -1 \pm j1\}$ by a state feedback $u = fx$? Explain the reason briefly.

(b) Can you place the closed-loop poles at $\{-1, -1 \pm j1\}$ by a state feedback $u = fx$? If you can, compute the state feedback gain f.

(c) When the input vector is changed to $b = [0 \ 1 \ 1]^T$, find the state feedback gain f that places the closed-loop poles at $\{-2, -1 \pm j1\}$.

6.5 Given a linear system

$$\dot{x}(t) = \begin{bmatrix} 2 & -1 \\ 1 & 2 \end{bmatrix} x(t) + \begin{bmatrix} 0 \\ 1 \end{bmatrix} u(t), \ \ y(t) = [1 \ 0]x(t).$$

(a) Compute the poles of system and check its stability.
(b) Check the controllability and observability.
(c) Find a state feedback $u(t) = fx(t)$ such that the closed-loop poles are placed at $\{-1, -2\}$.
(d) Design a full-order observer with poles $\{-3, -4\}$.
(e) Design a minimal order observer with a pole -3.

6.6 Discuss whether the system

$$\dot{x} = \begin{bmatrix} 0 & 1 \\ 1 & 0 \end{bmatrix} x + \begin{bmatrix} 0 \\ 1 \end{bmatrix} u, \quad y = [1 \ 1]x$$

can be stabilized by a static output feedback $u = ky$.

6.7 Given a linear system

$$\dot{x} = \begin{bmatrix} 0 & 1 \\ 0 & 0 \end{bmatrix} x + \begin{bmatrix} 0 \\ 1 \end{bmatrix} u, \quad y = [1 \ 0]x.$$

(a) When all states are measured, we hope to use state feedback $u = fx$ to place the poles of closed-loop system at $\{-1, -2\}$. Compute the feedback gain f.
(b) When only the output is measured, we want to use a full-order observer to estimate the state $x(t)$. Design an observer with poles of $\{-4, -5\}$.

6.8 In order to realize the state feedback control in Exercise 6.4(c), we want to design an observer.
(a) When the output equation is $y = [0 \ 0 \ 1]x$, use block diagram to explain whether a full-order observer can be designed.
(b) When the output equation is $y = [1 \ 0 \ 1]x$, design a full-order observer with poles $-4, -2 \pm j2$.
(c) When the output equation is $y = [1 \ 0 \ 1]x$, design a minimal order observer with poles $-2 \pm j2$.

6.9 Applying the state feedback $u = Fx$ to an MIMO system $\dot{x} = Ax + Bu + Dw$, $y = Cx$, the closed-loop system becomes

$$\dot{x}(t) = (A + BF)x(t) + Dw(t), \quad y(t) = Cx(t).$$

In this closed-loop system, prove first that for any disturbance $w(t)$ the output response is identically zero iff

$$C(sI - A - BF)^{-1}D \equiv 0.$$

Then, prove that this condition is equivalent to

$$CD = C(A + BF)D = \cdots = C(A + BF)^{n-1}D = 0.$$

Finally, consider an SISO system. When the relative degree of the transfer function of $u \mapsto y$ is r, prove that there is a state feedback gain f satisfying the condition iff

$$cd = cAd = cA^2d = \cdots = cA^{r-1}d = 0.$$

(This implies that the transfer function of $d \mapsto y$ has a relative degree higher than the transfer function of $u \mapsto y$. That is, the input u reaches the output y faster than the disturbance d, which explains why the disturbance can be perfectly cancelled.)

6.10 In Gopinath algorithm, prove that (A_{12}, A_{22}) is also observable when (C, A) is observable.

Notes and References

For more details on the design methods of state feedback and observer, refer to Refs [12, 44] and [2].

7

Parametrization of Stabilizing Controllers

In the design of control systems, a prerequisite is to ensure the stability of closed-loop system. In traditional control theories, no matter the classical control theory or the modern control theory, the central issue is to design a *single controller* capable of controlling the plant. However, in control system design we often hope to optimize the control performance in certain sense. This optimization problem basically boils down to shaping the closed-loop transfer matrix. At the same time, the stability of closed-loop system has to be ensured. However, in performance optimization it becomes an obstacle to ensure the stability of system. Therefore, a question arises naturally: is it possible to describe all controllers that stabilize the plant by a formula with a free parameter? If possible, the problem of performance optimization reduces to the optimization of the free parameter, and it may be expected that the optimization problem is greatly simplified.

This chapter proves that this expectation is possible. Parametrization of stabilizing controllers is a great progress in control theory. Here, we first introduce the notion of generalized closed-loop control system and several applications. Then, we show a formula for all controllers that can stabilize a fixed plant. This formula contains a free parameter, so it is called the *parametrization* of the controllers. After that, we will analyze the structures of the closed-loop transfer matrix and state-space realization in detail. At last, features of 2-degree-of-freedom systems are enclosed.

Robust Control: Theory and Applications, First Edition. Kang-Zhi Liu and Yu Yao.

7.1 Generalized Feedback Control System

7.1.1 Concept

Recall the two-mass–spring system in Example 1.2, where the state vector is $x = [\omega_L\ \phi\ \omega_M]^T$ and the state equation is

$$\dot{x} = \begin{bmatrix} -\dfrac{D_L}{J_L} & \dfrac{k}{J_L} & 0 \\ -1 & 0 & 1 \\ 0 & -\dfrac{k}{J_M} & -\dfrac{D_M}{J_M} \end{bmatrix} x + \begin{bmatrix} \dfrac{1}{J_L} \\ 0 \\ 0 \end{bmatrix} d + \begin{bmatrix} 0 \\ 0 \\ \dfrac{1}{J_M} \end{bmatrix} u$$

$$y_P = [0 \quad 0 \quad 1]x. \tag{7.1}$$

The performance specification is to suppress the influence of load torque disturbance d and ensure that ω_L tracks the reference input r. In this problem, the output that needs to be controlled is the speed error $r - \omega_L$ of load which is different from the measured signal ω_M; also, the torque disturbance d is different from the control input u in their properties and locations where they enter the system. For such control problems, in order to optimize the disturbance (or reference tracking) response directly in control design, new input/output description is needed. The new description introduced for this purpose is shown in Figure 7.1. In this figure, $G(s)$ contains the plant, signals for performance optimization and weighting functions, and is called the *generalized plant*. K is the controller. Further, we will use the following terms:

- z is the output vector used for specifying the control performance and model uncertainty[1], called the *performance output*.
- y is the input vector of the controller (e.g., outputs of sensors and tracking errors), called the *measured output*.
- w is the external input vector used for specifying the control performance and model uncertainty, called the *disturbance*.
- u is the command vector of actuators, called the *control input*.

The input/output relationships of the generalized plant and controller are, respectively,

$$\begin{bmatrix} \hat{z}(s) \\ \hat{y}(s) \end{bmatrix} = G(s) \begin{bmatrix} \hat{w}(s) \\ \hat{u}(s) \end{bmatrix} \tag{7.2}$$

$$\hat{u}(s) = K(s)\hat{y}(s). \tag{7.3}$$

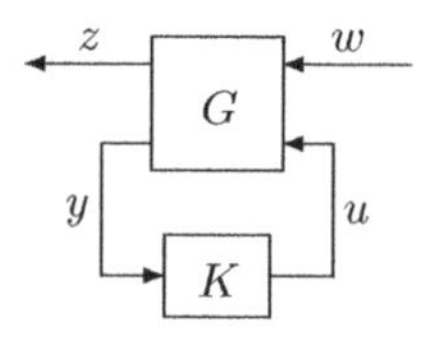

Figure 7.1 Generalized feedback system

[1] Refer to Chapter 11 for model uncertainty.

It should be noted that, the so-called measured output in the generalized feedback system is the input of controller $K(s)$, not necessarily the output of sensor. For example, in the reference tracking problem of one-degree-of-freedom (1-DOF in short) systems, the tracking error $r - y_P$ is the input of controller (y_P denotes the output of plant) which is the so-called measured output y; meanwhile, in the two-degree-of-freedom (2-DOF in short) system to be introduced next, the measured output y is the signal vector $\begin{bmatrix} r \\ y_P \end{bmatrix}$.

Let the state equation of generalized plant $G(s)$ be

$$\begin{bmatrix} \dot{x} \\ z \\ y \end{bmatrix} = \begin{bmatrix} A & B_1 & B_2 \\ C_1 & D_{11} & D_{12} \\ C_2 & D_{21} & 0 \end{bmatrix} \begin{bmatrix} x \\ w \\ u \end{bmatrix} \tag{7.4}$$

and partition the transfer matrix of generalized plant as

$$G(s) = \begin{bmatrix} G_{11} & G_{12} \\ G_{21} & G_{22} \end{bmatrix} = \left[\begin{array}{c|cc} A & B_1 & B_2 \\ \hline C_1 & D_{11} & D_{12} \\ C_2 & D_{21} & 0 \end{array}\right] \tag{7.5}$$

corresponding to the dimensions of input $\begin{bmatrix} w \\ u \end{bmatrix}$ and output $\begin{bmatrix} z \\ y \end{bmatrix}$. Then, the closed-loop transfer matrix of $w \mapsto z$ equals

$$H_{zw}(s) = G_{11} + G_{12}K(I - G_{22}K)^{-1}G_{21}. \tag{7.6}$$

In fact, not only the design of feedback control systems but also the design of feedforward systems like filters as well as the design of 2-DOF control systems can be handled in this framework. In the sequel, we will introduce a few examples: one is to derive the generalized plant based on transfer function; the rest are based on state space.

7.1.2 Application Examples

Example 7.1 (2-DOF control system) *In reference tracking problems, signals that can be used in control are the plant output y_P and the reference input r. However, in the traditional control structure shown in Figure 5.1 feedbacked is the difference $r - y_P$ of signals. This kind of control structure is called* 1-DOF control system . *1-DOF control system does not make full use of the available information, so it is difficult to improve the tracking performance. On the contrary, control structure with not only the feedback of y_P but also the feedforward of r is called* 2-DOF control system . *Its general structure is shown in Figure 7.2*

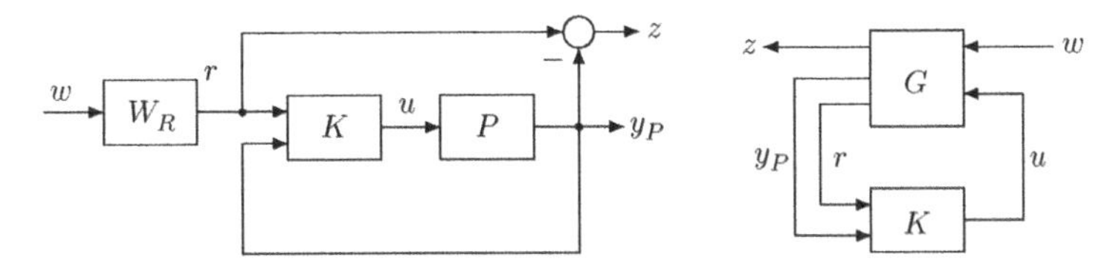

Figure 7.2 Reduction of 2-DOF system to generalized feedback system

where $K(s) = [K_F \quad K_B]$ contains two blocks $K_F(s)$ and $K_B(s)$, which correspond to the feedforward signal r and the feedback signal y_P, respectively.

Let the model of reference input be $W_R(s)$. Since the control purpose is to reduce the tracking error as small as possible, we choose the performance output as the tracking error $z = r - y_P$. Besides, noting that the disturbance is the impulse input w of $W_R(s)$ and the measured output is $\begin{bmatrix} r \\ y_P \end{bmatrix}$, we get the input/output relation:

$$\begin{bmatrix} \hat{z} \\ \hline \hat{r} \\ \hat{y}_P \end{bmatrix} = G(s) \begin{bmatrix} \hat{w} \\ \hat{u} \end{bmatrix} = \left[\begin{array}{c|c} W_R & -P \\ \hline W_R & 0 \\ 0 & P \end{array}\right] \begin{bmatrix} \hat{w} \\ \hat{u} \end{bmatrix} \tag{7.7}$$

$$\hat{u} = K(s) \begin{bmatrix} \hat{r} \\ \hat{y}_P \end{bmatrix} = K_F \hat{r} + K_B \hat{y}_P \tag{7.8}$$

in which the dashed line is used to distinguish input/output signals with different properties as well as the corresponding matrix blocks.

Example 7.2 *Consider specifically the generalized plant in the load torque disturbance control problem of two-mass–spring system. For convenience, denote the coefficient matrices of the state-space model in Section 7.1.1 with A, b_1, b_2, c_2, respectively. Then*

$$\begin{aligned} \dot{x} &= Ax + b_1 d + b_2 u \\ y_P &= c_2 x. \end{aligned} \tag{7.9}$$

According to the specification, we select the tracking error of load speed as the performance output:

$$z = r - x_1 = [-1 \quad 0 \quad 0]x + r = c_1 x + r.$$

Since the reference input r and motor speed y_P are known, the measured output is (2-DOF control)

$$y = \begin{bmatrix} r \\ y_P \end{bmatrix}.$$

In this problem, the disturbance[2] is $\overline{w} = [r \ \ d]^T$. Denote by $P(s)$ the generalized plant from $[\overline{w}^T \quad u]^T$ to $[z \ \ y^T]^T$. Reorganizing the preceding state equations according to the input and output of $P(s)$, we obtain the state-space model of $P(s)$:

$$\begin{bmatrix} \dot{x} \\ \hline z \\ \hline r \\ y_P \end{bmatrix} = \left[\begin{array}{c|cc|c} A & 0 & b_1 & b_2 \\ \hline c_1 & 1 & 0 & 0 \\ \hline 0 & 1 & 0 & 0 \\ c_2 & 0 & 0 & 0 \end{array}\right] \begin{bmatrix} x \\ \hline r \\ d \\ \hline u \end{bmatrix}. \tag{7.10}$$

[2] From the angle of tracking error z, the reference input r is equivalent to a disturbance.

Further, in order to take the dynamics of reference input into account, we denote its model by $W_R(s)$, while the disturbance model is set as $W_D(s)$. Substituting their input/output relationships

$$\hat{r}(s) = W_R(s)\hat{w}_1(s), \quad \hat{d}(s) = W_D(s)\hat{w}_2(s), \tag{7.11}$$

we eventually get the generalized plant of $[w_1 \quad w_2 \quad u]^T \mapsto [z \ y^T]^T$ *as*

$$G(s) = P(s) \times \text{diag}(W_R(s) \quad W_D(s) \quad 1). \tag{7.12}$$

Example 7.3 (Filter design) *Let us consider the problem of estimating a signal q consisting of some states (Figure 7.3) from the plant input and output. Assume that noise n exists in the plant and the state equation of plant is*

$$\begin{aligned} \dot{x} &= Ax + B_1 n + B_2 u \\ y_P &= Cx + D_1 n + D_2 u \\ q &= Hx. \end{aligned}$$

Here, we hope to process the known input/output signals (u, y_P) appropriately with a filter $F(s)$ so as to obtain an estimate $\overline{q}$ of signal q. As a rule of filter design, the estimation error $z = q - \overline{q}$ should be as small as possible. In terms of generalized feedback system, (n, u) is the disturbance, estimation error z is the performance output, (u, y_P) is the measured output, and $\overline{q}$ is the control input. The input/output relationship of the generalized feedback system is

$$\begin{bmatrix} \hat{z} \\ \hat{y}_P \\ \hat{u} \end{bmatrix} = \left[\begin{array}{c|cc:c} A & B_1 & B_2 & 0 \\ \hline H & 0 & 0 & -I \\ C & D_1 & D_2 & 0 \\ 0 & 0 & I & 0 \end{array}\right] \begin{bmatrix} \hat{n} \\ \hat{u} \\ \hat{\overline{q}} \end{bmatrix} = P(s) \begin{bmatrix} \hat{n} \\ \hat{u} \\ \hat{\overline{q}} \end{bmatrix} \tag{7.13}$$

$$\hat{\overline{q}} = F(s) \begin{bmatrix} \hat{y}_P \\ \hat{u} \end{bmatrix}. \tag{7.14}$$

Furthermore, when the noise n is a colored one with dynamics $W_n(s)$, then $\hat{n}(s) = W_n(s)\hat{w}(s)$ in which w is a white noise. Substitution of this relation yields the generalized plant with weighting function for the filter problem:

$$\begin{bmatrix} \hat{z} \\ \hat{y}_P \\ \hat{u} \end{bmatrix} = G \begin{bmatrix} \hat{w} \\ \hat{u} \\ \hat{\overline{q}} \end{bmatrix}, \quad G = P \begin{bmatrix} W_n & & \\ & I & \\ & & I \end{bmatrix}. \tag{7.15}$$

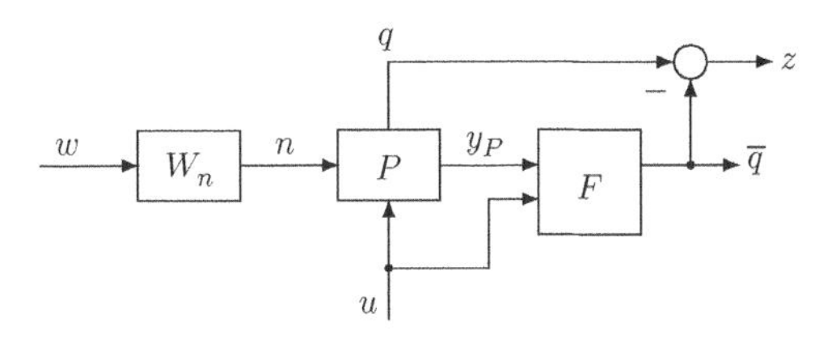

Figure 7.3 Filtering problem

From these examples, we conclude that any control problem can be converted to a control problem of the generalized feedback system in Figure 7.1. Therefore, we only analyze the system in Figure 7.1 hereafter.

7.2 Parametrization of Controllers

In Chapter 6, we discussed the issue of stabilizing a plant by using state feedback and observer and obtained a dynamic output feedback controller given by (6.48). However, in addition to this specific controller, there exist a lot of stabilizing controllers. The goal of this section is to derive a formula for all stabilizing controllers. It is called the parametrization of controllers because this formula contains a free parameter.

First of all, we show a basic property of the generalized feedback system.

Lemma 7.1 *$K(s)$ stabilizes $G(s)$ iff $K(s)$ stabilizes $G_{22}(s) = (A, B_2, C_2, 0)$.*

Proof. It suffices to prove that these two closed-loop systems have the same A matrix. Consider the closed-loop system (G_{22}, K) first. The state of G_{22} is x and the corresponding state equation is

$$\dot{x} = Ax + B_2u, \quad y = C_2x. \tag{7.16}$$

Further, the state equation of controller $K(s)$ is given by

$$\dot{x}_K = A_Kx_K + B_Ky, \quad u = C_Kx_K + D_Ky. \tag{7.17}$$

Substituting $y = C_2x$ into this equation, we get

$$\dot{x}_K = B_KC_2x + A_Kx_K, \quad u = D_KC_2x + C_Kx_K.$$

Then, substitution of this u into the equation of $\dot{x}$ gives

$$\dot{x} = (A + B_2D_KC_2)x + B_2C_Kx_K.$$

Therefore, the state equation of closed-loop system (G_{22}, K) becomes

$$\begin{bmatrix} \dot{x} \\ \dot{x}_K \end{bmatrix} = \begin{bmatrix} A + B_2D_KC_2 & B_2C_K \\ B_KC_2 & A_K \end{bmatrix} \begin{bmatrix} x \\ x_K \end{bmatrix}. \tag{7.18}$$

The difference between $G(s)$ and $G_{22}(s)$ is that a term B_1w about the disturbance is added to the equation of $\dot{x}$. But the disturbance term does not affect the A matrix of closed-loop system. So they are the same. •

Note that even if the state-space realization (7.4) of $G(s)$ is a minimal realization, the state-space realization $(A, B_2, C_2, 0)$ of $G_{22}(s)$ is not necessarily minimal.

7.2.1 *Stable Plant Case*

Let us consider stable $G(s)$ first. In this case, the formula of stabilizing controllers is particularly concise.

Theorem 7.1 *Assume that $G(s)$ is stable. Then, all stabilizing controllers are parameterized by*

$$K(s) = Q(I + G_{22}Q)^{-1} \tag{7.19}$$

in which $Q(s)$ is an arbitrary stable matrix with suitable dimension.

Proof. According to Lemma 7.1, we need only prove that $K(s)$ stabilizes $G_{22}(s)$. From the stability of $G(s)$, we see that $G_{22}(s)$ is also stable. By the definition of internal stability, if all the following four transfer matrices

$$(I - G_{22}K)^{-1}, \quad K(I - G_{22}K)^{-1}, \quad G_{22}K(I - G_{22}K)^{-1}, \quad (I - G_{22}K)^{-1}G_{22}$$

are stable, the closed-loop system (G_{22}, K) is internally stable. With the controller substituted, we have that these four transfer matrices are equal to

$$I + G_{22}Q, \ Q, \ G_{22}Q, \ (I + G_{22}Q)G_{22},$$

respectively. They are certainly stable. Conversely, when $K(s)$ is a stabilizing controller, $K(I - G_{22}K)^{-1} := Q(s)$ must be stable. Solving for $K(s)$, we see that it is described by $K(s) = Q(I + G_{22}Q)^{-1}$. •

Further, the next corollary follows immediately from this theorem. It corresponds to the case of $G_{22}(s) = -P(s)$.

Corollary 7.1 *Assume that the plant $P(s)$ in Figure 7.4 is stable. Then all controllers that stabilize the closed-loop system are parameterized by*

$$K(s) = Q(I - PQ)^{-1}$$

in which $Q(s)$ is any stable matrix with appropriate dimension.

Example 7.4 *Consider the single-input single-output (SISO) feedback system in Figure 7.4 where $P(s)$ is stable. Find all controllers that enable the asymptotic tracking of step reference input r.*

Solution Laplace transform of the tracking error is

$$\hat{e}(s) = \hat{r}(s) - \hat{y}(s) = \frac{1}{1 + PK}\hat{r}(s) = \frac{1}{1 + PK}\frac{1}{s}.$$

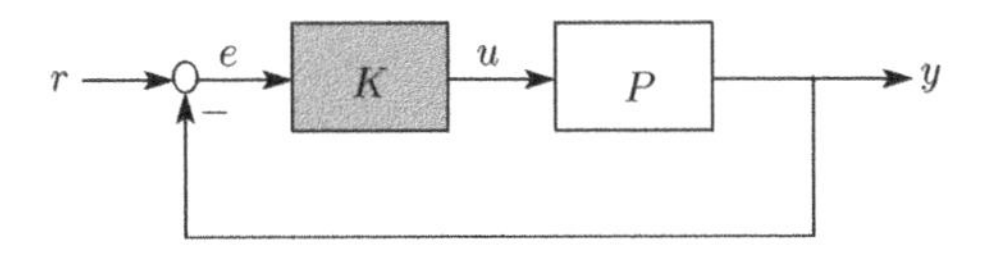

Figure 7.4 1-DOF feedback system

Substitution of the parametrization $K(s) = Q/(1 - PQ)$ leads to $\hat{e}(s) = (1 - PQ)\frac{1}{s}$. It follows from the final value theorem of Laplace transform that

$$e(\infty) = \lim_{s \to 0} s\hat{e}(s) = 1 - P(0)Q(0).$$

For $e(\infty) = 0$ to hold, there must be

$$P(0) \neq 0, \quad Q(0) = \frac{1}{P(0)}.$$

So, all controllers that can ensure the asymptotic tracking are given by the set

$$\left\{ K(s) = \frac{Q}{1 - PQ} \,\middle|\, Q \text{ is } \textit{stable and } Q(0) = \frac{1}{P(0)} \right\}.$$

In addition,

$$K(0) = \lim_{s \to 0} \frac{Q}{1 - PQ} \to \infty$$

holds. So, $K(s)$ contains at least one integrator $1/s$, which agrees with the internal model principle.

For instance, for the plant

$$P(s) = \frac{1}{(s+1)(s+2)},$$

one of the controllers that ensures the asymptotic tracking of step reference input is obtained as

$$K(s) = \frac{2(s+1)(s+2)}{s(s+3)}$$

when the free parameter is selected as $Q = 1/P(0) = 2$. ▽

Remark 7.1 *When we treat this example in the framework of generalized feedback system, the generalized plant with weighting function cannot be stabilized because the weighting function $1/s$ is not stable. So the result of this chapter cannot be applied directly. But a careful observation shows that there is no need to stabilize the state of weighting function; what is needed is to stabilize the real feedback system (P, K) and to ensure the stability of the closed-loop transfer function $\frac{1}{1+PK}\frac{1}{s}$ weighted by unstable function. This stability property is named as the comprehensive stability . The condition needed for comprehensive stability and the parametrization of its solutions are discussed in Refs [60, 61].*

Example 7.5 *Consider the SISO feedback system in Figure 7.5. Assume that $P(s)$ is stable and $P(0) \neq 0$. Find all controllers that are capable of asymptotic rejection of step disturbance d. Further, for $P(s) = 1/(s+1)$, select the free parameter as $Q(s) = P^{-1}(s)\frac{k}{1+\epsilon s}$ $(\epsilon > 0)$ and design a controller satisfying $\|y\|_2 \leq 0.1$.*

Solution Laplace transform of the disturbance response is

$$\hat{y}(s) = \frac{P}{1 + PK}\hat{d}(s) = \frac{P}{1 + PK}\frac{1}{s}.$$

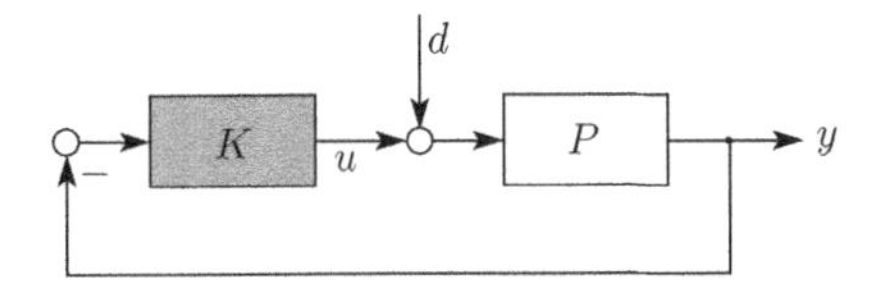

Figure 7.5 Disturbance control

Substitution of stabilizing controller $K = Q/(1 - PQ)$ yields

$$\hat{y}(s) = P(1 - PQ)\frac{1}{s}. \tag{7.20}$$

Similar to Example 7.4, from the final value theorem, it can be derived that all controllers guaranteeing zero steady-state output w.r.t. step disturbance d are given by

$$\left\{K = \frac{Q}{1 - PQ} : Q(s) \text{ is stable and } Q(0) = \frac{1}{P(0)}\right\}. \tag{7.21}$$

Each controller $K(s)$ contains at least one integrator $1/s$.

To ensure that the output norm $\|y\|_2$ of $P(s) = 1/(s+1)$ is bounded, $y(\infty) = 0$ must be satisfied. So, $k = Q(0) = 1/P(0) = 1$. Then, we obtain

$$\hat{y}(s) = \frac{\epsilon}{(s+1)(\epsilon s + 1)} = \frac{\epsilon}{1-\epsilon}\left(\frac{1}{s+1} - \frac{1}{s + 1/\epsilon}\right)$$
$$\Rightarrow\ y(t) = \frac{\epsilon}{1-\epsilon}(e^{-t} - e^{-t/\epsilon}), \quad t \geq 0$$

via substitution of the given $P(s), Q(s)$ into (7.20). So,

$$\|y\|_2^2 = \int_0^\infty y^2(t)\,dt = \frac{\epsilon^2}{2(1+\epsilon)} \leq 0.1^2 \Rightarrow \epsilon^2 - 0.02\epsilon - 0.02 \leq 0.$$

The solution of this inequality is $-0.131 \leq \epsilon \leq 0.151$. Considering the stability condition $\epsilon > 0$ on the free parameter Q, the range of solution finally obtained is $0 < \epsilon \leq 0.151$. Computing the corresponding controller, we get the following PI compensator:

$$K(s) = \frac{s+1}{\epsilon s} = \frac{1}{\epsilon} + \frac{1}{\epsilon s}.$$

▽

7.2.2 *General Case*

When G is not stable, the formula of its stabilizing controllers gets rather complicated. There are two kinds of methods on the parametrization: one using coprime factorization [90] over the ring of stable rational matrices and another using the state-space method directly. Here, we illustrate the latter.

Theorem 7.2 *Suppose that (A, B_2) is stabilizable and (C_2, A) is detectable. Let matrices F and L be such that stabilize $A + B_2F$ and $A + LC_2$, respectively. Then, all controllers*

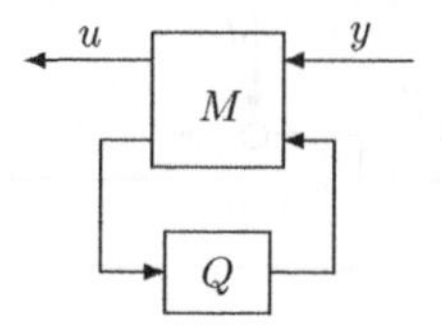

Figure 7.6 Parametrization of stabilizing controllers

stabilizing the generalized plant $G(s)$ of (7.4) are given by the transfer matrix $\mathcal{F}_\ell(M,Q)$ from y to u in Figure 7.6, where $Q(s)$ is any stable matrix with an appropriate dimension, and the coefficient matrix $M(s)$ is

$$M(s)=\left[\begin{array}{c|cc} A+B_2F+LC_2 & -L & B_2 \\ \hline F & 0 & I \\ -C_2 & I & 0 \end{array}\right].$$

Proof. The sufficiency can be proved through substitution of the stable transfer matrix $Q(s)=(A_Q,B_Q,C_Q,D_Q)$ and calculation of the closed-loop system A matrix. First of all, according to the linear fractional transformation (LFT) interconnection formula (4.63), the state-space realization of $K(s)=\mathcal{F}_\ell(M,Q)$ is obtained as

$$\begin{aligned}\mathcal{F}_\ell(M,Q) &= (A_K,\ B_K,\ C_K,\ D_K)\\ &= \left[\begin{array}{cc|c} A+B_2F+LC_2-B_2D_QC_2 & B_2C_Q & B_2D_QC_Q-L \\ -B_QC_2 & A_Q & B_Q \\ \hline F-D_QC_2 & C_Q & D_Q \end{array}\right].\end{aligned} \tag{7.22}$$

Substituting it into the transfer matrix of closed-loop system $H_{zw}=\mathcal{F}_\ell(G,K)$, and applying the LFT interconnection formula once again, we obtain the A matrix of closed-loop system as

$$A_c=\begin{bmatrix} A+B_2D_QC_2 & B_2F-B_2D_QC_2 & B_2C_Q \\ B_2D_QC_2-LC_2 & A+B_2F+LC_2-B_2D_QC_2 & B_2C_Q \\ B_QC_2 & -B_QC_2 & A_Q \end{bmatrix}. \tag{7.23}$$

To examine its stability, we try to convert A_c into a block triangular matrix via similarity transformation. The following similarity transformation is performed: add the second column to the first, subtract the first row from the second, and then exchange the second and third columns as well as the second and third rows[3]. Finally, we have that A_c is similar to the block triangular matrix:

$$\begin{bmatrix} A+B_2F & B_2C_Q & B_2F-B_2D_QC_2 \\ 0 & A_Q & -B_QC_2 \\ 0 & 0 & A+LC_2 \end{bmatrix}. \tag{7.24}$$

This matrix obviously is stable.

---

[3] $T_1=\begin{bmatrix} I & 0 & 0\\ I & I & 0\\ 0 & 0 & I\end{bmatrix}$, $T_2=\begin{bmatrix} I & 0 & 0\\ 0 & 0 & I\\ 0 & I & 0\end{bmatrix}$ are the transformation matrices of these two similarity transformations.

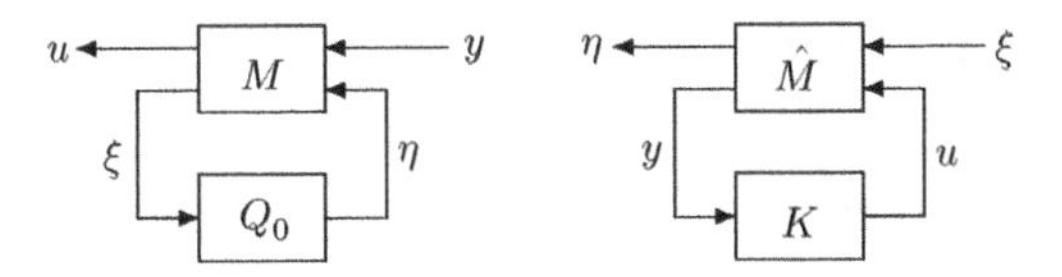

Figure 7.7 Input/output relations of $K = \mathcal{F}_\ell(M, Q_0)$ and $Q_0 = \mathcal{F}_\ell(\hat{M}, K)$

To prove the necessity, we need just prove that any stabilizing controller $K(s)$ can be described as $K(s) = \mathcal{F}_\ell(M, Q_0)$ with a stable $Q_0(s)$. Hence, we consider calculating $Q_0(s)$ from this formula and then verify its stability. According to the input/output relation in Figure 7.7, there hold

$$\begin{bmatrix} \hat{u} \\ \hat{\xi} \end{bmatrix} = M(s) \begin{bmatrix} \hat{y} \\ \hat{\eta} \end{bmatrix}, \quad \hat{u} = K(s)\hat{y}; \quad \begin{bmatrix} \hat{\eta} \\ \hat{y} \end{bmatrix} = \hat{M}(s) \begin{bmatrix} \hat{\xi} \\ \hat{u} \end{bmatrix}, \quad \hat{\eta} = Q_0(s)\hat{\xi}.$$

From these equations, we get that the relationship between $\hat{M}$ and M is

$$\hat{M}(s) = \begin{bmatrix} & I \\ I & \end{bmatrix} M^{-1} \begin{bmatrix} & I \\ I & \end{bmatrix}.$$

Calculation based on the formula of inverse system in Section 4.1.8 gives the state realization of $\hat{M}(s)$:

$$\hat{M}(s) = \left[\begin{array}{c|cc} A & -L & B_2 \\ \hline -F & 0 & I \\ C_2 & I & 0 \end{array}\right].$$

$\hat{M}(s)$ and $G(s)$ share the same (2, 2) block, namely, $\hat{M}_{22}(s) = G_{22}(s) = C_2(sI - A)^{-1}B_2$. So they both are stabilized by $K(s)$ (Lemma 7.1). Therefore, $Q_0(s) := \mathcal{F}_\ell(\hat{M}, K)$ is stable. •

Example 7.6 *Consider the stabilization of integrator $P(s) = 1/s := G_{22}$. One of its state realizations is $(0, 1, 1, 0)$. When $F = L = -1$ are chosen, $A + B_2F = A + LC_2 = -1$ are stable. From the coefficient matrix*

$$M(s) = \left[\begin{array}{c|cc} -2 & 1 & 1 \\ \hline -1 & 0 & 1 \\ -1 & 1 & 0 \end{array}\right] = \frac{1}{s+2}\begin{bmatrix} -1 & s+1 \\ s+1 & -1 \end{bmatrix},$$

we get the parametrization of stabilizing controllers:

$$K(s) = -\frac{1}{s+2} + \left(\frac{s+1}{s+2}\right)^2 Q(s)\left(1 + \frac{1}{s+2}Q(s)\right)^{-1}.$$

When $Q(s) = 0$, the controller is $K(s) = -1/(s+2)$. The characteristic polynomial of the closed-loop system is equal to the numerator polynomial $s(s+2) + 1 = (s+1)^2$ of $1 - PK$, so the closed-loop system is stable.

7.3 Youla Parametrization

Parametrization of stabilizing controllers stemmed from Youla's study on the strong stabilization problem [97, 90], which has a form different from the LFT given in the previous section. For some problems, Youla parametrization is more convenient. So, this section introduces this method.

We start from the input/output relation of $M(s)$:

$$\begin{bmatrix} \hat{u} \\ \hat{\xi} \end{bmatrix} = M(s) \begin{bmatrix} \hat{y} \\ \hat{\eta} \end{bmatrix} \tag{7.25}$$

(see Figure 7.7), arranging it as

$$\begin{bmatrix} \hat{u} \\ \hat{y} \end{bmatrix} = \Phi(s) \begin{bmatrix} \hat{\eta} \\ \hat{\xi} \end{bmatrix} \tag{7.26}$$

and

$$\begin{bmatrix} \hat{\eta} \\ \hat{\xi} \end{bmatrix} = \Phi^{-1}(s) \begin{bmatrix} \hat{u} \\ \hat{y} \end{bmatrix} := \Theta(s) \begin{bmatrix} \hat{u} \\ \hat{y} \end{bmatrix}. \tag{7.27}$$

To find the transfer matrix $\Phi(s)$, we transform the state equation of $M(s)$ so as to derive the state equation of $\Phi(s)$. The key is to focus on the input/output relation of $\Phi(s)$. Denoting the state of $M(s)$ by x_M, its state equation can be written as

$$\begin{aligned} \dot{x}_M &= (A + B_2F + LC_2)x_M - Ly + B_2\eta \\ u &= Fx_M + \eta \\ \xi &= -C_2x_M + y. \end{aligned} \tag{7.28}$$

From $\xi = -C_2x_M + y$, we get $y = C_2x_M + \xi$. Substituting it back into $\dot{x}_M$, we obtain the state equation of $\Phi(s)$:

$$\begin{aligned} \dot{x}_M &= (A + B_2F)x_M + B_2\eta - L\xi \\ u &= Fx_M + \eta \\ y &= C_2x_M + \xi. \end{aligned} \tag{7.29}$$

Hence, the transfer matrix $\Phi(s)$ is

$$\Phi(s) = \left[\begin{array}{c|cc} A + B_2F & B_2 & -L \\ \hline F & I & 0 \\ C_2 & 0 & I \end{array}\right] := \begin{bmatrix} D(s) & -Y(s) \\ N(s) & -X(s) \end{bmatrix}. \tag{7.30}$$

Further, calculating $\Phi^{-1}(s) = \Theta(s)$ based on the formula in Section 4.1.8, it is easy to get

$$\Theta(s) = \left[\begin{array}{c|cc} A + LC_2 & -B_2 & L \\ \hline F & I & 0 \\ C_2 & 0 & I \end{array}\right] := \begin{bmatrix} -\tilde{X}(s) & \tilde{Y}(s) \\ -\tilde{N}(s) & \tilde{D}(s) \end{bmatrix}. \tag{7.31}$$

Obviously,

$$\begin{bmatrix} -\tilde{X}(s) & \tilde{Y}(s) \\ -\tilde{N}(s) & \tilde{D}(s) \end{bmatrix} \begin{bmatrix} D(s) & -Y(s) \\ N(s) & -X(s) \end{bmatrix} = I \tag{7.32}$$

holds. After these preparations, we derive the Youla parametrization.

Theorem 7.3 *Suppose that (A, B_2) is stabilizable and (C_2, A) is detectable. F and L are matrices that stabilize $A + B_2F$ and $A + LC_2$, respectively. Then, the following statements hold:*

1. $G_{22}(s) = N(s)D^{-1}(s) = \tilde{D}^{-1}(s)\tilde{N}(s)$.
2. *All controllers stabilizing the generalized plant $G(s)$ of (7.5) are parameterized by*

$$K(s) = (\tilde{X} - Q\tilde{N})^{-1}(\tilde{Y} - Q\tilde{D}) = (Y - DQ)(X - NQ)^{-1} \tag{7.33}$$

where $Q(s)$ is any stable transfer matrix with appropriate dimension.

Proof. Statement 1 can be proved by calculating the product of transfer matrices using the realizations of $N(s), D(s), \tilde{N}(s), \tilde{D}(s)$ and then conducting a similarity transformation (Exercise 7.8). Statement 2 is proved here.

Noting that $\hat{\eta} = Q\hat{\xi}$ (Figure 7.7), (7.26) can be written as

$$\hat{u} = (DQ - Y)\hat{\xi}, \quad \hat{y} = (NQ - X)\hat{\xi}. \tag{7.34}$$

Removing $\hat{\xi}$, we get

$$\hat{u} = (Y - DQ)(X - NQ)^{-1}\hat{y} = K(s)\hat{y}. \tag{7.35}$$

On the other hand,

$$\hat{\eta} = \tilde{Y}\hat{y} - \tilde{X}\hat{u} = Q\hat{\xi}, \quad \hat{\xi} = \tilde{D}\hat{y} - \tilde{N}\hat{u} \tag{7.36}$$

follows from (7.27). Substituting the second equation into the first and removing $\hat{\xi}$ and $\hat{\eta}$, after some rearrangement we obtain the relationship between $\hat{u}$ and $\hat{y}$:

$$\hat{u} = (\tilde{X} - Q\tilde{N})^{-1}(\tilde{Y} - Q\tilde{D})\hat{y} = K(s)\hat{y}. \tag{7.37}$$

•

Statement (1) of this theorem shows that transfer matrix $G_{22}(s)$, either stable or unstable, can be described as a fraction of two stable transfer matrices $N(s)$ and $D(s)$. Moreover, from the $(1, 1)$ block of Eq. (7.32), we have

$$\tilde{Y}(s)N(s) - \tilde{X}(s)D(s) = I. \tag{7.38}$$

This equation implies that $N(s)$ and $D(s)$ do not have any common zero and right zero vector[4]. Therefore, this kind of factorization is called *coprime factorization* [90] based on stable transfer matrices.

It is easy to verify the following conclusions (Exercise 7.8):

- Zeros of $N(s)$ and $\tilde{N}(s)$ are the same as those of $G_{22}(s)$.
- Zeros of $D(s)$ and $\tilde{D}(s)$ are the same as the poles of $G_{22}(s)$.

[4] Suppose that z is a common zero with right zero vector $u \neq 0$, then $D(z)u = 0, N(z)u = 0$. Hence, postmultiplying (7.38) by u and substituting $s = z$ into it leads to a contradiction $0 = u$.

7.4 Structure of Closed-Loop System

In this section, we analyze the structure of the stabilized closed-loop system. Illustrated are two types of structures: one is the parametric structure of closed-loop system with controller parameter as the variable, and another is the structure of closed-loop transfer matrix with the free parameter of stabilizing controller as the variable.

7.4.1 *Affine Structure in Controller Parameter*

A controller

$$K(s) = \left[\begin{array}{c|c} A_K & B_K \\ \hline C_K & D_K \end{array}\right]$$

is used to control the generalized plant (7.4), as shown in Figure 7.1. We investigate the relationship between the coefficient matrix of generalized feedback system and that of the controller. Suppose that the state equation of controller $K(s)$ is

$$\begin{bmatrix} \dot{x}_K \\ u \end{bmatrix} = \begin{bmatrix} A_K & B_K \\ C_K & D_K \end{bmatrix} \begin{bmatrix} x_K \\ y \end{bmatrix}. \tag{7.39}$$

The state vector of closed-loop system is selected as $\begin{bmatrix} x \\ x_K \end{bmatrix}$. Substituting y into (7.39) and then substituting u back into (7.4), we obtain the state equation of closed-loop system as

$$\begin{bmatrix} \dot{x} \\ \dot{x}_K \\ \hdashline z \end{bmatrix} = \begin{bmatrix} A_c & B_c \\ C_c & D_c \end{bmatrix} \begin{bmatrix} x \\ x_K \\ \hdashline w \end{bmatrix} \tag{7.40}$$

where

$$\begin{bmatrix} A_c & B_c \\ C_c & D_c \end{bmatrix} = \left[\begin{array}{cc:c} A + B_2D_KC_2 & B_2C_K & B_1 + B_2D_KD_{21} \\ B_KC_2 & A_K & B_KD_{21} \\ \hdashline C_1 + D_{12}D_KC_2 & D_{12}C_K & D_{11} + D_{12}D_KD_{21} \end{array}\right]. \tag{7.41}$$

Now, we examine the relationship between the coefficient matrices of closed-loop system and controller. Firstly, it is noted that A_c can be written as

$$\begin{aligned} A_c &= \begin{bmatrix} A & 0 \\ 0 & 0 \end{bmatrix} + \begin{bmatrix} B_2D_KC_2 & B_2C_K \\ B_KC_2 & A_K \end{bmatrix} \\ &= \begin{bmatrix} A & 0 \\ 0 & 0 \end{bmatrix} + \begin{bmatrix} B_2 & 0 \\ 0 & I \end{bmatrix} \begin{bmatrix} D_K & C_K \\ B_K & A_K \end{bmatrix} \begin{bmatrix} C_2 & 0 \\ 0 & I \end{bmatrix}. \end{aligned}$$

That is, A_c is an affine function of the coefficient matrix of controller:

$$\mathcal{K} = \begin{bmatrix} D_K & C_K \\ B_K & A_K \end{bmatrix}. \tag{7.42}$$

Rewriting other coefficient matrices analogously, the coefficient matrices of closed-loop system can be written as

$$\begin{bmatrix} A_c & B_c \\ C_c & D_c \end{bmatrix} = \begin{bmatrix} \overline{A} & \overline{B}_1 \\ \overline{C}_1 & \overline{D}_{11} \end{bmatrix} + \begin{bmatrix} \overline{B}_2 \\ \overline{D}_{12} \end{bmatrix} \mathcal{K}[\overline{C}_2, \overline{D}_{21}]. \tag{7.43}$$

Obviously, they are affine functions of $\mathcal{K}$. The coefficient matrices are given by

$$\begin{bmatrix} \overline{A} & \overline{B}_1 & \overline{B}_2 \\ \overline{C}_1 & \overline{D}_{11} & \overline{D}_{12} \\ \overline{C}_2 & \overline{D}_{21} & \end{bmatrix} = \left[\begin{array}{cc|c|cc} A & 0 & B_1 & B_2 & 0 \\ 0 & 0 & 0 & 0 & I \\ \hline C_1 & 0 & D_{11} & D_{12} & 0 \\ \hline C_2 & 0 & D_{21} & & \\ 0 & I & 0 & & \end{array}\right]. \tag{7.44}$$

The closed-loop transfer matrix is a linear fractional function of the controller which is nonlinear. Meanwhile, in state space their coefficient matrices have an affine relation which is much simpler. This affine feature about the controller parameter is a very important character of linear closed-loop systems. It is because of this character that the state-space method is effective in various kinds of optimal control designs. For example, in the $\mathcal{H}_\infty$ control and multiobjective control to be introduced later on, this affine relationship plays a fundamental role in deriving the LMI solutions.

7.4.2 Affine Structure in Free Parameter

First, for simplicity define the following notations:

$$\begin{aligned} A_F &:= A + B_2F, \quad C_F := C_1 + D_{12}F \\ A_L &:= A + LC_2, \quad B_L := B_1 + LD_{21} \\ \hat{A} &:= A + B_2F + LC_2. \end{aligned} \tag{7.45}$$

Applying the stabilizing controller of Theorem 7.2, we get the closed-loop transfer matrix from w to z (Figure 7.8) as

$$H_{zw}(s) = \mathcal{F}_\ell(G, K) = \mathcal{F}_\ell(G, \mathcal{F}_\ell(M, Q)) = \mathcal{F}_\ell(N, Q).$$

Its coefficient matrix $N(s)$ can be obtained based on the state equations of $G(s)$(with state x) and $M(s)$(with state x_M) as follows: select the state vector of closed-loop system as $\begin{bmatrix} x \\ x-x_M \end{bmatrix}$ and then eliminate the intermediate variables (y, u). The specific result is as follows (refer to Exercise 7.11):

$$N(s) = \begin{bmatrix} N_{11} & N_{12} \\ N_{21} & N_{22} \end{bmatrix} = \left[\begin{array}{cc|cc} A_F & -B_2F & B_1 & B_2 \\ 0 & A_L & B_L & 0 \\ \hline C_F & -D_{12}F & D_{11} & D_{12} \\ 0 & C_2 & D_{21} & 0 \end{array}\right]. \tag{7.46}$$

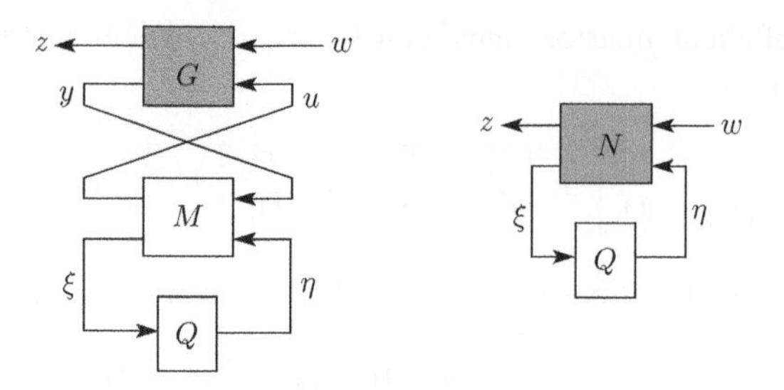

Figure 7.8 Closed-loop system

Expanding the transfer matrices in Eq. (7.46), we get

$$N_{12}(s) = \left[\begin{array}{c|c} A_F & B_2 \\ \hline C_F & D_{12} \end{array}\right], \quad N_{21}(s) = \left[\begin{array}{c|c} A_L & B_L \\ \hline C_2 & D_{21} \end{array}\right], \quad N_{22}(s) = 0.$$

Eventually, the closed-loop transfer matrix becomes

$$H_{zw}(s) = N_{11}(s) + N_{12}(s)Q(s)N_{21}(s). \tag{7.47}$$

Namely, $H_{zw}(s)$ is an affine function of $Q(s)$. This affine structure will be used in solving the $\mathcal{H}_2$ optimal control problem. Historically, the $\mathcal{H}_\infty$ control problem was also treated based on this formula at the beginning.

7.5 2-Degree-of-Freedom System

In this section, the 2-DOF system in Figure 7.9 is considered. We will analyze the structural property of the stabilized 2-DOF system in detail and discuss the implementation of 2-DOF control.

7.5.1 *Structure of 2-Degree-of-Freedom Systems*

The control objective of 2-DOF system shown in Figure 7.9 is to let the plant output $y(t)$ track the reference input $r(t)$ while suppressing the influence of disturbance $d(t)$ simultaneously. Here, take the tracking error

$$e(t) = r(t) - y_P(t) \tag{7.48}$$

as the performance output and assume that the state equation

$$\dot{x} = Ax + Hd + Bu \tag{7.49}$$

$$y_P = Cx \tag{7.50}$$

of the plant is stabilizable and detectable. Note that in practice the disturbance d may enter the closed-loop system at a location different from the control input u (for instance,

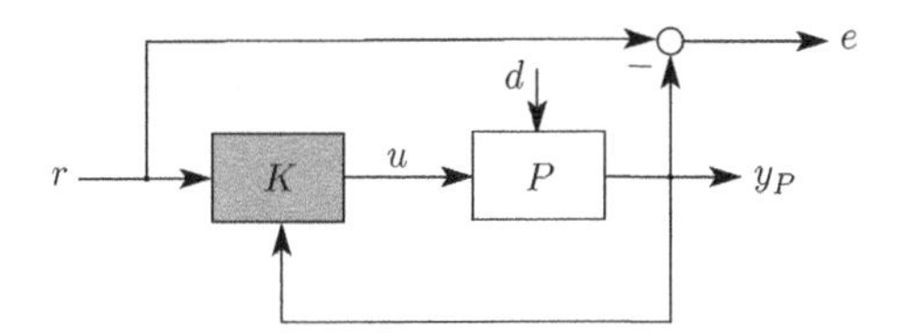

Figure 7.9 2-DOF control system

two-mass–spring system), so their coefficient matrices are set differently. Then, transfer matrix $P_u(s)$ of $u \mapsto y_P$ and transfer matrix $P_d(s)$ of $d \mapsto y_P$ are

$$P_u(s) = C(sI - A)^{-1}B, \quad P_d(s) = C(sI - A)^{-1}H, \tag{7.51}$$

respectively. In order to analyze the closed-loop structure of 2-DOF system based on the result of the last section, we regard the 2-DOF system as a generalized feedback system where the disturbance is $w = \begin{bmatrix} r \\ d \end{bmatrix}$ and the measured output is $y = \begin{bmatrix} r \\ y_P \end{bmatrix}$. Consequently, the state equation of generalized plant becomes

$$\begin{bmatrix} \dot{x} \\ e \\ y \end{bmatrix} = \left[\begin{array}{c|cc|c} A & 0 & H & B \\ \hline -C & I & 0 & 0 \\ \hline 0 & I & 0 & 0 \\ C & 0 & 0 & 0 \end{array}\right] \begin{bmatrix} x \\ w \\ u \end{bmatrix} = \left[\begin{array}{c|c|c} A & B_1 & B_2 \\ \hline C_1 & D_{11} & D_{12} \\ \hline C_2 & D_{21} & 0 \end{array}\right] \begin{bmatrix} x \\ w \\ u \end{bmatrix}.$$

Here, let matrix F be a gain that stabilizes $A_F := A + B_2F = A + BF$. Noting that $C_2 = \begin{bmatrix} 0 \\ C \end{bmatrix}$, the observer gain in Theorem 7.2 can be chosen as $[0 \quad L]$ where the matrix L stabilizes $A_L := A + [0 \quad L]C_2 = A + LC$. So the closed-loop transfer matrix from $\begin{bmatrix} r \\ d \end{bmatrix}$ to e is

$$[T_{er}(s) \quad T_{ed}(s)] = N_{11}(s) + N_{12}(s)Q(s)N_{21}(s). \tag{7.52}$$

Computing the matrices defined by (7.45), we obtain $B_L = [0 \quad H], C_F = -C$. So, according to (7.46) the following coefficient matrices are obtained:

$$N_{12}(s) = -C(sI - A_F)^{-1}B, \quad N_{21}(s) = \begin{bmatrix} I & 0 \\ 0 & C(sI - A_L)^{-1}H \end{bmatrix}$$
$$N_{11}(s) = \begin{bmatrix} I & -N_{12}(s)F(sI - A_L)^{-1}H - C(sI - A_F)^{-1}H \end{bmatrix}.$$

Finally, partition the free parameter $Q(s)$ into

$$Q(s) = [Q_F(s) \quad Q_B(s)] \tag{7.53}$$

according to the dimensions of $y = \begin{bmatrix} r \\ y_P \end{bmatrix}$. Then, the reference tracking transfer matrix T_{er} and disturbance suppression transfer matrix T_{ed} become

$$T_{er}(s) = I + N_{12}(s)Q_F(s) \tag{7.54}$$

$$\begin{aligned} T_{ed}(s) = {} & N_{12}(s)Q_B(s)C(sI - A_L)^{-1}H \\ & - N_{12}(s)F(sI - A_L)^{-1}H - C(sI - A_F)^{-1}H, \end{aligned} \tag{7.55}$$

respectively. Obviously, $T_{er}(s)$ is only related with $Q_F(s)$, while $T_{ed}(s)$ only depends on $Q_B(s)$. This means that $T_{er}(s)$ and $T_{ed}(s)$ can be designed independently.

Further, since the zeros of $N_{12}(s)$ are identical to those of $P_u(s)$, there exists a vector $u \neq 0$ satisfying $u^* N_{12}(z) = 0$ corresponding to an unstable zero z of $P_u(s)$. Then, $u^* T_{er}(z) = u^*$ always holds no matter what $Q_F(s)$ is. That is, the reference tracking transfer matrix $T_{er}(s)$ is constrained by the unstable zeros of $P_u(s)$[5]. Compared with this, the disturbance attenuation transfer matrix $T_{ed}(s)$ is constrained not only by the unstable zeros of plant $P_u(s)$ but also by the unstable zeros of the disturbance transfer matrix $P_d(s)$ (zeros of transfer matrix $C(sI - A_L)^{-1}H$ are identical to those of $P_d(s)$).

In particular, when the plant is stable, the feedback gains can be set as $F = 0, L = 0$. Then, the description of closed-loop transfer matrices simplifies to

$$T_{er}(s) = I - P_u(s)Q_F(s) \tag{7.56}$$

$$T_{ed}(s) = -P_u(s)Q_B(s)P_d(s) - P_d(s) \tag{7.57}$$

and the constraints on each transfer matrix as well as their design independence can be seen more clearly.

Example 7.7 *Consider the following first-order system:*

$$\dot{x} = -2x + u + d, \quad y_P = 2x.$$

Let the reference input r and the disturbance d be unit step signal $1(t)$. We design a 2-DOF control system to reduce the reference tracking error $e(t)$.

Here, the plant is stable and

$$P_u(s) = P_d(s) = \frac{2}{s+2}.$$

So, we choose the free parameters as follows:

$$Q_F(s) = P_u^{-1}(s)\frac{1}{\epsilon s + 1}, \quad Q_B(s) = -P_u^{-1}(s)\frac{1}{\tau s + 1}, \epsilon, \tau > 0.$$

A simple calculation gives

$$T_{er}(s) = 1 - P_u Q_F = \frac{s}{s + 1/\epsilon}$$

$$T_{ed}(s) = -(P_u Q_B + 1)P_d = -\frac{2s}{(s+2)(s+1/\tau)}.$$

So the tracking error equals

$$\hat{e}(s) = T_{er}\hat{r} + T_{ed}\hat{d} = \frac{1}{s + 1/\epsilon} - \frac{2}{(s+2)(s+1/\tau)}$$

$$\Rightarrow e(t) = e^{-t/\epsilon} - \frac{2\tau}{1 - 2\tau}(e^{-2t} - e^{-t/\tau}). \tag{7.58}$$

[5] Refer to Section 10.2 and Section 10.4 for details about such constraints.

The tracking error can be reduced by lowering ϵ, τ. The corresponding controller is computed based on the state-space realization of 2-DOF system and (7.19):

$$G_{22} = \begin{bmatrix} 0 \\ C \end{bmatrix} (sI - A)^{-1}B = \begin{bmatrix} 0 \\ 2 \end{bmatrix} (s+2)^{-1} \cdot 1 = \begin{bmatrix} 0 \\ P_u \end{bmatrix} \Rightarrow$$

$$K(s) = \frac{Q}{1+QG_{22}} = \frac{[Q_F \quad Q_B]}{1+Q_B P_u} = \frac{\tau s+1}{\tau s} \begin{bmatrix} \frac{s+2}{2(\epsilon s+1)} & -\frac{s+2}{2(\tau s+1)} \end{bmatrix}.$$

Obviously, the low-frequency gain of $K(s)$ increases when τ is reduced, while ϵ does not affect the low-frequency gain of $K(s)$. High gain of controller leads to an increase of control input. Note that the second term of (7.58) is the disturbance response. In order to realize signal tracking using an input as small as possible, we should better mainly use feedforward control (i.e., lowering ϵ only). The feedback should be strengthened only when the disturbance is strong (lowering both ϵ and τ).

7.5.2 Implementation of 2-Degree-of-Freedom Control

Here, we only discuss the implementation of 2-DOF control for stable SISO systems. Since $G_{22} = [0 \ \ P]^T, P(s) := C(sI - A)^{-1}B$, the controller becomes

$$K := [K_F \quad -K_B] = \frac{[Q_F \quad -Q_B]}{1 + [Q_F \quad -Q_B]G_{22}} = \frac{[Q_F \quad -Q_B]}{1 - PQ_B}. \tag{7.59}$$

Drawing the block diagram, we get the general form of 2-DOF system as shown in Figure 7.10 (P_0 stands for the plant model contained in the controller).

The feature of this system is that the input signal of $Q_B(s)$ becomes zero when $P = P_0$ and $d(t) = 0$. So the feedback controller $K_B(s)$ is not activated. In this case, the transfer function of $r \mapsto y_P$ becomes

$$H_{y_P r}(s) = P(s)Q_F(s).$$

To guarantee a good time response, a method is to make the closed-loop transfer function match or close to a reference model $M(s)$ with good performance. This is called *model-matching* . It is very easy to derive the feedforward compensator Q_F according to the model-matching condition:

$$PQ_F = M \Rightarrow Q_F(s) = \frac{M(s)}{P(s)}. \tag{7.60}$$

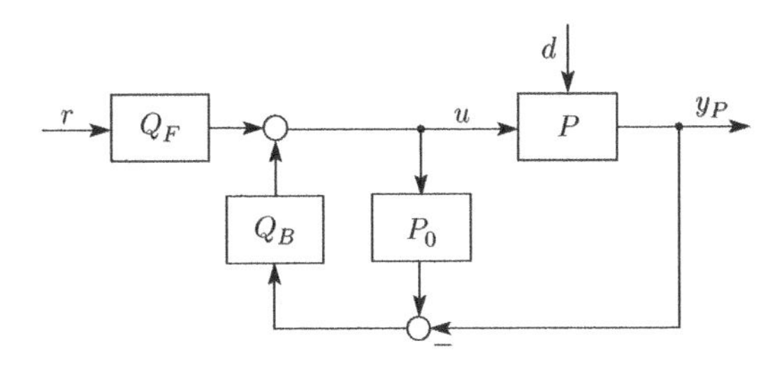

Figure 7.10 General form of 2-DOF systems

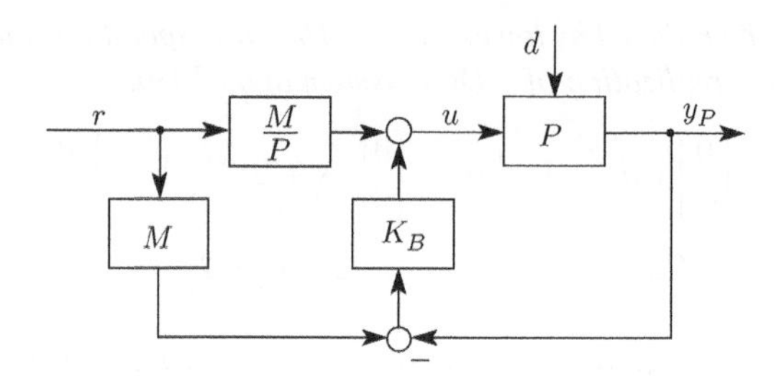

Figure 7.11 Another form of 2-DOF systems

Since the free parameter Q_F must be stable, this equation implies that when the plant $P(s)$ have unstable zeros, the model $M(s)$ must also contain the same zeros. That is, for a nonminimum phase plant, the output response cannot be improved arbitrarily.

Another implementation form of 2-DOF systems is shown in Figure 7.11, which can be verified as follows. Calculating the transfer function of $r \mapsto u$, we have

$$\begin{aligned} H_{ur}(s) &= \frac{M}{P} + MK_B = \frac{M}{P} + M\frac{Q_B}{1 - PQ_B} \\ &= \frac{M}{P}\frac{1 - PQ_B + PQ_B}{1 - PQ_B} = \frac{Q_F}{1 - PQ_B} \\ &= K_F. \end{aligned}$$

The physical implication of this implementation is that, when the output of actual plant is different from that of the reference model $M(s)$, the feedback controller is activated; when they are the same, the feedback controller stops working.

Example 7.8 *Consider the following plant with low damping:*

$$P(s) = \frac{4}{s^2 + s + 4} \quad \left(\zeta = \frac{1}{4}, \quad \omega_n = 2\right).$$

We hope that the closed-loop transfer function matches a reference model with a strengthened damping:

$$M(s) = \frac{4}{s^2 + 3s + 4} \quad (\zeta^* = 0.75, \quad \omega_n^* = 2).$$

Following the design method stated in the preceding text, we get the feedforward compensator

$$Q_F(s) = \frac{M}{P} = \frac{s^2 + s + 4}{s^2 + 3s + 4}.$$

The feedback controller is designed as

$$Q_B(s) = P^{-1}\frac{1}{(\epsilon s + 1)^2} \Rightarrow K_B(s) = \frac{s^2 + s + 4}{2\epsilon^2 s(s + 2/\epsilon)}.$$

Then, the sensitivity function becomes

$$S(s) = \frac{1}{1 + PK_B} = 1 - PQ_B = 1 - \frac{1}{(\epsilon s + 1)^2} = \frac{\epsilon s(\epsilon s + 2)}{(\epsilon s + 1)^2}$$

(MATLAB drill: set $\epsilon = 0.3$*, do simulations to compare the step responses of open-loop control and closed-loop control).*

Exercises

7.1 Given a stable plant $P = 1/(s + 2)$, a controller is desired that can track the step reference input $r(t) = 1\ (t \geq 0)$ asymptotically. Design such a first-order controller $K(s)$ using the formula of stabilizing controllers.

7.2 In Exercise 7.1, besides the asymptotic tracking of step reference input, the tracking error w.r.t. the ramp reference input $r(t) = t\ (t \geq 0)$ has to satisfy $|e(\infty)| \leq 0.05$. Let the free parameter be $Q(s) = P^{-1}/(\epsilon s + 1)$, find an $\epsilon > 0$ such that these two conditions are satisfied, and compute the corresponding controller $K(s)$.

7.3 Set the free parameter as $Q = (as + b)/(s + 1)$. Design a controller such that the output $y(t)$ of plant $P(s) = 1/(s + 2)$ tracks the ramp reference input $r(t) = t\ (t \geq 0)$ asymptotically.

7.4 The input of plant $P(s) = 5/(s + 5)$ is $u(t)$ and its output is $y(t)$. We want to ensure that output $y(t)$ asymptotically tracks unit step reference input $r(t) = 1(t)$ by using a stabilizing controller:

$$K(s) = \frac{Q}{1 - PQ}, \quad Q(s) = \frac{s + 5}{5(as + b)}, a, b > 0.$$

The tracking error is $e = r - y$. Let the interconnection relation of the system be $\hat{u}(s) = K(s)\hat{e}(s)$.

(a) Draw the block diagram of the closed-loop system.
(b) Describe the Laplace transform $\hat{e}(s)$ of tracking error using the free parameter $Q(s)$.
(c) To achieve the asymptotic tracking $e(\infty) = 0$, derive the conditions on the parameters (a, b).
(d) Further, derive the conditions on (a, b) for $\|e\|_2 \leq 0.1$.
(e) Select a set of (a, b) satisfying conditions (c) and (d) and calculate the corresponding controller $K(s)$.

7.5 The SISO closed-loop system in Figure 7.12 is called *IMC* (internal model control) , which is widely used in process control. Here, $P(s)$ stands for the actual system, $P_0(s)$ stands for its model, and both are strictly proper. $Q(s)$ is a parameter used for control and is a stable transfer function. Let the tracking error be $e = r - y$.

(a) When $P(s) = P_0(s)$ holds and both are stable, prove that the closed-loop system is stable for any $Q(s)$.
(b) When $P(s) = P_0(s)$ holds and both are minimum phase, propose a strategy that can improve the tracking performance for any reference input r.

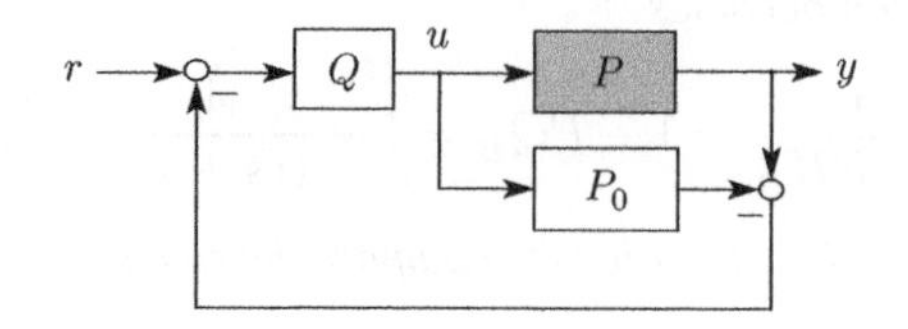

Figure 7.12 Structure of IMC

(c) When $P(s)$ and $P_0(s)$ contain a common unstable pole p, discuss the possibility of stabilization of the closed-loop system.

7.6 In the control system of Figure 7.5, the plant P and disturbance d are

$$P(s) = \frac{1}{s+1}, \quad \hat{d}(s) = \frac{1}{s}\hat{w}(s), \quad \|w\|_2 \leq 1.$$

To reduce the effect of disturbance, we hope that the closed-loop transfer function from w to y satisfies $\|H_{yw}\|_\infty < 1$. Examine whether this specification can be achieved by the following controller:

$$K(s) = \frac{Q}{1-PQ}, \quad Q(s) = \frac{s+1}{as+b}, \quad a > 0, \quad b > 0.$$

If possible, give the conditions on (a, b).

7.7 Following Examples 7.5 and 7.6, design a controller such that the response y w.r.t. the unit step disturbance acting on the input of plant $P(s) = 1/s$ satisfies $\|y\|_2 \leq 0.1$.

7.8 Prove statement 1 of Theorem 7.3 first. Then, prove that zeros of $N(s), \tilde{N}(s)$ are identical to those of $G_{22}(s)$ and zeros of $D(s), \tilde{D}(s)$ are identical to the poles of $G_{22}(s)$.

7.9 In the unity feedback system of Figure 7.4, find a controller such that the output of stable plant

$$P(s) = \frac{2}{(s+1)(s+2)}$$

can asymptotically track ramp reference input (Hint: set the free parameter as $Q = (as+b)/(s+1)$).

7.10 In the unity feedback system of Figure 7.5, the stable plant P satisfies $P(0) \neq 0$. Suppose that the input disturbance of P is a unit step signal $d(t) = 1(t)$:
(a) Use the formula of stabilizing controllers to design a controller $K(s)$ satisfying $y(\infty) = 0$. Here, fix the free parameter $Q(s)$ as a constant q.
(b) When $P(s) = 1/(s+2)$, concretely calculate the controller $K(s)$ satisfying the specification.

7.11 Derive the formula of coefficient matrix $N(s)$ given in Section 7.4.2 as follows: starting from the state equations of $G(s)$ and $M(s)$ and then eliminating the intermediate variables (y, u).

7.12 When $D_{22} \neq 0$, the parameter of closed-loop transfer matrix is no longer an affine function of the controller parameter $\begin{bmatrix} D_K & C_K \\ B_K & A_K \end{bmatrix}$. Find a way to transform the controller so that the parameter of the closed-loop transfer matrix is affine about the controller parameter after transformation (Hint: refer to Figure 7.13. Insert D_{22} into the controller and then consider the new controller $\hat{K} = K(I - D_{22}K)^{-1}$).

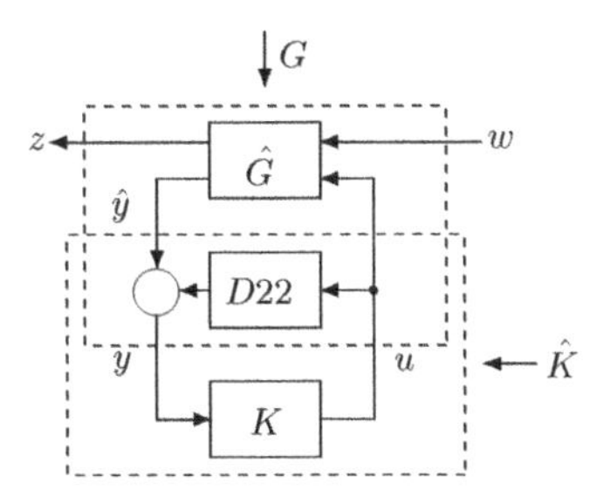

Figure 7.13 Treatment of nonzero D_{22}

7.13 In the unity feedback system of Figure 7.4, $P(s) = 1/(s-1)$. If we use controller $K = Q/(1 - PQ)$ to stabilize this unstable plant, what condition should $Q(s)$ satisfy?

7.14 Prove that for stable plant $P(s) = (s-1)/(s+1)(s+2)$, no matter what the controller K is, the sensitivity function $S = 1/(1 + PK)$ always satisfies

$$S(1) = S(\infty) = 1.$$

Then, prove $\min_K \|S\|_\infty = 1$ based on this fact and find a stabilizing controller that realizes the minimal sensitivity(Hint: since S is stable, the maximum of $|S(s)|$ in the closed right half-plane is equal to $\|S\|_\infty$. Use the formula of stabilizing controllers).

7.15 In the 2-DOF system in Figure 7.10, assume that the disturbance d is imposed at the same location as the control input u, namely, $y_P = P(u + d)$. Further, the plant is the same as its model:

$$P(s) = P_0(s) = \frac{1}{s+1}.$$

Design a controller satisfying the following specifications:

1. Design the free parameter $Q_F(s)$ such that transfer function of $r \mapsto y_P$ equals

$$M(s) = \frac{25}{s^2 + 5s + 25}.$$

2. When $r(t) = 0, d(t) = 1(t)$, design $Q_B(s) = P_0(s)^{-1}/(\epsilon s + 1)$ such that the output y_P satisfies $\|y_P\|_2 \leq 0.1$.

7.16 In Example 7.5, when the output response specification is changed to

$$(1) \quad \|y\|_1 \leq 0.1, \qquad (2) \quad \|y\|_\infty \leq 0.1,$$

find the allowable range of ϵ.

Notes and References

The parameterization of stabilizing controllers rooted in the study of strong stabilization [97]. More details can be found in books on robust control such as Refs [29, 25] and [100]. Liu and Mita [60, 61] extended it to optimal control problems with unstable weighting functions and can be applied to servo mechanism design problems [65, 70, 41].

8

Relation between Time Domain and Frequency Domain Properties

In the analysis and design of linear systems, the relationship between the time domain and frequency domain characteristics plays an especially important role. In system analysis, conditions for the robust stability and robust performance are described as specifications on the frequency response in most cases. However, in system design the state space in the time domain is more convenient. Therefore, it is indispensable to transform frequency domain conditions into equivalent time domain conditions in order to establish an effective and beautiful design theory. In this chapter, two important results are introduced on such relationships.

8.1 Parseval's Theorem

As the first classical principle, the famous Parseval's theorem reveals a very important relationship between the time domain and frequency domain characteristics of signals. Concretely speaking, Parseval's theorem gives the relationship between the squared integral of a time function and that of its Fourier transform, namely, the energy in the time domain is equal to the energy in the frequency domain.

In this chapter, all signals are defined in the time domain $t \geq 0$, that is, signals are zero when $t < 0$.

8.1.1 Fourier Transform and Inverse Fourier Transform

Before stating Parseval's theorem, let us recall the properties of Fourier transform. First of all, a scalar signal $f(t)$ is a real function. Its Fourier transform $\mathcal{F}[f(t)]$ is defined by

$$\hat{f}(j\omega) = \mathcal{F}[f(t)] = \int_0^\infty f(t)e^{-j\omega t}\,dt. \tag{8.1}$$

Robust Control: Theory and Applications, First Edition. Kang-Zhi Liu and Yu Yao.

Therefore, its conjugate $\hat{f}^*(j\omega)$ has the property

$$\hat{f}^*(j\omega) = \left(\int_0^\infty f(t)e^{-j\omega t}dt\right)^* = \int_0^\infty f(t)e^{j\omega t}dt = \hat{f}(-j\omega). \tag{8.2}$$

On the other hand, to recover a signal $f(t)$ from $\hat{f}(j\omega)$, the inverse Fourier transform $\mathcal{F}^{-1}[\hat{f}(j\omega)]$ is used, that is,

$$f(t) = \mathcal{F}^{-1}[\hat{f}(j\omega)] = \frac{1}{2\pi}\int_{-\infty}^\infty \hat{f}(j\omega)e^{j\omega t}d\omega. \tag{8.3}$$

8.1.2 Convolution

The convolution of $f_1(t)$ and $f_2(t)$ is defined as

$$f_1(t) * f_2(t) = \int_0^\infty f_1(t-\tau)f(\tau)d\tau. \tag{8.4}$$

Convolution is also called convolution integral. Similarly, the convolution between functions $\hat{f}_1(j\omega), \hat{f}_2(j\omega)$ in the frequency domain is defined as

$$\hat{f}_1(j\omega) * \hat{f}_2(j\omega) = \frac{1}{2\pi}\int_{-\infty}^\infty \hat{f}_1(j\omega - j\nu)\hat{f}_2(j\nu)d\nu. \tag{8.5}$$

In the properties about the convolution of time function, the following is well known:

$$\mathcal{F}[f_1(t) * f_2(t)] = \hat{f}_1(j\omega)\hat{f}_2(j\omega). \tag{8.6}$$

Similarly, the following property holds for the convolution of frequency domain functions:

$$\begin{aligned}
&\mathcal{F}[f_1(t)f_2(t)] = \frac{1}{2\pi}\hat{f}_1(j\omega) * \hat{f}_2(j\omega)\\
\Longleftrightarrow\ &f_1(t)f_2(t) = \frac{1}{2\pi}\mathcal{F}^{-1}[\hat{f}_1(j\omega) * \hat{f}_2(j\omega)].
\end{aligned} \tag{8.7}$$

This equation is proved as follows. First, due to (8.3) we have

$$\mathcal{F}^{-1}[\hat{f}_1(j\omega) * \hat{f}_2(j\omega)] = \frac{1}{2\pi}\int_{-\infty}^\infty \left[\int_{-\infty}^\infty \hat{f}_1(j\omega - j\nu)\hat{f}_2(j\nu)d\nu\right]e^{j\omega t}d\omega.$$

Then, by an exchange of the integration order and a change of variable $\omega = \nu + \mu$, we get

$$\begin{aligned}
\mathcal{F}^{-1}[\hat{f}_1(j\omega) * \hat{f}_2(j\omega)] &= \frac{1}{2\pi}\int_{-\infty}^\infty \left[\int_{-\infty}^\infty \hat{f}_1(j\omega - j\nu)e^{j\omega t}d\omega\right]\hat{f}_2(j\nu)d\nu\\
&= \frac{1}{2\pi}\int_{-\infty}^\infty \left[\int_{-\infty}^\infty \hat{f}_1(j\mu)e^{j\mu t}d\mu\right]\hat{f}_2(j\nu)e^{j\nu t}d\nu\\
&= 2\pi\left[\frac{1}{2\pi}\int_{-\infty}^\infty \hat{f}_1(j\mu)e^{j\mu t}d\mu\right]\left[\frac{1}{2\pi}\int_{-\infty}^\infty \hat{f}_2(j\nu)e^{j\nu t}d\nu\right]\\
&= 2\pi f_1(t)f_2(t).
\end{aligned}$$

Hence, (8.7) is proved.

8.1.3 Parseval's Theorem

Theorem 8.1 (Parseval's theorem) *Assume that signal vectors $f(t), f_1(t), f_2(t) \in \mathbb{R}^n$ have Fourier transforms $\hat{f}(j\omega), \hat{f}_1(j\omega), \hat{f}_2(j\omega)$, respectively. Then, there hold the following integral relationships:*

1. *The inner product in the time domain is equal to that in the frequency domain, that is,*

$$\int_0^\infty f_1^T(t)f_2(t)\,dt = \frac{1}{2\pi}\int_{-\infty}^\infty \hat{f}_1^*(j\omega)\hat{f}_2(j\omega)d\omega. \tag{8.8}$$

2. *The two-norm in the time domain is equal to that in the frequency domain:*

$$\int_0^\infty \|f(t)\|^2 dt = \frac{1}{2\pi}\int_{-\infty}^\infty \|\hat{f}(j\omega)\|^2 d\omega. \tag{8.9}$$

$\int_0^\infty \|f(t)\|^2 dt$ on the left-hand side of (8.9) represents the energy of signal $f(t)$. In this sense, $\|\hat{f}(j\omega)\|^2$ on the right-hand side can be regarded as the energy density at the frequency ω. Therefore, $\|\hat{f}(j\omega)\|^2$ is also called the *energy spectrum*.

Example 8.1 *An exponentially convergent signal*

$$f(t) = e^{-t} \quad \forall t \geq 0 \quad \Leftrightarrow \quad \hat{f}(j\omega) = \frac{1}{j\omega + 1}$$

is used to illustrate Parseval's theorem.

The left-hand side of (8.9) is

$$\int_0^\infty (e^{-t})^2 dt = \int_0^\infty e^{-2t} dt = \frac{1}{2}.$$

Meanwhile, its right-hand side is

$$\frac{1}{2\pi}\int_{-\infty}^\infty |\hat{f}(j\omega)|^2 d\omega = \frac{1}{2\pi}\int_{-\infty}^\infty \frac{1}{\omega^2+1}d\omega = \frac{1}{2\pi}\arctan\omega\Big|_{-\infty}^{\infty} = \frac{1}{2}.$$

Obviously, both sides are equal. In this example, the energy spectrum is $|\hat{f}(j\omega)|^2 = 1/(\omega^2+1)$. From Figure 8.1, it can be seen clearly that the energy mainly concentrates in the low frequency.

In control design, it is very important to fully grasp the energy spectrum of a signal. When the signal in the previous example is a disturbance, the closed-loop system gain needs to be rolled off in the frequency band where the energy spectrum of disturbance is big in order to attenuate its influence on the system output. In this example, this band is roughly $0 \leq \omega \leq 6$(rad/s). Note also that an energy spectrum is an even function of frequency, so we do not need to consider the negative frequency. Finally, an energy spectrum is the square of the gain of a signal's frequency response. So we can also acquire the characteristic of a signal from its gain of frequency response.

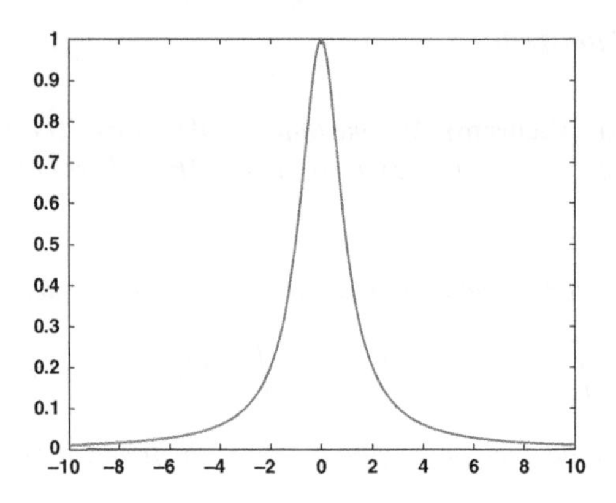

Figure 8.1 Energy spectrum

8.1.4 Proof of Parseval's Theorem

Expanding (8.7), we get

$$\int_0^\infty f_1(t)f_2(t)e^{-j\omega t}dt = \frac{1}{2\pi}\int_{-\infty}^\infty \hat{f}_1(j\omega - j\nu)\hat{f}_2(j\nu)d\nu.$$

When $\omega = 0$, the equation becomes

$$\int_0^\infty f_1(t)f_2(t)dt = \frac{1}{2\pi}\int_{-\infty}^\infty \hat{f}_1(-j\nu)\hat{f}_2(j\nu)d\nu.$$

Further, we obtain, by a substitution of the property $\hat{f}(-j\nu) = \hat{f}^*(j\nu)$ and a replacement of the integral variable ν by ω, that

$$\int_0^\infty f_1(t)f_2(t)dt = \frac{1}{2\pi}\int_{-\infty}^\infty \hat{f}_1^*(j\omega)\hat{f}_2(j\omega)d\omega. \tag{8.10}$$

For vectors $f_1(t), f_2(t) \in \mathbb{R}^n$, $f_{1i}(t), f_{2i}(t)$ are used to represent their ith elements. Then, their inner product can be written as

$$f_1^T f_2 = f_{11}f_{21} + \cdots + f_{1n}f_{2n}.$$

So, application of (8.10) to the scalar elements leads to statement (1). As for statement (2), it is obviously the result for $f(t) = f_1(t) = f_2(t)$. Thus, we have proved Parseval's theorem.

8.2 KYP Lemma

As a relationship bridging the time domain and frequency domain characteristics of a system, the so-called KYP lemma[1] is as famous as Parseval's theorem. This lemma stemmed from the Popov criterion [78] discovered by Popov in his study on Lur'e system. Later on, Kalman

[1] KYP is taken from the initials of Kalman, Yakubovich, and Popov, three masters in the control field.

[45] and Yakubovich [95] disclosed the positive real lemma, which reveals the relationship between the Popov criterion in the frequency domain and a Lyapunov function in the time domain. After numerous extensions, it eventually evolves into the present KYP lemma. As a powerful tool bridging the time domain and frequency domain properties of systems, KYP lemma plays a fundamental role in robust control.

Theorem 8.2 (KYP lemma) *Given matrices* $A \in \mathbb{R}^{n\times n}, B \in \mathbb{R}^{n\times m}, M = M^T \in \mathbb{R}^{(n+m)\times(n+m)}$. *Assume that* A *has no eigenvalues on the imaginary axis and* (A, B) *is controllable. Then the following statements are equivalent:*

1. *For all* ω *including the infinity, there holds the inequality:*

$$\begin{bmatrix} (j\omega I - A)^{-1}B \\ I \end{bmatrix}^* M \begin{bmatrix} (j\omega I - A)^{-1}B \\ I \end{bmatrix} \leq 0. \tag{8.11}$$

2. *There exists a symmetrical matrix* $P = P^T \in \mathbb{R}^{n\times n}$ *satisfying the following inequality:*

$$M + \begin{bmatrix} A^TP + PA & PB \\ B^TP & 0 \end{bmatrix} \leq 0. \tag{8.12}$$

Moreover, when both are strict inequalities, the equivalence is still true and (A, B) *needs not be controllable.*

"Elegant" is the best word for this equivalence. In the following subsections, we show several applications of KYP lemma.

8.2.1 Application in Bounded Real Lemma

Let us consider a gain property about a stable transfer matrix $G(s)$ (Figure 8.2):

$$G^*(j\omega)G(j\omega) < \gamma^2 I \quad \forall\omega \in [0,\ \infty]. \tag{8.13}$$

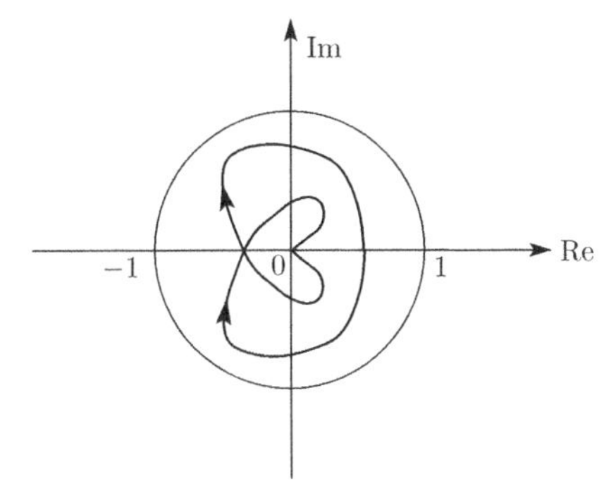

Figure 8.2 Nyquist diagram of a bounded real function ($\gamma = 1$)

This frequency property is equivalent to the $\mathcal{H}_\infty$ norm condition $\|G\|_\infty < \gamma$. Let

$$G(s) = C(sI - A)^{-1}B + D = [C\ \ D]\begin{bmatrix}(sI - A)^{-1}B \\ I\end{bmatrix}.$$

Then, this inequality can be written as

$$\begin{bmatrix}(j\omega I - A)^{-1}B \\ I\end{bmatrix}^* \begin{bmatrix}C^TC & C^TD \\ D^TC & D^TD\end{bmatrix}\begin{bmatrix}(j\omega I - A)^{-1}B \\ I\end{bmatrix} < \gamma^2 I \Leftrightarrow$$

$$\begin{bmatrix}(j\omega I - A)^{-1}B \\ I\end{bmatrix}^* \begin{bmatrix}C^TC & C^TD \\ D^TC & D^TD - \gamma^2 I\end{bmatrix}\begin{bmatrix}(j\omega I - A)^{-1}B \\ I\end{bmatrix} < 0. \tag{8.14}$$

Noting that the inequality is strict, application of KYP lemma yields the following equivalent condition, that is, there exists a symmetrical matrix P satisfying inequality:

$$\begin{bmatrix}C^TC & C^TD \\ D^TC & D^TD - \gamma^2 I\end{bmatrix} + \begin{bmatrix}A^TP + PA & PB \\ B^TP & 0\end{bmatrix} < 0. \tag{8.15}$$

This equivalence is called the *bounded real lemma*. In system design, it is not convenient to deal with products like C^TC. So, we further transform this inequality and obtain the following lemma.

Lemma 8.1 (Bounded real lemma) *Given $G(s) = (A, B, C, D)$, the following statements are equivalent:*

1. *A is stable and $\|G\|_\infty < \gamma$.*
2. *There exists a positive definite matrix P satisfying*

$$\begin{bmatrix}A^TP + PA & PB & C^T \\ B^TP & -\gamma I & D^T \\ C & D & -\gamma I\end{bmatrix} < 0. \tag{8.16}$$

Proof. Dividing both sides of inequality (8.15) with γ and renaming P/γ as P, (8.15) can be written as

$$\begin{bmatrix}A^TP + PA & PB \\ B^TP & -\gamma I\end{bmatrix} - \begin{bmatrix}C^T \\ D^T\end{bmatrix} \cdot (-\gamma I)^{-1} \cdot [C\ \ D] < 0.$$

It follows from Schur's lemma that (8.16) is equivalent to (8.15). Finally, we need only prove that the stability of A is equivalent to $P > 0$. Since the $(1, 1)$ block of (8.16) is

$$PA + A^TP < 0,$$

the equivalence is immediate by Lyapunov's stability theory (see Section 4.3). •

Example 8.2 *According to the Bode plot of transfer function*

$$G(s) = \frac{a}{s^2 + 2s + 2} = \left[\begin{array}{cc|c}0 & 1 & 0 \\ -2 & -2 & 1 \\ \hline a & 0 & 0\end{array}\right], \quad a > 0$$

its largest gain is $a/2$*. Thus, in order to ensure that the* $\mathcal{H}_\infty$ *norm of* $G(s)$ *is less than 1 (i.e.,* $\gamma = 1$*),* a *must be less than 2. Meanwhile, by using the LMI (8.16), we get a positive definite solution*

$$P = \begin{bmatrix} 3.8024 & 1.5253 \\ 1.5253 & 1.8265 \end{bmatrix}$$

when $a = 1.9$*. However, when* $a \geq 2$*, no positive definite solution exists for (8.16). This implies that we can use the bounded real lemma to calculate the* $\mathcal{H}_\infty$ *norm of transfer matrices. Generally speaking, in the calculation of the* $\mathcal{H}_\infty$ *norm of a stable transfer matrix* $G(s) = (A, B, C, D)$*, we need just to reduce* γ *in (8.16) gradually until there is no positive definite solution for (8.16). The last* γ *is the* $\mathcal{H}_\infty$ *norm of* $G(s)$*.*

8.2.1.1 Time Domain Interpretation of Bounded Real Lemma

Let x, u, y be the state, input, and output of a stable transfer matrix $G(s) = (A, B, C, D)$, respectively. Then

$$\dot{x} = Ax + Bu, \quad y = Cx + Du.$$

Next, consider a quadratic and positive definite function

$$V(x) = x^T P x. \tag{8.17}$$

Multiplying the inequality (8.15) by a nonzero vector $\left[\begin{smallmatrix} x \\ u \end{smallmatrix}\right]$ and its transpose from left and right, respectively, we get

$$\begin{aligned} 0 &> x^T(A^T P + PA)x + x^T PBu + u^T B^T Px + x^T C^T Cx \\ &\quad + x^T C^T Du + u^T D^T Cx + u^T(D^T D - I)u \\ &= x^T P(Ax + Bu) + (Ax + Bu)^T Px + (Cx + Du)^T(Cx + Du) - u^T u \\ &= x^T P\dot{x} + \dot{x}^T Px + y^T y - u^T u. \end{aligned}$$

Since $\dot{V} = x^T P\dot{x} + \dot{x}^T Px$, inequality

$$\dot{V}(x) < u^T u - y^T y \tag{8.18}$$

holds. After integration, we have

$$V(x(t)) < V(x(0)) + \int_0^t [u^T(\tau)u(\tau) - y^T(\tau)y(\tau)]d\tau. \tag{8.19}$$

$u^T u, y^T y$ are the input and output powers, respectively; their difference is the power supplied to the system. After integration it becomes the energy supplied to the system. Meanwhile, $V(x)$ can be regarded as a *storage function* of the system energy. Hence, this inequality implies that the variation of the energy stored in a system is less than the energy supplied by the input. That is, a bounded real system consumes a part of the energy supplied by the input. So, it is called a *dissipative system*.

8.2.2 Application in Positive Real Lemma

Stable transfer function satisfying the following inequality is called a *positive real matrix*:

$$G^*(j\omega) + G(j\omega) \geq 0 \quad \forall\omega \in [0, \ \infty]. \tag{8.20}$$

The numbers of input and output of a positive real matrix are equal. For a scalar transfer function, the left-hand side of the inequality is equal to twice of the real part of $G(j\omega)$. So this inequality means that the real part of $G(j\omega)$ is nonnegative, as illustrated in Figure 8.3. This is why $G(s)$ is called a positive real function. From the viewpoint of system, the phase angle of a positive real function is limited in $[-90^\circ, 90^\circ]$. For this reason, its relative degree does not exceed 1.

For example, it can be judged from the Nyquist diagram that transfer function $G(s) = 1/(s+1)$ is positive real. Furthermore, it is worth noting that some unstable systems may also have a frequency property like (8.20). For example,

$$G(s) = \frac{s-1}{s-2} \Rightarrow \Re[G(j\omega)] = \Re\left[\frac{j\omega - 1}{j\omega - 2}\right] = \frac{2+\omega^2}{4+\omega^2} > 0;$$

thus it satisfies the condition (8.20).

Next, let us derive the condition for $G(s) = (A, B, C, D)$ to be positive real. To this end, we rewrite the inequality (8.20) as

$$\begin{bmatrix} (j\omega I - A)^{-1}B \\ I \end{bmatrix}^* \begin{bmatrix} C^T \\ D^T \end{bmatrix} + [C \ \ D] \begin{bmatrix} (j\omega I - A)^{-1}B \\ I \end{bmatrix} \geq 0.$$

To convert it into a form compatible with KYP lemma, we use the identity $I = [((j\omega I - A)^{-1}B)^* \ \ I] \begin{bmatrix} 0 \\ I \end{bmatrix}$. Multiplying I^* and I to the left-hand side of the first term and the right-hand side of the second term in the inequality, the positive real condition turns into

$$-\begin{bmatrix} (j\omega I - A)^{-1}B \\ I \end{bmatrix}^* \begin{bmatrix} 0 & C^T \\ C & D + D^T \end{bmatrix} \begin{bmatrix} (j\omega I - A)^{-1}B \\ I \end{bmatrix} \leq 0$$

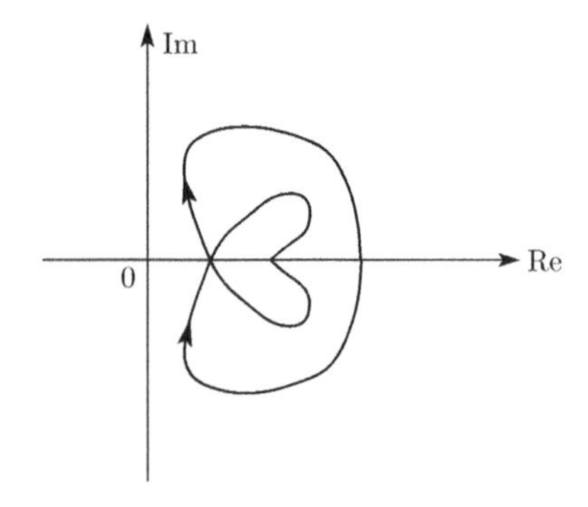

Figure 8.3 Nyquist diagram of a positive real function

after some rearrangement. When (A, B) is controllable, the following equivalent condition is obtained. That is, there is a symmetrical matrix P satisfying the matrix inequality:

$$\begin{bmatrix} A^T P + PA & PB \\ B^T P & 0 \end{bmatrix} - \begin{bmatrix} 0 & C^T \\ C & D + D^T \end{bmatrix} \leq 0. \tag{8.21}$$

Further, the positive real function here is stable and the stability of the matrix A will impose some constraint on the matrix P. With this condition added, we obtain a result called *positive real lemma.*

Lemma 8.2 (Positive real lemma) *Let (A, B, C, D) be a minimal realization of $G(s)$ and matrix A be stable. Then the following statements are equivalent:*

1. *$G(s)$ satisfies the positive real condition (8.20).*
2. *There is a positive definite matrix P satisfying inequality (8.21).*
3. *There is a positive definite matrix P and a full row rank matrix $[L\ W]$ satisfying*

$$\begin{bmatrix} A^T P + PA & PB \\ B^T P & 0 \end{bmatrix} - \begin{bmatrix} 0 & C^T \\ C & D + D^T \end{bmatrix} = - \begin{bmatrix} L^T \\ W^T \end{bmatrix} [L\ W]. \tag{8.22}$$

Proof. From the discussion in the preceding text, the positive real condition (8.20) is equivalent to that the inequality (8.21) has a symmetrical solution P. Since the matrix on the left-hand side of (8.21) is negative semidefinite, it can be factorized as the product of a full row rank matrix and its transpose. Then, the inequality (8.21) is equivalent to the Eq. (8.22). We prove that the stability of A requires $P > 0$.

From the stability of A, it is known that $A^T P + PA \leq 0$ requires $P \geq 0$ (Lemma 4.7). Therefore, we need just prove that P is nonsingular. Suppose that $x \in \text{Ker}\, P$. Multiplying the equation $A^T P + PA = -L^T L$ by x^T, x from the left and right, respectively, we get $Lx = 0$. Then, multiplying the equation $PB - C^T = -L^T W$ by x^T from the left, we obtain $Cx = 0$. Multiplying the equation $A^T P + PA = -L^T L$ again by x from the right, we have $PAx = 0$. This shows that the kernel space Ker P is A-invariant. Therefore, there is a vector $x \in \text{Ker}\, P$ that is an eigenvector of matrix A. Let the corresponding eigenvalue be λ:

$$Ax = \lambda x, \quad Cx = 0$$

hold simultaneously. This contradicts the observability of (C, A). Hence, P must be nonsingular. •

Example 8.3 *Consider a transfer function:*

$$G(s) = \frac{s + a}{s^2 + 2s + 2} = \left[\begin{array}{cc|c} 0 & 1 & 0 \\ -2 & -2 & 1 \\ \hline a & 1 & 0 \end{array}\right].$$

A simple calculation shows that the real part of its frequency response is

$$\Re[G(j\omega)] = \Re\left[\frac{a + j\omega}{2 - \omega^2 + j2\omega}\right] = \frac{2a + (2 - a)\omega^2}{(2 - \omega^2)^2 + 4\omega^2}.$$

For all ω, the condition for it to be positive is

$$2a > 0, 2 - a \geq 0 \Rightarrow 0 < a \leq 2.$$

On the other hand, solving the LMI (8.21) w.r.t. $a = 1$ and $a = 3$, we get a positive definite solution

$$P = \begin{bmatrix} 4.0 & 1.0 \\ 1.0 & 1.0 \end{bmatrix}$$

for $a = 1$, but no positive definite solution exists for $a = 3$. This shows that this transfer function is not positive real when $a = 3$.

Further, when the strict inequality

$$G^*(j\omega) + G(j\omega) > 0 \quad \forall \omega \in [0, \ \infty] \tag{8.23}$$

holds, $G(s)$ is called a *strongly positive real matrix*. Obviously, the strongly positive real condition requires that $G^*(j\infty) + G(j\infty) = D^T + D > 0$. That is, the relative degree of $G(s)$ must be zero. The transfer function in Example 8.3 does not satisfy the strongly positive real condition. But for transfer function $G(s) = (s+a)/(s+1)$,

$$G(j\omega) = \frac{j\omega + a}{j\omega + 1} = \frac{a + \omega^2 + j(1-a)\omega}{\omega^2 + 1}$$

holds. So, it is strongly positive real as long as $a > 0$. Another feature of a strongly positive real matrix is that its normal rank must be full. That is, for almost all complex numbers s, the matrix $G(s)$ should have full rank.

From KYP lemma, it is easy to derive the following strongly positive real lemma. Detailed proof is left for the reader to complete.

Lemma 8.3 (Strongly positive real lemma) *For transfer matrix $G(s) = (A, B, C, D)$, the following statements are equivalent:*

1. *Matrix A is stable and $G(s)$ is strongly positive real.*
2. *There is a positive definite matrix P satisfying the strict inequality*

$$\begin{bmatrix} A^T P + PA & PB \\ B^T P & 0 \end{bmatrix} - \begin{bmatrix} 0 & C^T \\ C & D + D^T \end{bmatrix} < 0. \tag{8.24}$$

Note that (A, B, C, D) does not need to be a minimal realization here.

Since a strictly proper transfer function $G(s)$ has the property $G(j\infty) = D = 0$ at $s = \infty$, it is not a strongly positive real function. But there are many transfer matrices satisfying $G^*(j\omega) + G(j\omega) > 0$ at all finite frequencies except the infinity, that is,

$$G^*(j\omega) + G(j\omega) > 0 \quad \forall \omega \in [0, \ \infty). \tag{8.25}$$

Such a stable transfer matrix is called a *strictly positive real matrix*. For example, the transfer function in Example 8.3 is strictly positive real when $0 < a \leq 2$.

What is the state-space condition for a transfer function to be strictly positive real? Unfortunately, this is still an open problem for the strictly positive real matrix defined in (8.25). In view of this fact, Narendra–Taylor [71] proposed to use the following frequency domain characteristics to replace (8.25).

Definition 8.1 (Modified strictly positive realness) *$G(s)$ is said to be strictly positive real if there exists a constant $\epsilon > 0$ such that $G(s-\epsilon)$ is stable and satisfies the positive real condition:*

$$G^*(j\omega - \epsilon) + G(j\omega - \epsilon) \geq 0 \quad \forall \omega \in [0,\ \infty]. \tag{8.26}$$

The following lemma gives the relationship between the frequency characteristics of $G(s-\epsilon)$ and $G(s)$. Refer to Ref. [18] for a detailed proof.

Lemma 8.4 (Strictly positive real lemma) *Suppose that a square transfer matrix $G(s) = (A, B, C, D)$ is stable and has full normal rank. Then the following two statements are equivalent:*

1. *There exists a constant $\epsilon > 0$ such that $G(s-\epsilon)$ is positive real.*
2. *$G^*(j\omega) + G(j\omega) > 0$ holds for any finite frequency ω and*

$$\lim_{\omega \to \infty} \omega^{2\rho} \det[G^*(j\omega) + G(j\omega)] > 0$$

where ρ is the dimension of the kernel space of constant matrix $D + D^T$, namely, $\rho = \dim\ (\mathrm{Ker}\ (D + D^T))$.

That is, the positive realness of $G(s-\epsilon)$ guarantees that $G(s)$ is strictly positive real.

We examine the transfer function of Example 8.3 once again. From the preceding discussion, we have known that it is strictly positive real when $0 < a \leq 2$.

Example 8.4 *Consider the transfer function*

$$G(s) = \frac{s+a}{s^2 + 2s + 2}.$$

For sufficiently small $\epsilon > 0$, $G(s-\epsilon)$ is still stable. Substituting $s = j\omega - \epsilon$ into $G(s)$, some simple calculation shows that its real part is

$$\begin{aligned}\Re[G(j\omega - \epsilon)] &= \Re\left[\frac{a - \epsilon + j\omega}{1 + (1-\epsilon)^2 - \omega^2 + j2(1-\epsilon)\omega}\right] \\ &= \frac{(a-\epsilon)[1 + (1-\epsilon)^2] + (2 - a - \epsilon)\omega^2}{[1 + (1-\epsilon)^2 - \omega^2]^2 + 4(1-\epsilon)^2\omega^2}.\end{aligned}$$

For all ω including the infinity, $\Re[G(j\omega - \epsilon)] \geq 0$ if

$$a - \epsilon \geq 0, \quad 2 - a - \epsilon \geq 0 \Rightarrow \epsilon \leq a \leq 2 - \epsilon.$$

There is obviously a small gap between this bound and the strictly positive real condition $0 < a \leq 2$. This gap shrinks as $\epsilon \to 0$.

8.2.2.1 Time Domain Interpretation of Positive Real Lemma

Suppose that the state equation of stable transfer matrix $G(s)$ is

$$\dot{x} = Ax + Bu, \quad y = Cx + Du$$

and the storage function of this system is

$$V(x) = x^T Px. \tag{8.27}$$

Multiplying to the left and right of the inequality (8.21) with a nonzero vector $[x^T \ \ u^T]$ and its transpose, we have

$$\begin{aligned} 0 &\le x^T(A^TP + PA)x + x^TPBu + u^TB^TPx - x^TC^Tu - u^TCx \\ &\quad - u^T(D^T + D)u \\ &= x^TP(Ax + Bu) + (Ax + Bu)^TPx - (Cx + Du)^Tu - u^T(Cx + Du) \\ &= x^TP\dot{x} + \dot{x}^TPx - y^Tu - u^Ty. \end{aligned}$$

Since y^Tu, u^Ty are scalars, they are equal. So there holds

$$\dot{V}(x) \le 2y^Tu. \tag{8.28}$$

An integration leads to

$$V(x(t)) \le V(x(0)) + 2\int_0^t y^T(\tau)u(\tau)d\tau. \tag{8.29}$$

If y^Tu is treated as the *supply rate* of the energy injected into the system, then this inequality shows that the energy stored in the system is less than the energy supplied by the input. Such a system is called a *passive system*.

Finally, to see why y^Tu is regarded as the supply rate of the energy injected into the system, let us examine the circuit consisting of an ideal voltage source and a load shown in Figure 8.4 in which u is the voltage of power source and y is the current of load impedance Z. Their product is apparently the power which the power source supplies to the load impedance Z.

The concepts of passivity and dissipativity are the time domain characterizations of positive realness and bounded realness (frequency domain) of linear systems. They can be extended to nonlinear systems. Interested readers may consult Refs [47, 68, 93].

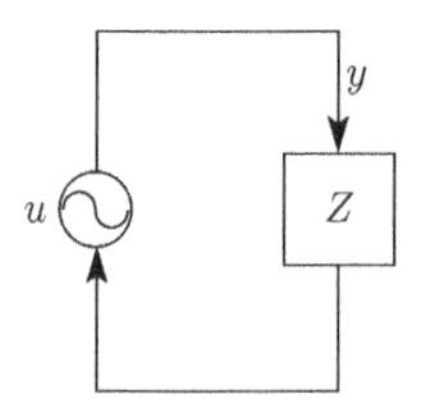

Figure 8.4 Energy supply rate for load impedance

8.2.3 Proof of KYP Lemma*

In this subsection, we present an elementary proof by Rantzer [80]. To understand this proof, one only needs some knowledge of linear algebra and the separating hyperplane. The idea of proof is briefly outlined here: firstly, transform the frequency domain property in statement (1) of KYP lemma equivalently to that certain two sets are disjoint; secondly, prove that the separation property still holds even if one of the two sets is extended to its convex hull; and finally, use the separating hyperplane theorem to prove its equivalence to the time domain property in statement (2) of KYP lemma.

Some preliminaries are introduced in the following which are necessary for the second step of the proof.

Lemma 8.5 *Suppose that F, G are complex matrices with the same dimension. The following statements are true:*

1. $FF^* = GG^*$ *iff there exists a unitary matrix* U *such that* $F = GU$.
2. $FF^* \leq GG^*$ *iff there exists a matrix* U *satisfying* $UU^* \leq I$ *such that* $F = UG$.
3. $FG^* + GF^* = 0$ *iff there exists a unitary matrix* U *such that* $F(I + U) = G(I - U)$.
4. $FG^* + GF^* \geq 0$ *iff there exists a matrix* U *satisfying* $UU^* \leq I$ *such that* $F(I + U) = G(I - U)$.

Proof.

1. $FF^* = GG^*$ implies that $\mathrm{Ker}\, F^* = \mathrm{Ker}\, G^*$. So we can use a row transformation matrix T to convert them into $TF = \begin{bmatrix} F_1 \\ 0 \end{bmatrix}$, $TG = \begin{bmatrix} G_1 \\ 0 \end{bmatrix}$ where both matrices F_1 and G_1 have full row rank. If we can use a matrix U to express F_1 as $F_1 = G_1 U$, then

$$TF = \begin{bmatrix} F_1 \\ 0 \end{bmatrix} = \begin{bmatrix} G_1 U \\ 0 \end{bmatrix} = \begin{bmatrix} G_1 \\ 0 \end{bmatrix} U = TGU \Rightarrow F = GU.$$

Therefore, we need only prove the full row rank case. Since $FF^* = GG^* := H$ is a positive definite, this Hermitian matrix H can be decomposed into the product of its square root, that is, $H = H^{1/2} H^{1/2}$. Set $U_{F1} = H^{-1/2} F$, $U_{G1} = H^{-1/2} G$ (both are wide matrices), and then by $FF^* = H^{1/2} H^{1/2} = GG^*$, it is obvious that $U_{F1} U_{F1}^* = I = U_{G1} U_{G1}^*$. Thus, $F = H^{1/2} U_{F1}, G = H^{1/2} U_{G1}$ hold. Moreover, using a matrix U_{F2} to make $U_F = \begin{bmatrix} U_{F1} \\ U_{F2} \end{bmatrix}$ unitary (such matrix always exists), we can write F as

$$F = \begin{bmatrix} H^{1/2} & 0 \end{bmatrix} U_F, \quad U_F = \begin{bmatrix} U_{F1} \\ U_{F2} \end{bmatrix}.$$

Similarly, G also can be written as

$$G = \begin{bmatrix} H^{1/2} & 0 \end{bmatrix} U_G, \quad U_G = \begin{bmatrix} U_{G1} \\ U_{G2} \end{bmatrix}.$$

Finally, $F = GU$ follows by setting $U = U_G^* U_F$, where U is unitary.

2. When $FF^* \leq GG^*$, the positive semidefinite matrix $GG^* - FF^* \geq 0$ can be decomposed into $GG^* - FF^* = HH^*$. So we get

$$FF^* + HH^* = GG^* \Rightarrow [F \ H][F \ H]^* = [G \ 0][G \ 0]^*.$$

Here, the dimensions of $[F \quad H]$ and $[G \quad 0]$ are equal. Then, according to statement 1 we know that

$$[F \quad H] = [G \quad 0]\begin{bmatrix} U & V \\ V^* & W \end{bmatrix}, \quad \begin{bmatrix} U & V \\ V^* & W \end{bmatrix}\begin{bmatrix} U & V \\ V^* & W \end{bmatrix}^* = I$$

holds for some unitary matrix. Then

$$F = GU, \quad UU^* = I - VV^* \leq I.$$

Statements (3) and (4) are derived by replacing F, G with $G - F, G + F$, respectively. •

Lemma 8.6 *Suppose that $f, g \in \mathbb{C}^n$ and $g \neq 0$. The following statements hold:*

1. *$fg^* + gf^* = 0$ is equivalent to $f = j\omega g$ for some $\omega \in \mathbb{R}$.*
2. *$fg^* + gf^* \geq 0$ is equivalent to $f = sg$ for some $s \in \mathbb{C}$ with positive real part.*

Proof. We use the following transformation:

$$U = \frac{1-s}{1+s} \Leftrightarrow s = \frac{1-U}{1+U}.$$

When $s = j\omega$, the amplitude of U is 1. When the real part of s is positive, $|U| \leq 1$ holds. Subject to the given conditions,

$$f = \frac{1-U}{1+U}g$$

holds owing to Lemma 8.5 (3,4). $|U| = 1$ in case (1) and $|U| \leq 1$ in case (2). Therefore, the conclusion follows immediately. •

Lemma 8.7 *Suppose that M, N are $n \times r$ complex matrices. If a matrix $W \geq 0$ satisfies*

$$NWM^* + MWN^* = 0,$$

then W can be written as $W = \sum_{k=1}^{r} w_k w_k^*$ with complex vectors $w_k \in \mathbb{C}^r$ $(k = 1, \ldots, r)$ satisfying

$$Nw_k w_k^* M^* + Mw_k w_k^* N^* = 0. \tag{8.30}$$

Proof. Factorize W as $W^{1/2}W^{1/2}$ and absorb them into M, N, respectively. Then, there exists a unitary matrix U satisfying

$$MW^{1/2}(I + U) = NW^{1/2}(I - U)$$

according to Lemma 8.5 (3). Using an orthonormal complex basis $\{u_k \in \mathbb{C}^r | k = 1, \ldots, r\}$ to form a basis $\{u_k u_k^* \mid k = 1, \ldots, r\}$ of dimension $r \times r$ for unitary matrices, U can be written as

$$U = \sum_{k-1}^{r} \alpha_k u_k u_k^*, \quad \alpha_k \in \mathbb{C}.$$

To solve for the coefficients α_k, we multiply this equation with u_k. From the orthonormal property $u_i^* u_j = \delta_{ij}$, we know that $U u_k = \alpha_k u_k$. That is, α_k, u_k are, respectively, eigenvalue and eigenvector of U. Further, we know that $|\alpha_k| = 1$ since $U^* U = I$. So, α_k can be described by some phase angle θ_k as $\alpha_k = e^{j\theta_k}$. Now, define a vector w_k as $w_k = W^{1/2} u_k$. Then,

$$\begin{aligned}\sum_{k-1}^{r} w_k w_k^* = \sum_{k-1}^{r} W^{1/2} u_k u_k^* W^{1/2} &= W^{1/2} \left(\sum_{k-1}^{r} u_k u_k^* \right) W^{1/2} = W^{1/2} \cdot I \cdot W^{1/2} \\ &= W\end{aligned}$$

holds in which we have used the property $\sum_{k-1}^{r} u_k u_k^* = I$ of orthonormal basis $\{u_k\}$. Moreover, for $k = 1, \ldots, r$, there also holds

$$\begin{aligned} M w_k (1 + e^{j\theta_k}) &= M W^{1/2} (I + U) u_k \\ &= N W^{1/2} (I - U) u_k \\ &= N w_k (1 - e^{j\theta_k}). \end{aligned}$$

Applying Lemma 8.5 (3), it follows that w_k satisfies (8.30). •

8.2.3.1 Proof of Theorem 8.2

As stated at the beginning of this section, the proof consists of three parts. We will use a set (a, A) that is composed of a scalar a and a symmetrical matrix A. In this space, the inner product is defined as

$$\langle (a, A), (b, B) \rangle = ab + \mathrm{Tr}(AB).$$

Step 1 Transform statement (1) equivalently into the separation condition of two sets.

According to the definition of positive semidefinite matrix, statement (1) is equivalent to that the inequality

$$\begin{bmatrix} x(\omega) \\ u \end{bmatrix}^* M \begin{bmatrix} x(\omega) \\ u \end{bmatrix} \le 0$$

holds for any vector $u \in \mathbb{C}^m$ and frequency ω where the vector $x(\omega) = (j\omega I - A)^{-1} Bu \in \mathbb{C}^n$ depends on the frequency ω and input u. Next, introduce the following two sets:

$$\begin{aligned} \mathcal{P} &= \{(r, 0) \mid r > 0\}, \\ \Theta(\omega) &= \left\{ \left(\begin{bmatrix} x(\omega) \\ u \end{bmatrix}^* M \begin{bmatrix} x(\omega) \\ u \end{bmatrix}, x(\omega)(Ax(\omega) + Bu)^* + (Ax(\omega) + Bu)x^*(\omega) \right) \right\}. \end{aligned}$$

Here u is arbitrary and $x(\omega)$ is the signal given in the preceding text. Since $x(\omega)$ satisfies $j\omega x(\omega) = Ax(\omega) + Bu$, we have $(Ax(\omega) + Bu)x^*(\omega) + x(\omega)(Ax(\omega) + Bu)^* = 0$. Obviously, these two sets do not intersect iff the previous inequality holds.

Step 2 Change $x(\omega)$ to a new x independent of ω and u and extend the set $\Theta(\omega)$ to

$$\Theta = \left\{ \left(\begin{bmatrix} x \\ u \end{bmatrix}^* M \begin{bmatrix} x \\ u \end{bmatrix}, x(Ax + Bu)^* + (Ax + Bu)x^* \right) \middle| \begin{bmatrix} x \\ u \end{bmatrix} \in \mathbb{C}^{n+m} \right\}.$$

Then since (x, u) is arbitrary, $x(Ax + Bu)^* + (Ax + Bu)x^* = 0$ is no longer guaranteed. That is, the inclusion relation $\Theta(\omega) \subset \Theta$ is true. The equivalence between $\Theta(\omega) \cap \mathcal{P} = \emptyset$ and $\Theta \cap \mathcal{P} = \emptyset$ is proved below.

When Θ is separated from $\mathcal{P}$, due to $\Theta(\omega) \subset \Theta$, obviously $\Theta(\omega)$ is also separated from $\mathcal{P}$. Therefore, we prove that $\Theta \cap \mathcal{P} = \emptyset$ holds if $\Theta(\omega) \cap \mathcal{P} = \emptyset$. To this end, we consider the convex hull $\mathbf{conv}\Theta$ of Θ and prove that when $\mathbf{conv}\Theta \cap \mathcal{P} \neq \emptyset$, a contradiction occurs. Since Θ is a cone[2], its convex hull is a convex cone. Therefore, for any natural number N

$$\sum_{k=1}^{N} \left(\begin{bmatrix} x_k \\ u_k \end{bmatrix}^* M \begin{bmatrix} x_k \\ u_k \end{bmatrix}, \quad x_k(Ax_k + Bu_k)^* + (Ax_k + Bu_k)x_k^* \right) \tag{8.31}$$

belongs to $\mathbf{conv}\Theta$. Noting the relation

$$\begin{bmatrix} x_k \\ u_k \end{bmatrix}^* M \begin{bmatrix} x_k \\ u_k \end{bmatrix} = \mathrm{Tr}\left(\begin{bmatrix} x_k \\ u_k \end{bmatrix} \begin{bmatrix} x_k \\ u_k \end{bmatrix}^* M \right)$$

$$(Ax_k + Bu_k)x_k^* = [A \quad B] \begin{bmatrix} x_k \\ u_k \end{bmatrix} \begin{bmatrix} x_k \\ u_k \end{bmatrix}^* \begin{bmatrix} I \\ 0 \end{bmatrix},$$

(8.31) can be summarized as

$$\left(\mathrm{Tr}(WM), [I \quad 0]W \begin{bmatrix} A^T \\ B^T \end{bmatrix} + [A \quad B]W \begin{bmatrix} I \\ 0 \end{bmatrix} \right)$$

where

$$W = \sum_{k=1}^{N} \begin{bmatrix} x_k \\ u_k \end{bmatrix} \begin{bmatrix} x_k \\ u_k \end{bmatrix}^* \geq 0$$

and its dimension is $(n + m) \times (n + m)$. If the convex hull convΘ intersects with $\mathcal{P}$, there exists a vector sequence $(x_1, u_1), \ldots, (x_N, u_N)$ satisfying

$$\mathrm{Tr}(WM) > 0, \quad [I \quad 0]W \begin{bmatrix} A^T \\ B^T \end{bmatrix} + [A \quad B]W \begin{bmatrix} I \\ 0 \end{bmatrix} = 0.$$

Applying Lemma 8.7 to the second equation, we can decompose W into $W = \sum_{k=1}^{n+m} \begin{bmatrix} x_k \\ u_k \end{bmatrix} \begin{bmatrix} x_k \\ u_k \end{bmatrix}^*$ by using suitably rearranged vector sequence $(x_1, u_1), \ldots, (x_{n+m}, u_{n+m})$ such that $x_k(Ax_k + Bu_k)^* + (Ax_k + Bu_k)x_k^* = 0$. Therefore,

[2] Multiplying $\theta \in \Theta$ by $t \geq 0$ and setting $x_1 = \sqrt{t}x, u_1 = \sqrt{t}u$, we have $t\theta \in \Theta$. So, the set Θ is a cone.

$x_k = (j\omega I - A)^{-1}Bu_k$ holds for some ω according to Lemma 8.6. That is, these (x_k, u_k) are vectors belonging to $\Theta \cap \Theta(\omega)$. Moreover, from the first inequality we have $\begin{bmatrix} x_k \\ u_k \end{bmatrix}^* M \begin{bmatrix} x_k \\ u_k \end{bmatrix} > 0$ for at least one k. Thus we have $\Theta(\omega) \cap \mathcal{P} \neq \emptyset$ which contradicts the given condition.

Step 3 Prove the equivalence between $\Theta \cap \mathcal{P} = \emptyset$ and statement 2.

When $\Theta \cap \mathcal{P} = \emptyset$, there exists a hyperplane separating them. Let the normal of the separating hyperplane be (p, P) in which p is a scalar and P is a symmetrical matrix. Θ is contained in the half-space defined by the following inequality:

$$0 \geq p \begin{bmatrix} x \\ u \end{bmatrix}^* M \begin{bmatrix} x \\ u \end{bmatrix} + \mathrm{Tr}(P[x(Ax+Bu)^* + (Ax+Bu)x^*])$$
$$= \begin{bmatrix} x \\ u \end{bmatrix}^* \left\{ pM + \begin{bmatrix} A^TP + PA & PB \\ B^TP & 0 \end{bmatrix} \right\} \begin{bmatrix} x \\ u \end{bmatrix}.$$

Since (x, u) is arbitrary, this inequality is equivalent to

$$pM + \begin{bmatrix} A^TP + PA & PB \\ B^TP & 0 \end{bmatrix} \leq 0. \tag{8.32}$$

Next we need to only prove that we can set $p = 1$. Due to the fact that the set $\mathcal{P}$ is contained in the opposite side of the half-space, we have $p \cdot r + \mathrm{Tr}(P \cdot 0) = pr \geq 0$. So, $p \geq 0$.

Without the loss of generality, we can set $P = \mathrm{diag}(P_1, 0), \det(P_1) \neq 0$. If $p = 0$, partition (A, B) as

$$A = \begin{bmatrix} A_{11} & A_{12} \\ A_{21} & A_2 \end{bmatrix}, \quad B = \begin{bmatrix} B_1 \\ B_2 \end{bmatrix}$$

in accordance with the partition of P. Then from matrix inequality (8.32), we get

$$\begin{bmatrix} A_{11}^TP_1 + P_1A_{11} & P_1A_{12} & P_1B_1 \\ A_{12}^TP_1 & 0 & 0 \\ B_1^TP_1 & 0 & 0 \end{bmatrix} \leq 0$$
$$\Rightarrow P_1B_1 = 0, P_1A_{12} = 0 \Rightarrow A_{12} = 0, B_1 = 0.$$

In this case, A_{11} becomes uncontrollable which contradicts to the controllability of (A, B). Hence $p > 0$ is proved. Finally, statement 2 is derived by dividing (8.32) with p and renaming P/p as P.

Further, the hyperplane with normal $(1, P)$ separates the set $\Theta(\omega)$ from the open set $\mathcal{P}$ when statement 2 is true. So, they cannot intersect because $\mathcal{P}$ is open by the inverse theorem of separating hyperplane.

Finally, the proof is similar for the strict inequality case. The difference lies in that the convex set $\mathcal{P}$ is replaced by

$$\mathcal{P} = \{(r, 0) | r \geq 0\}$$

and Θ needs to be compact in order to guarantee strict separation. Therefore, a new restriction

$$\|x\|^2 + \|u\|^2 \leq 1$$

is imposed on the vector (x, u). Then, the strict inequality is obtained in (8.32). Further, if $p = 0$ the block in the lower right corner of (8.32) is 0 and the strict inequality is not satisfied. So, $p \neq 0$ is obtained without the controllability condition.

Exercises

8.1 Calculate the inner products and the 2-norms for the following functions in the time domain and frequency domain, respectively, so as to validate Parseval's theorem:

$$x(t) = e^{-t} \sin t, \quad y(t) = e^{-2t}.$$

8.2 Constant-scaled bounded real lemma [3]: Given $G(s) = (A, B, C, D)$ prove that the following conditions are equivalent:
1. A is stable and $\|L^{1/2} G L^{-1/2}\|_\infty < \gamma$ for a matrix $L > 0$ with compatible dimension.
2. There exist positive definite matrices $P > 0, L > 0$ satisfying the inequality:

$$\begin{bmatrix} A^T P + PA & PB & C^T \\ B^T P & -\gamma L & D^T \\ C & D & -\gamma L^{-1} \end{bmatrix} < 0.$$

8.3 Prove the strongly positive real lemma.

8.4 Suppose that (A, B, C, D) is a minimal realization of $G(s)$ and the matrix A is stable. Prove that $(A + \epsilon I, B, C, D)$ is a minimal realization of $G(s - \epsilon)$ and $A + \epsilon I$ is stable for sufficiently small $\epsilon > 0$. Then prove that when $G(s - \epsilon)$ is positive real, $G^*(j\omega) + G(j\omega) > 0$ holds for any finite frequency ω.

8.5 Discuss the positive realness and strictly positive realness of transfer matrix

$$G(s) = \begin{bmatrix} 1 & \dfrac{1}{s+1} \\ -\dfrac{1}{s+1} & \dfrac{1}{s+1} \end{bmatrix}$$

through concrete calculation of its frequency response. Then verify the positive realness and strictly positive realness of this transfer matrix again by using the state-space characterizations.

8.6 Discuss the time domain property of strictly positive real systems from the viewpoint of energy dissipation, based on the strictly positive real lemma.

Notes and References

KYP lemma originated in the study of absolute stability [78] and the positive real lemma [45, 95]. The notion of strictly positive realness was from Refs [71, 86, 91], and [18] gave some slightly different strictly positive real conditions. This book adopts the proof of Ref. [80].

9

Algebraic Riccati Equation

The algebraic Riccati equation (ARE) plays an extremely important role in control system design theory. This chapter provides a concise treatment on it. Here, we focus on explaining the properties of ARE and its computing algorithm as well as some basic engineering applications.

Let $A, Q = Q^T, R = R^T$ be real-valued $n \times n$ matrices. The following matrix equation is called an *algebraic Riccati equation*, or simply ARE:

$$A^T X + XA + XRX + Q = 0. \tag{9.1}$$

This is a nonlinear algebraic matrix equation whose solution is not unique and difficult to solve directly. Even for the simplest 2×2 Riccati equation, it is not an easy job to solve the equation directly. For example, for a second-order case

$$A = \begin{bmatrix} 0 & 1 \\ -a_1 & -a_2 \end{bmatrix}, \quad R = rI_2, \quad Q = qI_2, \quad X = \begin{bmatrix} x_1 & x_2 \\ x_2 & x_3 \end{bmatrix},$$

expansion of the Riccati equation yields the following second-order simultaneous polynomial equations:

$$\begin{aligned} & x_1^2 + x_2^2 - 2a_1x_2 + q = 0 \\ & x_2^2 + x_3^2 + 2x_2 - 2a_2x_3 + q = 0 \\ & x_1x_2 + x_2x_3 + x_1 - a_2x_2 - a_1x_3 = 0. \end{aligned}$$

So, new perspective is necessary in order to solve Riccati equation, especially to get solutions with certain specific property.

9.1 Algorithm for Riccati Equation

Riccati equation is deeply tied to a $2n \times 2n$ matrix:

$$H := \begin{bmatrix} A & R \\ -Q & -A^T \end{bmatrix}. \tag{9.2}$$

Robust Control: Theory and Applications, First Edition. Kang-Zhi Liu and Yu Yao.

Matrix in this form is called a *Hamiltonian matrix*, which is used to solve the ARE (9.1).

To understand why the Hamiltonian matrix is related with Riccati equation, we note that the Riccati equation can be rewritten as

$$\begin{aligned}0 &= A^TX + XA + XRX + Q\\ &= \begin{bmatrix} X & -I \end{bmatrix}\begin{bmatrix} A & R \\ -Q & -A^T \end{bmatrix}\begin{bmatrix} I \\ X \end{bmatrix}.\end{aligned}$$

This means that

$$\mathrm{Im}\left(H\begin{bmatrix} I \\ X \end{bmatrix}\right) \subset \mathrm{Ker}\begin{bmatrix} X & -I \end{bmatrix}.$$

It is easy to see that $\mathrm{Ker}[X \quad -I] = \mathrm{Im}[\begin{smallmatrix} I \\ X \end{smallmatrix}]$. So

$$\mathrm{Im}\left(H\begin{bmatrix} I \\ X \end{bmatrix}\right) \subset \mathrm{Im}\begin{bmatrix} I \\ X \end{bmatrix} \tag{9.3}$$

holds. This shows that $\mathrm{Im}[\begin{smallmatrix} I \\ X \end{smallmatrix}]$ is an H-invariant subspace (see Section 2.8). So there exists a matrix $\Lambda \in \mathbb{R}^{n\times n}$ satisfying

$$H\begin{bmatrix} I \\ X \end{bmatrix} = \begin{bmatrix} I \\ X \end{bmatrix}\Lambda. \tag{9.4}$$

In fact, $\Lambda = A + RX$. Therefore, X certainly can be obtained by solving the eigenvalue problem of H.

Next we analyze the eigenvalue property of H. Note that a symmetric matrix is obtained by multiplying the first column of H by $-I$ and then exchanging the first and the second column, that is,

$$\begin{bmatrix} A & R \\ -Q & -A^T \end{bmatrix}\begin{bmatrix} 0 & -I \\ I & 0 \end{bmatrix} = \begin{bmatrix} R & -A \\ -A^T & Q \end{bmatrix}.$$

Denote this transformation matrix by

$$J := \begin{bmatrix} 0 & -I \\ I & 0 \end{bmatrix}.$$

Then, we have $HJ = (HJ)^T = -JH^T$. Hence, the equation

$$J^{-1}HJ = -H^T$$

follows immediately, and H and $-H^T$ are similar. So, if λ is an eigenvalue of H, due to the identity

$$\begin{aligned}|\lambda I - H| &= |J^{-1}(\lambda I - H)J| = |\lambda I - J^{-1}HJ|\\ &= |\lambda I + H^T| = (-1)^{2n}|((-\overline{\lambda})I - H)^*|\\ &= |(-\overline{\lambda})I - H|^*,\end{aligned}$$

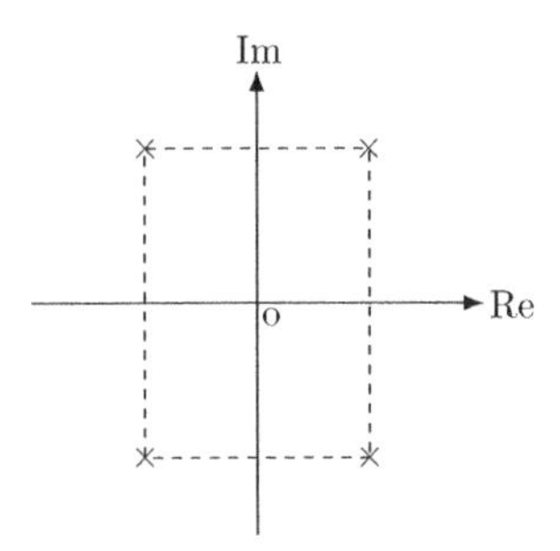

Figure 9.1 Eigenvalue location of Hamiltonian matrix

we see that $-\overline{\lambda}$ is also an eigenvalue of H. Further, since H is real valued, coefficients of its characteristic polynomial must be real. Therefore $(\overline{\lambda}, -\lambda)$, the conjugates of $(\lambda, -\overline{\lambda})$, are also eigenvalues of H. That is, the eigenvalue set $\sigma(H)$ of H is symmetrically placed not only about the real axis but also about the imaginary axis (see Figure 9.1).

The second property of Hamiltonian matrix H is $\sigma(\Lambda) \subset \sigma(H)$. This is easily proved as follows. Set $\Lambda u = \lambda u, u \neq 0$. Then, λ is an eigenvalue of Λ and u the corresponding eigenvector. There holds

$$H \begin{bmatrix} I \\ X \end{bmatrix} u = \begin{bmatrix} I \\ X \end{bmatrix} \Lambda u = \lambda \begin{bmatrix} I \\ X \end{bmatrix} u. \tag{9.5}$$

Since $\left[\begin{smallmatrix} I \\ X \end{smallmatrix}\right] u \neq 0$, λ becomes an eigenvalue of H too.

As aforementioned, the solution of Riccati equation is not unique. What the following theorem states is how to use the basis of H's invariant subspace to construct the solutions of ARE (9.1). Here, we use $\sigma(H|S)$ to denote the eigenvalue set of map H restricted in the invariant subspace S of H. Namely,

$$\sigma(H|S) := \{\lambda \mid Hu = \lambda u, \ 0 \neq u \in S\}. \tag{9.6}$$

So, the eigenvector u corresponding to any eigenvalue $\lambda \in \sigma(H|S)$ belongs to S.

Theorem 9.1 *Suppose that $S \subset \mathbb{C}^{2n}$ is an n-dimensional H-invariant subspace and $X_1, X_2 \in \mathbb{C}^{n\times n}$ are complex matrices satisfying*

$$S = \mathrm{Im} \begin{bmatrix} X_1 \\ X_2 \end{bmatrix}.$$

If X_1 is invertible, $X := X_2X_1^{-1}$ is a solution of ARE (9.1) and $\sigma(A + RX) = \sigma(H|S)$ holds. Moreover, X is independent of the selection of the basis of S.

Conversely, if $X \in \mathbb{C}^{n\times n}$ is a solution of ARE (9.1), then there exist $X_1, X_2 \in \mathbb{C}^{n\times n}$ with X_1 invertible such that $X = X_2X_1^{-1}$. Further, $S = \mathrm{Im}\left[\begin{smallmatrix} X_1 \\ X_2 \end{smallmatrix}\right]$ is an n-dimensional invariant subspace of H and satisfies $\sigma(A + RX) = \sigma(H|S)$.

Proof. Since S is an invariant subspace of H, there exists a matrix $\Lambda \in \mathbb{C}^{n\times n}$ satisfying

$$\begin{bmatrix} A & R \\ -Q & -A^T \end{bmatrix} \begin{bmatrix} X_1 \\ X_2 \end{bmatrix} = \begin{bmatrix} X_1 \\ X_2 \end{bmatrix} \Lambda.$$

Postmultiplying this equation by X_1^{-1}, we get

$$\begin{bmatrix} A & R \\ -Q & -A^T \end{bmatrix} \begin{bmatrix} I \\ X \end{bmatrix} = \begin{bmatrix} I \\ X \end{bmatrix} X_1 \Lambda X_1^{-1}. \tag{9.7}$$

Then, premultiplying Eq. (9.7) by $\begin{bmatrix} X & -I \end{bmatrix}$, we obtain

$$\begin{aligned} 0 &= \begin{bmatrix} X & -I \end{bmatrix} \begin{bmatrix} A & R \\ -Q & -A^T \end{bmatrix} \begin{bmatrix} I \\ X \end{bmatrix} \\ &= XA + A^T X + XRX + Q, \end{aligned}$$

that is, X is a solution of (9.1). Further, we have from (9.7) that

$$A + RX = X_1 \Lambda X_1^{-1},$$

that is, $\sigma(A + RX) = \sigma(\Lambda)$. According to its definition, Λ is the matrix description of the map $H|S$. So, $\sigma(A + RX) = \sigma(H|S)$ holds. Finally, any basis of S can be described as

$$\begin{bmatrix} X_1 \\ X_2 \end{bmatrix} P = \begin{bmatrix} X_1 P \\ X_2 P \end{bmatrix}$$

by a nonsingular matrix P. Since $(X_2 P)(X_1 P)^{-1} = X_2 X_1^{-1} = X$, X is independent of the selection of basis.

Conversely, if X is a solution of (9.1), then it satisfies (9.4) where $\Lambda := A + RX$. So, columns of $\left[\begin{smallmatrix} I \\ X \end{smallmatrix}\right]$ span an n-dimensional invariant subspace S of H, and $\sigma(A + RX) = \sigma(H|S)$ holds. The proof is completed by setting $X_1 = I$, $X_2 = X$. •

In this theorem, no restriction is placed on the property of $\sigma(H|S)$, not even the symmetry about the real axis. This is why we regard the invariant subspace S and X as complex so far. If more properties are imposed on $\sigma(H|S)$, then the solution X of Riccati equation will possess the corresponding properties. In the next section, we discuss Riccati equation with all eigenvalues of $\sigma(H|S)$ being located in the left half-plane.

9.2 Stabilizing Solution

When all eigenvalues of $\sigma(A + RX)$ are in the open left plane, X is called the *stabilizing solution* of Riccati equation. Often used in the control theory is the stabilizing solution. In this section, the existence condition for this solution and its property will be discussed.

According to the distribution of $\sigma(H)$, apparently the existence of stabilizing solution requires that no eigenvalue of H is on the imaginary axis. In this case, there are n eigenvalues of H in $\Re(s) < 0$ and $\Re(s) > 0$, respectively. Denote the invariant subspaces corresponding to the eigenvalues of H in the open left plane by $\mathcal{X}_-(H)$. After obtaining the base vectors of $\mathcal{X}_-(H)$, aligning them into a matrix, and decomposing the matrix, we get

$$\mathcal{X}_-(H) = \mathrm{Im} \begin{bmatrix} X_1 \\ X_2 \end{bmatrix}.$$

The eigenvalues of H in the open left plane are symmetrical about the real axis. Hence, if λ is an eigenvalue, so is its conjugate $\overline{\lambda}$. According to Section 2.8, the invariant subspace $\mathcal{X}_-(H)$ can be selected as real valued so that X_1, X_2 are real-valued matrices.

If X_1 is nonsingular, then the following two subspaces

$$\mathcal{X}_-(H), \quad \mathrm{Im}\begin{bmatrix} 0 \\ I \end{bmatrix} \tag{9.8}$$

are complementary subspaces. So, $X := X_2X_1^{-1}$ can be defined. From the fact that X is independent of the choice of bases, X is determined solely by H. That is, the mapping $H \mapsto X$ can be regarded as a function. Denoting this function by Ric, then X can be written as $X = Ric(H)$. Further, the domain of function Ric is denoted by $\mathrm{dom}(Ric)$. This domain consists of all Hamilton matrices H having the following two properties:

1. H has no eigenvalues on the imaginary axis.
2. The two subspaces in (9.8) are complementary.

Apparently, the stabilizing solution $X = Ric(H)$ of Riccati equation is real valued. Moreover, as shown in Theorem 9.1 in the previous section, this solution is independent of the selection of bases of $\mathcal{X}_-(H)$. So, it is unique.

The following theorem proves that H belongs to $\mathrm{dom}(Ric)$ if and only if Riccati equation has a real symmetric stabilizing solution.

Theorem 9.2 *$H \in \mathrm{dom}(Ric)$ iff there exists a matrix X satisfying the following conditions:*

1. X is real symmetric.
2. X satisfies the algebraic Riccati equation

$$A^TX + XA + XRX + Q = 0.$$

3. $A + RX$ is stable.

Further, $X = Ric(H)$ holds.

Proof. (Sufficiency) Obviously, when there exists X satisfying conditions 1–3,

$$\begin{bmatrix} A & R \\ -Q & -A^T \end{bmatrix}\begin{bmatrix} I \\ X \end{bmatrix} = \begin{bmatrix} I \\ X \end{bmatrix}(A + RX)$$

holds. According to the symmetric property of H's eigenvalues about the imaginary axis, there are no eigenvalues of H on the imaginary axis since $A + RX$ is stable. So

$$\mathcal{X}_-(H) = \mathrm{Im}\begin{bmatrix} I \\ X \end{bmatrix}$$

Holds, and $H \in \mathrm{dom}(Ric)$.

(Necessity) Let $X = Ric(H)$. As the proof of conditions 2 and 3 are exactly the same as that of Theorem 9.1, we only prove condition 1.

Suppose that $X_1, X_2 \in \mathbb{R}^{n\times n}$ satisfies $\mathcal{X}_-(H) = \mathrm{Im}\left[\begin{smallmatrix}X_1\\X_2\end{smallmatrix}\right]$. We first prove that $X_1^T X_2$ is symmetrical. Because $\mathcal{X}_-(H)$ is the invariant subspace corresponding to the eigenvalues of H in the open left plane, there exists a stable matrix $H_- \in \mathbb{R}^{n\times n}$ satisfying

$$H\begin{bmatrix}X_1\\X_2\end{bmatrix} = \begin{bmatrix}X_1\\X_2\end{bmatrix}H_-.$$

Here, H_- is the matrix description of the mapping $H|\mathcal{X}_-(H)$. Premultiplying this equation with $\left[\begin{smallmatrix}X_1\\X_2\end{smallmatrix}\right]J$, we get

$$\begin{bmatrix}X_1\\X_2\end{bmatrix}^T JH\begin{bmatrix}X_1\\X_2\end{bmatrix} = \begin{bmatrix}X_1\\X_2\end{bmatrix}^T J\begin{bmatrix}X_1\\X_2\end{bmatrix}H_-. \tag{9.9}$$

Since JH is symmetrical, the left and right sides of (9.9) are also symmetrical. So

$$\begin{aligned}&(-X_1^TX_2 + X_2^TX_1)H_- = H_-^T(-X_1^TX_2 + X_2^TX_1)^T = -H_-^T(-X_1^TX_2 + X_2^TX_1)\\ &\Rightarrow (-X_1^TX_2 + X_2^TX_1)H_- + H_-^T(-X_1^TX_2 + X_2^TX_1) = 0\end{aligned}$$

holds. This equation is a Lyapunov equation. Its unique solution is

$$-X_1^TX_2 + X_2^TX_1 = 0$$

since H_- is stable. That is, $X_1^TX_2$ is symmetrical. Next, owing to the nonsingularity of X_1 and $X = X_2X_1^{-1}$, premultiplying the previous equation by $(X_1^{-1})^T$ and postmultiplying it by X_1^{-1}, we have $X = (X_1^{-1})^T(X_2^TX_1)X_1^{-1}$. So X is also symmetrical. •

This theorem shows that, by the introduction of function *Ric*, Riccati equation has a real symmetric stabilizing solution iff Hamiltonian matrix H belongs to $\mathrm{dom}(Ric)$. Compared with proving the existence of its stabilizing solution and describing the results directly with Riccati equation, analyzing the eigenvalues of H is much more concise. Therefore, this book adopts this description.

The theorem that follows gives the necessary and sufficient condition for the existence of the unique solution of (9.1) in the case where R is definite. The proof is left as an exercise.

Theorem 9.3 *Suppose that H has no pure imaginary eigenvalues and R is either positive semidefinite or negative semidefinite. Then, $H \in \mathrm{dom}(Ric)$ iff (A, R) is stabilizable.*

Furthermore, when the matrix Q has certain special structure, we have the following conclusion.

Theorem 9.4 *Suppose that H is of the following form:*

$$H = \begin{bmatrix}A & -BB^T\\ -C^TC & -A^T\end{bmatrix}.$$

Then, $H \in \mathrm{dom}(Ric)$ iff (A, B) is stabilizable, and (C, A) has no unobservable eigenvalues on the imaginary axis. Further, if $H \in \mathrm{dom}(Ric)$, then $X = Ric(H) \geq 0$. In addition, $X > 0$ iff (C, A) has no stable unobservable eigenvalues.

Proof. From Theorem 9.3 we know that the stabilizability of (A, B) is necessary for $H \in \mathrm{dom}(Ric)$, and it becomes the sufficient condition if H has no poles on the imaginary axis. Therefore, we need just to prove that, when (A, R) is stabilizable, H has no eigenvalue on imaginary axis iff (C, A) has no unobservable eigenvalues on imaginary axis.

Suppose instead that $j\omega$ is an eigenvalue of H and the corresponding eigenvector is $\left[\begin{smallmatrix} x \\ z \end{smallmatrix}\right] \neq 0$. Then

$$Ax - BB^T z = j\omega x, \quad -C^T Cx - A^T z = j\omega z \tag{9.10}$$

hold. Premultiplying these two equations by z^*, x^*, respectively, we get

$$z^*(A - j\omega I)x = z^* BB^T z \tag{9.11}$$

$$-x^*(A - j\omega I)^* z = x^* C^T Cx \tag{9.12}$$

after rearrangement. Due to the conjugate symmetry of the right side of (9.12), we have $z^*(A - j\omega I)x = -x^* C^T Cx$. Combined with (9.11), we get

$$-\|Cx\|^2 = \|B^T z\|^2.$$

This means that $B^T z = 0$, $Cx = 0$. Substituting them into the two equations of (9.10), respectively, we have

$$(A - j\omega I)x = 0, \quad (A - j\omega I)^* z = 0.$$

Combining these four equations, we obtain

$$z^*[A - j\omega I \quad B] = 0, \quad \begin{bmatrix} A - j\omega I \\ C \end{bmatrix} x = 0.$$

$z = 0$ follows from the stabilizability of (A, B). So there must be $x \neq 0$, namely, $j\omega$ is an unobservable eigenvalue of (C, A) on the imaginary axis. Therefore, $j\omega$ is an eigenvalue of H iff it is an unobservable eigenvalue of (C, A).

Next, we prove that $X \geq 0$ when $X := Ric(H)$. Riccati equation

$$A^T X + XA - XBB^T X + C^T C = 0$$

can be written as

$$(A - BB^T X)^T X + X(A - BB^T X) + XBB^T X + C^T C = 0. \tag{9.13}$$

Noting that $A - BB^T X$ is stable (Theorem 9.2), we regard this equation as a Lyapunov equation and obtain an explicit expression of the solution

$$X = \int_0^\infty \mathrm{e}^{(A-BB^T X)^T t}(XBB^T X + C^T C)\mathrm{e}^{(A-BB^T X)t}\, dt. \tag{9.14}$$

Because the integral function on the right is positive semidefinite, so is X.

Finally, we prove that $\det(X) = 0$, that is, $\mathrm{Ker}(X)$ is not empty iff (C, A) has a stable unobservable eigenvalue. When $x \in \mathrm{Ker}(X)$, there holds $Xx = 0$. Premultiplying (9.13) by x^* and postmultiplying it by x, we get

$$\|Cx\|^2 = 0 \ \Rightarrow \ Cx = 0.$$

Postmultiplying (9.13) by x again, we get

$$XAx = 0.$$

That is, $\mathrm{Ker}(X)$ is A-invariant. When $\mathrm{Ker}(X) \neq \{0\}$, there exists $0 \neq x \in \mathrm{Ker}(X)$ and λ satisfying $\lambda x = Ax = (A - BB^TX)x$[1] and $Cx = 0$. $\Re(\lambda) < 0$ is true since $(A - BB^TX)$ is stable, that is, λ is a stable unobservable eigenvalue.

Conversely, when (C, A) has a stable unobservable eigenvalue λ, there exists a vector $x \neq 0$ satisfying $Ax = \lambda x$, $Cx = 0$. Premultiplying the Riccati equation by x^* and postmultiplying it by x, we have

$$2\Re(\lambda)x^*Xx - x^*XBB^TXx = 0.$$

$x^*Xx = 0$ must hold since $\Re(\lambda) < 0$, that is, X is not positive definite. •

The following corollary is very helpful in solving $\mathcal{H}_2$ control problems.

Corollary 9.1 *Suppose that matrix D has full column rank. Then $R = D^TD > 0$. Given a Hamiltonian matrix*

$$\begin{aligned} H &= \begin{bmatrix} A & 0 \\ -C^TC & -A^T \end{bmatrix} - \begin{bmatrix} B \\ -C^TD \end{bmatrix} R^{-1} \begin{bmatrix} D^TC & B^T \end{bmatrix} \\ &= \begin{bmatrix} A - BR^{-1}D^TC & -BR^{-1}B^T \\ -C^T(I - DR^{-1}D^T)C & -(A - BR^{-1}D^TC)^T \end{bmatrix}, \end{aligned}$$

$H \in \mathrm{dom}(Ric)$ iff (A, B) is stabilizable and $\begin{bmatrix} A-j\omega I & B \\ C & D \end{bmatrix}$ has full column rank for all ω. Further, when $H \in \mathrm{dom}(Ric)$, there holds $X = Ric(H) \geq 0$. Moreover, $X > 0$ iff $\begin{bmatrix} A-sI & B \\ C & D \end{bmatrix}$ has full column rank for all $\Re(s) \leq 0$.

Proof. Firstly, the stabilizability of (A, B) is equivalent to that of $(A - BR^{-1}D^TC, BR^{-1/2})$ (refer to Exercise 9.4). Next, $\begin{bmatrix} A-sI & B \\ C & D \end{bmatrix}$ has full column rank iff $((I - DR^{-1}D^T)C, A - BR^{-1}D^TC)$ has no unobservable eigenvalues at s (Exercise 9.4). Since $I - DR^{-1}D^T \geq 0$, this condition is equivalent to that $((I - DR^{-1}D^T)^{1/2}C,\ A - BR^{-1}D^TC)$ has no unobservable eigenvalues at s. So, the conclusion follows from Theorem 9.4. •

[1] Refer to Exercise 2.14 of Chapter 2.

9.3 Inner Function

When a stable transfer function $N(s)$ with full row rank satisfies

$$N^{\sim}(s)N(s) = I \quad \forall\ s, \tag{9.15}$$

it is called an *inner function*. Since $N^*(j\omega)N(j\omega) = I$ is true for all ω, $\|N(j\omega)q\| = \|q\|(\forall\omega)$ and $\|Nv\|_2 = \|v\|_2$ hold for any constant vector q with suitable dimension and any two-norm bounded vector function $v(s)$. That is, the norms of vector function and constant vector do not change after being mapped by an inner function. Such norm-preserving property is the essence of inner function. From the viewpoint of system engineering, the so-called inner function is actually an *all-pass transfer function* whose gain is 1 over the whole frequency domain. Refer to Section 10.1.2 for more discussions and examples about the scalar case.

The characterization of inner function is given by the following theorem.

Theorem 9.5 *Suppose that $N(s) = (A, B, C, D)$ is stable and (A, B) is controllable. Further, assume that X is the observability Gramian satisfying*

$$A^T X + XA + C^T C = 0.$$

Then, $N(s)$ is an inner function iff the following two conditions are satisfied:

1. $D^T C + B^T X = 0$
2. $D^T D = I$.

Proof. According to $N(-j\infty)^T N(j\infty) = D^T D = I$, condition 2 must be true. Next, a similarity transformation with transformation matrix $T = \begin{bmatrix} I & 0 \\ -X & I \end{bmatrix}$ on

$$N(-s)^T N(s) = \left[\begin{array}{cc|c} A & 0 & B \\ -C^T C & -A^T & -C^T D \\ \hline D^T C & B^T & D^T D \end{array}\right]$$

and a substitution of the Lyapunov equation lead to

$$N(-s)^T N(s) = \left[\begin{array}{cc|c} A & 0 & B \\ -(A^T X + XA + C^T C) & -A^T & -(XB + C^T D) \\ \hline B^T X + D^T C & B^T & D^T D \end{array}\right]$$

$$= \left[\begin{array}{cc|c} A & 0 & B \\ 0 & -A^T & -(XB + C^T D) \\ \hline B^T X + D^T C & B^T & I \end{array}\right].$$

To enforce $N^{\sim}(s)N(s) = I$, all poles must be cancelled. That is, $N^{\sim}(s)N(s)$ has to be both uncontrollable and unobservable. So, from the controllability of (A, B) we see that the statement 1 must be true. Apparently, if conditions 1 and 2 are both satisfied, then $N^{\sim}(s)N(s) = I$. •

Exercises

9.1 Following the procedure that follows to prove Theorem 9.3. Here, assume that $R \geq 0$ (the case of $R \leq 0$ can be proved similarly).

1. To show the sufficiency, we need just to prove that the spaces $\mathcal{X}_-(H)$ and $\mathrm{Im}[\begin{smallmatrix}0\\I\end{smallmatrix}]$ are complementary. That is, to prove that $\mathrm{Ker}(X_1) = \{0\}$ when

$$\mathcal{X}_-(H) = \mathrm{Im}\begin{bmatrix}X_1\\X_2\end{bmatrix}, \quad H\begin{bmatrix}X_1\\X_2\end{bmatrix} = \begin{bmatrix}X_1\\X_2\end{bmatrix}H_-.$$

   To this end, (i) prove that $\mathrm{Ker}(X_1)$ is H_--invariant using $X_2^T X_1 = X_1^T X_2$ and (ii) prove that X_1 is nonsingular by reduction to contradiction. That is, assume that $\mathrm{Ker}(X_1) \neq \{0\}$, and then derive a contradiction.
2. Prove the necessity based on the fact that X is a stabilizing solution when $H \in \mathrm{dom}(Ric)$.

9.2 Following the subsequent procedure to prove the following statement: When (A, B) is stabilizable and (C, A) is detectable, Riccati equation

$$A^T X + XA - XBB^T X + C^T C = 0$$

has a unique positive semidefinite solution. Moreover, this solution is the stabilizing solution.

(a) By Theorem 9.4, Riccati equation has a unique and positive semidefinite solution under the given condition. So, if we can prove that any positive semidefinite solution $X \geq 0$ is the stabilizing solution, then the uniqueness of positive semidefinite solution follows from that of the stabilizing solution.

(b) Use reduction to contradiction. That is, assume that although $X \geq 0$ satisfies Riccati equation, it is not the stabilizing solution. So, $A - BB^T X$ has an unstable eigenvalue λ with eigenvector x. Then prove that $Ax = \lambda x$, $Cx = 0$ based on Riccati equation, that is, to derive a result contradictory to the detectability of (C, A).

9.3 When (A, B) is stabilizable, prove that $(A - BR^{-1}D^T C, BR^{-1}B^T)$ is also stabilizable where $R > 0$.

9.4 When D has full column rank, prove that $[\begin{smallmatrix}A-sI & B\\C & D\end{smallmatrix}]$ has full column rank at a point s iff $((I - DR^{-1}D^T)C, A - BR^{-1}D^T C)$ has no unobservable eigenvalues at this point.

Notes and References

Among the references on Riccati equation, the most complete one is Ref. [51]. Zhou *et al.* [100] also contains a selection of important topics and is easy to read.

10

Performance Limitation of Feedback Control

The fundamental mechanism of control is feedback. It is only with feedback that we can attenuate disturbance and realize precise reference tracking. However, feedback is not omnipotent. The performance of transient response cannot be enhanced unlimitedly by feedback control. In fact, due to the special structure of feedback control, feedback systems embody an unsurpassable limit. For linear systems, this limit is closely related to the unstable poles and zeros of the plant and controller. The purpose of this chapter is to reveal such limit via quantitative analysis so as to provide a guidance for setting reasonable performance specifications in the design of plant structure and the design of feedback controller.

In order to understand qualitatively how the performance limitation of feedback control results in, let us look at an example. Consider the reference tracking of single-input single-output (SISO) systems. Coprimely factorize plant P and controller K with their numerator and denominator polynomials, respectively:

$$P(s) = \frac{N_P}{M_P}, \quad K(s) = \frac{N_K}{M_K}. \tag{10.1}$$

Then, the transfer function from reference input $r(t)$ to output $y_P(t)$ equals

$$T(s) = \frac{PK}{1 + PK} = \frac{N_P N_K}{M_P M_K + N_P N_K}. \tag{10.2}$$

This transfer function is called the *complementary sensitivity* . To guarantee the stability of the system, there cannot be unstable pole-zero cancelation between $P(s)$ and $K(s)$. So the unstable zeros of $P(s)$ and $K(s)$ are kept in $T(s)$ as its zeros. Recalling Chapter 5, in the tracking of step reference input, zeros of $T(s)$ either amplify overshoot or cause undershoot, thus limiting the performance of system. Moreover, since the unstable poles of $P(s)$ cannot be canceled by the zeros of $K(s)$, it limits the range of allowable controllers. This certainly will have a bad impact on the transient response. This example reveals that the most significant character of feedback control is that *although poles can be moved arbitrarily, unstable zeros cannot be moved* (stable zeros can be canceled by the poles of closed-loop system).

Robust Control: Theory and Applications, First Edition. Kang-Zhi Liu and Yu Yao.

In the subsequent sections, we analyze the performance limitation of feedback system quantitatively. For simplicity, we only introduce the results on SISO systems. In fact, the limitation of feedback control is the most prominent in SISO systems. For multiple-input multiple-output (MIMO) systems, the limitation of control performance also involves the zero vectors of plant and the input vectors of reference signal. For details, refer to Refs [15, 14, 69, 84].

10.1 Preliminaries

This section introduces some mathematical knowledge necessary for the analysis of system performance limitation.

10.1.1 Poisson Integral Formula

In complex analysis, a function $F(s)$ differentiable in a simply connected region is called an analytic function in this region. For a function $F(s)$ analytic on a closed curve $\partial\Omega$ and its interior Ω in the complex plane, there holds the following Cauchy theorem [1]:

$$\frac{1}{2\pi j}\oint_{\partial\Omega} F(s)ds = 0. \tag{10.3}$$

Further, the following Cauchy integral formula

$$F(s_0) = \frac{1}{2\pi j}\oint_{\partial\Omega} \frac{F(s)}{s - s_0}ds, \quad s_0 \in \Omega \tag{10.4}$$

is also true where the positive direction of integral path $\partial\Omega$ is *counterclockwise*. Next, we prove Poisson integral formula using it.

Lemma 10.1 (Poisson integral formula) *Suppose that the rational function $F(s)$ is analytic in the open right half plane and $|F(s)|$ bounded in this region. Then, at any point $s_0 = x_0 + jy_0$ $(x_0 > 0)$ in the open right half plane, the following equation holds*

$$F(s_0) = \frac{1}{\pi}\int_{-\infty}^{\infty} F(j\omega)\frac{x_0}{x_0^2 + (\omega - y_0)^2}d\omega. \tag{10.5}$$

Proof. From the assumption we know that $F(s)(\overline{s}_0 + s_0)/(s + \overline{s}_0)$ is also analytic in the open right half plane. Let $\partial\Omega$ be composed of a line $\Re(s) = \epsilon \quad (> 0)$ and an arc in the *right* half plane with infinite radius. From Cauchy integral formula, we have

$$F(s_0) = F(s_0)\frac{\overline{s}_0 + s_0}{s_0 + \overline{s}_0} = \frac{1}{2\pi j}\oint_{\partial\Omega} F(s)\frac{\overline{s}_0 + s_0}{(s - s_0)(s + \overline{s}_0)}ds = I_1 + I_2$$

in which I_1 is the integral along the line $\Re(s) = \epsilon$:

$$\begin{aligned} I_1 &= \frac{1}{2\pi j}\int_{\infty}^{-\infty} F(\epsilon + j\omega)\frac{\overline{s}_0 + s_0}{(\epsilon + j\omega - s_0)(\epsilon + j\omega + \overline{s}_0)}d(j\omega) \\ &\to \frac{1}{\pi}\int_{-\infty}^{\infty} F(j\omega)\frac{x_0}{x_0^2 + (\omega - y_0)^2}d\omega \text{ (as } \epsilon \to 0). \end{aligned}$$

Further, I_2 is the integral along the arc $s = Re^{j\theta}$ with radius $R(\to \infty)$ in the right half plane:

$$I_2 = \frac{1}{\pi} \lim_{R\to\infty} \int_{-\arccos(\epsilon/R)}^{\arccos(\epsilon/R)} F(Re^{j\theta}) \frac{x_0}{(Re^{j\theta} - s_0)(Re^{j\theta} + \bar{s}_0)} Re^{j\theta} d\theta.$$

Since $|F(Re^{j\theta})|$ is bounded, the integral kernel of I_2 converges to zero as $R \to \infty$. As the integral interval $(-\pi/2, \pi/2)$ is bounded, we have $I_2 = 0$. Thus, the conclusion is proved. •

This lemma implies that the value of a stable transfer function at a point s_0 in the right half plane is determined uniquely by its frequency response and the point s_0.

10.1.2 All-Pass and Minimum-Phase Transfer Functions

Stable transfer function $A(s)$ satisfying

$$|A(j\omega)| = 1 \quad \forall \omega \tag{10.6}$$

is called an *all-pass transfer function* . For instance, transfer functions

$$1, \quad \frac{1-s}{1+s}, \quad \frac{\bar{\lambda}-s}{\lambda+s} (\Re(\lambda) > 0), \quad \frac{s^2-2s+5}{s^2+2s+5} = \frac{(1-j2-s)(1+j2-s)}{(1+j2+s)(1-j2+s)}$$

are all-pass. By definition, the physical implication of an all-pass transfer function is that its gain is 1 over the whole frequency domain. So, it does not change the gain of input but delays the phase of input. From the previous examples, it is clear that the poles and zeros of all-pass transfer function are symmetrical about the imaginary axis, as illustrated in Figure 10.1.

For an arbitrary stable transfer function $H(s)$, owing to the property (10.6) of all-pass transfer function $A(s)$, the following relations

$$\|AH\|_\infty = \|H\|_\infty, \quad \|AH\|_2 = \|H\|_2 \tag{10.7}$$

apparently hold. That is, an all-pass transfer function has the property of preserving $\mathcal{H}_\infty, \mathcal{H}_2$ norms.

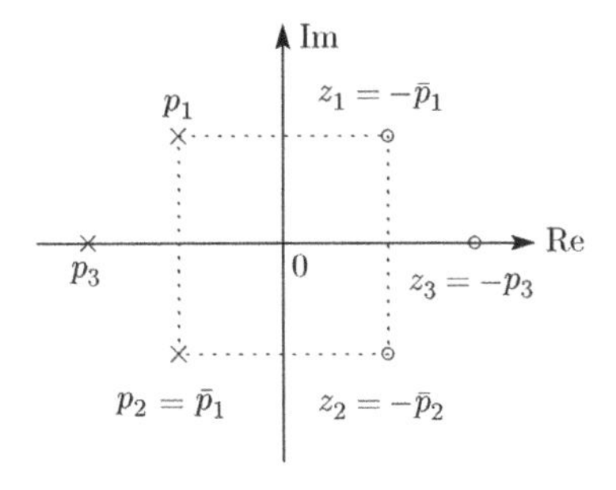

Figure 10.1 Location of poles and zeros of all-pass transfer function

On the other hand, stable transfer function without any zeros in the right half plane is called a *minimum-phase transfer function*. Some examples are

$$\frac{1}{s+1},\ \frac{s}{s+1},\ \frac{s+1}{s^2+2s+6},\ \frac{s^2+2s+6}{(s+2)^2}.$$

The reason why such a transfer function is called minimum phase is that its phase lag is minimal compared with other stable transfer functions with the same gain. For example, comparing the phases of transfer functions $s/(s+2)$ and $s(1-s)/(s+2)(s+1)$, the former is $\pi/2-\arctan(\omega/2)$ while the latter is $\pi/2-\arctan(\omega/2)-2\arctan\omega$.

On the contrary, transfer function with zeros in the right half plane is called a *nonminimum-phase transfer function* , such as

$$\frac{1-s}{1+s},\ \frac{1-s+s^2}{(s+5)(1+s+s^2)}.$$

In fact, all-pass transfer functions are nonminimum phase.

Obviously, any stable rational function $P(s)$ can be factorized as the product of an all-pass transfer function $A(s)$ and a minimum-phase transfer function $P_m(s)$. For example, transfer function $P(s)=(s^2-2s+5)/(s+1)(s^2+4s+6)$ with zeros on $1\pm j2$ can be factorized as

$$P(s)=\frac{1-j2-s}{1+j2+s}\times\frac{1+j2-s}{1-j2+s}\times\frac{s^2+2s+5}{(s+1)(s^2+4s+6)}.$$

10.2 Limitation on Achievable Closed-loop Transfer Function

As stated in Chapter 5, control performance can be quantified by the norm of closed-loop transfer function. From the control objective, the norm of closed-loop transfer function should be reduced as much as possible, or the closed-loop transfer function should get close to desired transfer function as much as possible. However, achievable closed-loop transfer functions are limited. This will be revealed by using the parametrization of stabilizing controllers in this section.

10.2.1 Interpolation Condition

In general, after the substitution of stabilizing controller, the closed-loop transfer function can be described as (see Subsection 7.4.2)

$$H(s)=N_{11}(s)+N_{12}(s)Q(s)N_{21}(s). \tag{10.8}$$

Here, we examine the case where $N_{12}(s)$ and $N_{21}(s)$ have unstable zeros. Make the following assumptions:

Assumption 1 z ($\Re(z)\geq 0$) is a zero of $N_{12}(s)$ with a multiplicity l.
Assumption 2 p ($\Re(p)\geq 0$) is a zero of $N_{21}(s)$ with a multiplicity r.

These conditions imply that $N_{12}(s)$ contains a factor $(s-z)^l$ and $N_{21}(s)$ contains a factor $(s-p)^r$. Mathematically, they are equivalent to

$$N_{12}^{(i)}(z)=0, \quad i=0,\ldots,l-1 \tag{10.9}$$

$$N_{21}^{(j)}(p)=0, \quad j=0,\ldots,r-1 \tag{10.10}$$

where $X^{(i)}(z)$ stands for the derivative $d^iX(s)/ds^i|_{s=z}$, and $X^{(0)}(s)=X(s)$.

Moving the term N_{11} in (10.8) to the left side, we get

$$H(s)-N_{11}(s)=N_{12}(s)Q(s)N_{21}(s).$$

Since $Q(s)$ is stable, unstable zeros of $N_{12}(s)$ and $N_{21}(s)$ will not be canceled by the poles of $Q(s)$. So $H(s)-N_{11}(s)$ has the same zeros. Thus, $H(s)$ must satisfy

$$(H-N_{11})^{(i)}(z)=0, \quad i=0,\ldots,l-1 \tag{10.11}$$

$$(H-N_{11})^{(j)}(p)=0, \quad j=0,\ldots,r-1. \tag{10.12}$$

Such constraint is called *interpolation condition*, which is independent of controller. The physical impact of such constraint will be illustrated in the example of sensitivity.

10.2.2 Analysis of Sensitivity Function

Assume that the plant $P(s)$ satisfies the following assumptions:

Assumption 1 z ($\Re(z)\geq 0$) is a zero of $P(s)$ with a multiplicity l.
Assumption 2 p ($\Re(p)\geq 0$) is a pole of $P(s)$ with a multiplicity r.

The sensitivity function $S(s)$ is defined as

$$S(s)=\frac{1}{1+P(s)K(s)}. \tag{10.13}$$

From Figure 10.2, we see that $S(s)$ is the transfer function from reference input r to tracking error $e=r-y_P$ in the closed-loop system. In the sequel, we use Youla parametrization to expand the sensitivity. In the present case, $G_{22}=-P$ in the generalized plant and can be coprimely factorized as

$$G_{22}(s)=-P(s)=N(s)D^{-1}(s)=\tilde{D}^{-1}(s)\tilde{N}(s). \tag{10.14}$$

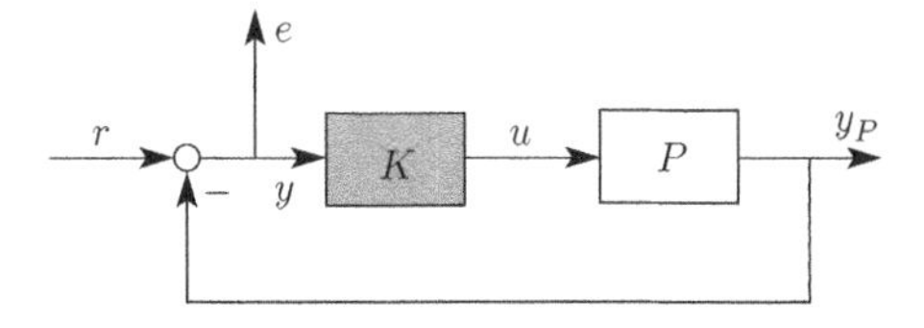

Figure 10.2 Sensitivity and reference tracking

Substituting Youla parametrization given in Theorem 7.3 into $S(s)$, we have

$$\begin{aligned} S(s) =& [1 - \tilde{D}^{-1}(s)\tilde{N}(s)(Y - DQ)(X - NQ)^{-1}]^{-1} \\ =& (X - NQ)[\tilde{D}X - \tilde{N}Y - (\tilde{D}N - \tilde{N}D)Q]^{-1}\tilde{D}. \end{aligned}$$

$\tilde{D}N = \tilde{N}D$ follows from (10.14). Further, from the $(2,2)$ block of (7.32), we have $\tilde{D}X - \tilde{N}Y = -1$. Eventually, the sensitivity $S(s)$ becomes

$$S(s) = -(X - NQ)\tilde{D}. \tag{10.15}$$

Moreover, changing the order of the two matrices on the left side of (7.32), their product is still an identity. From its $(2,2)$ block, we obtain $N\tilde{Y} - X\tilde{D} = 1$. So, substituting $-X\tilde{D} = 1 - N\tilde{Y}$ into the sensitivity S in (10.15), we obtain another expression:

$$S(s) = 1 - N\tilde{Y} + NQ\tilde{D} = 1 + N(-\tilde{Y} + Q\tilde{D}). \tag{10.16}$$

Zeros of $N(s)$ are the zeros of $P(s)$, and zeros of $\tilde{D}(s)$ are the poles of $P(s)$ (refer to Exercise 7.8 in Chapter 7). Thus, the sensitivity $S(s)$ must satisfy the following interpolation conditions:

$$S(z) = 1, \ S^{(i)}(z) = 0, \ i = 1, \ldots, l - 1 \tag{10.17}$$

$$S^{(j)}(p) = 0, \ j = 0, \ldots, r - 1. \tag{10.18}$$

That is, when there exist unstable zeros or poles in the plant, feedback control cannot achieve arbitrary shaping of sensitivity. Next, we illustrate the impact it brings to system performance by examples.

Example 10.1 *Let z be an unstable zero of $P(s)$. Substituting the coprime factorization $N(s)/M(s)$ of open-loop system $L(s) = P(s)K(s)$ into the sensitivity, we get*

$$S(s) = \frac{1}{1 + L(s)} = \frac{M(s)}{M(s) + N(s)}. \tag{10.19}$$

Thus, we see that the unstable poles $p_i (i = 1, \ldots, k)$ of $L(s)$ are the zeros of sensitivity. So, $S(s)$ can be factorized as the product of a minimum-phase transfer function $S_m(s)$ and an all-pass transfer function $A(s)$:

$$S(s) = S_m(s)A(s), \ A(s) = \prod_{i=1}^{k} \frac{p_i - s}{\overline{p}_i + s}. \tag{10.20}$$

In the minimization of sensitivity weighted by $W(s)$, $\|SW\|_\infty$ represents the tracking error corresponding to a reference input modeled by $W(s)$. So, the smaller it is the better. However, due to $S(z) = 1$, the invariance of $\mathcal{H}_\infty$ norm w.r.t. all-pass transfer function and the maximum modulus theorem[1], it can be proved that

$$\| WS\|_\infty = \|WS_m\|_\infty \geq |W(z)S_m(z)|$$

[1] A function analytical in the closed right half plane must take its maximum absolute value on the imaginary axis (see reference [1]).

$$=|W(z)S(z)A^{-1}(z)| = |W(z)A^{-1}(z)| \tag{10.21}$$

$$=|W(z)|\prod_{i=1}^{k}\left|\frac{\overline{p}_i + z}{p_i - z}\right|. \tag{10.22}$$

When there is a pole p_i close to z, the second factor will be very big, thus enlarging the tracking error. Similarly, when z comes close to a pole of $W(s)$, this norm cannot be reduced. For example, in the tracking of step signal $W(s) = 1/s$, if $|z| < 1$ then $\|SW\|_\infty > 1/|z| \gg 1$. Thus, good tracking performance cannot be achieved. This in fact corresponds to the well-known fact that the plant cannot have zeros at the origin in tracking step signals.

Deeper quantitative analyses on the tracking error will be elaborated in the following sections.

10.3 Integral Relation

The famous Bode integral relation on the sensitivity and Bode phase formula are presented in this section.

10.3.1 Bode Integral Relation on Sensitivity

Let the open-loop transfer function be $L(s) = P(s)K(s)$ and make the following assumption:

Assumption 3 The closed-loop system (P, K) is stable and $L(s)$ contains poles $p_i(i = 1, \dots, k)$ in the open right half plane.

As stated before, all poles of the open-loop transfer function $L(s)$ in the closed right half plane become the zeros of sensitivity function $S(s)$. Thus, by Assumption 3, $S(s)$ contains zeros $p_i(i = 1, \dots, k)$ in the open right half plane. Therefore, the sensitivity function $S(s)$ can be factorized as (10.20). Note that, since the relative degree of sensitivity $S(s)$ is zero, $S_m(s)$ does not have zeros in the closed right half plane including infinity except those coming from the poles of $L(s)$ on the imaginary axis. So, function $F(s) := \ln\ S_m(s)$ is analytical and bounded in the closed right half plane except the imaginary axis poles of $L(s)$. The reason is as follows: $d\ln S_m(s)/ds = S_m^{-1}(s)dS_m/ds$ in form, but $S_m^{-1}(s)$ is bounded in the closed right half plane except the imaginary axis poles of $L(s)$; further, dS_m/ds is also bounded in the closed right half plane due to the stability of $S_m(s)$.

Then according to Lemma 10.1, at an arbitrary point $s_0 = x_0 + jy_0$ in the closed right half plane,

$$\ln S_m(s_0) = \frac{1}{\pi}\int_{-\infty}^{\infty} \ln S_m(j\omega)\frac{x_0}{x_0^2 + (\omega - y_0)^2}d\omega \tag{10.23}$$

holds. Write $S_m(s)$ as $|S_m(s)|e^{j\arg S_m(s)}$, we have $\ln S_m(s) = \ln|S_m(s)| + j\arg S_m(s)$. Since $|A(j\omega)| = 1$, $\ln|S(j\omega)| = \ln|S_m(j\omega)| = \Re(\ln S_m(j\omega))$ holds. Taking the real parts on both sides of (10.23), we have[2]

$$\ln|S_m(s_0)| = \frac{1}{\pi}\int_{-\infty}^{\infty} \ln|S(j\omega)|\frac{x_0}{x_0^2+(\omega-y_0)^2}d\omega. \tag{10.24}$$

Based on this, we can derive the relation between the unstable poles of open-loop system and sensitivity.

Theorem 10.1 *Suppose that the relative degree γ of $L(s)$ is no less than 2. Subject to Assumption 3, the following Bode sensitivity integral relation holds:*

$$\int_0^{\infty} \ln|S(j\omega)|d\omega = \pi\sum_{i=1}^{k}\Re(p_i) > 0. \tag{10.25}$$

Moreover, this integral is zero when $L(s)$ is stable.

Proof. Let $y_0 = 0$ in (10.24). Multiplying both sides with x_0 and taking limit, we obtain

$$\lim_{x_0\to\infty} x_0 \ln|S_m(x_0)| = \frac{1}{\pi}\int_{-\infty}^{\infty} \ln|S(j\omega)|d\omega = \frac{2}{\pi}\int_0^{\infty} \ln|S(j\omega)|d\omega. \tag{10.26}$$

Next, we prove

$$\lim_{x_0\to\infty} x_0 \ln S(x_0) = 0. \tag{10.27}$$

First, as $x_0 \to \infty$, by the assumption on relative degree, $L(x_0) \to c/x_0^{\gamma}$, $S(x_0) \to 1$ where c is a constant, $\gamma \geq 2$. So $x_0^2 dL(x_0)/dx_0 \to -\gamma c/x_0^{\gamma-1} \to 0$. Then, since $dS(x_0)/dx_0 = -S^2(x_0)dL(x_0)/dx_0$, by De Lôpital's Theorem, it can be derived that

$$\begin{aligned}
\lim_{x_0\to\infty} x_0 \ln S(x_0) &= \lim_{x_0\to\infty} \frac{\ln S(x_0)}{1/x_0} = \lim_{x_0\to\infty} \frac{\frac{1}{S(x_0)}\frac{dS(x_0)}{dx_0}}{-1/x_0^2} \\
&= \lim_{x_0\to\infty} S(x_0)x_0^2\frac{dL(x_0)}{dx_0} = \lim_{x_0\to\infty} S(x_0) \lim_{x_0\to\infty} x_0^2\frac{dL(x_0)}{dx_0} \\
&= 0.
\end{aligned}$$

Therefore, taking the real part of (10.27) we get $\lim_{x_0\to\infty} x_0 \ln|S(x_0)| = 0$. Since $\ln|S(x_0)| = \ln|S_m(x_0)| + \ln|A(x_0)|$, there holds

$$\begin{aligned}
\lim_{x_0\to\infty} x_0 \ln|S_m(x_0)| &= -\lim_{x_0\to\infty} x_0 \ln|A(x_0)| \\
&= -\sum_{i=1}^{k} \lim_{x_0\to\infty} x_0 \ln\left|\frac{p_i - x_0}{\overline{p}_i + x_0}\right|.
\end{aligned}$$

Finally, we need only prove that $\lim_{x_0\to\infty} x_0 \ln\left|\frac{p_i - x_0}{\overline{p}_i + x_0}\right| = -2\Re(p_i)$. This is accomplished by applying De Lôpital theorem once again.

[2] Note that $|A(s_0)| \neq 1$. So the left is no equal to $\ln|S(s_0)|$.

Finally, when $L(s)$ is stable and $S(s)$ has no unstable zeros, $S(s) = S_m(s)$. Therefore, the integral is zero in this case. •

Note that $\ln|S(j\omega)| < 0$ as long as $|S(j\omega)| < 1$. Similarly, $\ln|S(j\omega)| > 0$ as long as $|S(j\omega)| > 1$. Theorem 10.1 indicates that, if the sensitivity is less than 1 in some frequency band, there is another frequency band in which the sensitivity is greater than 1. Further, when there are unstable zeros in the open-loop transfer function, the area in the frequency band where the sensitivity amplitude is greater than 1 is bigger than the area in the frequency band where the sensitivity amplitude is less than 1. This is known as the *water bed effect*. That is, when one place is pushed down, other places come up. This means that it is impossible to roll off the sensitivity uniformly over the whole frequency domain.

Example 10.2 *As an example, Figure 10.3 depicts the gain of sensitivity for the open-loop transfer function:*

$$L(s) = \frac{4(2-s)}{s(s+2)(s+10)}.$$

The water bed effect appears clearly in the figure.

Fortunately, in practice, the sensitivity represents the transfer function about reference tracking, and most reference inputs, such as step signal, have less high-frequency components. So, as long as the gain of sensitivity is low enough in the low-frequency band, tracking error is reduced substantially. A slightly higher sensitivity gain in other frequency bands will not result in much bad influence.

Furthermore, when the relative degree of open-loop transfer function is 1, it is known that the integral relation

$$\int_0^\infty \ln|S(j\omega)|d\omega = \pi\sum_{i=1}^{k}\Re(p_i) - \frac{\pi}{2}\lim_{s\to\infty} sL(s) \tag{10.28}$$

holds between the sensitivity and the open-loop transfer function [94].

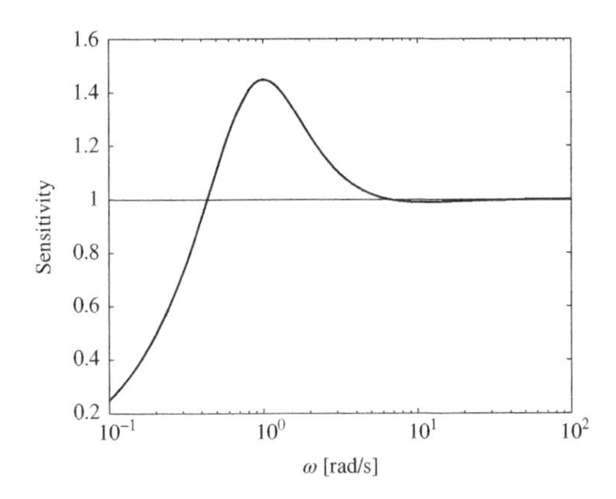

Figure 10.3 Water bed effect

10.3.2 Bode Phase Formula

In the loop shaping method of classic control, Bode phase formula on minimum-phase open-loop systems plays a very important role. The theorem described in the following is a slightly extended version of Bode phase formula.

Theorem 10.2 *Assume that the open-loop transfer function $L(s)$ has no poles and zeros in the open right half plane and $L(j0) > 0$. Then, at any point $j\omega_0$ ($\omega_0 > 0$) different from poles and zeros of $L(s)$ on the imaginary axis, the following relation holds*

$$\arg L(j\omega_0) = \frac{1}{\pi}\int_{-\infty}^{\infty} \frac{d\ln|L(j\omega_0 e^{\nu})|}{d\nu} \ln\coth\frac{|\nu|}{2} d\nu \tag{10.29}$$

where $\nu := \ln(\omega/\omega_0)$.

Note that this theorem is true even for improper transfer function $L(s)$.

The curve of function $\ln\coth\frac{|\nu|}{2}$ is shown in Figure 10.4. Clearly, this function begins rolling off sharply roughly from $\nu = 1 \Leftrightarrow \omega = \omega_0$. Its area can be approximated by the area in $\nu \in (-2.5, 2.5) \Leftrightarrow \omega \in (0.1\omega_0, 10\omega_0)$. It can be proved that the area of function $\ln\ \coth\frac{|\nu|}{2}$ is $\pi^2/2$. Next, we consider the case where ω_0 is the crossover frequency ω_c (satisfying $|L(j\omega_c)| = 1$). If the slope of open-loop gain is constant in a frequency band of 1 dec around ω_c, that is, $d\ln|L|/d\nu = d\ln|L(j\omega)|/d\ln\omega = -k$ (equivalent to $-20k$ dB/dec) for $\omega \in (0.1\omega_c, 10\omega_c)$, according to theorem 10.2 we have

$$\arg L(j\omega_c) \approx -\frac{\pi}{2}k. \tag{10.30}$$

Note that $\mathrm{PM} := \pi + \arg L(j\omega_c)$ is the phase margin of the closed-loop feedback system. Since the phase angle in ω_c equals

$$\arg L(j\omega_c) \approx \begin{cases} -90^\circ, & k = 1 \\ -180^\circ, & k = 2, \end{cases} \tag{10.31}$$

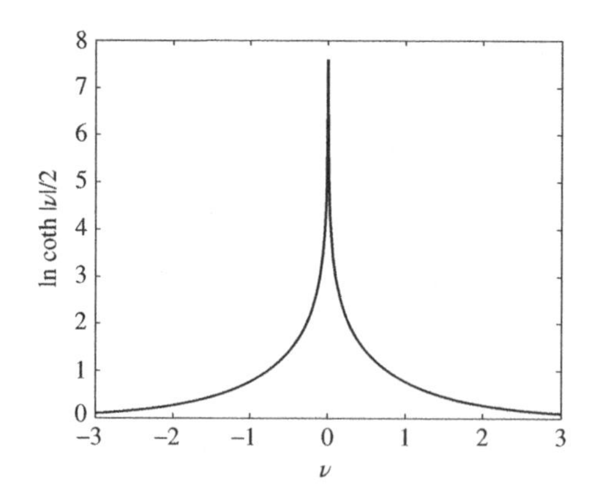

Figure 10.4 Curve of function $\ln\coth\frac{|\nu|}{2}$

to guarantee a sufficient phase margin ($40^\circ \leq \text{PM} \leq 60^\circ$), the slope of logarithmic gain needs be kept as -1 within a frequency band of 1 dec around the crossover frequency. This has been an important rule in classic control design.

10.3.2.1 Proof of Theorem 10.2

First of all, we compute the integral along a path consisting of the right part C_r of a circle centered at $(0, j\omega)$ $(\omega \neq \omega_0)$ with a radius r (refer to C_1, C_2 in Figure 10.5). $s - j\omega = re^{j\theta}$ on C_r, so we have

$$\int_{C_r} \frac{\ln\,(s - j\omega)}{s^2 + \omega_0^2} ds = \int_{-\pi/2}^{\pi/2} \frac{jr(\ln\ r + j\theta)e^{j\theta}}{(j\omega + re^{j\theta})^2 + \omega_0^2} d\theta.$$

Note that $r \ln r \to 0$ as $r \to 0$. So

$$\lim_{r\to 0} \int_{C_r} \frac{\ln\,(s - j\omega)}{s^2 + \omega_0^2} ds = 0.$$

Although the poles and zeros of $L(s)$ on the imaginary axis are the singularities of $\ln L(s)$, it follows from the previous equation that these singularities does not affect the integral of $\ln L(s)/(s^2 + \omega_0^2)$ along the imaginary axis. Besides, by assumption we see that $\ln L(s)$ is analytic in the open right half plane.

Then, setting $\partial\Omega$ as the curve shown in Figure 10.5, we get

$$\oint_{\partial\Omega} \frac{\ln L(s)}{s^2 + \omega_0^2} ds = 0 \tag{10.32}$$

from Cauchy integral formula. In this integral, the integral along the semicircle with infinite radius is zero because the integrand is zero. So,

$$0 = \int_{-j\infty}^{-j(\omega_0+r)} + \int_{C_1} + \int_{-j(\omega_0-r)}^{j(\omega_0-r)} + \int_{C_2} + \int_{j(\omega_0+r)}^{j\infty}$$

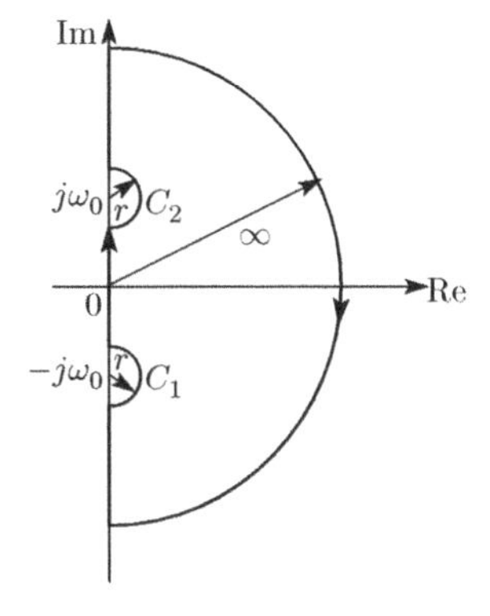

Figure 10.5 Integration path

holds. $s = -j\omega_0 + re^{j\theta}$ ($\theta \in [-\pi/2, \pi/2]$) on the semicircle C_1. Substituting it and taking limit $r \to 0$, we get

$$\int_{C_1} \frac{\ln L(s)}{s^2+\omega_0^2} ds = \int_{-\pi/2}^{\pi/2} \frac{\ln L(-j\omega_0 + re^{j\theta})}{re^{j\theta} - j2\omega_0} jd\theta \quad \to \quad -\frac{\pi}{2\omega_0} \ln L(-j\omega_0).$$

Similarly, when $r \to 0$, we have

$$\int_{C_2} \frac{\ln L(s)}{s^2+\omega_0^2} ds \to \frac{\pi}{2\omega_0} \ln L(j\omega_0).$$

So, by taking the limit $r \to 0$, we obtain

$$0 = j\int_{-\infty}^{\infty} \frac{\ln L(j\omega)}{\omega_0^2 - \omega^2} d\omega + \frac{\pi}{2\omega_0} \ln \frac{L(j\omega_0)}{L(-j\omega_0)}. \tag{10.33}$$

Since $L(s)$ is a real rational function and $L(j0) > 0$, there holds $|L(j\omega_0)| = |L(-j\omega_0)|$ and $\arg\ L(-j\omega_0) = -\arg\ L(j\omega_0)$. Therefore, the second term on the right side becomes $j\frac{\pi}{\omega_0} \arg\ L(j\omega_0)$. Comparing the imaginary parts of both sides, we obtain

$$\arg\ L(j\omega_0) = \frac{\omega_0}{\pi} \int_{-\infty}^{\infty} \frac{\ln |L(j\omega)|}{\omega^2 - \omega_0^2} d\omega = \frac{2\omega_0}{\pi} \int_0^{\infty} \frac{\ln |L(j\omega)|}{\omega^2 - \omega_0^2} d\omega. \tag{10.34}$$

(Although the integrand on the right is divergent at ω_0, the integral itself is bounded.)

Next, we convert the right side into the desired form. First, direct calculation shows that

$$\frac{d}{d\nu}\left(\ln \coth \frac{|\nu|}{2}\right) = -\frac{1}{\sinh \nu}. \tag{10.35}$$

Then, by a variable change $\nu = \ln(\omega/\omega_0)$, the integral interval $[0, \infty)$ turns into $(-\infty, \infty)$. Further, via substitution of $\omega/\omega_0 = e^\nu$ and $d\nu = d\omega/\omega$, the right side of (10.34) reduces to

$$\begin{aligned}
&\frac{2}{\pi} \int_0^{\infty} \frac{\ln |L(j\omega)|}{\omega/\omega_0 - \omega_0/\omega} \frac{d\omega}{\omega} = \frac{2}{\pi} \int_{-\infty}^{\infty} \frac{\ln |L(j\omega_0 e^\nu)|}{e^\nu - e^{-\nu}} d\nu \\
&= \frac{1}{\pi} \int_{-\infty}^{\infty} \frac{\ln |L(j\omega_0 e^\nu)|}{\sinh \nu} d\nu \\
&= -\frac{1}{\pi}\left[\ln |L(j\omega_0 e^\nu)| \ln\ \coth \frac{|\nu|}{2}\right]_{-\infty}^{\infty} + \frac{1}{\pi} \int_{-\infty}^{\infty} \frac{d \ln |L(j\omega_0 e^\nu)|}{d\nu} \ln\ \coth \frac{|\nu|}{2} d\nu
\end{aligned} \tag{10.36}$$

Since $|L(j\omega_0 e^\nu)|$ is not an even function of ν, it is not straightforward to prove that the first term on the right side is zero. But the proof is accomplished through the following steps:

1. It can be proved that $\coth \frac{|\nu|}{2} = \left|\frac{\omega+\omega_0}{\omega-\omega_0}\right|$. So the first term on the right side of the equation can be written as $-\frac{1}{\pi}\left[\ln |L(j\omega)| \ln \left|\frac{\omega+\omega_0}{\omega-\omega_0}\right|\right]_0^{\infty}$.

2. When $\omega \to \infty$ or 0, we always have $\ln\left|\frac{\omega+\omega_0}{\omega-\omega_0}\right| \to 0$. So, as long as $\ln|L(j\infty)|$ or $\ln|L(j0)|$ is bounded, the term corresponding to it is zero.
3. Note that $\frac{d\ln|L(j\omega)|}{d\ln\ \omega}$ is the slope of Bode gain and is bounded when $\omega \to 0$ or ∞. If $\ln|L(j\omega)|$ diverges as $\omega \to 0$ or ∞, it can be proved that its limit is zero by applying De Lôpital theorem to $\ln|L(j\omega)|\ln\left|\frac{\omega+\omega_0}{\omega-\omega_0}\right| = \ln|L(j\omega)|/(1/\ln\left|\frac{\omega+\omega_0}{\omega-\omega_0}\right|)$.

As such, we have obtained the conclusion. •

Further, when the open-loop transfer function $L(s)$ has poles and zeros in the right half plane, we may do factorization $L(s) = L_m(s)A(s)B(s)^{-1}$ to factorize the open-loop transfer function into the product of a minimum-phase transfer function $L_m(s)$ satisfying conditions in Theorem 10.2 and all-pass transfer functions $A(s) = \prod_{i=1}^{k}(z_i - s)/(\overline{z}_i + s)$, $B(s) = \prod_{j=1}^{l}(p_j - s)/(\overline{p}_j + s)$. Here, z_i, p_j are, respectively, the zero and pole of $L(s)$ in the open right half plane. So, $|L| = |L_m|$ and $\arg L = \arg L_m + \arg A - \arg B$ hold. Applying the Bode integral relation given by the theorem above to L_m and substituting it into the integral, Bode integral condition can be extended to general open-loop system as follows.

Corollary 10.1 *Suppose that the open-loop transfer function $L(s)$ satisfies $L(j0) > 0$, and its poles and zeros in the open right half plane are $z_i (i = 1, \ldots, k)$ and $p_j (j = 1, \ldots, l)$, respectively. Then,*

$$\begin{aligned}\arg L(j\omega_0) =& \frac{1}{\pi}\int_{-\infty}^{\infty}\frac{d\ln|L|}{d\nu}\ln\coth\frac{|\nu|}{2}d\nu + \sum_{i=1}^{k}\arg\frac{z_i - j\omega_0}{\overline{z}_i + j\omega_0}\\ &-\sum_{j=1}^{l}\arg\frac{p_j - j\omega_0}{\overline{p}_j + j\omega_0}\end{aligned} \tag{10.37}$$

and the phase angle of the second term on the right side is always negative.

10.4 Limitation of Reference Tracking

In this section, we prove that the infimum of tracking error area w.r.t. step reference signal is constrained by unstable poles and zeros of the plant. These results reveal quantitatively the limitation of feedback control.

In this section, the following basic assumption is made.

Assumption 4 $P(s)$ has no zeros on the imaginary axis and its zeros in the open right half plane are z_i $(i = 1, \ldots, k)$ (multiplicity counted).

10.4.1 1-Degree-of-Freedom System

The structure of 1-DOF system is shown in Figure 10.2. The closed-loop transfer function of $r \mapsto e$ is the sensitivity $S(s)$, and the tracking error is $\hat{e}(s) = S(s)\hat{r}(s)$.

10.4.1.1 Stable Plant Case

In this case, all stabilizing controllers are given by $K(s) = Q/(1 - PQ)$. Substituting it into the sensitivity $S(s)$, we get

$$S(s) = 1 - P(s)Q(s). \tag{10.38}$$

Further, the plant $P(s)$ can be factored as

$$P(s) = L(s)P_m(s),\ L(s) = \prod_{i=1}^{k} L_i(s),\ L_i(s) = \frac{\overline{z}_i}{z_i}\frac{z_i - s}{\overline{z}_i + s} \tag{10.39}$$

where $L_i(s)$ is an all-pass transfer function, and $P_m(s)$ is a minimum-phase transfer function without imaginary zeros. Note that $L_i(s)$ has been normalized as such that $L_i(0) = 1$.

The following theorem gives the infimum on the tracking error achievable by feedback control.

Theorem 10.3 *Suppose that $P(s)$ is stable and satisfies Assumption 4. Then, the infimum of tracking error w.r.t. unit step reference input is*

$$\inf\|e\|_2^2 = \sum_{i=1}^{k} \frac{2\Re(z_i)}{|z_i|^2}. \tag{10.40}$$

Further, $\inf\|e\|_2^2 = 0$ *when P has no unstable zeros.*

Proof. Laplace transform of the tracking error is

$$\hat{e}(s) = (1 - PQ)\hat{r}(s) = (1 - PQ)\frac{1}{s}. \tag{10.41}$$

For the two-norm of $e(t)$ to be bounded, $\hat{e}(s)$ must be stable. So, the free parameter $Q(s)$ must satisfy $1 - P(0)Q(0) = 0$. By this and $L_i(0) = 1$, we see that $1 - P_m(0)Q(0)\prod_{i=j\geq 1}^{k} L_i(0) = 0$ holds for all $1 \leq j \leq k$. So, by the norm-preserving property of all-pass transfer function, it can be proved that

$$\begin{aligned}\|e\|_2^2 &= \left\|\left(1 - P_mQ\prod_{i=1}^{k} L_i\right)\frac{1}{s}\right\|_2^2 = \left\|\left(L_1^{-1} - P_mQ\prod_{i=2}^{k} L_i\right)\frac{1}{s}\right\|_2^2 \\ &= \left\|(L_1^{-1} - 1)\frac{1}{s} + \left(1 - P_mQ\prod_{i=2}^{k} L_i\right)\frac{1}{s}\right\|_2^2.\end{aligned} \tag{10.42}$$

$(L_1^{-1} - 1)/s = 2\Re(z_1)/\overline{z}_1(z_1 - s)$ is antistable, meanwhile the second term is stable due to $1 - P_m(0)Q(0)\prod_{i=2}^{k} L_i(0) = 0$. So, the first term is orthogonal to the second. Hence, from Lemma 2.6 we get

$$\begin{aligned}\|e\|_2^2 &= \left\|(L_1^{-1} - 1)\frac{1}{s}\right\|_2^2 + \left\|\left(1 - P_mQ\prod_{i=2}^{k} L_i\right)\frac{1}{s}\right\|_2^2 \\ &= \frac{4(\Re(z_1))^2}{|z_1|^2}\left\|\frac{1}{z_1 - s}\right\|_2^2 + \left\|\left(1 - P_mQ\prod_{i=2}^{k} L_iP_mQ\right)\frac{1}{s}\right\|_2^2.\end{aligned} \tag{10.43}$$

Further, application of Lemma 2.7 leads to

$$\|e\|_2^2 = \frac{2\Re(z_1)}{|z_1|^2} + \left\| \left(1 - P_m Q \prod_{i=2}^{k} L_i\right) \frac{1}{s} \right\|_2^2. \tag{10.44}$$

Repeating this procedure, we get eventually

$$\|e\|_2^2 = \sum_{i=1}^{k} \frac{2\Re(z_i)}{|z_i|^2} + \left\| (1 - P_m Q) \frac{1}{s} \right\|_2^2. \tag{10.45}$$

Although $P_m^{-1}(s)$ is improper, $Q(s,\epsilon) := P_m^{-1}(s)/(\epsilon s + 1)^r$ is stable and proper in which $\epsilon > 0$ and r is the relative degree of $P(s)$. Further, $Q(s,\epsilon) \to P_m^{-1}(s)$ when $\epsilon \to 0$. That is, we can make $1 - P_m Q$ approach arbitrarily close to zero with a stable and proper Q. So, the infimum of the second term in the previous equation is zero. Therefore, we obtain

$$\inf \|e\|_2^2 = \sum_{i=1}^{k} \frac{2\Re(z_i)}{|z_i|^2}.$$

$P = P_m$ when P has no unstable zeros. From the last part of the proof, infimum of the two-norm of tracking error is zero. •

This theorem reveals that the precision of reference tracking for a stable plant is limited by its unstable zeros. Obviously, when there exists unstable zeros close to the origin, the tracking performance will be worsened inevitably.

Example 10.3 *Design a controller for stable plant* $P(s) = (s-2)/(s+1)(s+3)$ *such that the tracking error is as close to the infimum as possible.*
Solution Firstly, factorize $P(s)$ as the product of $P_m(s) = -(s+2)/(s+1)(s+3)$ and $L(s) = (2-s)/(2+s)$. Since there is an unstable zero $z = 2$, the infimum of tracking error is

$$\inf \|e\|_2^2 = \frac{2z}{z^2} = 1.$$

As seen from the proof of the theorem, we need just to design a controller such that $\|(1 - P_m Q)/s\|_2$ approaches zero. Here, we choose $Q(s)$ as

$$Q(s,\epsilon) = \frac{P_m^{-1}(s)}{\epsilon s + 1} = -\frac{(s+1)(s+3)}{(s+2)(\epsilon s + 1)}. \tag{10.46}$$

By a simple calculation, we get

$$\left\| (1 - P_m Q) \frac{1}{s} \right\|_2^2 = \left\| \frac{1}{s + \epsilon^{-1}} \right\|_2^2 = \frac{1}{2\epsilon^{-1}} = \frac{\epsilon}{2}$$

which approaches zero as $\epsilon \to 0$. The corresponding controller is

$$K(s,\epsilon) = -\frac{1}{2(1+\epsilon)} \times \frac{(s+1)(s+3)}{s(\frac{\epsilon}{2(1+\epsilon)} s + 1)}. \tag{10.47}$$

When $\epsilon \to 0$, the controller approaches the improper optimal controller $K_{\text{opt}}(s) = -(s+1)(s+3)/(2s)$. That is, although the infimum can be approached arbitrarily, there does not exist a proper controller achieving this infimum.

In addition, characteristic polynomial of the closed-loop system is

$$\begin{aligned}p(s) =& M_P M_K + N_P N_K \\ =& (s+1)(s+3)\left[\epsilon s^2 + (1+2\epsilon)s + 2\right] \\ =& \epsilon(s+1)(s+3)(s+2)\left(s + \frac{1}{\epsilon}\right).\end{aligned} \tag{10.48}$$

Its roots are $-1, -3, -2, and -1/\epsilon$ in which -1 and -3 stem from the cancelation of stable poles of plant and zeros of controller; another root -2 is the mirror image about imaginary axis of right half plane zero $z = 2$. The last root $-1/\epsilon$ approaches infinity as $\epsilon \to 0$. That is, the character of the suboptimal controller is that canceling all stable poles and zeros of the plant, placing poles of closed-loop system at the locations of unstable zero's mirror image, and finally moving all other poles to the negative infinity.

Moreover, observing the improper optimal controller $K_{\text{opt}}(s)$ carefully, we find that it takes the stable poles $p = -1, -3$ of the plant as its zeros and includes the infinite zeros of plant as its poles. This is why the optimal controller is improper.

Furthermore, calculating the optimal open-loop transfer function and sensitivity, we find that they are

$$L_{\text{opt}}(s) = -\frac{s-2}{2s}, \quad S_{\text{opt}}(s) = \frac{2s}{s+2}$$

respectively. The open-loop gain is $1/2$ at frequency $\omega > 2$ rad/s and does not rolls off enough. So, although this optimal controller can minimize the tracking error, it is very sensitive to sensor noise. This means that in control design it is rather dangerous to pay too much attention to some specific specification and ignore the others. ▽

10.4.1.2 Unstable Plant Case

For such plants, the following assumption is needed besides Assumption 4.

Assumption 5 $P(s)$ has no poles on the imaginary axis and has *distinct* poles p_j $(j = 1, \ldots, h)$ in the open right half plane.

Owing to the analysis in Section 10.2.2, $S(s)$ can be written as $S = -(X - NQ)\tilde{D} = 1 + N(-\tilde{Y} + Q\tilde{D})$. Under Assumptions 4 and 5, the following properties hold:

1. The only unstable zeros of $N(s)$ are z_i $(i = 1, \ldots, k)$.
2. The only unstable zeros of $\tilde{D}(s)$ are p_j $(j = 1, \ldots, h)$.
3. $S(z_i) = 1(\forall i)$ and $S(p_j) = 0(\forall j)$.

First, factorize $N(s), \tilde{D}(s)$ as products of all-pass transfer function and minimum-phase transfer function:

$$N(s) = LM, \quad L(s) = \prod_{i=1}^{k} \frac{\overline{z}_i}{z_i} \frac{z_i - s}{\overline{z}_i + s} \tag{10.49}$$

$$\tilde{D}(s) = gb, \quad b(s) = \prod_{j=1}^{h} \frac{\overline{p}_j}{p_j} \frac{p_j - s}{\overline{p}_j + s} \tag{10.50}$$

where $M(s)$ and $g(s)$ are minimum-phase transfer functions. Thus, the sensitivity can further be written as

$$S(s) = 1 + LM(-\tilde{Y} + Q\tilde{D}) = -(X - NQ)gb. \tag{10.51}$$

To ensure the boundedness of $\|e\|_2$, there must be $S(0) = 0$. This is equivalent to

$$1 - M(0)\tilde{Y}(0) + M(0)Q(0)\tilde{D}(0) = 0. \tag{10.52}$$

Further, $L(0) = b(0) = 1$.

It is proved in the following theorem that for unstable plant, the tracking precision is constrained not only by unstable zeros but also by unstable poles.

Theorem 10.4 *Suppose that $P(s)$ satisfies Assumptions 4 and 5. The infimum of tracking error w.r.t. unit step reference is*

$$\inf\|e\|_2^2 = \sum_{i=1}^{k} \frac{2\Re(z_i)}{|z_i|^2} + \sum_{i=1}^{h}\sum_{j=1}^{h} \frac{4\Re(p_i)\Re(p_j)}{(\overline{p}_i + p_j)p_i\overline{p}_j\overline{b}_i b_j}(1 - L^{-1}(p_i))^*(1 - L^{-1}(p_j)) \tag{10.53}$$

where the constant b_i is given by

$$b_i = \prod_{\substack{j \neq i \\ j=1}}^{h} \frac{\overline{p}_j}{p_j} \frac{p_j - p_i}{\overline{p}_j + p_i}.$$

Further, $\inf\|e\|_2^2 = 0$ *when $P(s)$ has no unstable zeros.*

The first term in the infimum is the constraint brought by unstable zeros. It is not so severe so long as the zeros are not too close to the origin. On the other hand, the second term indicates that unstable poles amplify the constraint of unstable zeros. Particularly, when there are unstable zeros and poles close to each other in $P(s)$, L^{-1} gets very big. Similarly, when there exist unstable poles close to each other, b_i^{-1} increases greatly. The previous theorem shows that the tracking error worsens whichever the case is.

However, when the plant has no unstable zeros, $L(s) = 1$ so that the second term of $\inf\|e\|_2^2$, namely, the term representing the influence of unstable poles, disappears. This means that *the constraint on reference tracking is essentially caused by unstable zeros. Unstable poles only amplify this effect.*

10.4.1.3 Proof of Theorem 10.4

By $L(0) = 1$, all poles of $(L^{-1} - 1)/s$ are in the open right half plane. In addition, owing to the boundedness condition (10.52) for $\|e\|_2$, $(1 - M\tilde{Y} + MQ\tilde{D})/s$ must be stable. So, applying Lemmas 2.5(4) and 2.6, we obtain

$$\begin{aligned}\|e\|_2^2 &= \left\|\left(1 - LM\tilde{Y} + LMQ\tilde{D}\right)\frac{1}{s}\right\|_2^2 = \left\|\left(L^{-1} - M\tilde{Y} + MQ\tilde{D}\right)\frac{1}{s}\right\|_2^2\\ &= \left\|(L^{-1} - 1)\frac{1}{s} + (1 - M\tilde{Y} + MQ\tilde{D})\frac{1}{s}\right\|_2^2\\ &= \left\|(L^{-1} - 1)\frac{1}{s}\right\|_2^2 + \left\|\left(1 - M\tilde{Y} + MQ\tilde{D}\right)\frac{1}{s}\right\|_2^2. \end{aligned} \tag{10.54}$$

Calculating the first term with a method similar to Theorem 10.3, we get

$$\|e\|_2^2 = \sum_{i=1}^{k}\frac{2\Re(z_i)}{|z_i|^2} + \left\|\left(1 - M\tilde{Y} + MQgb\right)\frac{1}{s}\right\|_2^2. \tag{10.55}$$

Therefore, we need to just consider the second term. For simplicity, define

$$J = \left\|\left(1 - M\tilde{Y} + MQgb\right)\frac{1}{s}\right\|_2^2 = \left\|\left(\frac{1 - M\tilde{Y}}{b} + MQg\right)\frac{1}{s}\right\|_2^2. \tag{10.56}$$

Before calculating J, some preparation is done. Firstly, from $S(p_j) = 0$ and (10.51) we know that $1 - L(p_j)M(p_j)Y(p_j) = 0$. So, $1 - M(p_j)Y(p_j) = 1 - L^{-1}(p_j)$ holds. Based on this, we expand $(1 - M\tilde{Y})/b$ into partial fraction

$$\frac{1 - M\tilde{Y}}{b} = \sum_{j=1}^{h} a_j\frac{p_j}{\overline{p}_j}\frac{\overline{p}_j + s}{p_j - s} + R(s) \tag{10.57}$$

in which $R(s)$ is a stable transfer function, and the coefficient a_j satisfies

$$a_j = \frac{\overline{p}_j}{p_j(p_j + \overline{p}_j)}\lim_{s\to p_j}(p_j - s)\frac{1 - M\tilde{Y}}{b} = \frac{1 - L^{-1}(p_j)}{b_j}. \tag{10.58}$$

Substituting this partial fraction expansion into J, we get

$$\begin{aligned} J &= \left\|\left(\sum_{j=1}^{h} a_j\frac{p_j}{\overline{p}_j}\frac{\overline{p}_j + s}{p_j - s} + R + MQg\right)\frac{1}{s}\right\|_2^2\\ &= \left\|\left(\sum_{j=1}^{h} a_j\left(\frac{p_j}{\overline{p}_j}\frac{\overline{p}_j + s}{p_j - s} - 1\right) + \sum_{j=1}^{h} a_j + R + MQg\right)\frac{1}{s}\right\|_2^2\\ &= \left\|\sum_{j=1}^{h} a_j\frac{2\Re(p_j)}{\overline{p}_j}\frac{1}{p_j - s} + \left(\sum_{j=1}^{h} a_j + R + MQg\right)\frac{1}{s}\right\|_2^2. \end{aligned}$$

By (10.57), it is easy to know that $\sum a_j + R(0) = 1 - M(0)\tilde{Y}(0)$. Then, $\sum a_j + (R + MQg)(0) = 0$ follows from (10.52) so that $(\sum a_j + R + MQg)/s$ is stable. Hence,

$$J = \left\| \sum_{j=1}^{h} a_j \frac{2\Re(p_j)}{\overline{p}_j} \frac{1}{p_j - s} \right\|_2^2 + \left\| \left(\sum_{j=1}^{h} a_j + R + MQg \right) \frac{1}{s} \right\|_2^2. \tag{10.59}$$

Since the inverses of $M(s), g(s)$ are stable. The infimum of the second term becomes zero. So,

$$\inf J = \left\| \sum_{j=1}^{h} a_j \frac{2\Re(p_j)}{\overline{p}_j} \frac{1}{p_j - s} \right\|_2^2. \tag{10.60}$$

On the other hand, by the properties of inner product (Lemmas 2.5 and 2.7), we have

$$\begin{aligned}
\inf J &= \sum_{i=1}^{h} \sum_{j=1}^{h} \left\langle a_i \frac{2\Re(p_i)}{\overline{p}_i} \frac{1}{p_i - s}, a_j \frac{2\Re(p_j)}{\overline{p}_j} \frac{1}{p_j - s} \right\rangle \\
&= \sum_{i=1}^{h} \sum_{j=1}^{h} \bar{a}_i \frac{2\Re(p_i)}{p_i} a_j \frac{2\Re(p_j)}{\overline{p}_j} \left\langle \frac{1}{p_i - s}, \frac{1}{p_j - s} \right\rangle \\
&= \sum_{i=1}^{h} \sum_{j=1}^{h} \bar{a}_i a_j \frac{4\Re(p_i)\Re(p_j)}{p_i \overline{p}_j} \frac{1}{\overline{p}_i + p_j}.
\end{aligned} \tag{10.61}$$

This proves the first half of the theorem.

When $P(s)$ has no unstable zeros, $N(s)$ becomes a minimum-phase system. So,

$$\|e\|_2^2 = \left\| (-X + NQ)\, gb \frac{1}{s} \right\|_2^2 = \left\| (-X + NQ)\, g \frac{1}{s} \right\|_2^2. \tag{10.62}$$

$N(s)$ has a stable inverse. When the free parameter is selected as $Q(s, \epsilon) = N^{-1}X/(\epsilon s + 1)^r$ (r is the relative degree of $N(s)$, namely, the relative degree of $P(s)$),

$$-X(s) + N(s)Q(s, \epsilon) = -X(s) \frac{(\epsilon s + 1)^r - 1}{(\epsilon s + 1)^r} \tag{10.63}$$

holds and possesses the following asymptotic properties:

$$-X(0) + N(0)Q(0, \epsilon) = 0, \ \lim_{\epsilon \to 0} (-X(s) + N(s)Q(s, \epsilon)) = 0. \tag{10.64}$$

Therefore, $\inf \|e\|_2^2 = 0$ is true.

10.4.2 2-Degree-of-Freedom System

In this case, according to Section 7.5, the closed-loop transfer function of $r \mapsto e = r - y_P$ is

$$T_{er}(s) = 1 + N_{12}(s)Q_F(s), \ N_{12}(s) = -C(sI - A_F)^{-1}B \tag{10.65}$$

in which $Q_F(s)$ is an arbitrary stable transfer function. Since the zeros of $N_{12}(s)$ coincide with those of $P(s)$, its result is of course the same as that of stable plant in 1-DOF systems. So, there holds the following theorem:

Theorem 10.5 *Suppose that Assumption 4 holds. In a 2-DOF system, the infimum of tracking error w.r.t. unit step reference input is*

$$\inf \|e\|_2^2 = \sum_{i=1}^{k} \frac{2\Re(z_i)}{|z_i|^2}. \tag{10.66}$$

Further, $\inf \|e\|_2^2 = 0$ *when* $P(s)$ *has no unstable zeros.*

That is, in 2-DOF systems, unstable poles of the plant are not an obstacle for reference tracking. This is a big advantage of 2-DOF systems. Moreover, the infimum of tracking error is the same as that of 1-DOF systems when the plant is stable. Paradoxically, for stable plants, the infimum cannot be reduced even if 2-DOF control is introduced, that is, it makes no contribution to the ultimate improvement of the limitation on reference response.

Exercises

10.1 Factorize transfer function $G(s) = s(s^2 - 5s + 6)/(s + 5)^2(s^2 + 2s + 5)$ as the product of an all-pass transfer function and a minimum-phase transfer function.

10.2 Prove that the zeros and relative degree of $C(sI - A - BF)^{-1}B$ and $C(sI - A)^{-1}B$ are the same.

10.3 When the plant $P(s)$ has an unstable zero z and two unstable poles p_1 and p_2, investigate the influence of their relative positions on the tracking performance.

10.4 For unstable plant $P(s) = 1/(s - 1)$, mimic the proof of Theorem 10.4 to compute the infimum of tracking error $\inf \|e\|_2$ w.r.t. unit step reference input and design a suboptimal controller.

Notes and References

Many contributions to the performance limitation study are owing to J. Chen [15, 14]. Some extensions are in Ref. [94]. For more details on the performance limitation, refer to Refs [84] and [31].

11

Model Uncertainty

In model-based control design, the mathematical model plays an extremely important role. However, the dynamics of a physical system cannot be described completely by a mathematical model obtained from physical laws or system identification. Only a part of it can be characterized by the model. In addition, a physical system is basically nonlinear and linear approximation is valid only around the operating point. When the range gets large or the operating point changes, the effect of nonlinearity appears. Therefore, in control design it is impossible to design a high-quality control system without considering the discrepancy between the model and the true system.

In this chapter, we focus on the model uncertainty and illustrate in detail the types of model uncertainty, its description, and the modeling method of its bound. Specifically, to be exposed are the uncertainty of dynamics, the variation of parameter, the uncertainty caused by linear approximation of nonlinear element, and the impact of model uncertainty on the control design. Finally, we will illustrate briefly the notions of robust stability and robust performance.

11.1 Model Uncertainty: Examples

In this section, we explore what kinds of uncertainty exist in real systems by examining some simple real system models.

Example 11.1 *A simplified speed control model for a car is given by*

$$P(s) = \frac{1}{Ms + \mu} \tag{11.1}$$

in which M is the mass of the car and μ is the coefficient of road friction. The mass changes with load and the friction coefficient changes with the road condition. In system design, we only know the ranges of these parameters. For example,

$$M_1 \leq M \leq M_2, \quad \mu_1 \leq \mu \leq \mu_2.$$

Robust Control: Theory and Applications, First Edition. Kang-Zhi Liu and Yu Yao.

Example 11.2 (IEEJ benchmark [87]) *In Chapter 1 we have touched on the uncertain resonant modes in the high-frequency band in hard disk drives (HDD). In general, the physical model is obtained as*

$$\tilde{P}(s) = \frac{K_p}{s^2} + \frac{A_1}{s^2 + 2\zeta_1\omega_1 s + \omega_1^2} + \frac{A_2}{s^2 + 2\zeta_2\omega_2 s + \omega_2^2} + \cdots \tag{11.2}$$

via finite element method and modal analysis. The problem lies in that the parameters of high-order resonant modes vary with manufacturing error, and hundreds of thousands of HDDs have to be controlled by the same controller in order to lower the price of products. So, the control design is carried out based on the rigid body model $P(s) = K_p/s^2$.

In the IEEJ (Institute of Electrical Engineers of Japan) benchmark, the parameters are as listed in Table 11.1. In this table, the numbers within the parentheses indicate the relative variation, and the angular frequency is $\omega_i = 2\pi f_i$. *The typical frequency response of HDD and that of the nominal model are illustrated in Figure 11.1(b).*

As shown by these examples, the mathematical model of a plant inevitably contains some uncertain parts. Even so, we still hope to design a control system, which can operate normally

Table 11.1 Parameters of HDD benchmark

i	f_i (Hz)	ζ_i	A_i
1	4 100 (±15%)	0.02	−1.0
2	8 200 (±15%)	0.02	1.0
3	12 300 (±10%)	0.02	−1.0
4	16 400 (±10%)	0.02	1.0
5	3 000 (±5%)	0.005	0.01 (−200% ∼ 0%)
6	5 000 (±5%)	0.001	0.03 (−200% ∼ 0%)
K_p			3.744×10^9

(a)

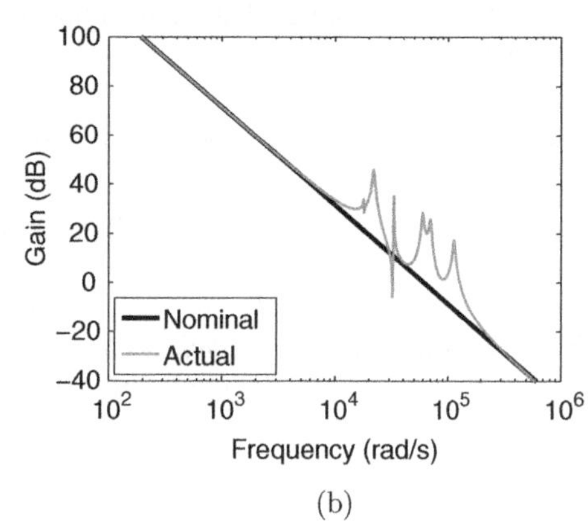

(b)

Figure 11.1 Hard disk drive and its frequency response (a) Photo, (b) Bode plot

and has a high performance, based on a model with uncertainty. The mission of robust control theory is to develop such an effective control design method for systems with uncertainty.

11.1.1 Principle of Robust Control

Since a real physical system cannot be modeled accurately, it is impossible to describe a real system using a single transfer function. Instead, we can determine a model which is used for design and then evaluate the difference between the real system and the model, that is, the uncertainty range. In this way, we can obtain a set of systems that includes the real system. The model selected for design is called the *nominal plant*. If the stability and control performance are guaranteed for the plant set in the control design, these properties carry over to the real system. This is the basic philosophy of robust control.

In other words, the achievable performance will be significantly limited by the worst-case uncertainty because we need to maintain the same level of performance for all plants in the set. Here, important are the description method of plant set and the modeling of uncertainty bounds. These two issues will be discussed, respectively, in the succeeding sections.

11.1.2 Category of Model Uncertainty

The model uncertainty can be roughly classified into parameter uncertainty and dynamic uncertainty.

11.1.2.1 Parameter Uncertainty

As shown in Examples 1.2 and 11.1, in this kind of uncertain systems, the model structure is known but the parameters are uncertain. The influence of parameter uncertainty on the transmission of signals mainly appears in the low- and middle-frequency bands.

11.1.2.2 Dynamic Uncertainty

This kind of model uncertainty can further be classified as

- Unmodeled high-frequency resonant modes, as shown in Figure 11.2
- Dynamics ignored deliberately for the simplification of system analysis and design, particularly the high-frequency dynamics

Remark 11.1 *Let us examine carefully the model uncertainty brought by system identification. A typical identification method for linear systems is to apply sinusoidal input to the system and then measure the steady-state response of the output. For unit sinusoidal input with a frequency* ω*, the steady-state response is still a sinusoid with the same frequency. Its amplitude* $K(\omega)$ *and phase angle* $\phi(\omega)$ *are*

$$K(\omega) = |G(j\omega)|, \quad \phi(\omega) = \arg G(j\omega). \tag{11.3}$$

That is, they are closely related with the frequency response of the transfer function. Changing the input frequency, the frequency response at a different frequency can be measured. Repeating this process, a set of gain–phase data can be obtained (Figure 11.2). Then, a transfer function is identified by finding a rational function whose frequency response matches the measured data. However, in reality vibration causes mechanical attrition, so the input frequency cannot be too high. That is, it is impossible to get the frequency response in the high-frequency band. Hence, model uncertainty in the high-frequency band is inevitable.

11.2 Plant Set with Dynamic Uncertainty

The uncertainties of plants may take various forms. But in order to establish an effective theory for analysis and design, the form of plant set description must be limited. It is especially important to ensure that the uncertainty description can be widely applied.

Dynamic uncertainty stands for the model uncertainty where the degrees of the real system and the nominal model are different. Its influence mainly appears in the high-frequency band. There are three major methods to characterize a dynamic uncertainty: using the range of its gain variation, using the range of its phase variation, and using both. The first method is introduced in this section, while the rest will be introduced in other chapters.

11.2.1 Concrete Descriptions

Historically, dynamic uncertainty is the first one studied and there are many ways to describe a plant set with dynamic uncertainty. Among them, the simplest way is to take the difference between the nominal plant P and the real system $\tilde{P}$ as uncertainty Δ, namely,

$$\Delta(s) = \tilde{P}(s) - P(s). \tag{11.4}$$

To achieve a high performance, the key in feedback control design is how to make use of information about the uncertainty as more as possible. This information can only be described as the boundary of the uncertainty set. So, the next question is how to set up a bounding model for the uncertainty Δ. Obviously, the gain of Δ is a useful information. Let us look at the Bode

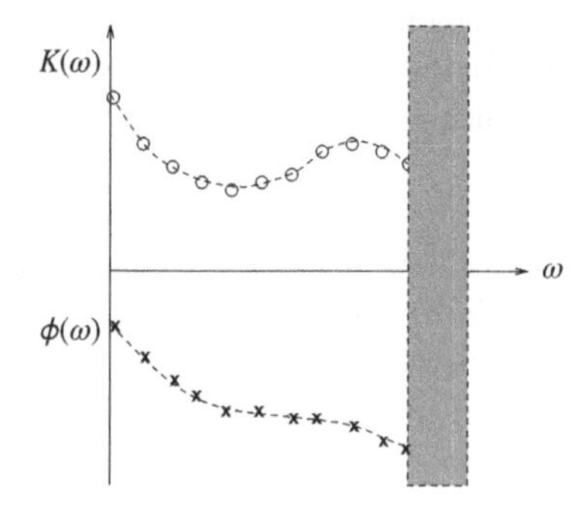

Figure 11.2 Identified frequency response with uncertainty in high frequency

plot of hard disk drive (Figure 11.1(b)) in Example 11.2. There is a significant discrepancy between the nominal plant and real system only in the frequency band of 10^4–10^5 rad/s. If this property is not reflected and the size of uncertainty is expressed simply as a constant, then the range of uncertainty is enlarged because this is equal to assuming that gain change of the uncertainty is the same over the whole frequency domain. The consequence is that the controller gain in low frequency has to be lowered so that high-performance control cannot be expected. Therefore, for dynamic uncertainty its gain bound must be described by a function of frequency.

Concretely, we carry out identification experiments under different conditions to get a number of models $P_i(j\omega)$ $(i = 1, 2, \cdots)$. Then, by calculating the gains of all $P_i(j\omega) - P(j\omega)$ and drawing curves, we can find a bounding function $W(s)$ satisfying

$$|\Delta_i(j\omega)| = |P_i(j\omega) - P(j\omega)| \le |W(j\omega)| \ \forall\omega,\ i. \tag{11.5}$$

This function is called the *weighting function*. For the sake of design, $W(s)$ is usually chosen as a low-order stable rational function. Further, assuming that *the set with bound $W(s)$ is filled with uncertainty* Δ (note that this assumption is conservative), the uncertainty can be expressed as

$$\Delta(s) = W(s)\delta(s), \quad |\delta(j\omega)| \le 1 \ \forall\omega \tag{11.6}$$

in which $\delta(s)$ is the normalized uncertainty. As such, we have obtained a plant set:

$$\{\tilde{P}(s) \,|\, \tilde{P} = P + W\delta, \quad \|\delta\|_\infty \le 1\}. \tag{11.7}$$

It is natural to assume that the real plant is contained in this set.

In the sequel, we summarize the commonly used descriptions of plant set.

11.2.1.1 Plant Set with Multiplicative Uncertainty

$$\tilde{P}(s) = (1 + \Delta W)P, \quad \|\Delta\|_\infty \le 1 \tag{11.8}$$

Here, $P(s)$ is the nominal plant, $\Delta(s)$ is the normalized uncertainty, and $W(s)$ is the weighting function that bounds the uncertainty set. The block diagram of this set is shown in Figure 11.3.

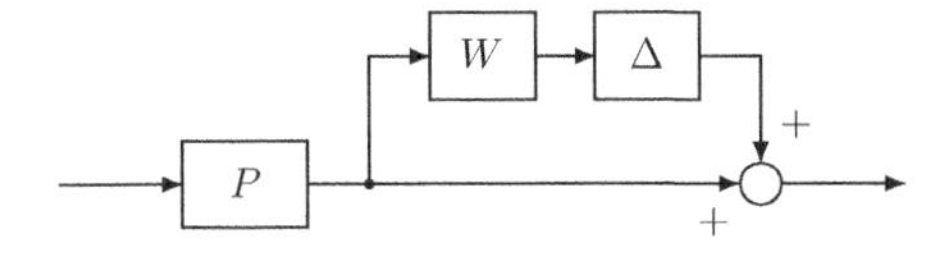

Figure 11.3 Plant set with multiplicative uncertainty

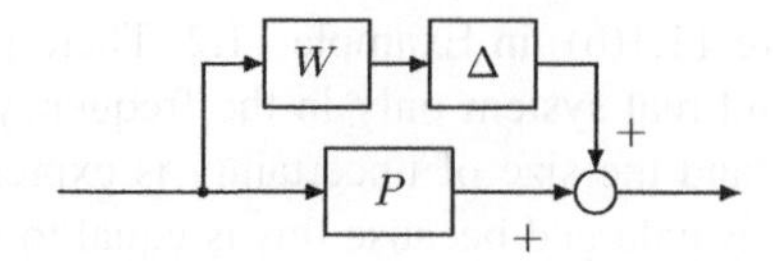

Figure 11.4 Plant set with additive uncertainty

11.2.1.2 Plant Set with Additive Uncertainty

$$\tilde{P}(s) = P + \Delta W, \quad \|\Delta\|_\infty \leq 1. \tag{11.9}$$

Figure 11.4 gives its block diagram.

11.2.1.3 Plant Set with Feedback Uncertainty

There are two forms of feedback uncertainty: type I and type II, which are described, respectively, as

$$\tilde{P}(s) = \frac{P}{1 + \Delta W P}, \quad \|\Delta\|_\infty \leq 1, \tag{11.10}$$

$$\tilde{P}(s) = \frac{P}{1 + \Delta W}, \quad \|\Delta\|_\infty \leq 1. \tag{11.11}$$

Their block diagrams are shown in Figures 11.5 and 11.6.

In these plant sets, the size of uncertainty is measured by the $\mathcal{H}_\infty$ norm. So it is called as *norm-bounded uncertainty*. In applications, we should select a plant set description that best incorporates all known information about the real plant.

Example 11.3 *As an example, we model the high-frequency resonant modes of the hard disk drive in Example 11.2 as a multiplicative uncertainty. The modeling process is as follows:*

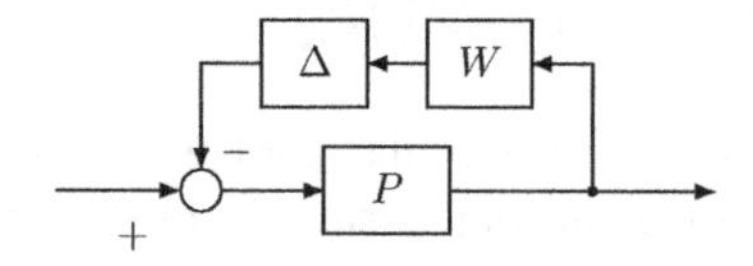

Figure 11.5 Plant set with type I feedback uncertainty

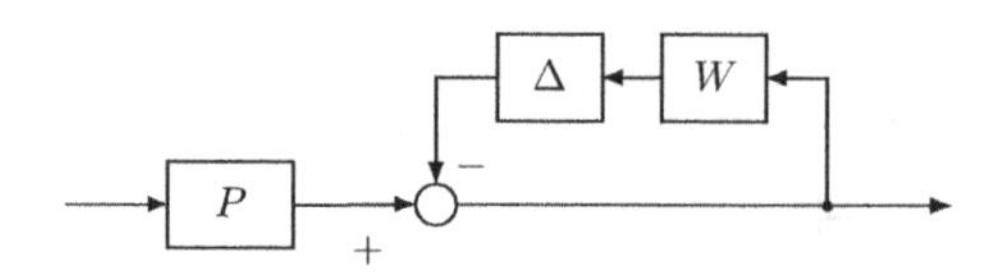

Figure 11.6 Plant set with type II feedback uncertainty

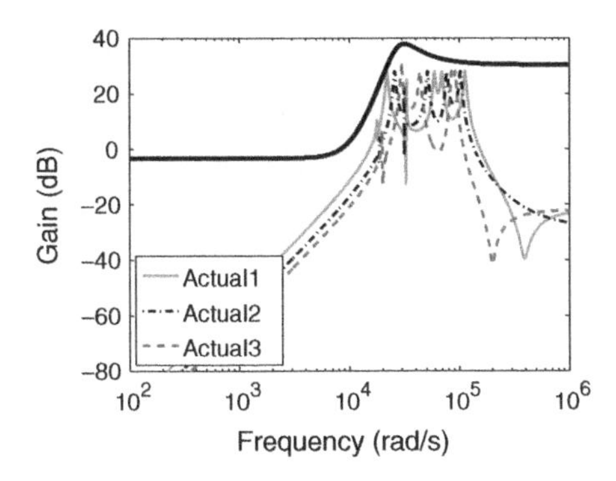

Figure 11.7 Uncertainty modeling of a hard disk drive

1. *Firstly, draw the relative error* $|1 - \frac{\tilde{P}(j\omega)}{P(j\omega)}|$ *between the real plant* $\tilde{P}(s)$ *and the nominal plant* $P(s)$ *in a Bode plot (Figure 11.7).*
2. *Apparently, the amplitude of relative error varies a lot at different frequencies. In this example, the model uncertainty mainly exists at high frequency. This property must be described by a frequency-dependent weighting function. That is, determine a minimum-phase weighting function such that the gain of its frequency response covers the relative errors. An example is the high-pass transfer function shown by the solid line in Figure 11.7.*
3. *In the end, the real plant* $\tilde{P}(s)$ *is contained in the plant set:*

$$\tilde{P}(s) = P(1 + \Delta W), \quad P(s) = \frac{K_p}{s^2}, \quad \|\Delta\|_\infty \leq 1.$$

Here, for the sake of design and implementation, low-order weighting function should be adopted. Further, to determine a weighting function $W(s)$, we may use an asymptote to determine roughly a transfer function. Then, the weighting function is obtained through a fine tuning of its coefficients.

In the case where no information is known on the structure of uncertainty, it is reasonable to use the description of additive or multiplicative uncertainty.

As seen from Figure 11.7, although the region below $W(s)$ (solid line) is treated as the uncertainty region, the real uncertainty comprises only a small part of it. In this sense, the foregoing plant set contains a lot of uncertainties that do not exist in reality. This inevitably enlarges the plant set and brings conservatism into robustness condition. The payoff is that the description of uncertainty is quite simple and suitable for analysis and design.

In a word, the range of true uncertainty will be enlarged more or less whatever description of plant set is used. To restrain the expansion of uncertainty range, the key is to use a weighting function that is as close to the true uncertainty as possible.

11.2.2 Modeling of Uncertainty Bound

In the frequency band where the uncertainty gain is big, precise control cannot be expected in principle. Therefore, in the modeling of uncertainty bound (weighting function), we should keep the gain of weighting function not far away from the upper bound of the uncertainty, at least in the low- and middle-frequency bands which have a strong impact on the control performance. This is because if the system model is more accurate in the control bandwidth, a better control performance may be achieved in the design.

In the sequel, we discuss the modeling method for the uncertainty bound according to the type of uncertainty.

11.2.2.1 System Identification Case

Calculate the difference between the frequency responses of true plant $\tilde{P}(j\omega)$ and the nominal model $P(j\omega)$ on a Bode plot and then find a weighting function $W(s)$ which covers $|\tilde{P}(j\omega) - P(j\omega)|$. An example is shown in Figure 11.8 where the solid line is $|\tilde{P}(j\omega) - P(j\omega)|$ and the dashed line is $|W(j\omega)|$.

11.2.2.2 Model Reduction Case

This is the case where a high-order plant $\tilde{P}(s)$ (including time delay) is approximated by a low-order model $P(s)$. A weighting function $W(s)$ is determined on a Bode plot that satisfies

$$|\tilde{P}(j\omega) - P(j\omega)| < |W(j\omega)| \quad \text{(additive uncertainty)}$$

or

$$\left|1 - \frac{\tilde{P}(j\omega)}{P(j\omega)}\right| < |W(j\omega)| \quad \text{(multiplicative uncertainty).}$$

Concretely speaking, we draw the curve of the left-hand side of the inequality and then find a rational transfer function $W(s)$ that covers this curve. This is how we determined the bound of multiplicative uncertainty in Example 11.3.

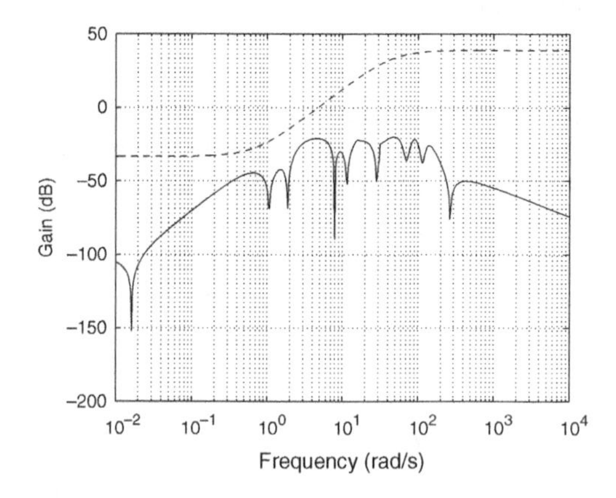

Figure 11.8 Weighting function of uncertainty

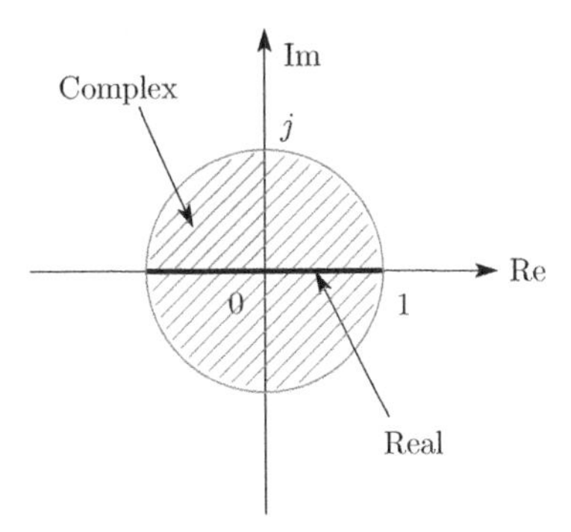

Figure 11.9 Dynamic uncertainty versus parameter uncertainty

11.3 Parametric System

The frequency response of a dynamic uncertainty forms a disk in the complex plane, as shown in Figure 11.9. Meanwhile, a parameter uncertainty is real, and its range is far less than the disk since it is simply the thick line segment contained in the disk. Therefore, for parameter uncertainty if a norm-bounded model is used, the range of plant set will be enlarged greatly and no meaningful result can be expected. So, in this section we discuss how to reasonably describe a plant set with parameter uncertainty. Such a plant set is named as a *parametric system.*

Example 11.4 *Consider the mass-spring-damper system shown in Figure 11.10(a). One end of the spring and damper is fixed on the wall. m is the mass, b the viscous friction coefficient of the damper, and k the spring constant. $y(t)$ is the displacement of mass m and $u(t)$ the external force. The state equation of this system is shown in Eq. (11.12) where the performance output is the displacement y:*

$$\dot{x} = \begin{bmatrix} 0 & 1 \\ -\dfrac{k}{m} & -\dfrac{b}{m} \end{bmatrix} x + \begin{bmatrix} 0 \\ \dfrac{1}{m} \end{bmatrix} u, \quad x = \begin{bmatrix} y \\ \dot{y} \end{bmatrix}. \tag{11.12}$$

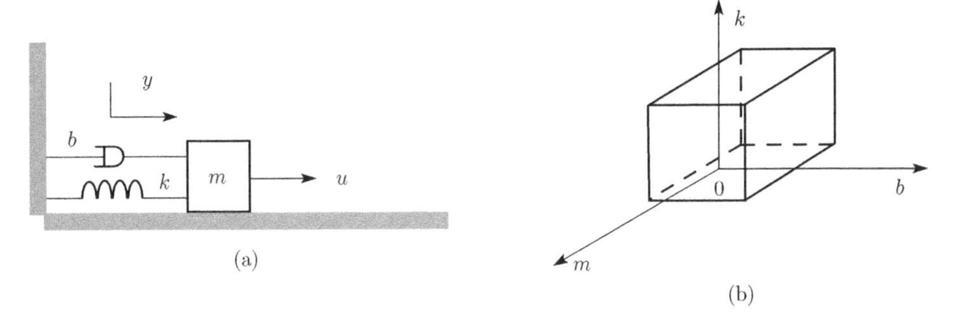

Figure 11.10 mass-spring-damper system (a) System configuration, (b) Parameter space

The uncertain parameters (m, b, k) *take their values in the following ranges:*

$$m_1 \le m \le m_2, \quad b_1 \le b \le b_2, \quad k_1 \le k \le k_2. \tag{11.13}$$

Putting these parameters into a vector, $[m \quad b \quad k]$ *forms a hexahedron with eight vertices in the three-dimensional space (Figure 11.10(b)).*

In this system, if each uncertain parameter is treated as a norm-bounded uncertainty, the uncertainty range would be enlarged significantly. Such a model is too conservative and impractical. So, we need to develop a new and more reasonable model for parameter uncertainty.

Example 11.5 *Linear motor is a motor which uses Lorentz force to offer a straight drive. Its principle is as follows: place many pairs of magnets with the same polarity in the order of* S, N*, respectively, on two parallel rails, and then change the polarity of electromagnetic windings installed in the bottom of the stage in an order of* $S \to N \to S \to N$ *periodically so as to generate a Lorentz force to drive the stage straightforward. The change of polarity of the electromagnetic windings is done by a servo amplifier. The input is the speed command and the performance output is the displacement of the stage. Usually, the dynamics of this system is designed as*

$$P(s) = \frac{K}{s(Ts+1)} = \frac{K/T}{s(s+1/T)}. \tag{11.14}$$

When the displacement and speed of the stage are taken as the states, the state equation becomes

$$\dot{x} = \begin{bmatrix} 0 & 1 \\ 0 & -\dfrac{1}{T} \end{bmatrix} x + \begin{bmatrix} 0 \\ \dfrac{K}{T} \end{bmatrix} u. \tag{11.15}$$

However, the linear motor has nonlinearities such as magnetic flux leakage, magnet saturation, and nonlinear friction with rails. So it cannot be described accurately by a linear model. Here, we regard the effect of these nonlinearities as the uncertainties of time constant T *and gain* K*. For example, by doing numerous experiments w.r.t. different speed commands, we can obtain many pairs of time constant and gain. An example is shown in Figure 11.11.*

Since the relationship between the time constant and the gain is not clear, a smart way is to enclose the experimental data using a minimum rectangle. That is, assuming that the parameters vary at random in the range of

$$T_1 \le T \le T_2, \quad K_1 \le K \le K_2.$$

(Refer to Figure 11.11(a)) However, a closer observation reveals that the measured data may be enclosed tightly with a polygon as shown in Figure 11.11(b). This polygon is just a portion of the rectangle in Figure 11.11(a).

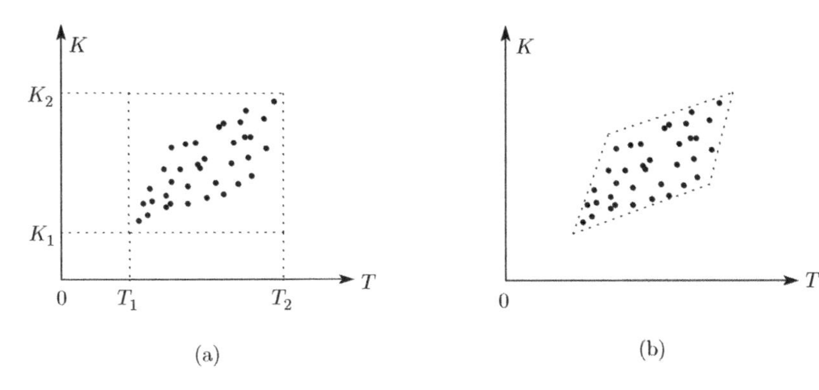

Figure 11.11 Measured parameters of linear motor (a) Rectangle approximation, (b) Tighter polytopic approximation

11.3.1 Polytopic Set of Parameter Vectors

How should we describe the uncertain parameter vector $[m \quad b \quad k]$ inside the cubic in Example 11.4? In order to search for a suitable method, let us examine the one parameter case.

Consider the case where the mass m takes values in an interval $[m_1, m_2]$. Note that m_1 and m_2, the two ends of the interval, are known. The question is how to use these vertices to express an arbitrary point in the interval. The simplest way is to take a vertex as a starting point and then add the variation relative to this starting point. This variation can be expressed as product of the interval length and a new variable $\lambda \in [0, 1]$ which represents the variation rate. So, any $m \in [m_1, m_2]$ can be written as

$$m = m_2 - \lambda(m_2 - m_1) = \lambda m_1 + (1 - \lambda)m_2.$$

The sum of the two coefficients in this equation is 1. Setting $\alpha_1 = \lambda, \alpha_2 = 1 - \lambda$, this equation can be written in a more compact form:

$$m = \alpha_1 m_1 + \alpha_2 m_2, \quad \alpha_1 + \alpha_2 = 1, \quad \alpha_i \geq 0. \tag{11.16}$$

In fact, this equation expresses a convex combination of the vertices m_1, m_2.

Next, when two parameters (m, b) change simultaneously, both $m \in [m_1, m_2]$ and $b \in [b_1, b_2]$ can be expressed as

$$\begin{aligned} m &= \alpha_1 m_1 + \alpha_2 m_2, \quad \alpha_1 + \alpha_2 = 1, \quad \alpha_i \geq 0 \\ b &= \beta_1 b_1 + \beta_2 b_2, \quad \beta_1 + \beta_2 = 1, \quad \beta_i \geq 0, \end{aligned}$$

respectively. As shown in Figure 11.12(a), the range of the vector $[m \quad b]^T$ is a rectangle with four vertices. Each vertex is named, respectively, as

$$\theta_1 = \begin{bmatrix} m_1 \\ b_1 \end{bmatrix}, \quad \theta_2 = \begin{bmatrix} m_1 \\ b_2 \end{bmatrix}, \quad \theta_3 = \begin{bmatrix} m_2 \\ b_1 \end{bmatrix}, \quad \theta_4 = \begin{bmatrix} m_2 \\ b_2 \end{bmatrix}. \tag{11.17}$$

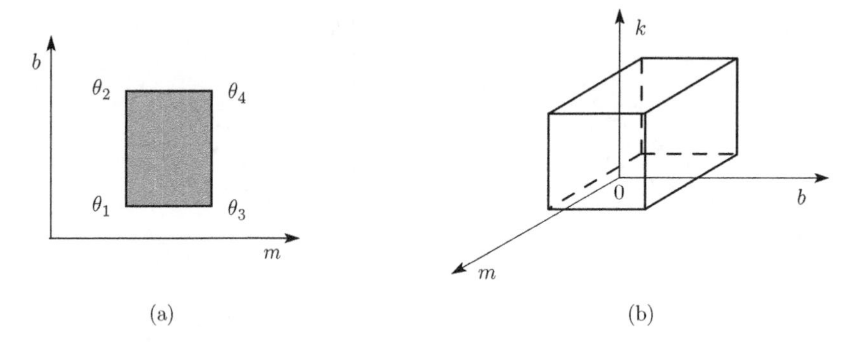

Figure 11.12 Polytope of parameter uncertainty (a) Two uncertain parameters, (b) Three uncertain parameters

Then we try to use these vertices to describe an arbitrary point $\theta = [m \quad b]$ in the rectangle. Simple calculation shows that θ can be written as

$$\theta = \begin{bmatrix} m \\ b \end{bmatrix} = \begin{bmatrix} (\beta_1 + \beta_2)(\alpha_1 m_1 + \alpha_2 m_2) \\ (\alpha_1 + \alpha_2)(\beta_1 b_1 + \beta_2 b_2) \end{bmatrix}$$
$$= \alpha_1\beta_1 \begin{bmatrix} m_1 \\ b_1 \end{bmatrix} + \alpha_1\beta_2 \begin{bmatrix} m_1 \\ b_2 \end{bmatrix} + \alpha_2\beta_1 \begin{bmatrix} m_2 \\ b_1 \end{bmatrix} + \alpha_2\beta_2 \begin{bmatrix} m_2 \\ b_2 \end{bmatrix}.$$

Renaming the coefficient of each vertex, we get the relation that

$$\lambda_1 = \alpha_1\beta_1, \quad \lambda_2 = \alpha_1\beta_2, \quad \lambda_3 = \alpha_2\beta_1, \quad \lambda_4 = \alpha_2\beta_2 \Rightarrow \lambda_i \geq 0$$
$$\lambda_1 + \lambda_2 + \lambda_3 + \lambda_4 = \alpha_1(\beta_1 + \beta_2) + \alpha_2(\beta_1 + \beta_2) = \alpha_1 + \alpha_2 = 1.$$

That is, a vector θ in a rectangle can always be described as a convex combination of the four vertices:

$$\theta = \lambda_1\theta_1 + \lambda_2\theta_2 + \lambda_3\theta_3 + \lambda_4\theta_4. \tag{11.18}$$

Finally, let us look at the case of three uncertain parameters. In this case, the parameter vector set forms a hexahedron with eight vertices (refer to Figure 11.12(b)). Similar to the two parameters case, after appropriate transformation the parameter vector can be expressed as a convex combination of all vertices:

$$\begin{bmatrix} m \\ b \\ k \end{bmatrix} = \lambda_1 \begin{bmatrix} m_1 \\ b_1 \\ k_1 \end{bmatrix} + \lambda_2 \begin{bmatrix} m_1 \\ b_1 \\ k_2 \end{bmatrix} + \lambda_3 \begin{bmatrix} m_1 \\ b_2 \\ k_1 \end{bmatrix} + \lambda_4 \begin{bmatrix} m_1 \\ b_2 \\ k_2 \end{bmatrix}$$
$$+ \lambda_5 \begin{bmatrix} m_2 \\ b_1 \\ k_1 \end{bmatrix} + \lambda_6 \begin{bmatrix} m_2 \\ b_1 \\ k_2 \end{bmatrix} + \lambda_7 \begin{bmatrix} m_2 \\ b_2 \\ k_1 \end{bmatrix} + \lambda_8 \begin{bmatrix} m_2 \\ b_2 \\ k_2 \end{bmatrix} \tag{11.19}$$

where $\lambda_i \geq 0$ for all i and $\sum_{i=1}^{8} \lambda_i = 1$. Through these examinations, we conclude that a point in a polytope can always be expressed as a convex combination of all vertices. This conclusion applies to parameter vectors with any dimension.

11.3.2 Matrix Polytope and Polytopic System

In practical system models, it is rare that all the parameters appear completely in linear form. For example, in the mass–spring–damper system

$$\dot{x} = \begin{bmatrix} 0 & 1 \\ -\dfrac{k}{m} & -\dfrac{b}{m} \end{bmatrix} x + \begin{bmatrix} 0 \\ \dfrac{1}{m} \end{bmatrix} u, \tag{11.20}$$

parameters appear in forms of product, ratio, and reciprocal. The question now is whether the coefficient matrix can be expressed as a convex combination of the matrices obtained via substitution of vector vertices into the matrix[1]?

Note that the mass m in the coefficient matrices appears only as a reciprocal $1/m$, which can be expressed as a convex combination of vertices as shown in the equation:

$$\frac{1}{m} = \alpha_1 \frac{1}{m_1} + \alpha_2 \frac{1}{m_2}, \quad \alpha_1 + \alpha_2 = 1, \quad \alpha_i \geq 0.$$

Substitution of this equation yields

$$\begin{bmatrix} 0 & 1 \\ -\dfrac{k}{m} & -\dfrac{b}{m} \end{bmatrix} = \alpha_1 \begin{bmatrix} 0 & 1 \\ -\dfrac{k}{m_1} & -\dfrac{b}{m_1} \end{bmatrix} + \alpha_2 \begin{bmatrix} 0 & 1 \\ -\dfrac{k}{m_2} & -\dfrac{b}{m_2} \end{bmatrix}$$

$$\begin{bmatrix} 0 \\ \dfrac{1}{m} \end{bmatrix} = \alpha_1 \begin{bmatrix} 0 \\ \dfrac{1}{m_1} \end{bmatrix} + \alpha_2 \begin{bmatrix} 0 \\ \dfrac{1}{m_2} \end{bmatrix}$$

since $\alpha_1 + \alpha_2 = 1$. A matrix set formed in this way is called a *matrix polytope*.

Another question is if the parameter product $b \times \frac{1}{m}$ still can be expressed as a convex combination of vertices when m, b both are uncertain. From the following transformation, we know that a parameter product can still be expressed as a convex combination of products of vertices

$$\begin{aligned} \frac{b}{m} &= \left(\alpha_1 \frac{1}{m_1} + \alpha_2 \frac{1}{m_2}\right)(\beta_1 b_1 + \beta_2 b_2) \\ &= \alpha_1 \beta_1 \frac{b_1}{m_1} + \alpha_1 \beta_2 \frac{b_2}{m_1} + \alpha_2 \beta_1 \frac{b_1}{m_2} + \alpha_2 \beta_2 \frac{b_2}{m_2} \\ &= \lambda_1 \frac{b_1}{m_1} + \lambda_2 \frac{b_2}{m_1} + \lambda_3 \frac{b_1}{m_2} + \lambda_4 \frac{b_2}{m_2} \end{aligned}$$

in which $\lambda_1 + \lambda_2 + \lambda_3 + \lambda_4 = 1, \lambda_i \geq 0$. So, the matrices can be described as matrix polytopes:

$$\begin{aligned} A(m, b) &= \lambda_1 A(b_1, m_1) + \lambda_2 A(b_2, m_1) + \lambda_3 A(b_1, m_2) + \lambda_4 A(b_2, m_2) \\ B(m) &= \lambda_1 B(m_1) + \lambda_2 B(m_1) + \lambda_3 B(m_2) + \lambda_4 B(m_2). \end{aligned}$$

The same is true for the parameter product $k \times \frac{1}{m}$.

[1] Such a matrix will be called a vertex matrix.

From these discussions we conclude that a product of different parameters can always be expressed as a convex combination of products of vertices. Now, can a power of a parameter, such as the square, still be expressed as a convex combination of its vertices? Let us look at the square of the mass:

$$m^2 = (\alpha_1 m_1 + \alpha_2 m_2)^2 = \alpha_1^2 m_1^2 + 2\alpha_1\alpha_2 m_1 m_2 + \alpha_2^2 m_2^2.$$

The cross product m_1m_2 appears in this square expansion. Apparently, m^2 cannot be described only by m_1^2 and m_2^2. This shows that a power of parameter over two is not equal to the convex combination of the vertices' power. Fortunately, the situation of parameter power is rare in real systems. This can be confirmed by the forgoing examples. However, in the design of robust control systems, this power problem will be a bottleneck. Details will be given in the forthcoming chapters.

In summary, if θ belongs to a polytope and its power higher than 2 does not exist in the system realization, then a system with uncertain parameter vector θ can be described by the model

$$\dot{x} = A(\theta)x + B(\theta)u \tag{11.21}$$

$$y = C(\theta)x \tag{11.22}$$

in which each coefficient matrix is a matrix polytope:

$$A(\theta) = \sum_{i=1}^{N} \lambda_i A(\theta_i), \quad B(\theta) = \sum_{i=1}^{N} \lambda_i B(\theta_i), \quad C(\theta) = \sum_{i=1}^{N} \lambda_i C(\theta_i) \tag{11.23}$$

$$\lambda_i \geq 0, \quad \sum_{i=1}^{N} \lambda_i = 1.$$

Here, θ_i is a known vertex of parameter vector. This sort of uncertain systems is called *polytopic systems*.

11.3.3 Norm-Bounded Parametric System

For the analysis of parametric systems, the polytopic model is a very useful model. Unfortunately, no satisfactory method has been found for the control design of polytopic systems up to now. Basically we do not know how to do the design except for some special problems such as the gain-scheduled control of Chapter 20. Therefore, in this section we will discuss other parametric models that are suitable for control design.

Recall the mass–spring–damper system:

$$\dot{x} = \begin{bmatrix} 0 & 1 \\ -\dfrac{k}{m} & -\dfrac{b}{m} \end{bmatrix} x + \begin{bmatrix} 0 \\ \dfrac{1}{m} \end{bmatrix} u.$$

Note that a feature of the coefficient matrices is that all uncertain parameters are in the second row. This makes it possible for us to put the uncertain parameters into a matrix and treat them as a matrix uncertainty. Set the nominal parameters as (m_0, k_0, b_0); the parameters can be written as

$$\frac{k}{m} = \frac{k_0}{m_0}(1 + w_1\delta_1), \quad \frac{b}{m} = \frac{b_0}{m_0}(1 + w_2\delta_2), \quad \frac{1}{m} = \frac{1}{m_0}(1 + w_3\delta_3), \quad |\delta_i| \leq 1.$$

Due to

$$\begin{bmatrix} 0 & 1 \\ -\frac{k}{m} & -\frac{b}{m} \end{bmatrix} = \begin{bmatrix} 0 & 1 \\ -\frac{k_0}{m_0} & -\frac{b_0}{m_0} \end{bmatrix} + \begin{bmatrix} 0 & 0 \\ -\frac{k_0}{m_0}w_1\delta_1 & -\frac{b_0}{m_0}w_2\delta_2 \end{bmatrix}$$

$$= \begin{bmatrix} 0 & 1 \\ -\frac{k_0}{m_0} & -\frac{b_0}{m_0} \end{bmatrix} + \begin{bmatrix} 0 \\ -1 \end{bmatrix} [\delta_1 \quad \delta_2 \quad \delta_3] \begin{bmatrix} \frac{k_0}{m_0}w_1 & 0 \\ 0 & \frac{b_0}{m_0}w_2 \\ 0 & 0 \end{bmatrix}$$

$$\begin{bmatrix} 0 \\ \frac{1}{m} \end{bmatrix} = \begin{bmatrix} 0 \\ \frac{1}{m_0} \end{bmatrix} + \begin{bmatrix} 0 \\ 1 \end{bmatrix} [\delta_1 \quad \delta_2 \quad \delta_3] \begin{bmatrix} 0 \\ 0 \\ \frac{1}{m_0}w_3 \end{bmatrix},$$

the state equation can be rewritten as

$$\dot{x} = (A + B_1\Delta C_1)x + (B_2 + B_1\Delta D_{12})u$$

where $\Delta = [\delta_1 \quad \delta_2 \quad \delta_3]$ and

$$A = \begin{bmatrix} 0 & 1 \\ -\frac{k_0}{m_0} & -\frac{b_0}{m_0} \end{bmatrix}, \quad B_1 = \begin{bmatrix} 0 \\ 1 \end{bmatrix}, \quad B_2 = \begin{bmatrix} 0 \\ \frac{1}{m_0} \end{bmatrix}$$

$$C_1 = -\begin{bmatrix} \frac{k_0}{m_0}w_1 & 0 \\ 0 & \frac{b_0}{m_0}w_2 \\ 0 & 0 \end{bmatrix}, \quad D_{12} = \begin{bmatrix} 0 \\ 0 \\ \frac{1}{m_0}w_3 \end{bmatrix}.$$

Naturally, the size of uncertainty vector Δ should be measured by vector norm. So, this system is called a *norm-bounded parametric system*.

However, a deeper thinking reveals that the present model has a defect. This defect lies in that, since there are products of parameters in the state equation, there exists a danger of enlarging the uncertainty range when these products are treated as independent parameters. For simplicity, let us look at the coefficients $\frac{k}{m}$ and $\frac{1}{m}$ related to parameters (m, k). Suppose that (m, k) take values in the following ranges, respectively:

$$k_{\text{m}} \leq k \leq k_{\text{M}}, \quad m_{\text{m}} \leq m \leq m_{\text{M}}. \tag{11.24}$$

The range of parameters is illustrated by the shaded square in Figure 11.13. But in state equation it is necessary to treat $\alpha = \frac{k}{m}$ and $\frac{1}{m}$ as uncertain parameters. Then, apparently the

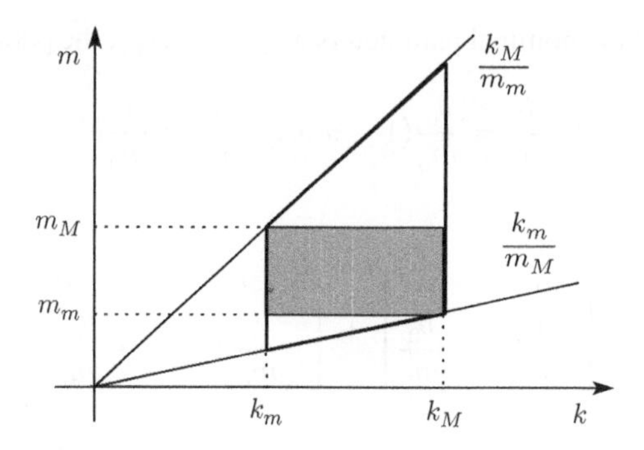

Figure 11.13 Parameter space of k and m

range of α is

$$\frac{k_{\rm m}}{m_{\rm M}} \leq \alpha \leq \frac{k_{\rm M}}{m_{\rm m}}. \tag{11.25}$$

When this range is converted back into the parameter space (k, m), from $km_m/k_M \leq m = k/\alpha \leq km_M/k_m$ it is known that the corresponding range is the area surrounded by the four thick lines in Figure 11.13, which is obviously bigger than the real range of physical parameters. This happens because the physical parameters are not treated independently.

However, when we multiply the second row of the state equation by m, the equation becomes

$$\begin{bmatrix} 1 & 0 \\ 0 & m \end{bmatrix} \dot{x} = \begin{bmatrix} 0 & 1 \\ -k & -b \end{bmatrix} x + \begin{bmatrix} 0 \\ 1 \end{bmatrix} u$$

in which each parameter appears independently. This model is called a *descriptor form.* Let each parameter take value in

$$m = m_0(1 + w_1\delta_1), \quad k = k_0(1 + w_2\delta_2), \quad b = b_0(1 + w_3\delta_3), \quad |\delta_i| \leq 1.$$

Dividing the second row of the state equation by m_0, we get

$$\begin{aligned} &(I + B_1\Delta C_1)\dot{x} = (A + B_1\Delta C_2)x + B_2u \\ &\Leftrightarrow \dot{x} = (I + B_1\Delta C_1)^{-1}(A + B_1\Delta C_2)x + (I + B_1\Delta C_1)^{-1}B_2u \end{aligned}$$

where $\Delta = [\delta_1 \ \ \delta_2 \ \ \delta_3]$ and the constant matrices are

$$A = \begin{bmatrix} 0 & 1 \\ -\dfrac{k_0}{m_0} & -\dfrac{b_0}{m_0} \end{bmatrix}, \quad B_1 = \begin{bmatrix} 0 \\ 1 \end{bmatrix}, B_2 = \begin{bmatrix} 0 \\ \dfrac{1}{m_0} \end{bmatrix}$$

$$C_1 = \begin{bmatrix} 0 & w_1 \\ 0 & 0 \\ 0 & 0 \end{bmatrix}, \quad C_2 = -\begin{bmatrix} 0 & 0 \\ \dfrac{k_0}{m_0}w_2 & 0 \\ 0 & \dfrac{b_0}{m_0}w_3 \end{bmatrix}.$$

Each coefficient matrix of this state equation is obviously a linear fractional transformation (LFT) of the uncertainty matrix Δ. In order to find the coefficient matrices of this LFT, we set $v = C_1\dot{x}$, $z = -v + C_2x$, and $w = \Delta z$. Then,

$$\dot{x} = Ax + B_1w + B_2u$$

holds so that $v = C_1Ax + C_1B_1w + C_1B_2u$. Substitution of this v into z leads to

$$z = (C_2 - C_1A)x - C_1B_1w - C_1B_2u.$$

Finally, we discuss the norm of matrix uncertainty Δ and its relationship with the parameter uncertainty range. For simplicity, let us just consider the case of two-parameter uncertainties (δ_1, δ_2). This uncertainty vector forms a square with the origin as the center. Meanwhile, two-dimensional vectors with a bounded two-norm form a disk. So, when a parameter uncertainty vector is treated as a norm-bounded uncertainty, in order not to enlarge the uncertainty range too much, the minimum disk containing the square shown in Figure 11.14 should be selected, whose radius is obviously $\sqrt{1^2 + 1^2} = \sqrt{2}$.

In this example, the uncertainty matrix is a three-dimensional vector, so its norm bound equals $\sqrt{3}$. In system analysis and design, uncertainty norm is usually normalized to 1. This is done by absorbing $\sqrt{3}$ into the signal z.

In summary, a general model for norm-bounded parametric systems is

$$G\begin{cases} \dot{x} = Ax + B_1w + B_2d + B_3u \\ z = C_1x + D_{11}w + D_{12}d + D_{13}u \\ e = C_2x + D_{21}w + D_{22}d + D_{23}u \\ y = C_3x + D_{31}w + D_{32}d \end{cases} \tag{11.26}$$

$$w = \Delta z, \|\Delta\|_2 \leq 1. \tag{11.27}$$

Here, newly included are a disturbance d and a performance output e used for control performance optimization, as well as a measured output y. The block diagram of this uncertain system is Figure 11.15.

The parameter uncertainty in this model can even be extended to a time-varying parameter matrix, namely,

$$\Delta(t) \in \mathbb{R}^{p\times q}, \quad \|\Delta(t)\|_2 \leq 1 \quad \forall t \geq 0. \tag{11.28}$$

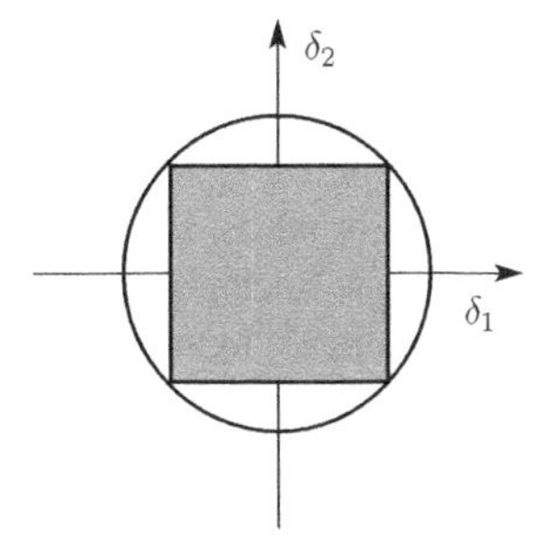

Figure 11.14 Range of parameter uncertainty: square vs disk

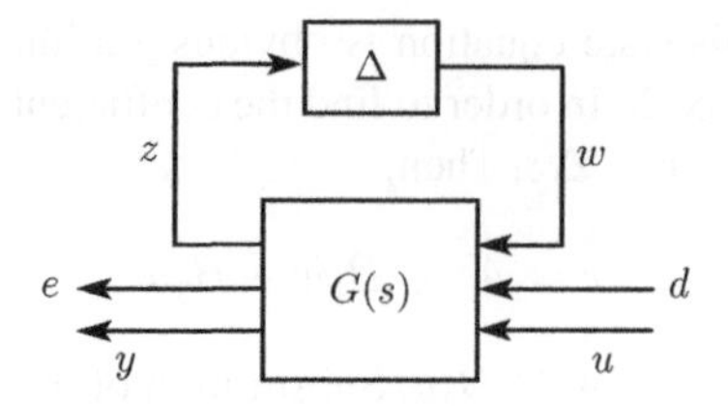

Figure 11.15 General form of norm-bounded parametric system

It should be noted that not all uncertain parameters in a parametric system can be collected into a vector like this example. For instance, in the two-mass–spring system in Example 1.2, when the load inertia moment J_L and the spring constant k are uncertain, we can only put them into a diagonal matrix:

$$\Delta = \begin{bmatrix} \delta_1 & \\ & \delta_2 \end{bmatrix}.$$

The detail is given in the next section.

11.3.4 Separation of Parameter Uncertainties

Let us reexamine the two-mass–spring system in Example 1.2. For simplicity, assume that the motor's inertia moment is 1 and all damping coefficients are 0. Then, its descriptor form becomes

$$\begin{bmatrix} J_L & & \\ & 1 & \\ & & 1 \end{bmatrix} \dot{x} = \begin{bmatrix} 0 & k & 0 \\ -1 & 0 & 1 \\ 0 & -k & 0 \end{bmatrix} x + \begin{bmatrix} 1 \\ 0 \\ 0 \end{bmatrix} d + \begin{bmatrix} 0 \\ 0 \\ 1 \end{bmatrix} u \tag{11.29}$$

and each parameter appears independently. As is clear from the coefficient matrices, the uncertain parameters are no longer in the same row or column, and they cannot be put into a vector.

The situation shown in this example applies to almost all mechanical systems. Therefore, next we consider an uncertain system with r uncertain parameters $\delta_1, \cdots, \delta_r$:

$$\begin{aligned} \left(E + \sum_{i=1}^{r} \delta_i E_i\right) \dot{x} &= \left(A + \sum_{i=1}^{r} \delta_i A_i\right) x + \left(B + \sum_{i=1}^{r} \delta_i B_i\right) u + Hd \\ e &= Mx + Nd + Vu \\ y &= Cx + Qd. \end{aligned} \tag{11.30}$$

Here, $x \in \mathbb{R}^n$ is the state, $u \in \mathbb{R}^{n_u}$ the input, $y \in \mathbb{R}^{n_y}$ the performance output, $e \in \mathbb{R}^{n_e}$ the performance output, and $d \in \mathbb{R}^{n_d}$ the disturbance. n_u denotes the dimension of input u; the rest of the subscripts have similar meanings. Each coefficient matrix is a constant matrix with appropriate dimension. Finally, E is assumed nonsingular, that is, the degree of system is not affected by the change of parameters. Let us figure out how to pull out all parameters independently.

First, carry out a matrix factorization:

$$E^{-1}[E_i \quad A_i \quad B_i] = L_i[R_i \quad W_i \quad Z_i], \quad i = 1, \cdots, r. \tag{11.31}$$

Let the rank of the matrix on the left be q_i. In the factorization, we should ensure that the matrix L_i has full column rank and $[R_i \quad W_i \quad Z_i]$ has full row rank. That is, to factorize a non-full rank matrix into the product of two full rank matrices. Such factorization is called the maximum rank factorization and its solution is not unique.

Next, we define the following matrices:

$$L = [L_1, \cdots, L_r], \quad R = \begin{bmatrix} R_1 \\ \vdots \\ R_r \end{bmatrix}, \quad W = \begin{bmatrix} W_1 \\ \vdots \\ W_r \end{bmatrix}, \quad Z = \begin{bmatrix} Z_1 \\ \vdots \\ Z_r \end{bmatrix}. \tag{11.32}$$

Then, the uncertain parameters can be separated independently as shown in the following lemma. Refer to Ref. [40] for the proof.

Lemma 11.1 *The uncertain system in (11.30) can be transformed into the system in Figure 11.15*

$$\begin{bmatrix} e \\ y \end{bmatrix} = \mathcal{F}_u(G, \Delta) \begin{bmatrix} d \\ u \end{bmatrix}$$

in which Δ is a diagonal matrix $\Delta = \mathrm{diag}(\delta_1 I_{q_1}, \cdots, \delta_r I_{q_r})$ and

$$G = \left[\begin{array}{c|ccc} \hat{A} & L & \hat{H} & \hat{B} \\ \hline W - R\hat{A} & -RL & -R\hat{H} & Z - R\hat{B} \\ M & 0 & N & V \\ C & 0 & Q & 0 \end{array}\right]$$

$$\hat{A} = E^{-1}A, \quad \hat{B} = E^{-1}B, \quad \hat{H} = E^{-1}H.$$

Example 11.6 *In the two-mass–spring example, assume that $J_L = J_0 + \delta_1 \Delta_J$, $k = k_0 + \delta_2 \Delta_k$. Then*

$$E = \mathrm{diag}(J_0, 1, 1), \quad E_1 = \mathrm{diag}(\Delta_J, 0, 0), \quad E_2 = 0$$

$$A = \begin{bmatrix} 0 & k_0 & 0 \\ -1 & 0 & 1 \\ 0 & -k_0 & 0 \end{bmatrix}, \quad A_1 = 0, \quad A_2 = \begin{bmatrix} 0 & \Delta_k & 0 \\ 0 & 0 & 0 \\ 0 & -\Delta_k & 0 \end{bmatrix}$$

$$B = [0 \quad 0 \quad 1]^T, \quad B_1 = 0, \quad B_2 = 0.$$

It is easy to see that the matrix ranks in (11.31) are $q_1 = q_2 = 1$ so that the parameter uncertainties are collected into a diagonal matrix $\Delta = \mathrm{diag}(\delta_1, \delta_2)$. An example of maximum rank factorization is given by

$$L = \begin{bmatrix} \sqrt{\frac{\Delta_J}{J_0}} & \sqrt{\frac{\Delta_k}{J_0}} \\ 0 & 0 \\ 0 & -\sqrt{\Delta_k} \end{bmatrix}, \quad R = \begin{bmatrix} \sqrt{\frac{\Delta_J}{J_0}} & 0 & 0 \\ 0 & 0 & 0 \end{bmatrix}$$

$$W = \begin{bmatrix} 0 & 0 & 0 \\ 0 & \sqrt{\Delta_k} & 0 \end{bmatrix}, \quad Z = \begin{bmatrix} 0 \\ 0 \end{bmatrix}.$$

11.4 Plant Set with Phase Information of Uncertainty

In practical systems static nonlinearity such as friction, dead zone, and saturation of actuator may be found everywhere, which usually cannot be modeled accurately. In dealing with such uncertainty, it is possible to achieve a high-performance robust control design if we can find some way to abstract its characteristics. Further, in flexible structures with high-order resonant modes, the resonant modes contain not only gain information but also phase information. In the sequel, we study the properties of such uncertainties through examples and discuss what kind of model can describe their properties.

The first example is the saturation function $\phi(\cdot)$ shown in Figure 11.16(a). $\phi(\cdot)$ can be enclosed by two lines with slopes of k and 0 as shown in the figure and can be described by a piecewise inequality as

$$0 \leq \phi(u) \leq ku \quad \forall u \geq 0; \quad -ku \leq \phi(u) \leq 0 \quad \forall u < 0. \tag{11.33}$$

This piecewise inequality can be summarized simply as

$$0 \leq u\phi(u) \leq ku^2 \quad \forall u. \tag{11.34}$$

The relationship between the magnetic flux and current in an electromagnet is exactly such a saturation. A more important character is that $\phi(\cdot)$ is located in the first and the third quadrants, namely, its slope changes between 0 and k. This shows that the phase angle of $\phi(\cdot)$ is zero, while its gain changes in the interval $[0, k]$. Particularly useful is the information on the phase angle.

If we need only consider inputs in a finite range, then the two lines surrounding the saturation $\phi(\cdot)$ can be ku and hu as shown in the figure. In this way, the range of saturation can be described more tightly. Then, the bound of saturation turns into

$$hu^2 \leq u\phi(u) \leq ku^2 \quad \forall u. \tag{11.35}$$

In addition, the so-called Stribeck friction shown in Figure 11.16(b) also has a similar character, only the range of gain changes to $[k, \infty]$.

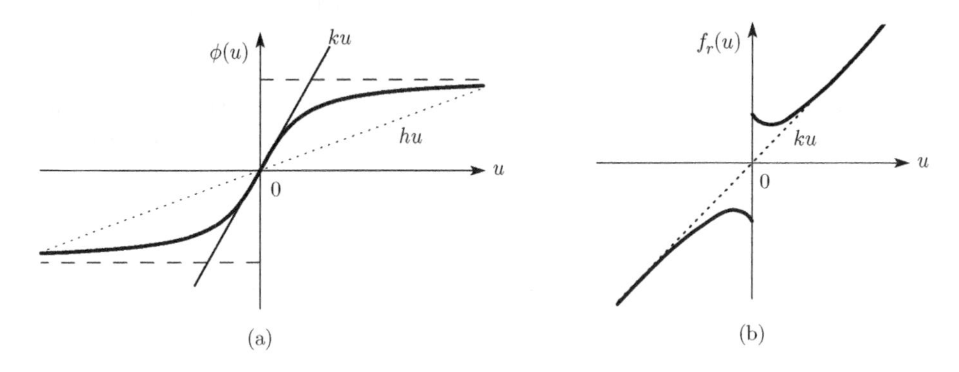

Figure 11.16 Examples of passive nonlinearity (a) Saturation, (b) Stribeck friction

Next, we discuss the character of a flexible space structure[2]: high-order resonant modes. The model of high-rise buildings also has the same property. When the input is a torque and the output is an angular velocity, the model is

$$P(s) = \frac{k_0 s}{s^2 + 2\zeta_0\omega_0 s + \omega_0^2} + \frac{k_1 s}{s^2 + 2\zeta_1\omega_1 s + \omega_1^2} + \cdots + \frac{k_n s}{s^2 + 2\zeta_n\omega_n s + \omega_n^2} \tag{11.36}$$

in which all gains k_i are positive (this kind of transfer function is called in-phase). The first-order resonant mode $\frac{k_0 s}{s^2+2\zeta_0\omega_0 s+\omega_0^2}$ can be identified accurately, but the second- and higher order resonant modes can hardly be identified correctly. Here, we treat the sum of these high-order resonant modes as a dynamic uncertainty. Although this uncertainty can be characterized by its gain information, we hope to find a model that is more suitable in describing its characteristic. To this end, let us analyze one of the resonant modes:

$$\frac{s}{s^2 + 2\zeta_i\omega_i s + \omega_i^2}.$$

Its frequency response is

$$\frac{j\omega}{\omega_i^2 - \omega^2 + j2\zeta_i\omega_i\omega} = \frac{2\zeta_i\omega_i\omega^2 + j\omega(\omega_i^2 - \omega^2)}{(\omega_i^2 - \omega^2)^2 + (2\zeta_i\omega_i\omega)^2}.$$

The real part is nonnegative when $\zeta_i > 0$ and 0 when $\zeta_i = 0$. This implies that the phase angle takes its value between $-\pi/2$ rad and $\pi/2$ rad (the Nyquist diagram of such a transfer function is shown in Figure 11.17), which is exactly the characteristic of positive-real function. Therefore, the sum of high-order resonant modes

$$\frac{k_1 s}{s^2 + 2\zeta_1\omega_1 s + \omega_1^2} + \cdots + \frac{k_n s}{s^2 + 2\zeta_n\omega_n s + \omega_n^2}$$

also has the same property. That is, the sum of high-order resonant modes is still positive-real which is an important character of flexible structures.

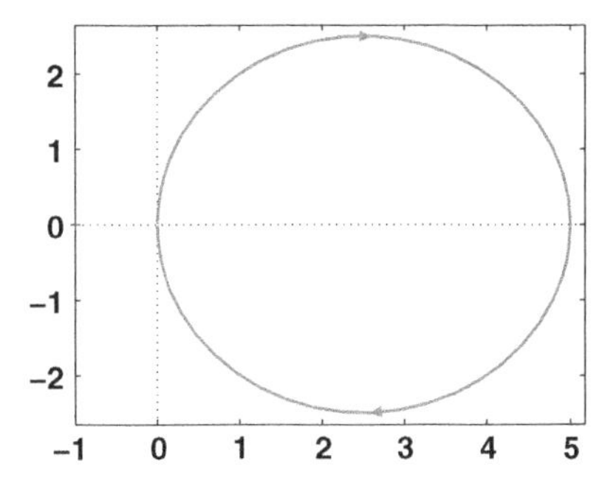

Figure 11.17 Nyquist diagram of a positive-real transfer function

[2] The most typical example is the solar battery panel used to supply power for satellite or space stations.

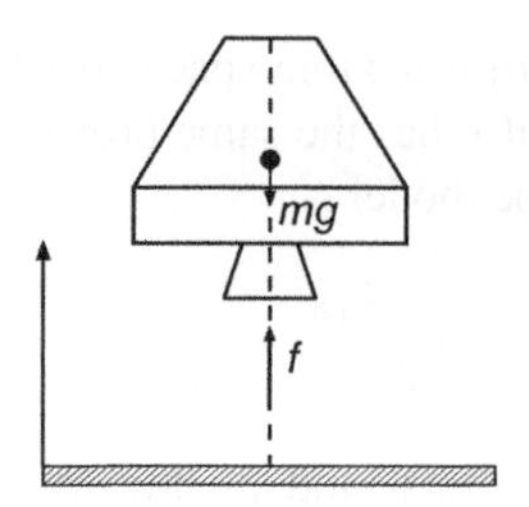

Figure 11.18 Lunar module

11.5 LPV Model and Nonlinear Systems

11.5.1 LPV Model

Consider a lunar module shown in Figure 11.18 which is used in the soft landing on the moon. m is the mass of the module including the fuel, and the thrust is $f = k\dot{m}$. Set the control input as $u = \dot{m}$ and the height of the module as y, and let g be the gravity constant on the moon surface. The state equation about state vector $x = [y \quad \dot{y} \quad m]^T$ is given by

$$\dot{x}_1 = x_2, \quad \dot{x}_2 = -g + \frac{k - x_2}{x_3}u, \quad \dot{x}_3 = u. \tag{11.37}$$

This model is derived according to Newton's second law $F = d(m\dot{y})/dt$. Although the state variable $x_3 = m$ decreases with the combustion of fuel, its range is less than the initial mass of fuel and is known. Secondly, the range of speed $x_2 = \dot{y}$ may be estimated in advance. Therefore, we can regard $p(t) = \frac{k-x_2}{x_3}$ as a time-varying parameter. Then, the nonlinear model can be written as

$$\dot{x} = \begin{bmatrix} 0 & 1 & 0 \\ 0 & 0 & 0 \\ 0 & 0 & 0 \end{bmatrix} x + \begin{bmatrix} 0 \\ p \\ 1 \end{bmatrix} u + \begin{bmatrix} 0 \\ -g \\ 0 \end{bmatrix}.$$

Though not a linear model, it is an affine model with a time-varying coefficient.

In general, when a nonlinear system works in a narrow range, a linear approximation is good enough for the control design. But in global control, or when the operating range is large, a linear approximation can no longer be trusted. Meanwhile, the need is very strong to use techniques similar to linear control in the control of nonlinear systems. So there arises a new challenge how to transform a nonlinear system to a quasilinear (viz., in a linear form) system. One way is to treat nonlinear functions as time-varying coefficients like the previous example. The *LPV* (linear parameter-varying) model to be illustrated below is the general form

$$\dot{x} = A(p(t))x + B(p(t))u \tag{11.38}$$

$$y = C(p(t))x. \tag{11.39}$$

Here, $p(t)$ is a time-varying parameter vector and each matrix is an affine function of $p(t)$.

For example, in the case of two time-varying parameters $p(t) = [p_1(t) \quad p_2(t)]$, each coefficient matrix can be written as

$$A(p(t)) = A_0 + p_1(t)A_1 + p_2(t)A_2, B(p(t)) = B_0 + p_1(t)B_1 + p_2(t)B_2$$
$$C(p(t)) = C_0 + p_1(t)C_1 + p_2(t)C_2$$

where all matrices are known except the time-varying parameter vector $p(t)$.

11.5.2 From Nonlinear System to LPV Model

Many nonlinear systems can be converted into LPV models by fully utilizing its dynamics. Here, we use a *single-machine infinite-bus power system*, shown in Figure 11.19, as an example to illustrate how to convert a nonlinear system into an LPV model. Infinite bus is a term in power systems which means that the entire power system connected with the generator is regarded as an ideal voltage source.

The nonlinear model of this power system is given by the following differential equation [50, 66]:

$$\dot{\delta} = \omega - \omega_0 \tag{11.40}$$

$$\dot{\omega} = \frac{\omega_0}{M} P_M - \frac{\omega_0}{M} P_e - \frac{D}{M}(\omega - \omega_0) \tag{11.41}$$

$$\dot{E}_q' = -\frac{1}{T_d'} E_q' + \frac{x_d - x_d'}{T_{d0} x_{d\Sigma}'} V_s \cos\delta + \frac{1}{T_{d0}} V_f. \tag{11.42}$$

In this model, δ is the rotor angle, ω is the angular velocity of rotor, E_q' is the q-axis transient voltage, V_s is the bus voltage, and P_M stands for the mechanical power from the turbine. Further, the active power of generator equals the following nonlinear function of states:

$$P_e = \frac{E_q' V_s}{x_{d\Sigma}'} \sin\delta - \frac{V_s^2}{2} \frac{x_d - x_d'}{x_{d\Sigma}' x_{d\Sigma}} \sin 2\delta. \tag{11.43}$$

The rest are known parameters. The control input is the excitation voltage V_f. Here, we consider the transient stability problem related to short-circuit fault of the transmission line. Short-circuit fault, especially a short-circuit fault occurring near the terminal of generator,

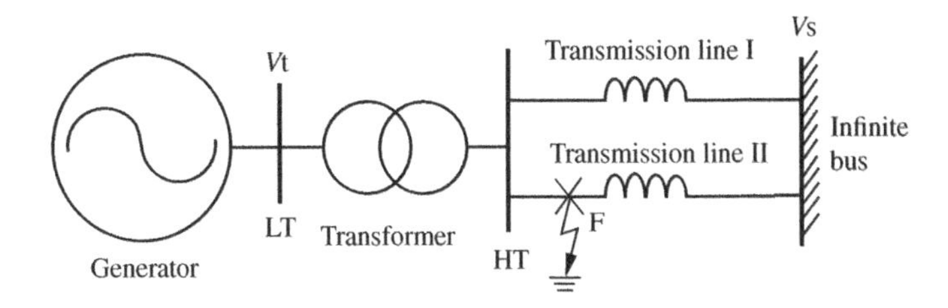

Figure 11.19 Single-machine infinite-bus power system

will make the active power drop instantly, thus causing a steep acceleration of the rotor (see Eq.(11.41)). If no excitation control is activated quickly right after clearing the fault by opening the short-circuit transmission line, the generator will lose synchronism. In the fault phase, all states deviate from the equilibrium significantly so that the linear approximation fails.

The equilibrium of power system is the state at which the mechanical power is equal to the active power, which is set as $(\delta_0, \omega_0, E'_{q0}, V_{f0})$. The goal of control is to restore the deviated states to the equilibrium. So it is equivalent to the stabilization of error states given by

$$x_1 = \delta - \delta_0, \quad x_2 = \omega - \omega_0, \quad x_3 = E'_q - E'_{q0}, \quad u = V_f - V_{f0}.$$

The state equation about the error state x is given by

$$\begin{aligned} \dot{x}_1 &= x_2 \\ \dot{x}_2 &= d_1 \sin\delta \cdot x_3 + d_1 E'_{q0}(\sin\delta - \sin\delta_0) + d_2 x_2 \\ \dot{x}_3 &= d_3 x_3 + d_4(\cos\delta - \cos\delta_0) + d_5 x_4. \end{aligned} \tag{11.44}$$

Next, we convert it into an LPV model. Focusing on the boundedness of $(\sin\delta - \sin\delta_0)/(\delta - \delta_0)$ and $(\cos\delta - \cos\delta_0)/(\delta - \delta_0)$, we can transform the nonlinear functions in the state equation into quasilinear functions with time-varying coefficients:

$$\sin\delta - \sin\delta_0 = \frac{\sin\delta - \sin\delta_0}{\delta - \delta_0} x_1, \quad \cos\delta - \cos\delta_0 = \frac{\cos\delta - \cos\delta_0}{\delta - \delta_0} x_1. \tag{11.45}$$

The following parameters are functions of the rotor angle δ:

$$p_1(\delta) = \frac{\sin\delta - \sin\delta_0}{\delta - \delta_0}, \quad p_2(\delta) = \sin\delta, \quad p_3(\delta) = \frac{\cos\delta - \cos\delta_0}{\delta - \delta_0}. \tag{11.46}$$

In this way, the state equation has been converted into the LPV model of Eq. (11.47):

$$\begin{aligned} \dot{x}_1 &= x_2 \\ \dot{x}_2 &= d_1 p_2(\delta) x_3 + d_1 E'_{q0} p_1(\delta) x_1 + d_2 x_2 \\ \dot{x}_3 &= d_3 x_3 + d_4 p_3(\delta) x_1 + d_5 x_4. \end{aligned} \tag{11.47}$$

Note that as long as the rotor angle $\delta(t)$ is measured, the time-varying parameter vector $p(t) = [p_1\ p_2\ p_3]^T$ can always be calculated online. Putting the LPV model (11.39) into a vector form, we get

$$\dot{x} = A(p)x + Bu \tag{11.48}$$

$$A(p) = \begin{bmatrix} 0 & 1 & 0 \\ d_1 E'_{q0} p_1(\delta) & d_2 & d_1 p_2(\delta) \\ d_4 p_3(\delta) & 0 & d_3 \end{bmatrix}, \quad B = \begin{bmatrix} 0 \\ 0 \\ d_5 \end{bmatrix}.$$

Moreover, the matrix $A(p)$ can be written as the following affine form:

$$A(p) = A_0 + p_1 A_1 + p_2 A_2 + p_3 A_3 \tag{11.49}$$

$$A_1 = \begin{bmatrix} 0 & 0 & 0 \\ d_1 E'_{q0} & 0 & 0 \\ 0 & 0 & 0 \end{bmatrix}, \quad A_2 = \begin{bmatrix} 0 & 0 & 0 \\ 0 & 0 & d_1 \\ 0 & 0 & 0 \end{bmatrix}, \quad A_3 = \begin{bmatrix} d_4 & 0 & 0 \\ 0 & 0 & 0 \\ 0 & 0 & 0 \end{bmatrix}.$$

In converting a nonlinear system into an LPV model, we need to ensure that the state-dependent time-varying parameters are bounded. Why? This is because robust control design can only treat bounded parameters with sufficiently small uncertainty range. In this sense, not all nonlinear systems can be converted into LPV models.

Another example is about the saturation $\phi(u)$ of an actuator (refer to Figure 11.16(a)). When $\phi(u)$ is a smooth function, then it can be written as a linear function with time-varying coefficients. For example, in

$$\phi(u) = \arctan u = \frac{\arctan u}{u} u = g(u)u,$$

the time-varying parameter $g(u) = \arctan u/u$ is bounded.

11.6 Robust Stability and Robust Performance

As aforementioned, the nominal plant P_0 is just an approximation of the actual system P and model uncertainty always exists in reality. However, mathematical models have to be used in the design of control systems. In order to ensure that the designed controllers can achieve the expected performance when implemented to the true system, we should guarantee the same level of stability and control performance for all plants in a plant set which contains the true system. See Figure 11.20 for an illustration.

In this book, we use the following terminologies.

Definition 11.1 *Given a plant set $\mathbb{P}$ and a corresponding performance specification, let $P \in \mathbb{P}$ be the nominal plant and K the controller. Then we say that*

1. *the closed-loop system is* nominally stable *if the controller K stabilizes the nominal plant P,*
2. *the closed-loop system is* robustly stable *if the controller K stabilizes all plants in the set $\mathbb{P}$,*

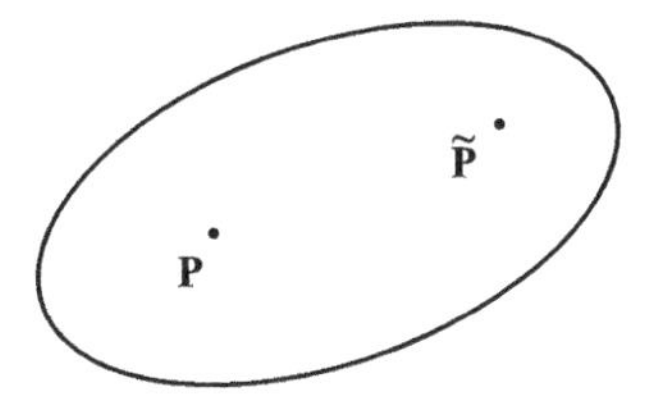

Figure 11.20 Plant set $\mathbb{P}$

3. *the closed-loop system achieves the* nominal performance *if the controller K satisfies the performance specification for the nominal plant P,*
4. *the closed-loop system achieves the* robust performance *if the controller K satisfies the performance specification for all plants in the set $\mathbb{P}$.*

The ultimate goal of robust control is to realize high-quality robust performance!

Exercises

11.1 Consider the following flexible system[3]

$$P(s) = K\left(\frac{1}{s^2} + \frac{A}{s^2 + 2\zeta_1\omega_1 s + \omega_1^2}\right) \tag{11.50}$$

in which, $K = 3.74 \times 10^9, A = 0.4 \sim 1.0, \zeta_1 = 0.02 \sim 0.6, \omega_1 = 4000 \sim 4200[\text{Hz}]$. Set the rigid body model

$$P_0(s) = \frac{K}{s^2}$$

as the nominal plant and treat the resonant mode

$$\Delta(s) = \frac{KA}{s^2 + 2\zeta_1\omega_1 s + \omega_1^2}$$

as a multiplicative uncertainty. Find an upper bound (weighting function) $W(s)$ for the uncertainty gain and establish the corresponding plant set.

11.2 Consider a plant with parameter uncertainty:

$$P(s) = \frac{k}{\tau s + 1}, \quad 0.8 \le k \le 1.2, \quad 0.7 \le \tau \le 1.3. \tag{11.51}$$

Here, the nominal plant is set as

$$P_0(s) = \frac{k_0}{\tau_0 s + 1}. \tag{11.52}$$

The model uncertainty is regarded as a multiplicative one $P(s)/P_0(s) = 1 + \Delta(s)$. The smaller the uncertainty is, the more possible it is for us to design a control system with a better performance. In order to minimize the gain of the frequency response of uncertainty $\Delta(s)$, how should we determine the nominal parameters (k_0, τ_0)? Compare with the case where the nominal values are set as the mean values.

11.3 For the two approximations of parameter range shown in Figure 11.11(a,b) in Example 11.5, calculate the corresponding matrix polytopes.

[3] In engineering practice, for a vibration system we can only obtain the range of frequency response of the system through system identification, not the range of parameters as shown in this exercise. It is solely for the convenience of presentation to assign ranges of parameters in this exercise.

11.4 The state-space model of the two-mass–spring system (Example 1.2 with friction ignored) is given by

$$\dot{x} = \begin{bmatrix} 0 & \frac{k}{J_L} & 0 \\ -1 & 0 & 1 \\ 0 & -\frac{k}{J_M} & 0 \end{bmatrix} x + \begin{bmatrix} \frac{1}{J_L} \\ 0 \\ 0 \end{bmatrix} d + \begin{bmatrix} 0 \\ 0 \\ \frac{1}{J_M} \end{bmatrix} u$$

$$y = [0 \quad 0 \quad 1]x$$

in which J_M, J_L, k are, respectively, the inertial moments of the motor and load and the spring constant of the shaft. Suppose that J_L, k take value in the following ranges:

$$J_1 \leq J_L \leq J_2, \quad k_1 \leq k \leq k_2.$$

Compute the polytopic description for this system.

11.5 Assume that $m \in [1, 2]$. Firstly, draw the area of vector $[m \;\; m^2]^T$. Then, use the following two methods to approximate the area of this parameter vector and compute the corresponding polytopic descriptions:

$$(1) \text{ rectangle}, \quad (2) \text{ triangle}.$$

Can you find a minimal trapezoid which covers the area of this uncertain vector?

11.6 In the two-mass–spring system in Exercise 11.4, let the uncertainty ranges of the load inertia moment J_L and the spring constant k be

$$J_L = J_{L0}(1 + w_1\delta_1), \quad k = k_0(1 + w_2\delta_2), \quad |\delta_i| \leq 1.$$

Transform the state equation into the form of (11.26) with norm-bounded parameter uncertainty where

$$\Delta = \text{diag}(\delta_1 \quad \delta_2).$$

11.7 In the single-machine infinite-bus power system of Subsection 11.5.2, the nonlinear term in Eq.(11.42) can be cancelled by the input $V_f = -\frac{x_d - x_d'}{x_{d\Sigma}'} V_s \cos\delta + u$. Derive a two-parameter LPV model w.r.t. the new input u.

Notes and References

Zhou *et al.* [100] provides a detailed treatment on the norm-bounded model uncertainty and Ref. [37] contains some discussion on the positive real uncertainty. Tits *et al.* [88] and Liu [53, 63, 56, 57] respectively proposed uncertainty models with both gain bound and phase bound. The corresponding robustness conditions were also derived. For more details on the power system involved in the LPV example, refer to Refs [50, 66] and [38].

12

Robustness Analysis 1: Small-Gain Principle

In this chapter, we analyze the robustness of system with dynamic uncertainty and try to find the conditions needed for ensuring robust stability and robust performance. This chapter treats dynamic uncertainties characterized by the range of gain variation, that is, only the information of uncertainty gain is used. The base of robustness analysis for such dynamic uncertainty is the small-gain theorem. After illustrating this theorem in detail, we will derive the robust stability criteria for various kinds of system sets and some sufficient conditions for robust performance. These conditions are all given in terms of $\mathcal{H}_\infty$ norm inequality about the closed-loop transfer matrix with weighting function.

12.1 Small-Gain Theorem

In the system shown in Figure 12.1, M is a given system and Δ is an uncertainty. Both are matrices of stable rational functions. We first consider the robust stability condition for this simplified closed-loop system. Then, we will give some examples to illustrate that, for all norm-bounded plant sets handled in Chapter 11, their robust stability criteria boil down to that of the system in Figure 12.1 through equivalent block diagram transformation.

Theorem 12.1 (Small-Gain Theorem) *Assume that $M(s), \Delta(s)$ are stable. The closed-loop system of Figure 12.1 is robustly stable iff one of the following conditions is true:*

1. *When $\|\Delta\|_\infty \leq 1$, there holds $\|M\|_\infty < 1$.*
2. *When $\|\Delta\|_\infty < 1$, there holds $\|M\|_\infty \leq 1$.*

Proof. Here, we give a detailed proof for single-input single-output (SISO) systems. As for multiple-input multiple-output (MIMO) systems, only an outline is provided at the end of

Robust Control: Theory and Applications, First Edition. Kang-Zhi Liu and Yu Yao.

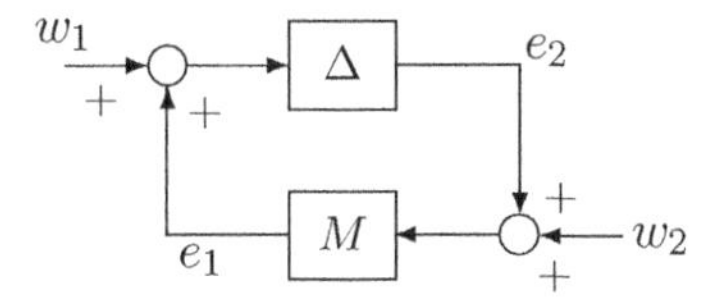

Figure 12.1 Small-gain theorem

proof. Refer to Ref. [100] for further details. Further, statement 2 can be proved by properly modifying the proof. So, we only prove statement 1.

The *return difference* of this feedback system is

$$R(s) = 1 - M(s)\Delta(s).$$

The denominator of all closed-loop transfer functions is $R(s)$. Since both M and Δ are stable, unstable pole–zero cancellation does not occur in their cascade connection. So, for the stability of closed-loop system, it suffices to ensure that $R(s)$ has no zero in the closed right half-plane. However,

$$\begin{aligned}|R(s)| &= |1 - M(s)\Delta(s)| \\ &\geq 1 - |M(s)||\Delta(s)| \\ &\geq 1 - \|M\|_\infty \|\Delta\|_\infty \\ &> 0\end{aligned}$$

holds at any point s in the closed right half-plane, in which the second inequality follows from the maximum modulus theorem [1] in complex analysis, while the last inequality is from the given condition $\|M\|_\infty \|\Delta\|_\infty < 1$. Therefore, the closed-loop system has no poles in the closed right half-plane and is stable.

The idea for the proof of necessity is to prove that we can find an uncertainty in the given uncertainty set Δ which destabilizes the closed-loop system if $\|M\|_\infty \geq 1$. Due to $\|M\|_\infty \geq 1$ and the continuity of frequency response, there must be a frequency $\omega_0 \in [0, \infty]$ at which $|M(j\omega_0)| = 1$ holds. Then, $M(j\omega_0)$ can be written as

$$M(j\omega_0) = e^{j\theta} \quad \text{or} \quad -e^{j\theta}, \quad \theta \in [0, \pi).$$

In the sequel, Δ is constructed only for the case of positive sign. In the case of negative sign, we just need reversing the sign of Δ shown in the succeeding text.

If we can find an uncertainty $\Delta(s)$ satisfying

$$\Delta(j\omega_0) = e^{-j\theta}, \quad \|\Delta\|_\infty \leq 1,$$

then

$$R(j\omega_0) = 1 - M(j\omega_0)\Delta(j\omega_0) = 0$$

holds. Since zeros of $R(s)$ are the poles of the closed-loop system, the closed-loop system has an unstable pole $j\omega_0$. Next, we will construct stable rational uncertainties satisfying this condition case by case.

$M(j\omega_0)$ is 1 when $\theta = 0$. The corresponding uncertainty is taken as $\Delta = 1$. On the other hand, when $\theta \in (0, \pi)$,

$$\Delta(s) = \frac{a - s}{a + s}, \quad a = \frac{\omega_0}{\tan\ \theta/2} > 0$$

satisfies $\Delta(j\omega_0) = e^{-j\theta}$ and $\|\Delta\|_\infty = 1$ simultaneously. The uncertainties constructed here are all stable and rational, and they belong to the given uncertainty set since their $\mathcal{H}_\infty$ norms are 1.

For MIMO systems, $\|M\|_\infty \geq 1$ implies that $\sigma_1 := \sigma_{\max}(M(j\omega_0)) = 1$ holds at certain frequency ω_0. Let the singular value decomposition of $M(j\omega_0)$ be

$$M(j\omega_0) = U\Sigma V^* = [u_1 \ \cdots \ u_p] \begin{bmatrix} \sigma_1 & & \\ & \sigma_2 & \\ & & \ddots \end{bmatrix} \begin{bmatrix} v_1^* \\ \vdots \\ v_q^* \end{bmatrix}.$$

If there exists an uncertainty satisfying

$$\Delta(j\omega_0) = \frac{1}{\sigma_1} v_1 u_1^* \ \Rightarrow \sigma_{\max}(\Delta(j\omega_0)) = 1,$$

then

$$\det(I - M(j\omega_0)\Delta(j\omega_0)) = \det\left(I - U\Sigma V^* \times \frac{1}{\sigma_1} v_1 u_1^*\right) = 1 - \frac{1}{\sigma_1} u_1^* U\Sigma V^* v_1 = 0$$

holds. So, the closed-loop system has an unstable pole $j\omega_0$.

Set

$$u_1 = [u_{11} e^{j\theta_1} \ \ \cdots \ ]^*, \quad v_1 = [v_{11} e^{j\phi_1} \ \ \cdots \ ]^*.$$

As the SISO case, we can find a stable rational matrix

$$\Delta(s) = \frac{1}{\sigma_1} \begin{bmatrix} v_{11} \dfrac{\alpha_1 - s}{\alpha_1 + s} \\ \vdots \end{bmatrix} \begin{bmatrix} u_{11} \dfrac{\beta_1 - s}{\beta_1 + s} & \cdots \end{bmatrix}$$

which satisfies the earlier condition at ω_0, thus destroying the stability of the closed-loop system. Further, this uncertainty satisfies $\|\Delta\|_\infty = 1$, hence belonging to the given uncertainty set. •

The small-gain theorem can also be interpreted as follows. At any point s in the right half-plane,

$$\|M(s)\Delta(s)\|_2 \leq \|M(s)\|_2 \|\Delta(s)\|_2 \leq \|M\|_\infty \|\Delta\|_\infty < 1$$

holds so that the infinite sum

$$I + M\Delta + (M\Delta)^2 + \cdots$$

converges to $(I - M\Delta)^{-1}$ in the closed right half-plane. This implies that $(I - M\Delta)^{-1}$ is stable. Moreover, since there is no pole–zero cancellation between M and Δ, the closed-loop system is also stable.

Physically, small-gain theorem corresponds to the fact that the external input is attenuated every time it circulates in the closed loop. More importantly, small-gain theorem is not only sufficient but also necessary. That is, for the norm-bounded uncertainty set under consideration, small-gain theorem is not conservative. However, we should understand this necessity correctly. It is true only when the phase of uncertainty can change freely, which seldom happens in practice. That is, for real-world systems small-gain theorem is simply sufficient, not necessary.

We will illustrate later on that, for any plant set containing norm-bounded uncertainty, its robust stability criterion can be derived from small-gain theorem. That is, small-gain theorem forms one of the foundations of robust control. It is also worth mentioning that, although small-gain theorem is illustrated only for linear systems here, it is also true for nonlinear time-varying systems [22, 47, 90, 98].

Example 12.1 *Let us test the validity of small-gain theorem on the system shown in Figure 12.2(a). Here, for the uncertainty* δ, *we consider two cases, respectively: gain uncertainty and phase uncertainty. The block diagram can be equivalently transformed into Figure 12.2(b). So, when* δ *is a gain uncertainty, the characteristic polynomial of closed-loop system is* $p(s) = s + 1 + \delta$, *and the closed-loop pole is* $s = -(1 + \delta) < 0$ $(\forall \delta > -1)$. *So, for* $-1 < \delta < 1$, *the closed-loop system is stable.*

Further, the Nyquist diagram of $\frac{1}{s+1}$ *is shown in Figure 12.2(c). From this figure we see that, even if the system contains a dynamic uncertainty* δ, *as long as* $|\delta| < 1$, *the Nyquist diagram*

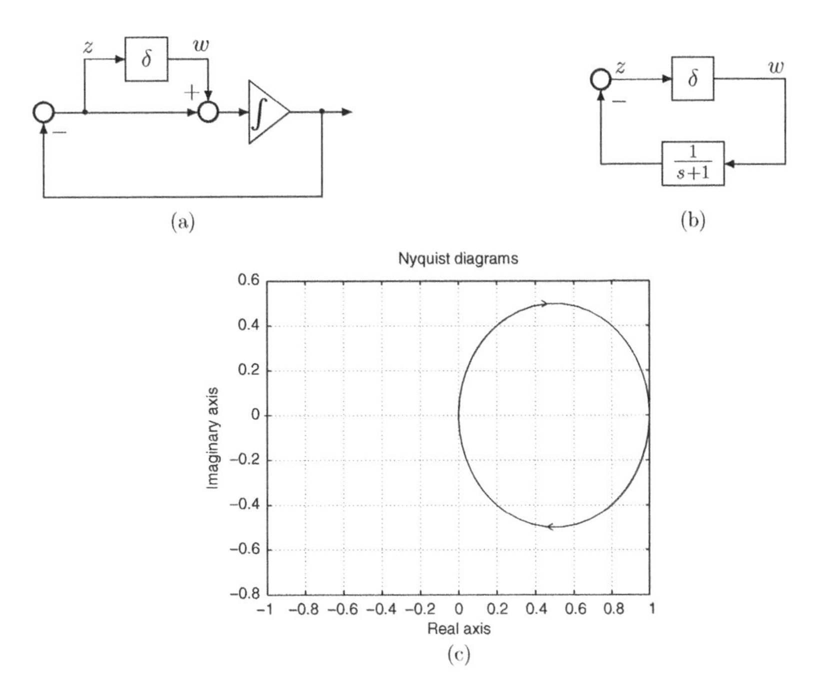

Figure 12.2 Stability margin problem (a) Original system (b) Transformed system (c) Nyquist diagram of nominal open-loop system

of the open-loop system $\delta\frac{1}{s+1}$ will not encircle the critical point $(-1, j0)$. So the closed-loop system is still stable. Here, $\left\|\frac{1}{s+1}\right\|_\infty = 1$ holds. So the result we get here coincides with that of small-gain theorem.

Next, let us examine what change occurs to the stability of closed-loop system in the case of $|\delta| \geq 1$. Set the uncertainty as $\delta = -1$. Then the closed-loop pole becomes $s = 0$, which is unstable. So, it has been verified that small-gain theorem is also necessary in this example.

12.2 Robust Stability Criteria

The robust stability conditions for uncertain systems with various sorts of descriptions are summarized in Table 12.1, where K is the controller and the sensitivity S and complementary sensitivity T are defined, respectively, as

$$S(s) = (I + PK)^{-1}, \quad T(s) = (I + PK)^{-1}PK. \tag{12.1}$$

Example 12.2 *As an example, let us derive the result for the closed-loop system with a feedback uncertainty shown in Figure 12.3(a). In this proof, we only need to conduct an equivalent transformation on the block diagram of closed-loop system with uncertainty so as to separate the uncertainty from others and then apply the small-gain theorem.*

Step 1 *When $\Delta(s) = 0$, the system must be nominally stable.*
Step 2 *In Figure 12.3(a), denote the input and output of uncertainty Δ as z, w, respectively.*

Table 12.1 Robust stability criteria

$W(s)$ and $\Delta(s)$ are stable, $\|\Delta\|_\infty \leq 1$	
Plant set $\mathbb{P}$	Robust stability criterion
$(I + \Delta W)P$	Nominal stability and $\|WT\|_\infty < 1$
$(I + \Delta W)^{-1}P$	Nominal stability and $\|WS\|_\infty < 1$
$P + \Delta W$	Nominal stability and $\|WKS\|_\infty < 1$
$P(I + \Delta WP)^{-1}$	Nominal stability and $\|WSP\|_\infty < 1$

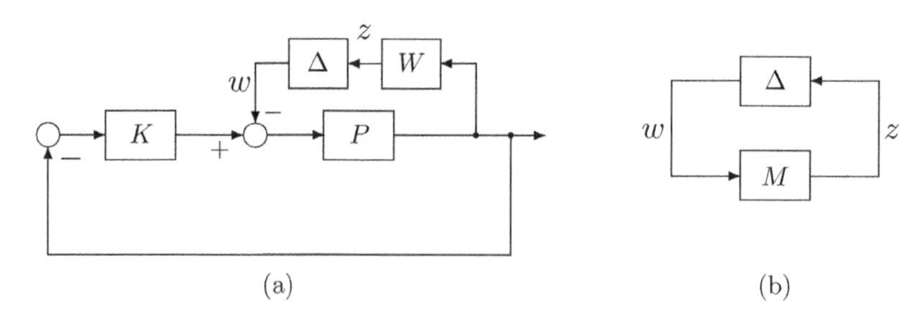

Figure 12.3 Equivalent transformation of block diagrams (a) Closed-loop system with feedback uncertainty (b) Transformed closed-loop system

Step 3 *Compute the transfer matrix from w to z:*

$$z = Mw, \quad M = -WSP.$$

Then, transform the closed-loop system to the system of Figure 12.3(b).

Step 4 *In the end, by small-gain theorem it is clear that the robust stability condition is that the nominal closed-loop system (P, K) is stable and satisfies*

$$1 > \|M\|_\infty = \|-WSP\|_\infty = \|WSP\|_\infty.$$

12.3 Equivalence between $\mathcal{H}_\infty$ Performance and Robust Stability

When dealing with robust control problems, there are two different kinds of specifications: robust stability and robust performance. In control design, it is very important to convert different specifications into the same type mathematically. Otherwise, it is very difficult to carry out effective control design. Fortunately, when the performance is measured in terms of $\mathcal{H}_\infty$ norm of transfer function, it is equivalent to a certain robust stability problem. This conclusion is illustrated by two examples in the succeeding text.

Example 12.3 *Consider the robust stability problem about additive uncertainty first (Figure 12.4(a), $\|\Delta\|_\infty \le 1$).*

$$\begin{aligned}\text{Robust stability } &\Leftrightarrow (P, K) \text{ is internal stable and } \|WKS\|_\infty < 1\\ &\quad\text{(according to Table 12.1)}\\ &\Leftrightarrow \text{Suppress the effect of virtual disturbance } w\\ &\quad\text{to plant input } u \text{ (Figure 12.4(b))}\end{aligned}$$

That is, mathematically the robust stability problem is equivalent to a performance problem of suppressing the input signal z of uncertainty Δ when the output w of uncertainty Δ is regarded as a virtual disturbance (Figure 12.4(b)).

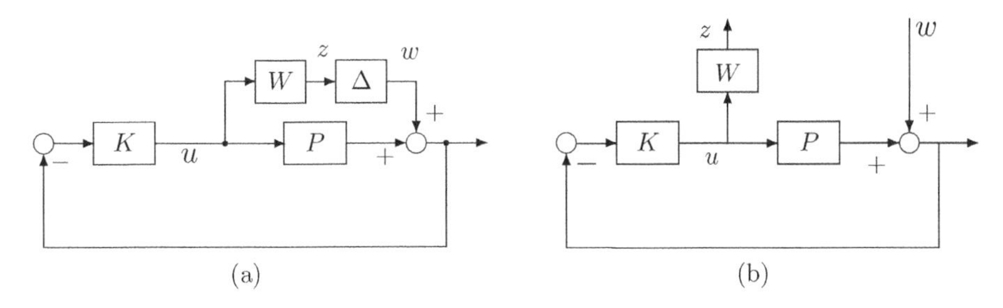

Figure 12.4 Equivalence between robust stability and disturbance attenuation (a) Robust stability problem (b) Equivalent disturbance attenuation problem

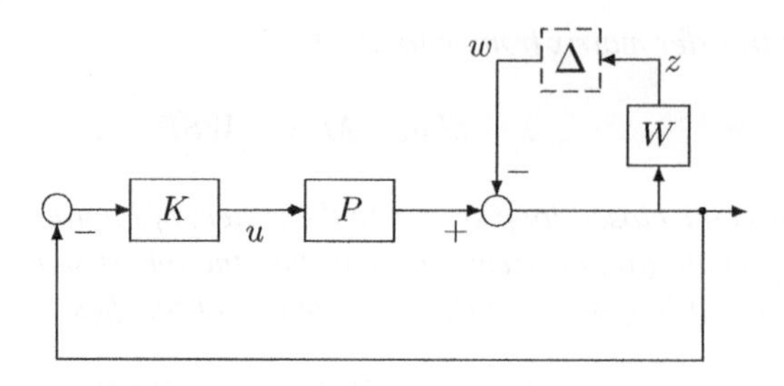

Figure 12.5 Equivalence between sensitivity reduction and robust stability

Example 12.4 *Sensitivity reduction (Figure 12.5)*

(P, K) is stable and $\| WS\|_\infty < 1 \Leftrightarrow$

Robustly stabilize plant set $\left\{\tilde{P} = \dfrac{P}{1+\Delta W}, \ \|\Delta\|_\infty \leq 1\right\}$ (Table 12.1)

That is, the problem of nominal $\mathcal{H}_\infty$ performance is equivalent to the robust stability problem of the closed-loop system formed by inserting a virtual uncertainty Δ between the input and output of the closed-loop transfer function, as shown in Figure 12.5.

This equivalence is true when the uncertainty is norm bounded, and the performance is measured by the $\mathcal{H}_\infty$ norm of closed-loop transfer function. As for other performance indices, it is not yet clear whether there exists such equivalence. These examples can be extended to the general case.

Theorem 12.2 *In the closed-loop system of Figure 12.6, the following statements hold:*

1. *The closed-loop system containing stable uncertainty $\|\Delta\|_\infty \leq 1$ is stable iff the nominal closed-loop system is stable and the $\mathcal{H}_\infty$ norm of transfer matrix $\mathcal{F}_\ell(G, K)$ is less than 1.*
2. *The nominal closed-loop system is stable and the $\mathcal{H}_\infty$ norm of transfer matrix $\mathcal{F}_\ell(G, K)$ is less than 1 iff the closed-loop system formed by inserting an arbitrary virtual stable uncertainty $\|\Delta\|_\infty \leq 1$ between its input w and output w is stable.*

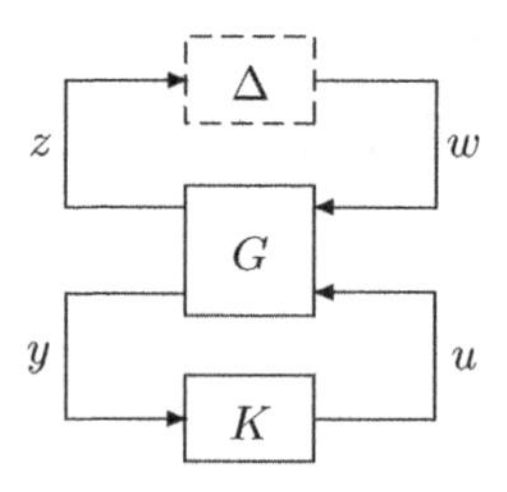

Figure 12.6 Equivalence between nominal performance and robust stability

12.4 Analysis of Robust Performance

The ultimate objective of control is to achieve required performance specifications (signal tracking, disturbance suppression, rapid response, etc.). Stability is simply the premise. Usually, before a control system turns unstable, its performance has significantly degraded. Therefore, the performance of uncertain plant should also be ensured, that is, it is necessary to guarantee the robust performance.

Then, what condition does the robust performance require? Let us illustrate it by examples.

Example 12.5 *Consider the tracking of a reference input with a model $W_S(s)$ (Figure 12.7(a)). The plant is an SISO system with additive uncertainty:*

$$\tilde{P} = P + \Delta W, \quad \|\Delta\|_\infty \leq 1.$$

The control purpose is to reduce the tracking error as much as possible, that is, to realize

$$\left\| W_S \frac{1}{1 + (P + \Delta W)K} \right\|_\infty < 1. \tag{12.2}$$

Firstly, the nominal system must satisfy this condition, namely,

$$\|W_S S\|_\infty < 1.$$

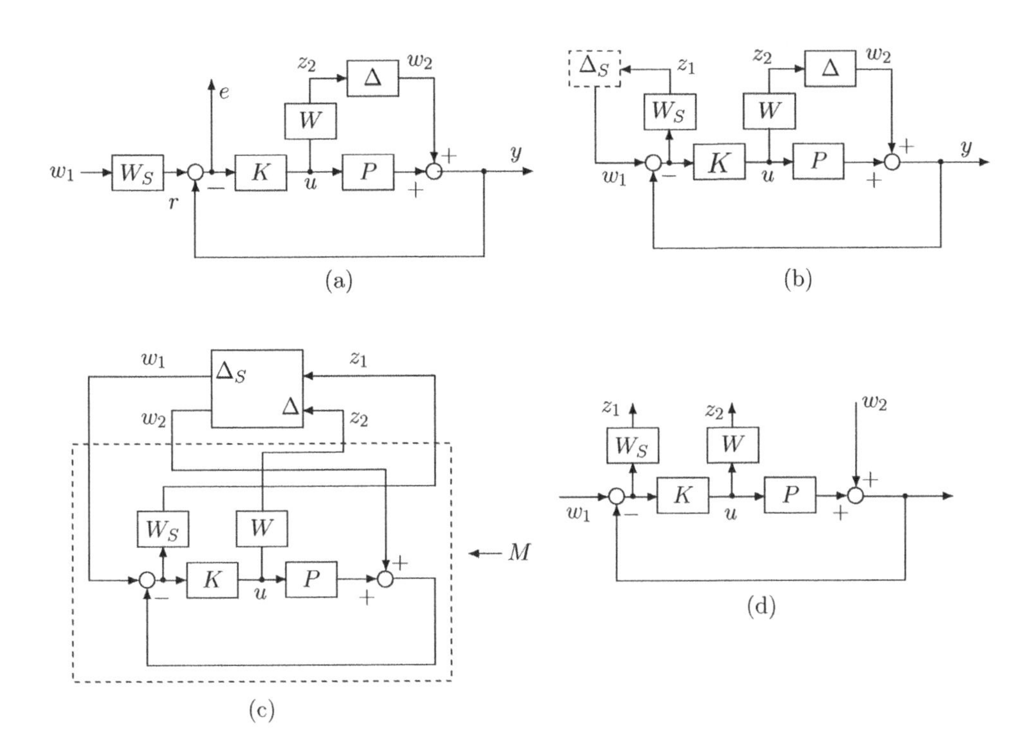

Figure 12.7 Robust tracking problem (a) Original problem (b) Equivalent robust stability problem (c) Separation of uncertainty (d) Conversion to disturbance control problem

Secondly, in order to ensure robust stability, there must be (Table 12.1)

$$\| WKS\|_\infty < 1.$$

Next, let us examine whether the nominal performance and robust stability can guarantee the robust performance. For this, we note that $W_S \frac{1}{1+(P+\Delta W)K}$ *can be rewritten as*

$$W_S \frac{1}{1+PK} \times \frac{1}{1+\Delta WKS} = W_S S(1+\Delta WKS)^{-1}.$$

$\Delta(\|\Delta\|_\infty \leq 1)$ *can take any complex value. So even if* $\| WKS\|_\infty < 1$, *that is, when the system is robustly stable, such a frequency still can be found at which* $|1+\Delta WKS| \ll 1$ *holds*[1]. *From the preceding equation, for this uncertainty* Δ, *the tracking performance deteriorates significantly. Therefore, no matter how good the nominal performance* ($\|W_S S\|_\infty$) *and robust stability are, the robust performance cannot be ensured.*

12.4.1 Sufficient Condition for Robust Performance

In the previous example, the condition for robust performance

$$\|W_S S(1+\Delta WKS)^{-1}\|_\infty < 1$$

contains an uncertainty. In order to carry out control design, it is necessary to turn it into some conditions without uncertainty. However, new concept is needed in order to derive the sufficient and necessary condition for robust performance (refer to Chapter 17). Here, we temporarily derive some simple sufficient conditions and look into its principle.

Example 12.6 *A condition for the preceding robust tracking problem can be derived via the following steps:*

Step 1 *According to Theorem 12.2, the robust performance problem of (12.2) is equivalent to the robust stability problem when a virtual uncertainty* $\Delta_S(\|\Delta_S\|_\infty \leq 1)$ *is inserted as in Figure 12.7(b)*[2].

Step 2 *After transforming Figure 12.7(b) into Figure 12.7(c), the problem is reduced to the robust stability problem of a closed-loop system containing an uncertainty with diagonal structure:*

$$\begin{bmatrix} w_1 \\ w_2 \end{bmatrix} = \begin{bmatrix} \Delta_S & \\ & \Delta \end{bmatrix} \begin{bmatrix} z_1 \\ z_2 \end{bmatrix}.$$

[1] For example, consider the case of $1 - \| WKS\|_\infty = \varepsilon \ll 1$. At the frequency ω_0 satisfying $WKS(j\omega_0) = \| WKS\|_\infty e^{j\phi}$, an uncertainty satisfying $\Delta(j\omega_0) = -e^{-j\phi}/(1+\varepsilon)$ makes $|1+\Delta WKS(j\omega_0)| = 1 - \| WKS\|_\infty/(1+\varepsilon) = 2\varepsilon/(1+\varepsilon) \ll 1$.

[2] We have already shifted the weighting function W_S to the tracking error port. Shifting W_S like this does not change the problem. But in the sufficient condition obtained without such shifting, matrix M contains SW_S and $WKSW_S$. When W_S is an unstable weighting function like $1/s$, they cannot be stable simultaneously except the special case where P has $1/s$. This is the reason why we shift W_S. For MIMO systems, the weighting matrix W_S usually cannot be shifted simply like this example and detailed analysis is needed. Refer to Ref. [59] for the specific conditions.

Step 3 *Since* $\left\|\begin{bmatrix}\Delta_S & \\ & \Delta\end{bmatrix}\right\|_\infty \leq 1$, *according to small-gain theorem a sufficient condition for the robust stability about uncertainty* $\begin{bmatrix}\Delta_S & \\ & \Delta\end{bmatrix}$ *is that the closed-loop transfer matrix* $M(s)$ *from the disturbance input* $[w_1 \ \ w_2]^T$ *to output* $[z_1 \ \ z_2]^T$ *shown in Figure 12.7(d) satisfies the* $\mathcal{H}_\infty$ *norm condition:*

$$\|M\|_\infty < 1$$

where

$$\begin{bmatrix}z_1\\z_2\end{bmatrix} = M\begin{bmatrix}w_1\\w_2\end{bmatrix}, \quad M(s) = \begin{bmatrix}W_S S & -W_S S\\ WKS & -WKS\end{bmatrix}.$$

It should be noted that, although for an uncertainty matrix whose elements can change freely[3], *small-gain theorem is necessary and sufficient for the robust stability, but for an uncertainty with a diagonal structure*[4], *small-gain theorem is only a sufficient condition, not necessary. To understand this point, the reader may refer to the uncertainty structure we constructed in the proof of the necessity of small-gain theorem.*

12.4.2 Introduction of Scaling

In the robust tracking problem of Example 12.6, the uncertainty in the equivalent robust stability problem is a diagonal matrix, so the condition obtained from small-gain theorem is only sufficient and is very conservative usually. In order to reduce this conservatism, we may bring in a *scaling matrix* before applying the small-gain theorem, in view of the diagonal structure of uncertainty.

Example 12.7 *Continued from Example 12.6.*

Step 4 *Since the poles and zeros of a minimum-phase transfer function are stable, as shown in Figure 12.8(a), the stability of closed-loop system does not change even though stable minimum-phase scaling functions are inserted in the closed-loop system consisting of* M *and* $\text{diag}(\Delta_s, \Delta)$[5]. *Then, we conduct the block diagram transformations shown in Figures 12.8(b) and 12.8(c). From Figure 12.8(c), it is seen that the following scaled norm condition can guarantee the robust performance:*

$$\|D^{-1}MD\|_\infty < 1, \quad D(s) = \begin{bmatrix}D_1(s) & \\ & D_2(s)\end{bmatrix}. \tag{12.3}$$

By selecting the scaling matrix D properly, it is possible to make $\|D^{-1}MD\|_\infty$ less than $\|M\|_\infty$. That is, the robust performance condition given by the former is less conservative than the latter.

For example, when the performance specification gets stricter (i.e., raising the gain of performance weighting function W_S), there will not be any controller satisfying $\|M\|_\infty < 1$. But the

[3] It is called *unstructured uncertainty*.

[4] It is called *structured uncertainty*.

[5] If $D(s)$ is nonminimum phase, then $D^{-1}(s)$ is unstable, and there is unstable pole-zero cancellation between D, D^{-1} so that the stability of closed-loop system fails.

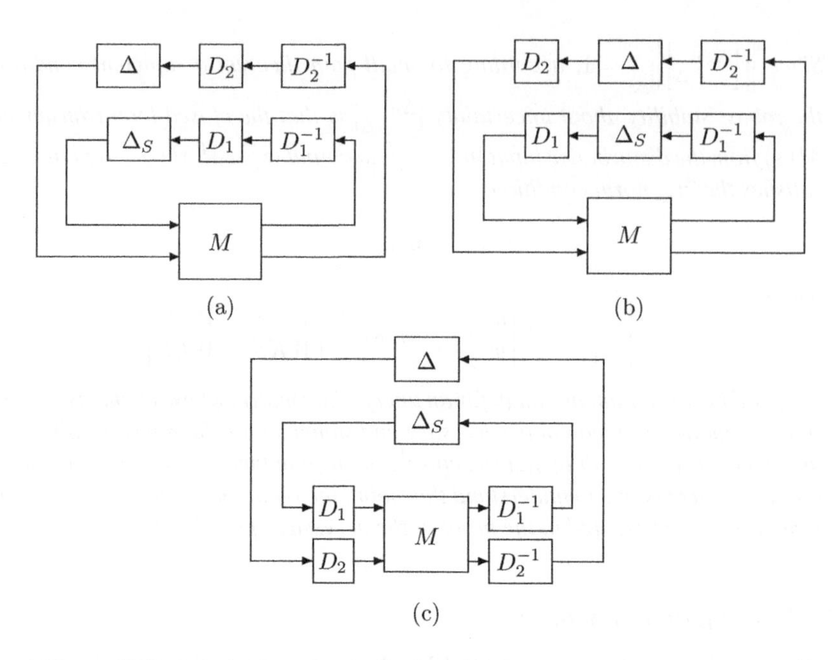

Figure 12.8 Introduction of scaling (a) Phase one (b) Phase two (c) Phase three

condition $\|M\|_\infty < 1$ is just a sufficient condition, not necessary. Its failure does not mean that there is no controller ensuring the robust performance. For the performance weighting function W_S, there may be a scaling function D and a controller K satisfying $\|D^{-1}MD\|_\infty \leq 1$. Therefore, systems with better performance may be designed by bringing scaling matrix into the control design.

This sort of problem is called the *scaled $\mathcal{H}_\infty$ problem.*

12.5 Stability Radius of Norm-Bounded Parametric Systems

Look at the following uncertain system

$$\dot{x} = (A + B\Delta(I - D\Delta)^{-1}C)x \tag{12.4}$$

where $\Delta \in \mathbb{R}^{p\times q}$ is a norm-bounded uncertain parameter matrix (see Subsection 11.3.3). Set

$$z = Cx + Dw, \quad w = \Delta z.$$

Then, the state equation can be rewritten as

$$\dot{x} = Ax + Bw.$$

That is, system (12.4) is equivalent to the closed-loop system consisting of a nominal system

$$M(s) = (A, B, C, D) \tag{12.5}$$

and a parameter uncertainty Δ, as shown in Figure 12.9.

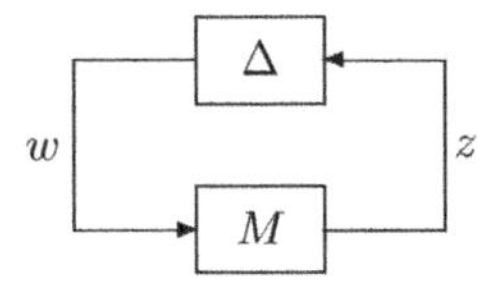

Figure 12.9 Stability radius problem of parametric systems

The question now is how to calculate the stability margin of this system. Since this margin is measured by norm, it is called *stability radius*. Small-gain theorem is necessary and sufficient only for norm-bounded dynamic uncertainty (complex). It is very conservative for parameter uncertainty (real), hence unable to answer the question now. This problem is resolved by Qiu *et al.* completely and beautifully [79]. The conclusion is as follows.

Theorem 12.3 (Qiu's Theorem) *The uncertain system (12.4) is robustly stable iff the parameter uncertainty matrix* Δ *satisfies*

$$\frac{1}{\|\Delta\|_2} > \sup_{\omega} \inf_{\gamma \in (0,1]} \sigma_2 \left(\begin{bmatrix} \Re(M(j\omega)) & -\gamma \Im(M(j\omega)) \\ \frac{1}{\gamma} \Im(M(j\omega)) & \Re(M(j\omega)) \end{bmatrix} \right). \tag{12.6}$$

Here, $\sigma_2(X)$ *denotes the second largest singular value of matrix* X.

For the analysis of system stability, Qiu's theorem is a very effective tool. However, it is not clear at present how to apply it to robust control design.

Exercises

12.1 In the transfer function

$$\tilde{P}(s) = \frac{1}{Ms + \mu}$$

of a car, let the mass M be a constant, while the friction coefficient varies in the range of

$$\mu = \mu_0 + c\Delta, \quad |\Delta| \le 1.$$

We hope to describe $\tilde{P}$ in a form of

$$\tilde{P} = \frac{P}{1 + \Delta W}, \quad P = \frac{1}{Ms + \mu_0}.$$

Find the weighting function $W(s)$.

12.2 Prove Table 12.1 based on the small-gain theorem and block diagram transformation.

12.3 In the plant set

$$\tilde{P} = (1 + \Delta W)^{-1} P,$$

suppose that the uncertainty $\Delta(s)$ is stable and satisfies $\|\Delta\|_\infty \leq 1$, the nominal model and weighting function are

$$P(s) = \frac{1}{s+1}, \quad W = 2.$$

Discuss whether this plant set can be robustly stabilized. (Hint: note that $P(j\infty) = 0$.)

12.4 We use controller $K(s) = 4$ to stabilize a plant set $\tilde{P}(s) = 1/(s+a)$:
(a) Find the range of a such that the closed-loop system is stable.
(b) Choose $P(s) = \frac{1}{s+10}$ as the nominal plant which corresponds to $a = 10$ and regard a as an uncertain parameter. Firstly, find a weighting function $W(s)$ such that

$$\tilde{P}(s) = \frac{1}{s+a} = \frac{P}{1+\Delta WP}$$

holds for $\Delta(s) = 1$. Then, compute the range of a which ensures the robust stability for all $\|\Delta\|_\infty \leq 1$, based on the robust stability condition. Further, compare it with the range of a obtained in (a).

12.5 Let the plant $P(s)$ and controller $K(s)$ be, respectively,

$$P(s) = \frac{1}{s-1}, \quad K(s) = \frac{1}{s} + k.$$

1. Find the allowable range of k for $K(s)$ to stabilize $P(s)$.
2. Prove that the output of plant can track step reference input asymptotically.
3. When the plant changes from $P(s)$ to a set

$$\tilde{P}(s) = \frac{1}{s-1+\alpha}, \quad |\alpha| < 1,$$

find the constraint on the gain k such that the controller can robustly stabilize $\tilde{P}(s)$.

12.6 In Example 11.1, consider the case that the car is subject to a disturbance $d(t)$ such as wind resistance. Suppose that the disturbance $\hat{d}(s)$ satisfies

$$|\hat{d}(j\omega)| < |W_d(j\omega)| \quad \forall\omega.$$

The control purpose is to suppress the disturbance and achieve the tracking of speed reference (Figure 12.10). Let the model of car speed reference $r(t)$ be $W_r(s)$. Find a sufficient condition for this robust performance.

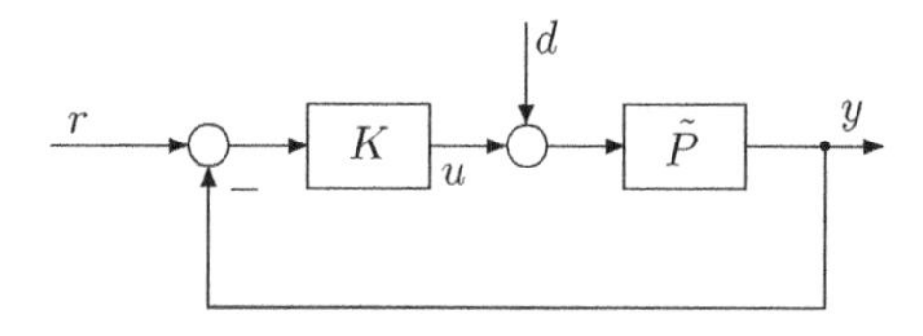

Figure 12.10 Speed tracking control of a car

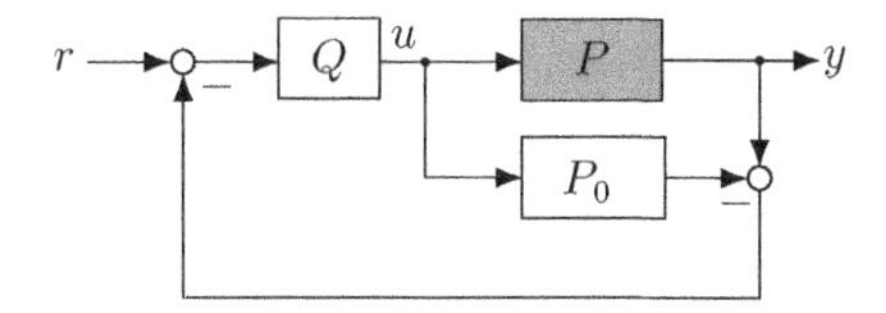

Figure 12.11 Structure of IMC

12.7 In the IMC control system shown in Figure 12.11, $P(s)$ is the actual system and $P_0(s)$ is the model, both being stable. The stable transfer function $Q(s)$ in the figure is the control parameter; the tracking error of reference input is $e = r - y$. Suppose that

$$\left| \frac{P(j\omega) - P_0(j\omega)}{P_0(j\omega)} \right| \leq |W(j\omega)|$$

for all frequencies.

1. Derive the robust stability condition for the closed-loop system.
2. Let the model of $r(t)$ be $W_R(s)$. Derive a condition that ensures the robust tracking performance $\|\hat{e}\|_\infty < \varepsilon$.

12.8 Consider the following typical process control system:

$$P(s) = \frac{1}{s+1} e^{-\theta s}, \quad 0 \leq \theta \leq 0.1.$$

When the nominal model is chosen as $P_0(s) = 1/(s+1)$, the uncertainty weighting function is given by

$$W(s) = \frac{0.2s}{0.1s+1}.$$

Design a control parameter $Q(s)$ satisfying the following specifications based on Exercise 12.7:

1. Robust stability of closed-loop system.
2. The output y robustly and asymptotically tracks the unit step reference input $r(t) = 1$.
3. For the nominal plant $P_0(s)$, guarantees the tracking performance $\|e\|_2 < 0.5$.

12.9 Let a transfer function $P_0(s)$ be factorized as $P_0(s) = n_0(s)/m_0(s)$ by using stable rational functions $n_0(s), m_0(s)$ that have no common unstable zeros. Suppose that the plant perturbs to

$$P(s) = \frac{n_0(s) + \Delta_n(s)}{m_0(s) + \Delta_m(s)}, \quad \Delta = [\Delta_n \quad \Delta_m] = W(s)\Delta_I(s)$$

in which both $W(s)$ and $\Delta_I(s)$ are stable transfer functions, and $\|\Delta_I\|_\infty \leq 1$. Let the controller be $K(s)$. The block diagram of the perturbed system is illustrated in Figure 12.12:

1. Derive the condition for robust stabilization of the uncertain system based on small-gain theorem.

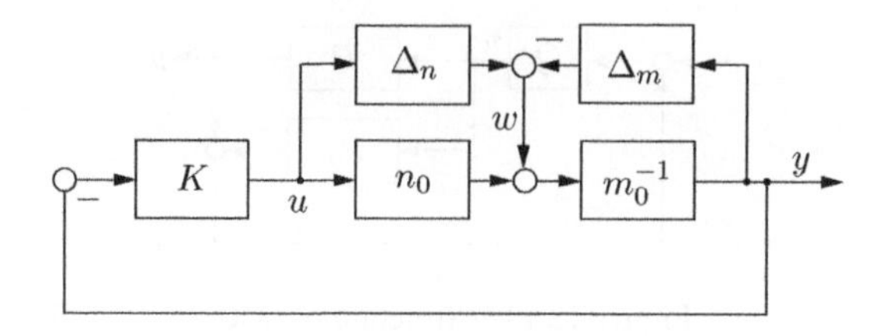

Figure 12.12 System with coprime uncertainty

2. When the nominal plant $P_0(s)$ is stable, express the derived condition as a function of free parameter $Q(s)$ by using the parameterization of stabilizing controllers.
3. Suppose that $P_0(s)$ is stable. Derive the condition on $Q(s)$ such that the output y robustly and asymptotically tracks unit step reference input.
4. Design a controller satisfying all these conditions. In numerical design, use

$$n_0(s) = \frac{1}{s+1}, \quad m_0(s) = 1, \quad W(s) = 0.2.$$

12.10 Control a plant set containing multiplicative uncertainty

$$P(s) = P_0(s)[1 + W_2(s)\Delta(s)], \quad \|\Delta\|_\infty \leq 1 \tag{12.7}$$

with controller $K(s)$. Prove that to ensure robust sensitivity performance

$$\|W_1 S\|_\infty < 1, \quad S(s) = \frac{1}{1 + P(s)K(s)}, \tag{12.8}$$

the necessary and sufficient condition is

$$|W_1(j\omega)S_0(j\omega)| + |W_2(j\omega)P_0(j\omega)| < 1 \ \forall\omega \tag{12.9}$$

where $S_0 = \frac{1}{1+P_0K}, T_0 = \frac{P_0K}{1+P_0K}$. (Hint: Note that the phase angle of uncertainty $\Delta(j\omega)$ is arbitrary. Do analysis at each frequency.)

12.11 In the 2-DOF system shown in Figure 12.13, *d* is an input disturbance and the plant set is given by

$$P(s) = P_0(s) + W(s)\Delta(s), \quad \|\Delta\|_\infty \leq 1.$$

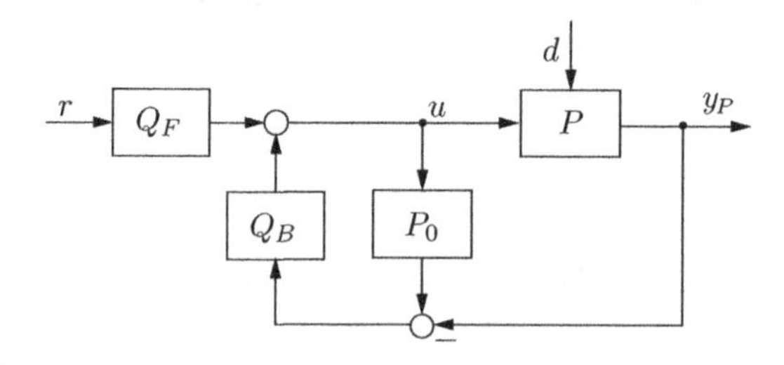

Figure 12.13 Two-degree-of-freedom system

1. Derive the robust stability condition.
2. Let the disturbance model be $W_D(s)$. For all uncertain plants, find a condition to ensure that the transfer function $H_{y_P d}(s)$ from disturbance d to output y_P satisfies $\|H_{y_P d}\|_\infty < 1$.
3. Let the reference input model be $W_R(s)$. For all uncertain plants, find a condition to ensure that the transfer function $H_{er}(s)$ from reference input r to tracking error $e = r - y_P$ satisfies $\|H_{er}\|_\infty < 1$.

Notes and References

The small gain principle rooted in the study of nonlinear and time-varying systems by Zames [98]. Zames [99] argued in favor of the use of $\mathcal{H}_\infty$ norm in robust control. Doyle and Stein [28] proposed the small gain condition for linear systems having multiplicative uncertainty while [13] gave a rigorous proof. More materials about the small gain principle may be found in Refs [100, 20, 25] and [68]. Tits *et al.* [88] and Liu [53, 63, 56, 57] studied the robust control problem of uncertain systems with known gain and phase bounds and derived the corresponding conditions. The robust steady-state performance was studied in Ref. [59].

13

Robustness Analysis 2: Lyapunov Method

In essence, the small-gain approach is tailored for dynamic uncertainty in which both gain and phase are uncertain. On the other hand, for a static uncertainty such as parameter variation, its phase angle is fixed. Disregarding this feature, the derived robustness conditions will be too conservative to be useful in the applications. In this chapter, we introduce the Lyapunov method which is able to make use of the feature that a parameter uncertainty is a real number in the robustness analysis of parametric systems.

This method, rooted in the study of Barmish [6], extends the Lyapunov method which was originally used for the stability analysis of nonlinear systems to uncertain linear systems. In this chapter, after a brief review of Lyapunov theory, we will describe in detail how to use this method to analyze robust stability, with a focus on the quadratic stability.

13.1 Overview of Lyapunov Stability Theory

Lyapunov stability theory was put forward by Lyapunov, an outstanding Russian mathematician in the late nineteenth century. This is a stability analysis method for nonlinear systems. This method not only has laid the foundation for the control of nonlinear systems but also has become one of the basic principles of robust control.

Consider a nonlinear system with state vector $x \in \mathbb{R}^n$:

$$\dot{x} = f(x), \quad x(0) \neq 0. \tag{13.1}$$

$f(x)$ is a nonlinear vector function satisfying $f(0) = 0$. Clearly, the origin $x = 0$ is an equilibrium. If the trajectory starting from an arbitrary point near the origin can always converge to the origin, the origin is said to be asymptotically stable w.r.t. the nonlinear system. The issue now is how to find a condition to ensure the asymptotic stability.

For this stability analysis problem, Lyapunov's idea is not to directly investigate the state trajectory, but to examine the variation of energy in the system instead. This idea can be understood as follows: since there is no external supply of energy in the autonomous system (13.1),

Robust Control: Theory and Applications, First Edition. Kang-Zhi Liu and Yu Yao.

the motion must stop when the internal energy becomes zero. On the contrary, the state does not stop moving as long as the internal energy is not zero. This means that so long as we know whether the internal energy converges to zero, we are definitely able to judge whether the state converges to the origin or not.

Concretely speaking, as a quantity representing energy, we use the so-called Lyapunov function $V(x)$, a positive definite function:

$$V(x) > 0 \quad \forall\ x \neq 0. \tag{13.2}$$

If its time derivative satisfies

$$\dot{V}(x) < 0 \quad \forall x \neq 0, \tag{13.3}$$

then it is guaranteed that

$$\lim_{t \to \infty} x(t) = 0.$$

Since this book focuses on linear systems, rigorous statement for nonlinear systems is omitted and only the brief description above is provided. The interested reader may refer to literatures such as Ref. [47].

For linear systems, the asymptotic stability condition can be proved as follows. Consider the linear system below:

$$\dot{x} = Ax, \quad x(0) \neq 0. \tag{13.4}$$

Set a Lyapunov function $V(x)$ as the following positive definite quadratic function:

$$V(x) = x^T Px > 0 \quad \forall x \neq 0. \tag{13.5}$$

Obviously $V(0) = 0$. The positive definiteness of $x^T Px$ is equivalent to that of the symmetric matrix P. If the trajectory of state $x(t)$ satisfies

$$\dot{V}(x) < 0 \quad \forall x(t) \neq 0, \tag{13.6}$$

then $V(x(t))$ decreases strictly monotonically w.r.t. the time t. $V(x)$ is bounded from below since $V(x) \geq 0$, so it has a limit. It can be seen from Eq. (13.6) that $V(x)$ continue decreasing as long as $x(t) \neq 0$. Therefore, $x(\infty) = 0$ must be true.

13.1.1 Asymptotic Stability Condition

Here we derive a stability condition for the linear system (13.4). The quadratic Lyapunov function in (13.5) is used for this purpose. Differentiation of $V(x) = x^T Px$ along the trajectory of $\dot{x} = Ax$ yields

$$\begin{aligned} \dot{V}(x) &= \dot{x}^T Px + x^T P\dot{x} \\ &= (Ax)^T Px + x^T P(Ax) \\ &= x^T(A^T P + PA)x. \end{aligned} \tag{13.7}$$

For all nonzero state x, the condition for $\dot{V}(x) < 0$ is

$$A^T P + PA < 0. \tag{13.8}$$

So, when this LMI (linear matrix inequality) has a positive definite solution P, asymptotic stability of the system is ensured. From this discussion, we obtain the following theorem:

Theorem 13.1 *The linear system (13.4) is asymptotically stable iff there exists a positive definite matrix P satisfying (13.8).*

Proof. The preceding discussion shows the sufficiency. For the proof of necessity, refer to Theorem 4.7(3) in Section 4.3. •

13.1.2 Condition for State Convergence Rate

Although the LMI (13.8) guarantees the stability of system, no guarantee on the convergence rate of state is provided. In order to ensure a convergence rate, we may use the following LMI condition:

$$A^T P + PA + 2\sigma P < 0, \quad \sigma > 0. \tag{13.9}$$

When this LMI has a positive definite solution P, there holds

$$\dot{V}(x) = x^T(A^T P + PA)x < x^T(-2\sigma P)x = -2\sigma V(x).$$

Noting that the solution of differential equation $\dot{y} = -2\sigma y$ is $y(t) = e^{-2\sigma t}y(0)$, according to the comparison principle [47], we know that $V(x)$ satisfies the inequality:

$$V(x(t)) < e^{-2\sigma t}V(x(0)).$$

Further, by using $\lambda_{\min}(P)\|x\|^2 \le x^T Px \le \lambda_{\max}(P)\|x\|^2$, we get

$$\lambda_{\min}(P)\|x(t)\|^2 \le x^T(x)Px(t) < e^{-2\sigma t}x^T(0)Px(0) \le e^{-2\sigma t}\lambda_{\max}(P)\|x(0)\|^2.$$

Therefore,

$$\|x(t)\| < \sqrt{\frac{\lambda_{\max}(P)}{\lambda_{\min}(P)}}\|x(0)\|e^{-\sigma t}, \tag{13.10}$$

that is, the state vector $x(t)$ converges to zero at a rate higher than σ.

13.2 Quadratic Stability

In the previous section, we have discussed the stability of linear nominal systems in the framework of Lyapunov theory. In this section, we consider the stability of a plant set with parameter uncertainty. The system under consideration is

$$\dot{x} = A(\theta)x, \quad x(0) \ne 0 \tag{13.11}$$

in which $\theta \in \mathbb{R}^p$ is a bounded vector of uncertain parameters.

For example, when the input $u = 0$ and no external force is imposed, the dynamics of the mass–spring–damper system is

$$\dot{x} = \begin{bmatrix} 0 & 1 \\ -\dfrac{k}{m} & -\dfrac{b}{m} \end{bmatrix} x = A(m, b, k)x$$

where the parameter vector is $\theta = [m\ b\ k]^T$.

When the parameters change, by nature the Lyapunov function should also change accordingly. However, we do not know in general the relation between Lyapunov function and the parameter vector θ. Based on this fact, Barmish suggested to use a fixed quadratic function $V = x^T Px$ to investigate the stability for the entire system set (13.11). That is, checking the stability of the uncertain system via a quadratic function $V(x)$ satisfying

$$V(x) = x^T Px > 0\ \forall x \neq 0; \quad \dot{V}(x, \theta) < 0\ \forall x \neq 0, \theta. \tag{13.12}$$

When this is possible, the system set is said to be *quadratically stable*. Since a common Lyapunov function is used for all systems in the set, it may be easier to find a stability condition.

Certainly, the quadratic stability is a very strong specification for stability. However, numerous applications indicate that it is quite effective in engineering applications.

13.2.1 *Condition for Quadratic Stability*

From $\dot{V}(x, \theta) = x^T(A^T(\theta)P + PA(\theta))x$, we know that the condition for the quadratic stability is that there is a positive definite matrix P satisfying the inequality:

$$A^T(\theta)P + PA(\theta) < 0\ \forall \theta. \tag{13.13}$$

Example 13.1 *Consider the following first-order system:*

$$\dot{x} = -(2 + \theta)x, \quad \theta > -2.$$

Since $A^T(\theta)P + PA(\theta) = -(2 + \theta)P - P(2 + \theta) = -2(2 + \theta)P$,

$$A^T(\theta)P + PA(\theta) = -2(2 + \theta) < 0 \quad \forall \theta \in (-2, \infty)$$

always holds w.r.t. $P = 1$. *Therefore, the stability is guaranteed. This conclusion is consistent with that obtained from the pole analysis.*

The next question is how to calculate a solution P for the matrix inequality (13.13). No general solution exists because $A(\theta)$ depends on the uncertain parameter vector θ. However, when the matrix $A(\theta)$ has some specific structures about the parameter vector θ, it is possible to solve the inequality. We will discuss this issue in detail in the next subsection.

13.2.2 Quadratic Stability Conditions for Polytopic Systems

Consider a system set whose coefficient matrix is a matrix polytope:

$$\dot{x} = \left(\sum_{i=1}^{N} \lambda_i A_i\right) x, \quad x(0) \neq 0 \tag{13.14}$$

in which λ_i $(i = 1, \ldots, N)$ is an uncertain parameter and satisfies $\lambda_i \geq 0$, $\sum_{i=1}^{N} \lambda_i = 1$.

For this system, the quadratic stability condition (13.13) can be rewritten as

$$\left(\sum_{i=1}^{N} \lambda_i A_i\right)^T P + P\left(\sum_{i=1}^{N} \lambda_i A_i\right) < 0 \ \ \forall \lambda_i$$

$$\Leftrightarrow \sum_{i=1}^{N} \lambda_i (A_i^T P + PA_i) < 0 \ \ \forall \lambda_i. \tag{13.15}$$

First of all, this inequality must hold at all vertices of the polytope. Hence,

$$A_i^T P + PA_i < 0 \ \ \forall i = 1, \ldots, N \tag{13.16}$$

must be true. Here, $A_i^T P + PA_i < 0$ is the condition for $\lambda_i = 1, \lambda_j = 0 \ (j \neq i)$, that is, at the ith vertex. As all λ_i are nonnegative and their sum is 1, at least one of them must be positive. So when (13.16) holds, the inequality

$$\sum_{i=1}^{N} \lambda_i (A_i^T P + PA_i) < 0$$

is always true. This shows that the LMI conditions (13.16) at all vertices are equivalent to the quadratic stability condition (13.13).

Example 13.2 *Let us examine the mass–spring–damper system with zero input, whose dynamics is*

$$\dot{x} = \begin{bmatrix} 0 & 1 \\ -\dfrac{k}{m} & -\dfrac{b}{m} \end{bmatrix} x.$$

Suppose that each parameter takes value in

$$1 \leq m \leq 2, \quad 10 \leq k \leq 20, \quad 5 \leq b \leq 10.$$

The parameter vector $\theta = [m \ b \ k]^T$ forms a cube with eight vertices. Solving the quadratic stability condition (13.16), we get a solution

$$P = \begin{bmatrix} 1.9791 & -2.8455 \\ -2.8455 & 14.2391 \end{bmatrix}.$$

Its eigenvalues are 1.35, 14.86, so P is positive definite. This means that the system is quadratically stable. This conclusion is very natural in view of the fact that the damping ratio b is positive.

On the other hand, when the damping ratio ranges over $0 \le b \le 5$, *the solution of (13.16) becomes*

$$P = \begin{bmatrix} 0.85 & 0.9 \\ 0.9 & 10.26 \end{bmatrix} \times 10^{-11} \approx 0$$

which is not positive definite. So we cannot draw the conclusion that this system is quadratically stable. In fact, this system set includes a case of zero damping. So the system set is not quadratically stable.

Generally speaking, the quadratic stability is conservative since it uses a common Lyapunov function to test the stability of all uncertain systems. Now, let us try to search for a less conservative stability condition by changing the Lyapunov function with the uncertain parameters. Look at a simple example:

$$\dot{x} = A(\theta)x = (A_0 + \theta A_1)x, \quad \theta \in [\theta_m, \theta_M].$$

Focusing on the affine structure of the matrix $A(\theta)$, we attempt to impose the same structure on the matrix $P(\theta)$:

$$P(\theta) = P_0 + \theta P_1.$$

This may lead to a less conservative condition. However, in this case, squared term θ^2 of the uncertain parameter appears in the stability condition. This can be seen clearly from

$$P(\theta)A(\theta) = (P_0 + \theta P_1)(A_0 + \theta A_1) = P_0 A_0 + \theta^2 P_1 A_1 + \theta(P_1 A_0 + P_0 A_1).$$

Due to θ^2, the affine structure, that is, the polytopic structure, is destroyed so that the stability condition cannot be reduced to the vertex conditions. In LMI approach, so far there is no good solution for problems like this.

The following outlines a method from [34]. The idea is to use the property of convex functions. Consider a quadratic Lyapunov function:

$$V(x, \theta) = x^T P(\theta) x, \quad P(\theta) > 0.$$

Its derivative

$$\begin{aligned} \dot{V}(x, \theta) &= x^T (A(\theta)^T P(\theta) + P(\theta) A(\theta)) x \\ &= x^T [(A_0^T P_0 + P_0 A_0) + \theta^2 (A_1^T P_1 + P_1 A_1) \\ &\quad + \theta(P_1 A_0 + P_0 A_1 + A_0^T P_1 + A_1^T P_0)] x \end{aligned}$$

is a quadratic function of parameter θ. As long as $\dot{V}(x, \theta)$ is a convex function of θ, the vertex conditions

$$A(\theta_m)^T P(\theta_m) + P(\theta_m) A(\theta_m) < 0, \quad A(\theta_M)^T P(\theta_M) + P(\theta_M) A(\theta_M) < 0$$

ensure that $\dot{V}(x, \theta)$ is negative definite (see Subsection 3.1.5). Meanwhile, a condition for convexity is

$$\frac{d^2}{d\theta^2} \dot{V}(x, \theta) = 2x^T (A_1^T P_1 + P_1 A_1) x \ge 0 \Rightarrow A_1^T P_1 + P_1 A_1 \ge 0.$$

Finally, the positive definiteness of $P(\theta)$ is guaranteed by the vertex conditions

$$P(\theta_m) > 0, \quad P(\theta_M) > 0.$$

13.2.3 Quadratic Stability Condition for Norm-Bounded Parametric Systems

As seen above, the polytopic model of uncertain systems is very effective in robustness analysis. Unfortunately, control design of polytopic systems is very difficult. So in this section, we discuss the quadratic stability of norm-bonded parametric systems. A general model for norm-bounded parametric systems is given in Subsection 11.3.3. In the inspection of system stability, the input u, disturbance d, measured output y, and performance output e may be ignored. Therefore, we analyze the following simplified parametric system:

$$M\begin{cases} \dot{x} = Ax + Bw \\ z = Cx + Dw \end{cases} \tag{13.17}$$

$$w = \Delta z, \quad \|\Delta(t)\|_2 \leq 1. \tag{13.18}$$

In this case, this time-varying system is as shown in Figure 13.1. State equation of the closed-loop system formed by nominal system $M(s)$ and parameter uncertainty $\Delta(t)$ is

$$\dot{x} = (A + B\Delta(I - D\Delta)^{-1}C)x, \quad \|\Delta(t)\|_2 \leq 1. \tag{13.19}$$

It is easy to see that when $\Delta(t)$ varies freely in the range of $\|\Delta(t)\|_2 \leq 1$, the invertibility condition for the matrix $I - D\Delta$ is $\|D\|_2 < 1$ (Exercise 13.2).

Applying the time-varying version of the small-gain theorem (refer to Exercise 13.3), we see that if there is a positive definite matrix P satisfying

$$\begin{bmatrix} A^TP + PA & PB & C^T \\ B^TP & -I & D^T \\ C & D & -I \end{bmatrix} < 0, \tag{13.20}$$

the system (M, Δ) is quadratically stable w.r.t. Lyapunov function $V(x) = x^TPx$.

It can be proven that this condition is necessary and sufficient for the quadratic stability of system (13.19).

Theorem 13.2 *The time-varying system (13.19) is quadratically stable iff there exists a positive definite matrix P satisfying (13.20).*

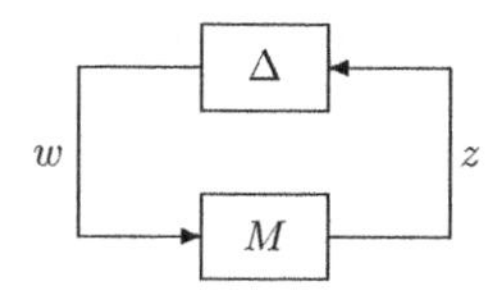

Figure 13.1 Parametric system

Proof. We use the so-called S-procedure[9]. By definition, quadratic stability implies the existence of $P > 0$ such that the Lyapunov function $V(x) = x^T P x$ satisfies $\dot{V}(x) < 0$ ($\forall x \neq 0$). Differentiation of $V(x)$ along the trajectory of system $M(s)$ gives

$$\begin{aligned}\dot{V}(x) &= \dot{x}^T P x + x^T P \dot{x} = (Ax + Bw)^T P x + x^T P (Ax + Bw) \\ &= \begin{bmatrix} x \\ w \end{bmatrix}^T \begin{bmatrix} A^T P + PA & PB \\ B^T P & 0 \end{bmatrix} \begin{bmatrix} x \\ w \end{bmatrix}. \end{aligned} \tag{13.21}$$

Meanwhile, $\|\Delta(t)\|_2 \leq 1$ implies that its input and output (z, w) satisfy $w^T w = z^T \Delta^T \Delta z \leq z^T z$. Substituting $z = Cx + Dw$, we get a condition

$$U(x,w) = \begin{bmatrix} x \\ w \end{bmatrix}^T \left\{ \begin{bmatrix} 0 & 0 \\ 0 & I \end{bmatrix} - \begin{bmatrix} C^T \\ D^T \end{bmatrix} [C \ D] \right\} \begin{bmatrix} x \\ w \end{bmatrix} \leq 0 \ \forall \begin{bmatrix} x \\ w \end{bmatrix} \neq 0 \tag{13.22}$$

which is equivalent to $\|\Delta(t)\|_2 \leq 1$.

As the invertibility of $I - D\Delta$ requires $\|D\|_2 < 1$, if $x = 0$ and $w \neq 0$, then $U(x,w) = w^T(I - D^T D)w > 0$. This is a contradiction. So, in any nonzero vector $\left[\begin{smallmatrix} x \\ w \end{smallmatrix}\right]$ satisfying the condition $U(x,w) \leq 0$, there must be $x \neq 0$. Invoking Schur's lemma, we see that (13.20) is equivalent to

$$\begin{aligned} 0 &> \begin{bmatrix} A^T P + PA & PB \\ B^T P & -I \end{bmatrix} + \begin{bmatrix} C^T \\ D^T \end{bmatrix} [C \ D] \\ &= \begin{bmatrix} A^T P + PA & PB \\ B^T P & 0 \end{bmatrix} - \left\{ \begin{bmatrix} 0 & 0 \\ 0 & I \end{bmatrix} - \begin{bmatrix} C^T \\ D^T \end{bmatrix} [C \ D] \right\}. \end{aligned}$$

Multiplying this inequality by a nonzero vector $\left[\begin{smallmatrix} x \\ w \end{smallmatrix}\right]$ and its transpose, we have

$$\dot{V}(x) < U(x,w) \leq 0.$$

From the discussion above, we know that $x \neq 0$ in this case, so the quadratic stability is proved. Conversely, when the system is quadratically stable, the two inequalities

$$\dot{V}(x) < 0, \quad U(x,w) \leq 0$$

hold simultaneously for nonzero state x. Consider a bounded $\left[\begin{smallmatrix} x \\ w \end{smallmatrix}\right]$. Then $\dot{V}(x)$ and $U(x,w)$ are also bounded. Enlarging $\dot{V}(x)$ suitably by a factor $\rho > 0$, we have

$$\rho \dot{V}(x) < U(x,w) \quad \forall x \neq 0.$$

Finally, absorbing ρ into the positive definite matrix P and renaming ρP as P, we obtain

$$\dot{V}(x) - U(x,w) < 0 \quad \forall \begin{bmatrix} x \\ w \end{bmatrix} \neq 0.$$

This inequality is equivalent to (13.20). •

Example 13.3 *We revisit the mass–spring–damper system in Example 13.2 starting from a norm-bounded model. Assume that the uncertain parameters are*

$$m = m_0(1 + w_1\delta_1), \quad k = k_0(1 + w_2\delta_2), \quad b = b_0(1 + w_3\delta_3), \quad |\delta_i| \le 1$$

in which the nominal parameters (m_0, k_0, b_0) *are the average of each uncertain parameter. The weights are*

$$w_1 = \frac{m_{\max}}{m_0} - 1, \quad w_2 = \frac{k_{\max}}{k_0} - 1, \quad w_3 = \frac{b_{\max}}{b_0} - 1.$$

According to Section 11.3.3, after normalizing the uncertainty matrix $\Delta = [\delta_1\ \delta_2\ \delta_3]$ *in the norm-bounded parametric system (13.19), the coefficient matrices become*

$$A = \begin{bmatrix} 0 & 1 \\ -\dfrac{k_0}{m_0} & -\dfrac{b_0}{m_0} \end{bmatrix}, \quad B = \begin{bmatrix} 0 \\ 1 \end{bmatrix}$$

$$C = -\sqrt{3}\begin{bmatrix} \dfrac{k_0}{m_0}w_1 & \dfrac{b_0}{m_0}w_1 \\ \dfrac{k_0}{m_0}w_2 & 0 \\ 0 & \dfrac{b_0}{m_0}w_3 \end{bmatrix}, \quad D = -\sqrt{3}\begin{bmatrix} w_1 \\ 0 \\ 0 \end{bmatrix}.$$

When the parameters take value in

$$1 \le m \le 2, \quad 10 \le k \le 20, \quad 5 \le b \le 10,$$

the solution to the quadratic stability condition (13.20) and $P > 0$ *is*

$$P = \begin{bmatrix} 1.9791 & -2.8455 \\ -2.8455 & 14.2391 \end{bmatrix}.$$

This shows that the system is quadratically stable.

On the other hand, when the damping ratio ranges over $0 \le b \le 5$, *no solution exists for (13.20) and* $P > 0$.

13.3 Lur'e System

The feedback system shown in Figure 13.2 consists of a linear system $G(s)$ and a static nonlinearity Φ. Let the realization of $G(s)$ be

$$\dot{x} = Ax + Bu, \quad y = Cx$$
$$x \in \mathbb{R}^n, \quad u \in \mathbb{R}^m, \quad y \in \mathbb{R}^m. \tag{13.23}$$

Its input u is supplied by

$$u = -\Phi(y) \tag{13.24}$$

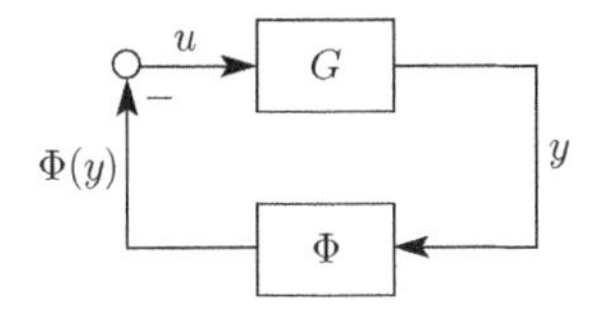

Figure 13.2 Lur'e system

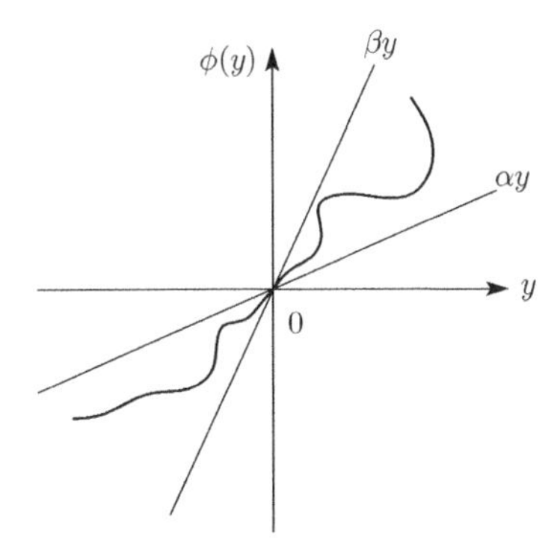

Figure 13.3 Static nonlinearity in a cone

in which the nonlinearity Φ satisfies

$$[\Phi(y) - K_{\min}y]^T[\Phi(y) - K_{\max}y] \leq 0. \tag{13.25}$$

Here, it is assumed that the coefficient matrices satisfy $K_{\max} - K_{\min} > 0$. For single-input single-output (SISO) systems, the nonlinearity is written as ϕ which satisfies

$$(\phi(y) - \alpha y)(\phi(y) - \beta y) \leq 0, \quad 0 \leq \alpha < \beta$$
$$\Rightarrow \alpha y \leq \phi(y) \leq \beta y,\ y \geq 0; \quad \beta y \leq \phi(y) \leq \alpha y, \quad y < 0.$$

As shown in Figure 13.3, ϕ is located in a sector bounded by straight lines with slopes of α, β. For simplicity, we call such region as a *sector* $[\alpha, \beta]$. Popov named the property wherein the asymptotic stability of closed-loop system is maintained for all static nonlinearities in a sector as *absolute stability*. In a modern viewpoint, $g(y) = \phi(y)/y$ can be regarded as a time-varying gain corresponding to the nonlinearity ϕ. This gain takes value in

$$\alpha \leq g(y) = \frac{\phi(y)}{y} \leq \beta. \tag{13.26}$$

So, Lur'e system can also be treated as a system with uncertain time-varying gain. Its absolute stability is exactly the robust stability w.r.t. the uncertain gain $g(y)$.

Now, what is the stability condition for Lur'e system? As mentioned in Chapter 11, a positive gain uncertainty always has zero phase angle. Keeping this feature in mind, we may guess intuitively the stability condition of Lur'e system based on Nyquist stability criterion. To illustrate it, let us look at the case where the nonlinear gain of a SISO system takes value in

$$0 \leq g(y) \leq K.$$

When the linear system $G(s)$ is stable, Nyquist stability criterion says that the closed-loop system is stable iff the return difference

$$1 + G(j\omega)g(y)$$

does not encircle the origin. As a closed curve encircling the origin must cross the imaginary axis, stability condition of the closed-loop system is equivalent to that $\Re[1 + G(j\omega)g(y)] \neq 0$ holds for all frequencies ω and all gains $g(y)$. Note that $\Re[1 + G(j\omega)g(y)] = 1 > 0$ when $g(y) = 0$. So long as

$$\Re[1 + G(j\omega)g(y)] = \Re[1 + G(j\omega)K] > 0 \quad \forall\omega$$

holds true for $g(y) = K$, then $\Re[1 + G(j\omega)g(y)] > 0$ is also true for any $g(y) \in [0, K]$ since this function $g(y)$ is affine. This condition implies the strongly positive realness of $1 + KG(s)$.

This condition can also be derived from another point of view. Note that the essence of this condition is to use the phase angle to ensure that Nyquist stability criterion is satisfied. Here, the key is that as long as the phase angle of the loop gain $G(s)g(y)$ is not $\pm 180°$, Nyquist stability criterion is met regardless of the gain. The gain $g(y) = \phi(y)/y$ of nonlinearity ϕ is finite. So, we try extending the nonlinear gain to the infinity. When $g(y)$ varies in the interval $[0, K]$, this is accomplished by the following transformation:

$$0 \leq g_N = \frac{g}{1 - g/K} < \infty.$$

Further, $g_N \to \infty$ as $g \to K$. Based on this fact, we make a transformation of block diagram as shown in Figure 13.4. After the transformation, the gain of the new nonlinearity $\phi_N(y_N)$ becomes g_N while the phase angle keeps zero. On the other hand, the new linear system turns into

$$G_N(s) = G(s) + \frac{1}{K} = \frac{1}{K}(1 + KG(s)).$$

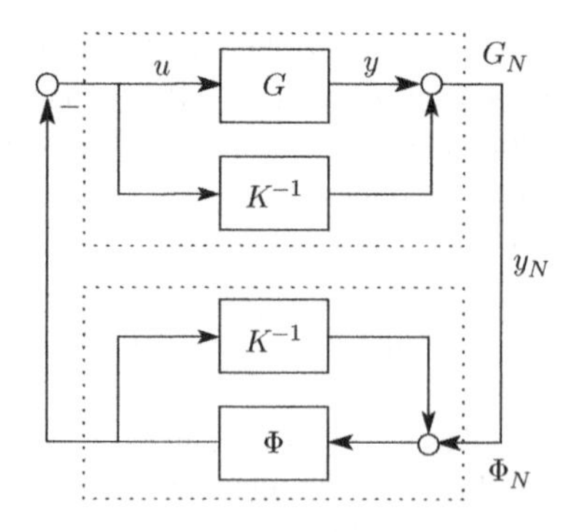

Figure 13.4 Equivalent transformation of Lur'e system

Then, the phase angle of the new loop gain $G_N(s)g_N(y)$ will never be $\pm 180°$ so long as $1 + KG(s)$ is positive-real. As such, stability of the closed-loop system is guaranteed.

Of course, the system under consideration is nonlinear, so the preceding discussion is not rigorous. A rigorous proof is done by using Lyapunov stability theory. In fact, the following conclusion holds:

Lemma 13.1 *Assume that A is stable in the realization (13.23) of m input m output linear system $G(s)$ and the nonlinearity Φ satisfies*

$$\Phi^T(\Phi - Ky) \leq 0.$$

Then, the closed-loop system is asymptotically stable if the transfer matrix

$$Z(s) = I + KG(s)$$

is strongly positive-real.

Proof. $Z^*(j\omega) + Z(j\omega)$ can be written as

$$Z^*(j\omega) + Z(j\omega) = \begin{bmatrix} (j\omega I - A)^{-1}B \\ I \end{bmatrix}^* \begin{bmatrix} 0 & (KC)^T \\ KC & 2I \end{bmatrix} \begin{bmatrix} (j\omega I - A)^{-1}B \\ I \end{bmatrix}.$$

According to KYP lemma, when $Z(s)$ is strongly positive-real, there is a symmetric matrix P satisfying

$$-\begin{bmatrix} 0 & (KC)^T \\ KC & 2I \end{bmatrix} + \begin{bmatrix} A^TP + PA & PB \\ B^TP & 0 \end{bmatrix} < 0. \tag{13.27}$$

Further, $P > 0$ follows from the $(1,1)$ block $A^TP + PA < 0$ and the stability of A. We construct a Lyapunov function

$$V(x) = x^TPx$$

using this positive definite matrix P and use it to prove the lemma. This $V(x)$ is radially unbounded. Differentiating $V(x)$ along the trajectory of system (13.23), we obtain that

$$\begin{aligned} \dot{V}(x) &= x^T(A^TP + PA)x + x^TPBu + u^TB^TPx \\ &= \begin{bmatrix} x \\ u \end{bmatrix}^T \begin{bmatrix} A^TP + PA & PB \\ B^TP & 0 \end{bmatrix} \begin{bmatrix} x \\ u \end{bmatrix} \\ &< \begin{bmatrix} x \\ u \end{bmatrix}^T \begin{bmatrix} 0 & (KC)^T \\ KC & 2I \end{bmatrix} \begin{bmatrix} x \\ u \end{bmatrix} \\ &= 2[u^T(KCx) + u^Tu] \\ &= 2[u^TKy + u^Tu] \end{aligned}$$

as long as $x \neq 0$. Further substitution of $u = -\Phi(y)$ yields

$$\dot{V}(x) < 2\Phi^T[\Phi - Ky] \leq 0, x \neq 0.$$

This implies the asymptotic stability of the closed-loop system according to Lyapunov stability theorem. •

Example 13.4 *Consider a linear system*

$$G(s) = \frac{6}{(s+1)(s+2)(s+3)}.$$

The nonlinearity ϕ is the ideal saturation function below:

$$\phi(y) = y, \ |y| < 1; \quad \phi(y) = \frac{y}{|y|}, \quad |y| \geq 1.$$

We examine the stability of this closed-loop system. The saturation is contained in the sector $[0, 1]$, that is, $K = 1$. Since this system is SISO, the strongly positive-real condition of $Z(s) = 1 + G(s)$ becomes

$$\Re[G(j\omega)] > -1.$$

The Nyquist plot of $G(j\omega)$ is shown in Figure 13.5. Clearly, this Nyquist plot is located on the right-hand side of the straight line $\Re[s] = -1$ and satisfies the stability condition of Lemma 13.1.

13.3.1 Circle Criterion

For a general sector $[K_{\min}, K_{\max}]$, we need to just transform the sector equivalently into $[0, K_{\max} - K_{\min}]$, as shown in Figure 13.6. Then, the stability condition can be derived similarly. In this case, the linear system turns into $G(I + K_{\min}G)^{-1}$.

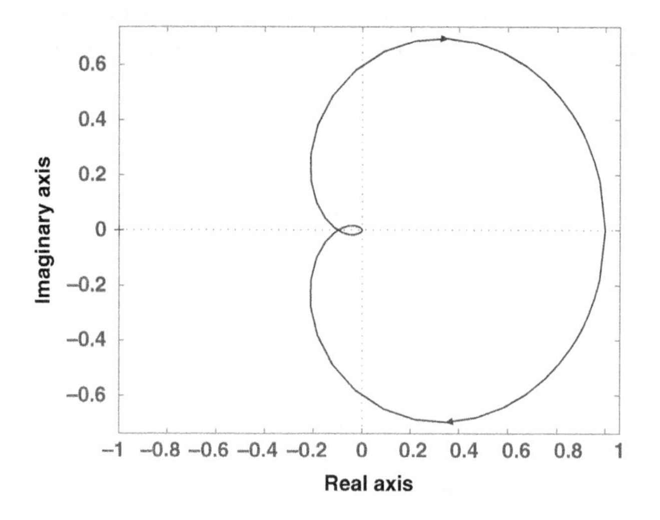

Figure 13.5 Nyquist plot

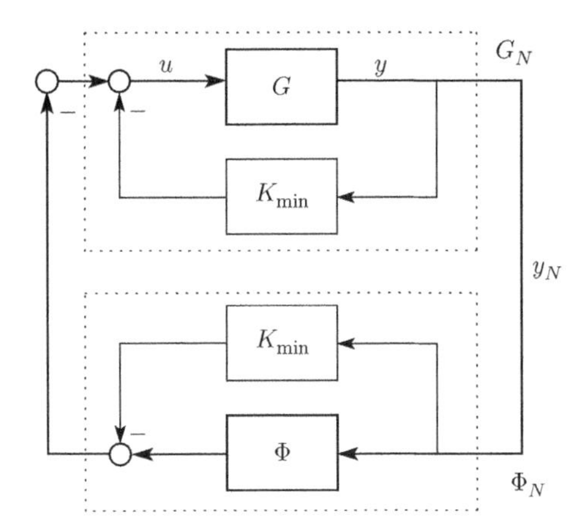

Figure 13.6 Equivalent transformation of Lur'e system: Circle criterion

Theorem 13.3 *In the closed-loop system composed of the linear system $G(s)$ of (13.23) and nonlinearity Φ of (13.25), if*

$$G_N(s) = G(s)[I + K_{\min}G(s)]^{-1}$$

is stable and

$$Z_N(s) = [I + K_{\max}G(s)][I + K_{\min}G(s)]^{-1}$$

is strongly positive-real, then the closed-loop system is asymptotically stable.

Proof. After block diagram transformation, the nonlinearity is in the sector $[0, K_{\max} - K_{\min}]$. So we can apply Lemma 13.1. Here, G becomes G_N and the $Z(s)$ turns into

$$I + (K_{\max} - K_{\min})G(s)[I + K_{\min}G(s)]^{-1} = [I + K_{\max}G(s)][I + K_{\min}G(s)]^{-1},$$

that is, $Z_N(s)$. So the conclusion follows immediately from Lemma 13.1. •

The SISO version of this theorem is the famous *circle criterion* below.

Theorem 13.4 (Circle criterion) *Consider an SISO Lur'e system. Assume that the nonlinearity ϕ is in the sector $[\alpha, \beta]$. Also define a disk:*

$$D(\alpha, \beta) = \left\{ z \in \mathbb{C} \mid \left| z + \frac{\alpha + \beta}{2\alpha\beta} \right| \leq \left| \frac{\beta - \alpha}{2\alpha\beta} \right| \right\}.$$

Then, the closed-loop system is asymptotically stable if one of the following statements holds.

1. *If $0 < \alpha < \beta$, the Nyquist plot of $G(j\omega)$ does not enter the disk $D(\alpha, \beta)$ and encircle it p times counterclockwise, where p is the number of unstable poles of $G(s)$.*

2. *If* $0 = \alpha < \beta$, $G(s)$ *is stable and the Nyquist plot of* $G(j\omega)$ *satisfies*

$$\Re[G(j\omega)] > -\frac{1}{\beta}.$$

3. *If* $\alpha < 0 < \beta$, $G(s)$ *is stable and the Nyquist plot of* $G(j\omega)$ *lies in the interior of the disk* $D(\alpha, \beta)$.

Proof. First, when $G/(1 + \alpha G)$ is stable, $1 + \alpha G(j\omega) \neq 0$ for all the frequencies ω. Now, $Z_N(s)$ is

$$Z_N(s) = \frac{1 + \beta G(s)}{1 + \alpha G(s)}.$$

It is strongly positive-real if $Z_N^*(j\omega) + Z_N(j\omega) > 0$ for all frequency ω. When $\alpha\beta \neq 0$, this is equivalent to

$$\begin{aligned}
0 &< (1 + \beta G^*(j\omega))(1 + \alpha G(j\omega)) + (1 + \alpha G^*(j\omega))(1 + \beta G(j\omega)) \\
&= 2 + (\alpha + \beta)(G^*(j\omega) + G(j\omega)) + 2\alpha\beta G^*(j\omega)G(j\omega) \\
&= 2\alpha\beta \left\{ G^*(j\omega)G(j\omega) + \frac{\alpha + \beta}{2\alpha\beta}(G^*(j\omega) + G(j\omega)) + \frac{1}{\alpha\beta} \right\} \\
&= 2\alpha\beta \left\{ \left| G(j\omega) + \frac{\alpha + \beta}{2\alpha\beta} \right|^2 - \left(\frac{\beta - \alpha}{2\alpha\beta} \right)^2 \right\}.
\end{aligned}$$

In the case of statement 1, strongly positive-real condition becomes

$$\left| G(j\omega) + \frac{\alpha + \beta}{2\alpha\beta} \right| > \left| \frac{\beta - \alpha}{2\alpha\beta} \right|.$$

It requires that the Nyquist plot of $G(j\omega)$ does not enter the disk $D(\alpha, \beta)$. On the other hand, the stability condition of $G/(1 + \alpha G)$ is that the Nyquist plot of $G(j\omega)$ must encircle the point $(-1/\alpha, j0)$ p times counterclockwise according to Nyquist criterion. Since this point is on the boundary of the disk $D(\alpha, \beta)$, the contour must encircle the disk p times counterclockwise.

In the case of statement 3, the aforementioned strongly positive-real condition becomes

$$\left| G(j\omega) + \frac{\alpha + \beta}{2\alpha\beta} \right| < \left| \frac{\beta - \alpha}{2\alpha\beta} \right|.$$

That is, the Nyquist plot of $G(j\omega)$ must lie in the interior of the disk $D(\alpha, \beta)$. So, the Nyquist plot of $G(j\omega)$ does not encircle the point $(-1/\alpha, j0)$. Therefore, $G(s)$ must be stable in order to guarantee the stability of $G/(1 + \alpha G)$.

Finally, in the case of statement 2, each part after transformation becomes

$$G_N = \frac{G}{1 + \alpha G} = G, \quad Z_N = 1 + \beta G.$$

Therefore, the stability of $G(s)$ is needed. Further, $Z_N(s) = 1 + \beta G(s)$ is strongly positive-real iff $\Re[G(j\omega)] > -\frac{1}{\beta}$. •

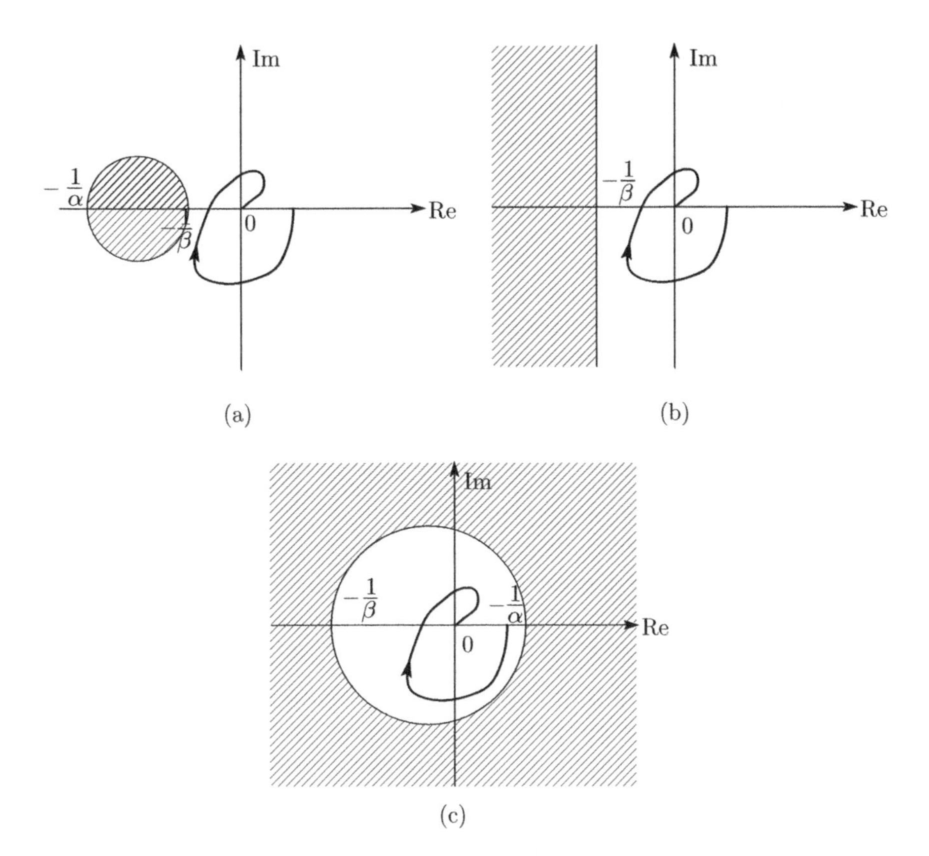

Figure 13.7 Circle criterion (a) Case 1: $0 < \alpha < \beta$ (b) Case 2: $0 = \alpha < \beta$ (c) Case 3: $\alpha < 0 < \beta$

The circle criterion for each case is illustrated in Figure 13.7(a)–(c).

Example 13.5 *Let the linear system be*

$$G(s) = \frac{6}{(s+1)(s+2)(s+3)}$$

and the nonlinearity ϕ be one of the following sectors:

$$(1)\ [1,3],\quad (2)\ [0,2],\quad (3)\ [-1,1]$$

We examine the stability of the closed-loop system.

The circle criterion for each sector is given by

$$(1)\ |G(j\omega) + \frac{2}{3}| \geq \frac{1}{3},\quad (2)\ \Re[G(j\omega)] \geq -\frac{1}{2},\quad (3)\ |G(j\omega)| \leq 1.$$

From the Nyquist plot of $G(s)$ shown in Figure 13.5, it is clear that all these conditions are met. Therefore, the closed-loop system is stable for static nonlinearity in any of these sectors.

13.3.2 Popov Criterion

In practical applications, basically the gain of static nonlinearity Φ does not change sign. Therefore, such a nonlinearity may be regarded as an uncertainty with zero phase angle; only its gain varies. Lemma 13.1 provides a stability condition which makes use of this feature. From the discussion before Lemma 13.1, we know that the essence of this lemma is to guarantee the stability of the closed-loop system only using the phase condition. However, in the block diagram transformation of Figure 13.4, although the transformed nonlinearity has a positive gain which takes value over a wide range, the phase angle of the linear part is still limited to $[-90^\circ, 90^\circ]$. Looking from the viewpoint that the phase angle of the open-loop system can get arbitrarily close to $\pm 180^\circ$, there is still a large room for improvement. So here we dig deeply into this idea.

There are many approaches to expand the range of phase angle of the open-loop system. However, from the viewpoint of theoretical simplicity, it is better to choose such a characteristic of phase angle that can be easily combined with KYP lemma. To this end, let us consider an open-loop transformation which makes the nonlinear part positive-real and the linear part strongly positive-real.

Assume that the nonlinearity satisfies

$$\Phi(y)^T[\Phi(y) - Ky] \leq 0, \quad K > 0. \tag{13.28}$$

We want to find a transformation so that the phase angle of the nonlinearity is in the range of $\pm 90^\circ$ and its gain may vary from zero to infinity, while the linear part is strongly positive-real. One transformation is shown in Figure 13.8 which meets these requirements. The gain of the transformed nonlinearity is

$$g_N = \frac{g\dfrac{1}{1+\eta s}}{1 - K^{-1}g\dfrac{1}{1+\eta s}} = \frac{g}{1 - K^{-1}g + \eta s} \tag{13.29}$$

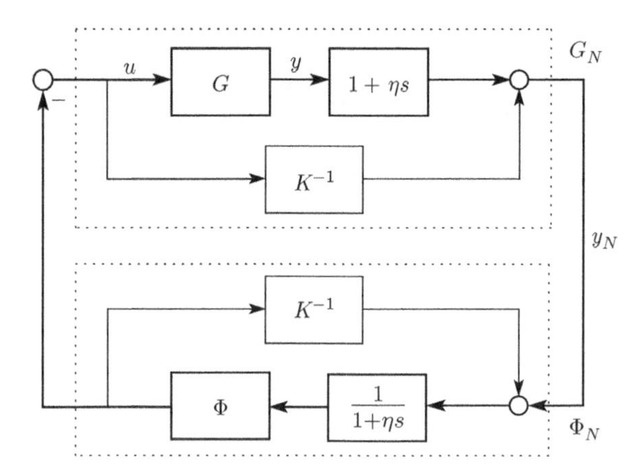

Figure 13.8 Equivalent transformation of Lur'e system: Popov criterion

in which $\eta > 0$. When $\omega \geq 0$, its phase angle changes from $0°$ to $-90°$. When $g \to K$ and $\omega \to 0$, the gain approaches $+\infty$. Meanwhile, when $\omega \to \infty$ or $g \to 0$, the gain converges to 0. Namely, g_N has the property of positive-real function. Therefore, intuitively the closed-loop system is stable as long as $G_N = (1+\eta s)G + K^{-1}$ is strongly positive-real.

More precisely, the so-called *Popov criterion* below holds.

Theorem 13.5 (Popov Criterion) *In the realization (13.23) of m-input m-output linear system $G(s)$, assume that A is stable. Then, the closed-loop system is asymptotically stable if there is a positive-real number η such that the transfer matrix*

$$Z(s) = I + (1+\eta s)KG(s)$$

is strongly positive-real.

Proof. Expanding $Z(s)$ using $s(sI-A)^{-1} = I + A(sI-A)^{-1}$, we have

$$\begin{aligned} Z(s) &= I + \eta KCB + KC(I+\eta A)(sI-A)^{-1}B \\ &= \begin{bmatrix} KC(I+\eta A) & I+\eta KCB \end{bmatrix} \begin{bmatrix} (sI-A)^{-1}B \\ I \end{bmatrix}. \end{aligned}$$

So, $Z^*(j\omega) + Z(j\omega)$ can be written as

$$\begin{bmatrix} (j\omega I-A)^{-1}B \\ I \end{bmatrix}^* \begin{bmatrix} 0 & (KC(I+\eta A))^T \\ KC(I+\eta A) & 2(I+\eta KCB) \end{bmatrix} \begin{bmatrix} (j\omega I-A)^{-1}B \\ I \end{bmatrix}.$$

According to KYP Lemma, there is a symmetric matrix P satisfying the inequality

$$-\begin{bmatrix} 0 & (KC(I+\eta A))^T \\ KC(I+\eta A) & 2(I+\eta KCB) \end{bmatrix} + \begin{bmatrix} A^TP+PA & PB \\ B^TP & 0 \end{bmatrix} < 0 \tag{13.30}$$

since $Z(s)$ is strongly positive-real. Further, we get $P > 0$ from the $(1,1)$ block $A^TP + PA < 0$ and the stability of A. In the sequel, we use a Lur'e–Postnikov-type Lyapunov function[1]

$$V(x) = x^TPx + 2\eta \int_0^y \Phi(v)^TKdv \tag{13.31}$$

to prove the theorem. Obviously, this $V(x)$ is positive definite and radially unbounded. Differentiating $V(x)$ along trajectory of Eq. (13.23), we obtain

$$\begin{aligned} \dot{V}(x) &= x^T(A^TP+PA)x + x^TPBu + u^TB^TPx + 2\eta\Phi(y)^TK\dot{y} \\ &= x^T(A^TP+PA)x + x^TPBu + u^TB^TPx - 2\eta u^TKC(Ax+Bu) \\ &= \begin{bmatrix} x \\ u \end{bmatrix}^T \left(\begin{bmatrix} A^TP+PA & PB \\ B^TP & 0 \end{bmatrix} - \begin{bmatrix} 0 & (\eta KCA)^T \\ \eta KCA & 2\eta KCB \end{bmatrix} \right) \begin{bmatrix} x \\ u \end{bmatrix} \end{aligned}$$

[1] Due to the property $\Phi(y)^TKy \geq \Phi(y)^T\Phi(y) \geq 0$, the second term is nonnegative. Regarding $\Phi(y)$ as a force and y as a displacement, then the integral represents a work.

$$
\begin{aligned}
&< \begin{bmatrix} x \\ u \end{bmatrix}^T \begin{bmatrix} 0 & (KC)^T \\ KC & 2I \end{bmatrix} \begin{bmatrix} x \\ u \end{bmatrix} \\
&= 2[u^T(KCx) + u^T u] \\
&= 2[u^T Ky + u^T u]
\end{aligned}
$$

as long as $x \neq 0$. Further, substituting $u = -\Phi(y)$ and invoking the property of nonlinearity Φ, we have

$$\dot{V}(x) < 2\Phi^T[\Phi - Ky] \leq 0 \quad \forall x \neq 0.$$

Therefore, the closed-loop system is asymptotically stable according to Lyapunov stability criterion. •

For a SISO system, Popov criterion can be checked graphically, as shown below. The strongly positive-real condition

$$\Re[1 + (1 + j\eta\omega)KG(j\omega)] > 0 \quad \forall\omega$$

of $Z(s)$ is equivalent to

$$\frac{1}{K} + \Re[G(j\omega)] - \eta\omega\Im[G(j\omega)] > 0 \quad \forall\omega.$$

In the Cartesian coordinate $(x, y) = (\Re[G(j\omega)], \omega\Im[G(j\omega)])$, this condition becomes

$$\frac{1}{K} + x(\omega) > \eta y(\omega) \quad \forall\omega.$$

This means that the trajectory $(x(\omega), y(\omega))$ is located below a line passing through $(-1/K, 0)$ and with a slope $1/\eta$, as shown in Figure 13.9. This figure is called *Popov plot.*

Example 13.6 *We revisit the example in the previous subsection. The linear system is*

$$G(s) = \frac{6}{(s+1)(s+2)(s+3)}.$$

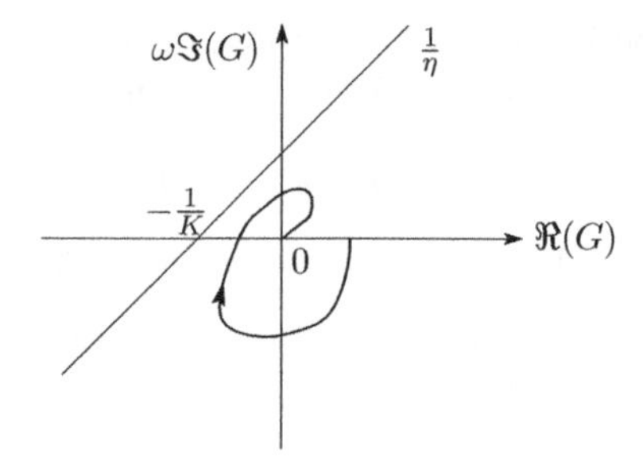

Figure 13.9 Popov criterion

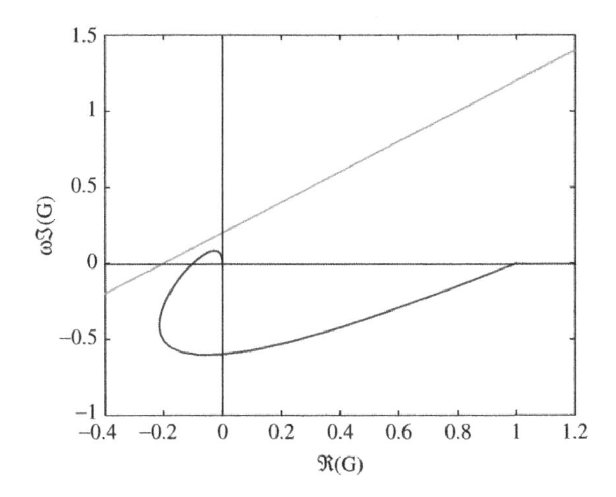

Figure 13.10 Popov plot of Example 13.6: $K = 5, \eta = 1$

When the sector where the nonlinearity ϕ lies is expanded to $[0, 5]$, the vertical line $-1/\beta = -0.2$ intersects the Nyquist plot of Figure 13.5 so that the circle criterion is not satisfied. However, when we draw the Popov plot, it is clear that Popov criterion is met (Figure 13.10). Therefore, the closed-loop system remains stable even in this case.

From this example, it is verified that Popov criterion is weaker than circle criterion. So, it has a wider field of applications.

13.4 Passive Systems

A system is called a *passive system* if its transfer function is either positive-real, or strongly positive-real, or strictly positive-real. In this section, we analyze the stability when a positive-real system is negatively feedback connected with a strongly positive-real or strictly positive-real system. It will be revealed that such closed-loop system is always stable. The closed-loop system under consideration is shown in Figure 13.11, in which the uncertainty $\Delta(s)$ is a positive-real transfer matrix, while the nominal closed-loop system $M(s)$ is either strongly positive-real or strictly positive-real. Intuitively, the phase angle of a positive-real system is limited to $[-90°, 90°]$ and that of a strongly positive-real system is restricted to $(-90°, 90°)$ (the infinite frequency is excluded for a strictly positive-real system). So the

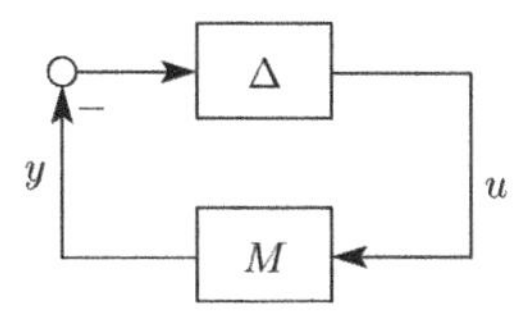

Figure 13.11 Closed-loop system with a positive-real uncertainty

phase angle of the open-loop system is always not $\pm 180^\circ$. In other words, the phase condition of Nyquist stability criterion is satisfied. So, the stability of closed-loop system may be expected.

Below, we give a rigorous description and proof for this inference. The following two conclusions are true:

Theorem 13.6 *In the closed-loop system of Figure 13.11, assume that the uncertainty $\Delta(s)$ is stable and positive-real. Then, the closed-loop system is asymptotically stable if the nominal system $M(s)$ is stable and strongly positive-real.*

Proof. From the stability assumption on M and Δ, we need to only discuss the case of minimal realization. So, the minimal state-space realizations of them are set as

$$\Delta(s) = (A_1, B_1, C_1, D_1), \quad M(s) = (A_2, B_2, C_2, D_2),$$

respectively. The corresponding state vectors are x_1, x_2. Noting the input–output relation, we have

$$\dot{x}_1 = A_1 x_1 + B_1(-y), \quad u = C_1 x_1 + D_1(-y)$$
$$\dot{x}_2 = A_2 x_2 + B_2 u, \quad y = C_2 x_2 + D_2 u.$$

According to positive-real lemma and strongly positive-real lemma, there exist positive definite matrices P, Q satisfying

$$\begin{bmatrix} A_1^T P + PA_1 & PB_1 \\ B_1^T P & 0 \end{bmatrix} - \begin{bmatrix} 0 & C_1^T \\ C_1 & D_1 + D_1^T \end{bmatrix} \leq 0 \tag{13.32}$$

$$\begin{bmatrix} A_2^T Q + QA_2 & QB_2 \\ B_2^T Q & 0 \end{bmatrix} - \begin{bmatrix} 0 & C_2^T \\ C_2 & D_2 + D_2^T \end{bmatrix} < 0 \tag{13.33}$$

under the given condition. Further, we define two Lyapunov functions:

$$V_1(x_1) = x_1^T P x_1, \quad V_2(x_2) = x_2^T Q x_2. \tag{13.34}$$

Both are positive definite. Multiplying a vector $[x_1^T \ \ -y^T] \neq 0$ and its transpose to (13.32), we get

$$\begin{aligned}
0 \geq & x_1^T (A_1^T P + PA_1) x_1 - x_1^T PB_1 y - y^T B_1^T P x_1 + x_1^T C_1^T y + y^T C_1 x_1 \\
& - y^T (D_1 + D_1^T) y \\
= & x_1^T P (A_1 x_1 - B_1 y) + (A_1 x_1 - B_1 y)^T P x_1 + y^T (C_1 x_1 - D_1 y) \\
& + (C_1 x_1 - D_1 y)^T y \\
= & x_1^T P \dot{x}_1 + \dot{x}_1^T P x_1 + u^T y + y^T u \\
\Rightarrow & \dot{V}_1(x_1) \leq -u^T y - y^T u.
\end{aligned}$$

Firstly, from (13.33) we have similarly that the inequality

$$\dot{V}_2(x_2) < u^T y + y^T u$$

holds for any nonzero vector $[x_2^T \ u^T]$. Secondly, $[x_1^T \ -y^T] \neq 0$, $[x_2^T \ u^T] \neq 0$ implies $[x_1^T \ x_2^T] \neq 0$. Take the quadratic positive definite function $V(x_1, x_2) = V_1(x_1) + V_2(x_2)$ as a Lyapunov candidate of the closed-loop system. Then

$$\dot{V}(x_1, x_2) = \dot{V}_1(x_1) + \dot{V}_2(x_2) < 0$$

holds for any nonzero vector $[x_1^T \ x_2^T]$. Therefore, the closed-loop system is asymptotically stable. •

Another conclusion is as follows:

Theorem 13.7 *Assume that the uncertainty $\Delta(s)$ is stable and positive-real in the closed-loop system of Figure 13.11. The closed-loop system is asymptotically stable if the nominal system $M(s)$ is stable, having full normal rank, and there is a constant $\epsilon > 0$ such that $M(s-\epsilon)$ is positive-real.*

Proof. Let the state vectors be x_1, x_2 and minimal realizations be $\Delta(s) = (A_1, B_1, C_1, D_1)$, $M(s) = (A_2, B_2, C_2, 0)$, respectively. From the input–output relation, we have

$$\dot{x}_1 = A_1 x_1 + B_1(-y), \quad u = C_1 x_1 + D_1(-y)$$
$$\dot{x}_2 = A_2 x_2 + B_2 u, \quad y = C_2 x_2.$$

Due to positive-real lemma, strictly positive-real lemma, and $M(s-\epsilon) = (A_2 + \epsilon I, B_2, C_2, 0)$, there exist positive definite matrices P, Q satisfying

$$\begin{bmatrix} A_1^T P + P A_1 & P B_1 \\ B_1^T P & 0 \end{bmatrix} - \begin{bmatrix} 0 & C_1^T \\ C_1 & D_1 + D_1^T \end{bmatrix} \leq 0 \tag{13.35}$$

$$\begin{bmatrix} (A_2 + \epsilon I)^T Q + Q(A_2 + \epsilon I) & Q B_2 \\ B_2^T Q & 0 \end{bmatrix} - \begin{bmatrix} 0 & C_2^T \\ C_2 & 0 \end{bmatrix} \leq 0. \tag{13.36}$$

Define two Lyapunov functions as follows:

$$V_1(x_1) = x_1^T P x_1, \quad V_2(x_2) = x_2^T Q x_2. \tag{13.37}$$

Multiplying (13.35) with nonzero vector $[x_1^T \ -y^T]$ and its transpose, we have

$$\dot{V}_1(x_1) \leq -u^T y - y^T u.$$

Similarly, multiplying (13.36) by nonzero vector $[x_2^T \ u^T]$ and its transpose, one gets that

$$\begin{aligned} 0 \geq & x_2^T (A_2^T Q + Q A_2 + 2\epsilon Q) x_2 + x_2^T Q B_2 u + u^T B_2^T Q x_2 \\ & - x_2^T C_2^T u - u^T C_2 x_2 \\ = & x_2^T Q \dot{x}_2 + \dot{x}_2^T Q x_2 + 2\epsilon x_2^T Q x_2 - y^T u - u^T y \\ \Rightarrow & \ \dot{V}_2(x_2) \leq u^T y + y^T u - 2\epsilon x_2^T Q x_2 \end{aligned}$$

holds for any nonzero vector $[x_2^T\ u^T]$. $[x_1^T\ x_2^T] \neq 0$ follows from $[x_1^T\ y^T] \neq 0$ and $[x_2^T\ u^T] \neq 0$. Take the quadratic function $V(x_1, x_2) = V_1(x_1) + V_2(x_2)$ as a Lyapunov candidate for the closed-loop system, which is obviously positive definite. From the preceding inequalities, we get that

$$\dot{V}(x_1, x_2) = \dot{V}_1(x_1) + \dot{V}_2(x_2) \leq -2\epsilon x_2^T Q x_2$$

holds for any nonzero vector $[x_1^T\ x_2^T]$. When x_2 is not identically zero, $V(x_1, x_2)$ strictly decreases. When $x_2(t) \equiv 0$, its derivative is also zero. So

$$0 = B_2 u, \quad y = 0.$$

Since $M(s)$ has full normal rank, B_2 must have full column rank. Therefore, $u(t) \equiv 0$. Substituting $u = 0, y = 0$ into the state equation of x_1, we have

$$\dot{x}_1 = A_1 x_1, \quad 0 = C_1 x_1.$$

$x_1(t) = 0$ is obtained because A_1 is stable and (C_1, A_1) is controllable. Therefore, the closed-loop system is asymptotically stable. •

Exercises

13.1 Consider a plant with uncertain parameters:

$$P(s) = \frac{k}{\tau s + 1}, \quad 0.8 \leq k \leq 1.2, \quad 0.7 \leq \tau \leq 1.3. \tag{13.38}$$

Its state equation is given by

$$\dot{x} = Ax + Bu = -\frac{1}{\tau}x + \frac{k}{\tau}u, \quad y = x. \tag{13.39}$$

For the state feedback $u = fx$, express the quadratic stability condition as the vertex conditions about the parameter vector $[k\ \tau]$. Then find a specific solution.

13.2 Assume that the matrix Δ varies in $\|\Delta\|_2 \leq 1$. Prove that the matrix $I - D\Delta$ is invertible iff $\|D\|_2 < 1$ by using the singular value decomposition of Δ.

13.3 Assume that the uncertainty $\Delta(t)$ in Figure 13.1 is a time-varying matrix and satisfies $\|\Delta(t)\|_2 \leq 1$. Further, let transfer matrix $M(s) = (A, B, C, D)$ be stable and satisfy $\|M\|_\infty < 1$. Denote the state of $M(s)$ by x. Starting from the time-domain interpretation of bounded-real lemma given in Subsection 8.2.1, prove that the closed-loop system is asymptotically stable by using the Lyapunov function $V(x) = x^T P x$.

13.4 Prove that the output of stable system $\dot{x} = Ax, y = Cx$ satisfies

$$\int_0^\infty y^T(t) y(t)\, dt = x^T(0) X x(0)$$

in which the matrix X is the solution to the Lyapunov equation below:

$$A^T X + XA + C^T C = 0.$$

13.5 Prove Theorem 13.6 for SISO systems based on Nyquist stability criterion. (Hint: Consider the phase angle range of $M(j\omega)\Delta(j\omega)$.)

13.6 Design a state feedback $u = Fx$ such that the uncertain system

$$\dot{x} = (A + \Delta A)x + (B + \Delta B)u$$
$$[\Delta A\ \Delta B] = E\Delta[C\ D], \quad \|\Delta(t)\|_2 \leq 1\ \forall t$$

is quadratically stabilized by following the procedure below:
Step 1 Transform the state feedback system into that of Figure 13.1 in which the state equation of the nominal system $M(s)$ is given by

$$\dot{x} = (A + BF)x + Ew$$
$$z = (C + DF)x.$$

Step 2 Apply Theorem 13.2 to derive the condition on F.
Step 3 Apply either the variable elimination or the variable change method.

13.7 Repeat the design problem of Exercise 13.6 w.r.t. the output feedback case. The plant is

$$\dot{x} = (A + B_2\Delta C_1)x + (B_1 + B_2\Delta D_{12})u, \quad y = C_2x$$
$$\|\Delta(t)\|_2 \leq 1\ \forall t$$

and the controller is $K(s) = (A_K, B_K, C_K, D_K)$.

13.8 Repeat the quadratic stabilization problem w.r.t. a parametric system:

$$\dot{x} = (\lambda_1 A_1 + \lambda_2 A_2)x + (\lambda_1 B_1 + \lambda_2 B_2)u, \quad y = Cx$$
$$\lambda_1, \lambda_1 \geq 0, \lambda_1 + \lambda_2 = 1.$$

Consider both the state feedback and the output feedback cases.

Notes and References

Refer to Refs [47, 90, 22, 45, 78, 93] for more details about Lyapunov theory, Lur'e systems, and passive systems. Particularly, [47] is a widely read classic book on this topic. The quadratic stability notion was proposed in Ref. [6]. Gahinet *et al.* [34] discussed how to deal with parametric systems by using a Lyapunov function with uncertain parameters.

14

Robustness Analysis 3: IQC Approach

Integral quadratic constraint (IQC) theory is a robustness analysis method proposed by Megretski–Rantzer [67]. The idea is to describe an uncertainty using an integral quadratic constraint (IQC) on its input and output. Based on this, one can find the robust stability condition for the closed-loop systems. In this chapter, we consider the system shown in Figure 14.1. $G(s)$ is a known linear system, and Δ is an uncertainty that includes static nonlinearity. Let the input of uncertainty Δ be z and the output be w. Here, we focus on the BIBO stability of the closed-loop system.

In this chapter, the same notation is used to denote both a time signal and its Fourier transform for convenience.

14.1 Concept of IQC

To illustrate the meaning and motivation of IQC, let us look at the characterizations of several kinds of uncertainty:

(1) For a norm-bounded uncertainty $\Delta(s)$ ($\|\Delta\|_\infty \leq 1$), inequality $\Delta^*(j\omega)\Delta(j\omega) \leq I$ holds for any frequency ω. Hence, multiplying both sides of the inequality by z^* and z, respectively, we obtain that

$$
\begin{aligned}
& w^*(j\omega)w(j\omega) = z^*(j\omega)\Delta^*(j\omega)\Delta(j\omega)z(j\omega) \leq z^*(j\omega)z(j\omega) \\
& \Rightarrow 0 \leq z^*(j\omega)z(j\omega) - w^*(j\omega)w(j\omega) = \begin{bmatrix} z(j\omega) \\ w(j\omega) \end{bmatrix}^* \begin{bmatrix} I & 0 \\ 0 & -I \end{bmatrix} \begin{bmatrix} z(j\omega) \\ w(j\omega) \end{bmatrix}
\end{aligned}
$$

holds for arbitrary ω. Of course, we have

$$
\int_{-\infty}^{\infty} \begin{bmatrix} z(j\omega) \\ w(j\omega) \end{bmatrix}^* \begin{bmatrix} I & 0 \\ 0 & -I \end{bmatrix} \begin{bmatrix} z(j\omega) \\ w(j\omega) \end{bmatrix} d\omega \geq 0.
$$

Robust Control: Theory and Applications, First Edition. Kang-Zhi Liu and Yu Yao.

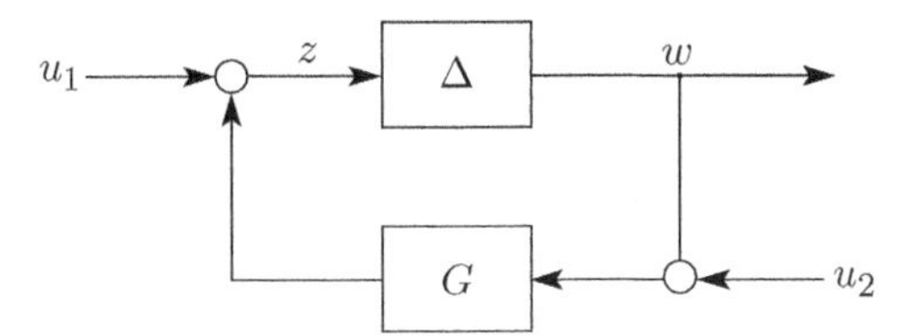

Figure 14.1 Closed-loop system

(2) For any frequency ω, a positive real uncertainty $\Delta(s)$ satisfies the inequality $\Delta^*(j\omega) + \Delta(j\omega) \geq 0$. Therefore, multiplication of both sides by z^* and z, respectively, yields

$$\begin{aligned}
0 \leq & z^*(j\omega)\Delta^*(j\omega)z(j\omega) + z^*(j\omega)\Delta(j\omega)z(j\omega) \\
= & w^*(j\omega)z(j\omega) + z^*(j\omega)w(j\omega) \\
\Rightarrow 0 \leq & \begin{bmatrix} z(j\omega) \\ w(j\omega) \end{bmatrix}^* \begin{bmatrix} 0 & I \\ I & 0 \end{bmatrix} \begin{bmatrix} z(j\omega) \\ w(j\omega) \end{bmatrix} \\
\Rightarrow & \int_{-\infty}^{\infty} \begin{bmatrix} z(j\omega) \\ w(j\omega) \end{bmatrix}^* \begin{bmatrix} 0 & I \\ I & 0 \end{bmatrix} \begin{bmatrix} z(j\omega) \\ w(j\omega) \end{bmatrix} d\omega \geq 0.
\end{aligned}$$

(3) The input and output of a real parameter uncertainty block $\Delta = \delta I$ $(|\delta| \leq 1)$ meet the relationship $w = \delta z$. Therefore, for a skew-symmetric matrix $Y = -Y^T$, the identity

$$\begin{aligned}
\begin{bmatrix} z(j\omega) \\ w(j\omega) \end{bmatrix}^* \begin{bmatrix} 0 & Y \\ Y^T & 0 \end{bmatrix} \begin{bmatrix} z(j\omega) \\ w(j\omega) \end{bmatrix} &= z^*(j\omega)Yw(j\omega) + w^*(j\omega)Y^T z(j\omega) \\
&= \delta z^*(j\omega)(Y + Y^T)z(j\omega) \\
&= 0
\end{aligned}$$

holds. So, for any matrix $X = X^T \geq 0$, we have

$$\begin{aligned}
\begin{bmatrix} z(j\omega) \\ w(j\omega) \end{bmatrix}^* \begin{bmatrix} X & Y \\ Y^T & -X \end{bmatrix} \begin{bmatrix} z(j\omega) \\ w(j\omega) \end{bmatrix} &= \begin{bmatrix} z(j\omega) \\ w(j\omega) \end{bmatrix}^* \begin{bmatrix} X & 0 \\ 0 & -X \end{bmatrix} \begin{bmatrix} z(j\omega) \\ w(j\omega) \end{bmatrix} \\
&= (1 - \delta^2)z^*(j\omega)Xz(j\omega) \geq 0 \\
&\Rightarrow \int_{-\infty}^{\infty} \begin{bmatrix} z(j\omega) \\ w(j\omega) \end{bmatrix}^* \begin{bmatrix} X & Y \\ Y^T & -X \end{bmatrix} \begin{bmatrix} z(j\omega) \\ w(j\omega) \end{bmatrix} d\omega \geq 0.
\end{aligned}$$

In fact, even if X, Y are extended to functions of frequency, the aforementioned inequality still holds.

As shown previously, many uncertainties can be described by the following integral inequality about the input z and output w of the uncertainty: [1]

$$\int_{-\infty}^{\infty} \begin{bmatrix} z(j\omega) \\ w(j\omega) \end{bmatrix}^* \Pi(j\omega) \begin{bmatrix} z(j\omega) \\ w(j\omega) \end{bmatrix} d\omega \geq 0, \quad \Pi(j\omega) = \Pi^*(j\omega). \tag{14.1}$$

[1] In fact, a quadratic inequality is satisfied at each frequency ω in the foregoing examples. The integral quadratic inequality about uncertainty is used simply because it is more convenient for the stability discussion of the closed-loop system.

This condition is called an *IQC conditions specified by the matrix* Π. However, since the uncertainty Δ under consideration contains static nonlinearity, we regard Δ as an operator.

14.2 IQC Theorem

Consider the closed-loop system shown in Figure 14.1 which is composed of negatively feedback connected transfer matrix $G(s)$ and uncertainty Δ. A sufficient condition for the robust stability of this system is given by the *IQC theorem* in the following text.

Theorem 14.1 **((IQC theorem))** *Let $G(s)$ be an $l \times m$ stable matrix and Δ be an $m \times l$ bounded causal operator. The closed-loop system (G, Δ) is BIBO stable when all of the following conditions are satisfied:*

1. *For any $\tau \in [0,1]$, $(I - \tau G\Delta)$ has a causal inverse.*
2. *For any $\tau \in [0,1]$, $\tau\Delta$ satisfies the IQC condition about Π.*
3. *There is an $\epsilon(> 0)$ satisfying*

$$\begin{bmatrix} G(j\omega) \\ I \end{bmatrix}^* \Pi(j\omega) \begin{bmatrix} G(j\omega) \\ I \end{bmatrix} \leq -\epsilon I \ \forall\omega. \tag{14.2}$$

Let us explain the meaning of *causality* first. An operation is called a *truncation* that chops off the part of signal $u(t)$ after the time instant T (refer to Figure 14.2). It is denoted by an operator P_T. That is,

$$P_T u(t) = \begin{cases} u(t), & t \in [0, T] \\ 0, & t > T \end{cases} \tag{14.3}$$

When a system G (may contain nonlinearity) has the property of

$$P_T G u(t) = P_T G P_T u(t) \tag{14.4}$$

for any signal $u(t)$, it is called a *causal system*. This equation means that the output of the system does not depend on future values of the input. When a system G has a causal inverse G^{-1}, it is called *causally invertible*.

The conditions (2) and (3) in IQC Theorem are only related with one of Δ and $G(s)$, while the condition (1) is related to both. Therefore, before checking condition 1, the other

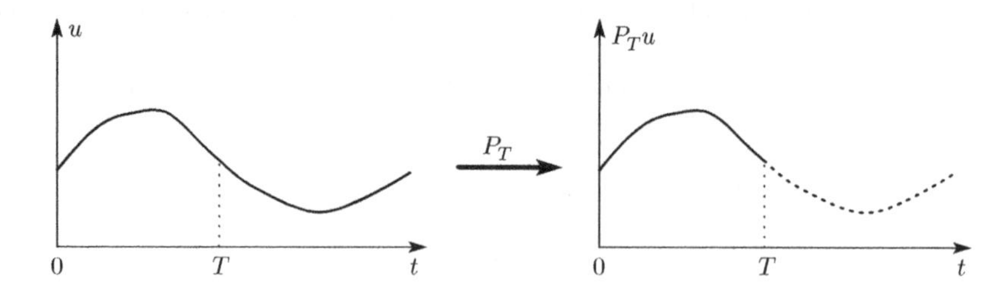

Figure 14.2 Truncation of a signal

two conditions should be satisfied. When $G(s)$ is strictly proper, condition 1 is automatically satisfied.

Next, we use this theorem to derive the robust stability conditions for several kinds of uncertainty.

Example 14.1 *As mentioned earlier, the IQC for a norm-bounded uncertainty* ($\|\Delta\|_\infty \leq 1$) *is specified by*

$$\Pi = \begin{bmatrix} I & 0 \\ 0 & -I \end{bmatrix}.$$

Therefore, condition 3 of Theorem 14.1 becomes

$$\begin{bmatrix} G(j\omega) \\ I \end{bmatrix}^* \begin{bmatrix} I & 0 \\ 0 & -I \end{bmatrix} \begin{bmatrix} G(j\omega) \\ I \end{bmatrix} = G^*(j\omega)G(j\omega) - I \leq -\epsilon I \iff \|G\|_\infty < 1.$$

Then, for any $\tau \in [0, 1]$, *obviously* $\det(I - \tau G(j\infty)\Delta(j\infty)) \neq 0$ *holds true due to the norm conditions of* G, Δ. *Therefore, condition 1 is satisfied. Next,* $w = \tau\Delta z$ *obviously satisfies* $z^*z - w^*w \geq 0$ *since* $\|\tau\Delta\|_\infty = \tau\|\Delta\|_\infty \leq \tau \leq 1$. *So condition 2 is also satisfied. Hence, the small-gain condition* $\|G\|_\infty < 1$ *ensures the robust stability. In fact, this proves the sufficiency of the small-gain theorem.*

Example 14.2 *The IQC about a positive real uncertainty (*$\Delta(j\omega) + \Delta^*(j\omega) \geq 0$*) is specified by*

$$\Pi = \begin{bmatrix} 0 & I \\ I & 0 \end{bmatrix}.$$

So, condition 3 of Theorem 14.1 becomes

$$\begin{bmatrix} G(j\omega) \\ I \end{bmatrix}^* \begin{bmatrix} 0 & I \\ I & 0 \end{bmatrix} \begin{bmatrix} G(j\omega) \\ I \end{bmatrix} = G^*(j\omega) + G(j\omega) \leq -\epsilon I < 0.$$

That is, $-G(s)$ *must be strongly positive real. To verify the condition 1, we need only prove that* $\det(I - \tau G(j\infty)\Delta(j\infty)) \neq 0$ *holds for any* $\tau \in [0, 1]$.

Assume that $[I - \tau G(j\infty)\Delta(j\infty)]z = 0$ *holds for a vector* $z \neq 0$ *when* $\tau \neq 0$. *Setting* $w = \tau\Delta(j\infty)z$, *then* $z = G(j\infty)w$ *and* $w \neq 0$ *hold by assumption. Based on these relations and the features of* G *and* Δ, *a contradiction*

$$\begin{aligned} 0 &\leq z^*(\tau\Delta(j\infty) + \tau\Delta^*(j\infty))z = z^*w + w^*z \\ &= w^*(G(j\infty) + G^*(j\infty))w \leq -\epsilon w^*w \\ &< 0 \end{aligned}$$

occurs so long as $\tau \neq 0$. *In addition, it is clear that* $I - \tau G(j\infty)\Delta(j\infty) = I$ *when* $\tau = 0$.

Next, the condition 2 follows from

$$w^*z + z^*w = \tau z^*(\Delta + \Delta^*)z \geq 0$$

in which $w = \tau\Delta z$. *From the IQC theorem, strongly positive realness of* $-G(s)$ *guarantees the robust stability of this system. As such, we have derived the positive real theorem of Chapter 13.*

14.3 Applications of IQC

In the previous section, we have shown the applications of IQC theory to systems with norm-bounded or positive real uncertainty. In this section, more examples are illustrated. It should be noted that different from Chapter 13, in the IQC framework, we consider systems with positive feedback rather than negative feedback. For the relation to the results in Chapter 13, we need just replace G by$-G$ in following examples:

Example 14.3 *Nonlinear time-varying block $\delta(t)I$ ($|\delta(t)| \leq 1$):*
As described in the constant scalar block, the IQC is specified by matrix

$$\Pi = \begin{bmatrix} X & Y \\ Y^T & -X \end{bmatrix}, \quad X = X^T \geq 0, \quad Y + Y^T = 0.$$

So, as long as $G(s)$ satisfies

$$\begin{bmatrix} G(j\omega) \\ I \end{bmatrix}^* \begin{bmatrix} X & Y \\ Y^T & -X \end{bmatrix} \begin{bmatrix} G(j\omega) \\ I \end{bmatrix} < 0 \quad \forall \omega,$$

the closed-loop system is stable.

Example 14.4 *Static nonlinearity in a sector $[\alpha, \beta]$ (Figure 14.3):*
This static nonlinearity locates in a sector bounded by two straight lines whose slopes are α and β. Here, $w = \Delta z = \phi(z)$ satisfies

$$(\phi(z) - \alpha z)(\beta z - \phi(z)) \geq 0$$

for arbitrary z. The inequality can be rewritten as

$$\begin{aligned} 0 &\leq \begin{bmatrix} z & \phi(z) \end{bmatrix} \begin{bmatrix} -\alpha \\ 1 \end{bmatrix} \begin{bmatrix} \beta & -1 \end{bmatrix} \begin{bmatrix} z \\ \phi(z) \end{bmatrix} \\ &= \begin{bmatrix} z & \phi(z) \end{bmatrix} \begin{bmatrix} \beta \\ -1 \end{bmatrix} \begin{bmatrix} -\alpha & 1 \end{bmatrix} \begin{bmatrix} z \\ \phi(z) \end{bmatrix}. \end{aligned}$$

Finally, in order to express it as an IQC about an Hermitian matrix Π, we add up both sides in the previous equation. Then, we get the following matrix Π that specifies the IQC:

$$\Pi = \begin{bmatrix} -2\alpha\beta & \alpha + \beta \\ \alpha + \beta & -2 \end{bmatrix}.$$

Expansion of condition 3 of IQC theorem shows that the inequality

$$\begin{aligned} 0 &> \begin{bmatrix} G(j\omega) \\ I \end{bmatrix}^* \begin{bmatrix} -2\alpha\beta & \alpha + \beta \\ \alpha + \beta & -2 \end{bmatrix} \begin{bmatrix} G(j\omega) \\ I \end{bmatrix} \\ &= -2\alpha\beta G^* G + (\alpha + \beta)(G^* + G) - 2 \end{aligned}$$

holds for any frequency ω. When $\alpha\beta \neq 0$, this means that

$$0 > -\alpha\beta \left| G - \frac{\alpha + \beta}{2\alpha\beta} \right|^2 + \alpha\beta \left| \frac{\alpha - \beta}{2\alpha\beta} \right|^2.$$

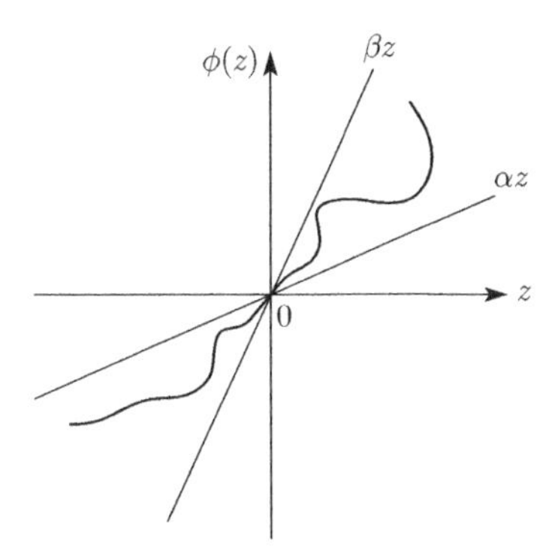

Figure 14.3 Static nonlinearity in a sector

If $\beta > \alpha > 0$, then

$$\left| G(j\omega) - \frac{\alpha+\beta}{2\alpha\beta} \right| < \left| \frac{\alpha-\beta}{2\alpha\beta} \right| \quad \forall\omega.$$

On the other hand, if $\alpha < 0 < \beta$, then

$$\left| G(j\omega) - \frac{\alpha+\beta}{2\alpha\beta} \right| > \left| \frac{\alpha-\beta}{2\alpha\beta} \right| \quad \forall\omega.$$

Finally, in the case of $\beta > \alpha = 0$, the inequality becomes

$$0 > \beta(G^* + G) - 2 \Rightarrow \Re[G(j\omega)] < \frac{\beta}{2} \ \forall\omega.$$

These conditions are exactly the same as the circle criterion.

Example 14.5 *Popov-type IQC: An essential feature of skew-symmetric matrix is that its quadratic form is zero. Noting this, we can always add an Hermitian matrix composed of a skew-symmetric matrix to the Hermitian matrix Π that specifies the IQC:*

$$\Pi_{\text{asym}} = \begin{bmatrix} 0 & H^*(j\omega) \\ H(j\omega) & 0 \end{bmatrix}, H(j\omega) + H^*(j\omega) = 0.$$

For example, for the conic nonlinearity discussed in Example 14.4, the Popov criterion may be derived as follows by setting $\alpha = 0$ in the Hermitian matrix Π and adding $\eta\beta\Pi_{\text{asym}}$ ($\eta > 0$) with $H(j\omega) = j\omega$.

Expansion of condition 3 of IQC theorem yields

$$\begin{aligned} 0 &> \beta G^* + \beta G - 2 + \eta\beta(HG + G^*H^*) \\ &= \beta(G + G^*) - 2 + \eta\beta[j\omega G + (j\omega G)^*] \\ &= 2\beta\,\Re[(1 + j\eta\omega)G] - 2. \end{aligned}$$

From it, the following Popov criterion is obtained:

$$\Re[(1 + j\eta\omega)G] < \frac{1}{\beta}.$$

Example 14.6 *Quadratic stability for polytopic uncertainty: Assume that the uncertainty set $\mathcal{D}$ is a polytope whose vertices are $\Delta_1, \ldots, \Delta_N \in \mathbb{R}^{m\times l}$ and the origin. We try to derive the stability conditions for the time-varying system:*

$$\dot{x} = (A + B\Delta(t)C)x, \quad \Delta(t) \in \mathcal{D}. \tag{14.5}$$

According to Chapter 13, the result obtained from a common Lyapunov function $V(x) = x^T Px$ is that the system is quadratically stable if there is a positive definite matrix P satisfying the following condition:

$$P(A + B\Delta_i C) + (A + B\Delta_i C)^T P < 0 \quad \forall i = 1, \ldots, N \tag{14.6}$$

at all vertices Δ_i. In this condition, the number of variables is $n(n+1)/2$ and the dimension of LMI is equal to Nn.

Meanwhile, different sufficient conditions may be derived via application of IQC theory. The basic idea is containing the matrix polytope with an ellipsoid first and then deriving the robust stability condition for all uncertainties inside the ellipsoid. Note that an open ellipsoid of matrices can be described by the following inequality:

$$Q + S\Delta + \Delta^T S^T + \Delta^T R\Delta > 0 \quad \forall \Delta \in \mathcal{D}. \tag{14.7}$$

To ensure that it corresponds to an ellipsoid, there must be $R < 0$. Then, the foregoing inequality is equivalently transformed to

$$\begin{bmatrix} Q + S\Delta + \Delta^T S^T & \Delta^T \\ \Delta & -R^{-1} \end{bmatrix} > 0. \tag{14.8}$$

Since this inequality is affine in Δ, it is satisfied for all $\Delta \in \mathcal{D}$ iff inequality

$$Q + S\Delta_i + \Delta_i^T S^T + \Delta_i^T R\Delta_i > 0 \ \forall i = 1, \ldots, N \tag{14.9}$$

is true at all vertices of $\mathcal{D}$. Obviously, uncertainty Δ that satisfies the matrix inequality (14.7) also satisfies an IQC specified by

$$\Pi = \begin{bmatrix} Q & S \\ S^T & R \end{bmatrix}.$$

Further, the current system is equivalent to the closed-loop system consisting of positively feedback connected linear system $G(s) = (A, B, C, 0)$ and uncertainty Δ. It is clear that condition 1 of IQC theorem is satisfied due to $G(j\infty) = 0$. Secondly, condition 2 can be proved by multiplying (14.8) with $\tau \in [0, 1]$. Therefore, by Theorem 14.1 we know that the condition

$$\begin{aligned} &\begin{bmatrix} G(j\omega) \\ I \end{bmatrix}^* \Pi \begin{bmatrix} G(j\omega) \\ I \end{bmatrix} \\ &= \begin{bmatrix} (j\omega I - A)^{-1}B \\ I \end{bmatrix}^* \begin{bmatrix} C^T & 0 \\ 0 & I \end{bmatrix} \Pi \begin{bmatrix} C & 0 \\ 0 & I \end{bmatrix} \begin{bmatrix} (j\omega I - A)^{-1}B \\ I \end{bmatrix} \\ &< 0 \ \forall \omega \end{aligned} \tag{14.10}$$

guarantees the robust stability. Finally, applying KYP lemma, we obtain a new robust stability condition:

$$\begin{bmatrix} C^TQC & C^TS \\ S^TC & R \end{bmatrix} + \begin{bmatrix} A^TP + PA & PB \\ B^TP & 0 \end{bmatrix} < 0. \tag{14.11}$$

In this condition, the variables are matrices (P, Q, S, R) *which have* $n(n+1)/2 + (n+m)(n+m+1)/2$ *elements. In this sense, it is unfavorable compared with the condition (14.6). However, the dimension of this LMI is* $n + m + Nl$*, so it is more beneficial than (14.6) when the number* N *of vertices is very large.*

By the way, it is proved in Ref. [67] that these two quadratic stability conditions are equivalent in theory.

14.4 Proof of IQC Theorem*

We prove the IQC theorem in this section. Here, signal norm used is only 2-norm and system norm is the norm $\|G\| = \sup_u \|y\|_2/\|u\|_2$ (u and y are, respectively, the input and output of system G) induced by signal 2-norm. For convenience, we omit the subscript of norm. In addition, the set of n-dimensional signals having bounded 2-norm is denoted by $\mathbb{L}_2^n$.

By assumptions, $G(s)$ and Δ are both bounded. So we need only prove that $(I - G\Delta)^{-1}$ is bounded. The proof is divided into three steps:

Step 1: Prove that for any $z \in \mathbb{L}_2^l$ and $\tau \in [0, 1]$, there is a finite constant $c_0 > 0$ satisfying

$$\|z\| \le c_0\|z - \tau G\Delta(z)\|.$$

Here, for simplicity we use the notation

$$\sigma(z, w) = \int_{-\infty}^{\infty} \begin{bmatrix} z(j\omega) \\ w(j\omega) \end{bmatrix}^* \Pi(j\omega) \begin{bmatrix} z(j\omega) \\ w(j\omega) \end{bmatrix} d\omega$$

and decompose the Hermitian matrix $\Pi(j\omega)$ into

$$\Pi(j\omega) = \begin{bmatrix} \Pi_{11}(j\omega) & \Pi_{12}(j\omega) \\ \Pi_{12}^*(j\omega) & \Pi_{22}(j\omega) \end{bmatrix}.$$

The norm $\|\Pi_{ij}\|$ of each block is denoted as m_{ij}.

The following equation can be confirmed via direct expansion:

$$\begin{bmatrix} z+\delta \\ w \end{bmatrix}^* \Pi \begin{bmatrix} z+\delta \\ w \end{bmatrix} - \begin{bmatrix} z \\ w \end{bmatrix}^* \Pi \begin{bmatrix} z \\ w \end{bmatrix} = \delta^*\Pi_{11}\delta + 2\Re(z^*\Pi_{11}\delta + w^*\Pi_{12}\delta).$$

So, for all quadratically integrable signals z, δ, w, inequality

$$\begin{aligned} |\sigma(z+\delta, w) - \sigma(z, w)| &= \left| \int_{-\infty}^{\infty} \{\delta^*\Pi_{11}\delta + 2\Re(z^*\Pi_{11}\delta + w^*\Pi_{12}\delta)\} d\omega \right| \\ &\le \int_{-\infty}^{\infty} \{|\delta^*\Pi_{11}\delta| + 2|\Re(z^*\Pi_{11}\delta)| + 2|\Re(w^*\Pi_{12}\delta)|\} d\omega \\ &\le m_{11}\|\delta\|^2 + 2\|\delta\|m_{11}\|z\| + 2\|\delta\|m_{12}\|w\| \end{aligned}$$

holds. Further, applying the inequality $2ab = 2(\frac{a}{\sqrt{\epsilon}})(\sqrt{\epsilon}b) \leq a^2/\epsilon + \epsilon b^2$ to product terms $2m_{11}\|\delta\| \cdot \|z\|$ and $2m_{12}\|\delta\| \cdot \|w\|$, we get

$$|\sigma(z+\delta, w) - \sigma(z, w)| \leq c(\epsilon)\|\delta\|^2 + \epsilon(\|z\|^2 + \|w\|^2)$$

in which $c(\epsilon) = m_{11} + m_{11}^2/\epsilon + m_{12}^2/\epsilon$. Then, multiplying both sides of (14.2) by $w^*(j\omega), w(j\omega)$ and integrating, we have

$$\sigma(Gw, w) \leq -\epsilon\|w\|^2 \ \forall w \in \mathbb{L}_2^m.$$

Let $\tau \in [0,1]$ and $w = \tau\Delta(z)$, $z \in \mathbb{L}_2^l$, $\epsilon_1 = \frac{\epsilon}{2+4\|G\|^2}$. As $\tau\Delta$ satisfies the IQC about Π and $\sigma(z,w) - \sigma(Gw,w) \leq |\sigma(z,w) - \sigma(Gw,w)|$, the following inequality holds:

$$\begin{aligned} 0 \leq \sigma(z,w) &= \sigma(Gw,w) + \sigma(z,w) - \sigma(Gw,w) \\ &\leq -\epsilon\|w\|^2 + c(\epsilon_1)\|z - Gw\|^2 + \epsilon_1(\|z\|^2 + \|w\|^2). \end{aligned}$$

Moreover, since $\|z\|^2 = \|Gw + (z - Gw)\|^2 \leq 2\|z - Gw\|^2 + 2\|Gw\|^2$ (parallelogram law) and $\|Gw\| \leq \|G\|\|w\|$, we have

$$\begin{aligned} 0 &\leq -\epsilon\|w\|^2 + (c(\epsilon_1) + 2\epsilon_1)\|z - Gw\|^2 + \epsilon_1(2\|Gw\|^2 + \|w\|^2) \\ &\leq -\epsilon\|w\|^2 + (c(\epsilon_1) + 2\epsilon_1)\|z - Gw\|^2 + \epsilon_1(2\|G\|^2\|w\|^2 + \|w\|^2) \\ &\leq -\frac{\epsilon}{2}\|w\|^2 + (c(\epsilon_1) + 2\epsilon_1)\|z - Gw\|^2. \end{aligned}$$

So $\|w\| \leq c\|z - Gw\|$. Thus, there holds

$$\|z\| \leq \|Gw\| + \|z - Gw\| \leq \|G\|\|w\| + \|z - Gw\| \leq c_0\|z - Gw\|.$$

Step 2: Prove that if $(I - \tau G\Delta)^{-1}$ is bounded for a $\tau \in [0,1]$, then $(I - \nu G\Delta)^{-1}$ remains bounded for all $\nu \in [0,1]$ satisfying $|\nu - \tau| < 1/(c_0\|G\|\|\Delta\|)$.

$(I - \tau G\Delta)^{-1}$ is causal due to condition 2. Define a truncated signal as

$$z_T = (I - \tau G\Delta)^{-1} P_T(z - \tau G\Delta(z)).$$

From causality we have $P_T z = P_T z_T$. Applying the result of Step 1 to z_T, we get

$$\begin{aligned} \|P_T z\| &= \|P_T z_T\| \leq \|z_T\| \\ &\leq c_0\|z_T - \tau G\Delta(z_T)\| = c_0\|P_T(z - \tau G\Delta(z))\| \\ &= c_0\|P_T(z - \nu G\Delta(z)) + (\nu - \tau)P_T G\Delta(z)\| \\ &\leq c_0\|P_T(z - \nu G\Delta(z))\| + c_0\|G\Delta\||\nu - \tau| \cdot \|P_T z\|, \end{aligned}$$

in which the triangle inequality and the submultiplicativity of norm have been used. Since $|\nu - \tau| < 1/(c_0\|G\|\|\Delta\|) \leq 1/(c_0\|G\Delta\|)$, we have

$$\begin{aligned} &\|P_T(z - \nu G\Delta(z))\| \geq \frac{(1 - c_0\|G\Delta\||\nu - \tau|)}{c_0}\|P_T z\| \\ \Rightarrow\ &\|z - \nu G\Delta(z)\| \geq \frac{(1 - c_0\|G\Delta\||\nu - \tau|)}{c_0}\|z\|. \end{aligned}$$

Let a signal be $u = (I - \nu G\Delta)z$. Then $z = (I - \nu G\Delta)^{-1}u$. This implies that u, z are the input and output of $(I - \nu G\Delta)^{-1}$. As $1 - c_0\|G\Delta\||\nu - \tau| > 0$, the aforementioned inequality implies that z must be bounded as long as u is bounded. Namely, the system $(I - \nu G\Delta)^{-1}$ is bounded.

Step 3: Cover the interval $[0, 1]$ by a finite number of intervals.

Let the length of each interval be less than$1/(c_0\|G\|\|\Delta\|)$. When $\tau = 0$, $(I - \tau G\Delta)^{-1} = I$ is obviously bounded. Thus, $(I - \tau G\Delta)^{-1}$ is still bounded in the first closed interval. Since the left vertex of the second closed interval is contained in the first one, the boundedness is true at this point. Next, repeat the same argument for every interval behind the second. Eventually, we get that $(I - \tau G\Delta)^{-1}$ is bounded for all $\tau \in [0, 1]$. So the boundedness of $(I - G\Delta)^{-1}$ at the point $\tau = 1$ is proved.

Notes and References

The materials of this chapter are taken from Ref. [67].

15

$\mathcal{H}_2$ Control

In control systems, the transient response is an extremely important specification. The $\mathcal{H}_2$ norm of a signal is its squared area. So the response of a signal is faster when its $\mathcal{H}_2$ norm is smaller. Namely, $\mathcal{H}_2$ norm is a very suitable indicator for judging the pros and cons of transient responses. Further, a disturbance or reference input can be regarded as the impulse response of their dynamics. So if this dynamics is connected to the closed-loop transfer function as a weighting function, the response of disturbance or reference input can be regarded as the impulse response of the weighted closed-loop transfer function. In the following section we will prove that the 2-norm of the impulse response of a transfer function is equal to the $\mathcal{H}_2$ norm of the transfer function itself. Therefore, in order to improve the quality of response to disturbance and reference input, we should make the $\mathcal{H}_2$ norm of the closed-loop transfer function as small as possible in design. This chapter discusses how to carry out such control design systematically. A case study is provided to illustrate the $\mathcal{H}_2$ control design and the importance of actuator/sensor location.

15.1 $\mathcal{H}_2$ Norm of Transfer Function

Firstly, we consider how to calculate the $\mathcal{H}_2$ norm of a stable transfer matrix, as well as the relation between the $\mathcal{H}_2$ norm of a transfer matrix and its input–output. Assume that the transfer matrix is given by

$$G(s) = (A,\ B,\ C,\ D). \tag{15.1}$$

Its $\mathcal{H}_2$ norm is defined as

$$\|G\|_2 := \sqrt{\frac{1}{2\pi}\int_{-\infty}^{\infty} \mathrm{Tr}[G^*(j\omega)G(j\omega)]d\omega}. \tag{15.2}$$

Denote the inverse Laplace transform of $G(s)$ (viz., its impulse response) by $g(t)$[1]. Since the impulse input is nonzero only at $t = 0$, the impulse response $g(t)$ of a stable transfer function

[1] Note that in general this is not a scalar, but a matrix.

satisfies $g(t) = 0$ $(\forall t < 0)$. Thus, according to Parseval's theorem, the $\mathcal{H}_2$ norm of transfer matrix $G(s)$ can be calculated by

$$\|G\|_2 = \|g\|_2 = \sqrt{\int_0^\infty \mathrm{Tr}[g^T(t)g(t)]\ dt}. \tag{15.3}$$

15.1.1 Relation with Input and Output

Consider the scalar case first. In this case, the output signal to the unit impulse input is $y(t) = g(t)$. From (15.3), we know that the $\mathcal{H}_2$ norm of the transfer function is equal to the 2-norm of its impulse response.

Now, what about the multiple-input multiple-output case? To analyze it in detail, we consider a transfer matrix with m inputs and assume that $\{u_i\}$ $(i = 1, \ldots, m)$ is an orthonormal set of vectors in the real space $\mathbb{R}^m$, namely,

$$U^T U = UU^T = I, \quad U = [u_1, \ldots, u_m].$$

Obviously, the response to the unit impulse input $w_i(t) = u_i\delta(t)$ is $y_i(t) = g(t)u_i$. Then, there holds

$$\begin{aligned}
\sum_{i=1}^m \|y_i\|_2^2 &= \sum_{i=1}^m \int_0^\infty y_i^T(t)y_i(t)dt = \sum_{i=1}^m \int_0^\infty u_i^T g^T(t)g(t)u_i dt \\
&= \int_0^\infty \sum_{i=1}^m \mathrm{Tr}\ (g^T(t)g(t)u_i u_i^T)dt = \int_0^\infty \mathrm{Tr}\ (g^T(t)g(t)UU^T)dt \\
&= \int_0^\infty \mathrm{Tr}\ (g^T(t)g(t))dt \\
&= \|G\|_2^2.
\end{aligned}$$

The properties of matrix trace, $\mathrm{Tr}\ (AB) = \mathrm{Tr}\ (BA)$ and $\sum \mathrm{Tr}\ (A_i) = \mathrm{Tr}\ (\sum A_i)$, have been used. This equation shows that the square of $\mathcal{H}_2$ norm of a transfer matrix is equal to the sum of the squared areas of all the responses corresponding to input in the orthonormal set of impulse vectors. Secondly, it is clear that

$$\|G\|_2^2 = \int_0^\infty \mathrm{Tr}\ (g(t)g^T(t))dt$$

also holds.

As another interpretation, we may prove that the square of $\mathcal{H}_2$ norm of a transfer matrix is equal to the variance of its steady-state response to a unit white noise vector[2]. That is, when the expectation and covariance of an input $u(t)$ are

$$\mathrm{E}[u(t)] = 0 \quad \forall t \tag{15.4}$$

[2] The signal $u(t)$ satisfying (15.4) and (15.5) is called a unit white noise vector.

and

$$\mathrm{E}[u(t)u^T(\tau)] = \delta(t-\tau)I, \tag{15.5}$$

respectively, the expectation $\mathrm{E}[y(t)]$ of its output y is also zero. The variance of y at t is[3]

$$\begin{aligned}
\mathrm{E}[y^T(t)y(t)] &= \mathrm{E}\left[\int_0^t\int_0^t (g(t-\alpha)u^T(\alpha))g(t-\beta)u(\beta)d\alpha d\beta\right] \\
&= \mathrm{E}\left[\int_0^t\int_0^t \mathrm{Tr}\left[g^T(t-\alpha)g(t-\beta)u(\beta)u^T(\alpha)\right) d\alpha d\beta\right] \\
&= \int_0^t\int_0^t \mathrm{Tr}\ (g^T(t-\alpha)g(t-\beta)\mathrm{E}(u(\beta)u^T(\alpha)))d\alpha d\beta \\
&= \int_0^t\int_0^t \mathrm{Tr}\ (g^T(t-\alpha)g(t-\beta)\delta(\beta-\alpha)I)d\alpha d\beta \\
&= \int_0^t \mathrm{Tr}\ (g^T(t-\beta)g(t-\beta))d\beta \\
&= \int_0^t \mathrm{Tr}\ (g^T(\tau)g(\tau))d\tau\ \ (\tau = t-\beta).
\end{aligned}$$

So, there holds the following equation:

$$\lim_{t\to\infty} \mathrm{E}[y^T(t)y(t)] = \|G\|_2^2. \tag{15.6}$$

This shows that the variance of the steady-state output is equal to the square of the $\mathcal{H}_2$ norm of the system.

For this reason, in filter design where the purpose is to remove noise from the measured data, the $\mathcal{H}_2$ method is a very effective tool. For related details, refer to Ref. [2].

15.1.2 Relation between Weighting Function and Dynamics of Disturbance/Noise

In general, a disturbance is not an impulse and a noise is not white. They all have certain dynamics, that is, possessing a certain frequency characteristic. Now assume that the frequency characteristic of disturbance d is $W(s)$. From Figure 15.1 we know that the output $y(t)$ is the impulse response of weighted transfer function $G(s)W(s)$. To attenuate the disturbance response, we should minimize

$$\|y\|_2 = \|GW\|_2. \tag{15.7}$$

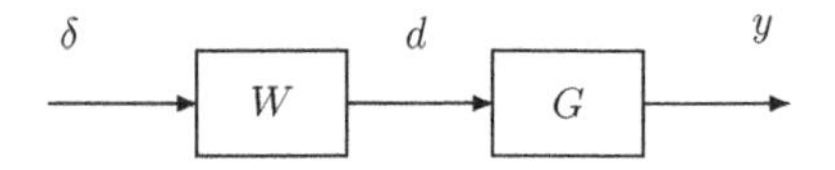

Figure 15.1 Response of disturbance with a frequency characteristic $W(s)$

[3] In the deduction, the property of impulse function, $\int_0^t f(\tau)\delta(\tau-\alpha)d\tau = f(\alpha)\ (\alpha\in[0,t])$, is used.

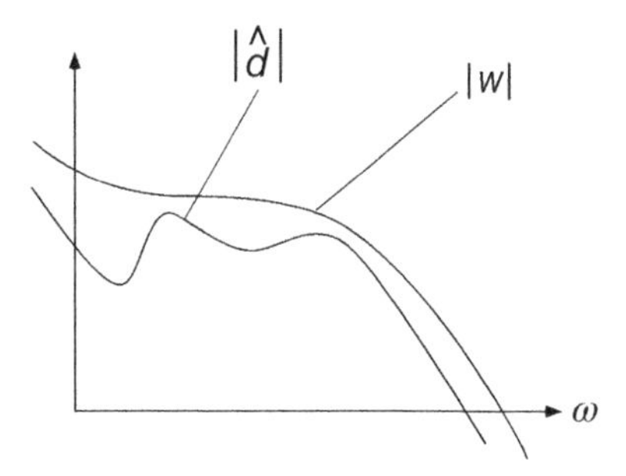

Figure 15.2 Frequency characteristics of weighting function and disturbance

Such a problem is called a control problem with weighting function, and $W(s)$ is called a *weighting function*. Further, even if the frequency characteristic of a disturbance is not known accurately, its upper bound can be used as the weighting function. In fact, when the amplitude of frequency response of a disturbance $d(t)$ is less than $|W(j\omega)|$ (refer to Figure 15.2), that is,

$$|\hat{d}(j\omega)| \leq |W(j\omega)| \ \forall \ \omega,$$

then since

$$\|G\hat{d}\|_2 \leq \|GW\|_2,$$

the disturbance response is suppressed if $\|GW\|_2$ is minimized.

In dealing with practical problems, disturbance of the system must have certain dynamics (frequency response). In the design of feedback systems, a better result can be obtained when this disturbance characteristic is taken into account. So, control designs using weighting function are better consistent with the engineering practice. In a word, the disturbance model must be used as a weighting function in practical designs.

15.1.3 Computing Methods

In this section, we consider how to calculate the $\mathcal{H}_2$ norm of a transfer matrix. Two more methods are introduced besides (15.3), the impulse response-based method.

First, for the existence of $\mathcal{H}_2$ norm of transfer matrix $G(s) = (A, B, C, D)$, the DC term D must be zero. If not, the impulse response of $G(s)$ contains an impulse term. But the squared area of an impulse signal is not bounded. So the squared area of the output of $G(s)$ also diverges. Mathematically, this can be understood as follows. In a sufficiently high-frequency band, since $(j\omega - A)^{-1} \approx 0$,

$$G(j\omega) \approx D, \quad \omega_N \leq \omega < \infty$$

holds. However, integrating $\mathrm{Tr}(D^T D) \neq 0$ in the band $[\omega_N, \infty)$, the integral diverges. So, we need only consider strictly proper, stable transfer matrix.

As $G(s)$ is strictly proper, $\lim_{R\to\infty} R \cdot G^{\sim}(Re^{j\theta})G(Re^{j\theta}) = 0$ holds. Therefore, the integral $\lim_{R\to\infty} \int_{\pi/2}^{3\pi/2} \mathrm{Tr}(G^{\sim}(Re^{j\theta})G(Re^{j\theta}))d(Re^{j\theta})$ along the semicircle with infinite radius

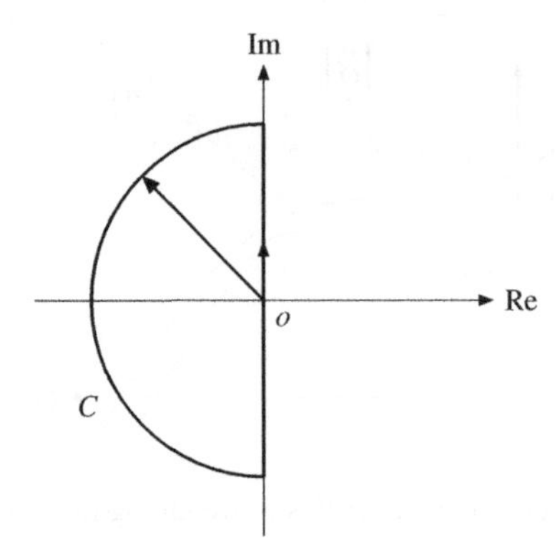

Figure 15.3 Integration path

is also zero. So we have

$$\begin{aligned}\|G\|_2^2 &= \frac{1}{2\pi}\int_{-\infty}^{\infty} \mathrm{Tr}(G^*(j\omega)G(j\omega))d\omega \\ &= \frac{1}{2\pi j}\oint_C \mathrm{Tr}(G^\sim(s)G(s))\,ds.\end{aligned}$$

As shown in Figure 15.3, the last line integral is along the imaginary axis and the semicircle on the left half-plane whose radius is infinite. According to the residue theorem, $\|G\|_2^2$ is equal to the sum of residues of $\mathrm{Tr}(G^\sim(s)G(s))$ at all of its poles in the left half-plane. Hence, in principle the calculation of $\mathcal{H}_2$ norm reduces to that of residue. This method is suitable for manual calculation, but not good for computer calculation.

Example 15.1 *Let the input of stable transfer function $P(s) = 1/(s+10)$ be $u(t)$ and the output be $y(t)$.*

1. *Calculate the norm $\|P\|_2$ of the transfer function.*
2. *When the input is $u(t) = 0.1e^{-t}$, compute the norm $\|y\|_2$ of output.*

Solution (1) First, the rational function $P(-s)P(s) = 1/(10-s)(10+s)$ has a pole $p = -10$ in the closed left half-plane. The residue at this pole is

$$\lim_{s\to -10}(s+10)P(-s)P(s) = \frac{1}{20}.$$

So $\|P\|_2 = \sqrt{5}/10$.

(2) As $u(t)$ is the unit impulse response of transfer function $W(s) = 0.1/(s+1)$, $y(t)$ can be viewed as the unit impulse response of $G = WP = 0.1/(s+1)(s+10)$. Thus, we have $\|y\|_2 = \|WP\|_2$. $G(-s)G(s) = 0.01/(1-s)(10-s)(1+s)(10+s)$ has two poles

$p = -1, -10$ in the closed left half-plane. Its residues at these two poles are

$$\lim_{s\to -1}(s+1)G(-s)G(s) = \frac{10^{-2}}{2\times 11\times 9}$$

$$\lim_{s\to -10}(s+10)G(-s)G(s) = -\frac{10^{-3}}{2\times 11\times 9}.$$

So $\|y\|_2 = \sqrt{5/11}\times 10^{-2}$ (compare with the impulse response-based calculation in Example 2.20). ▽

To find an algorithm suitable for computer, we consider how to make full use of the state-space description of $G(s)$. The following lemma gives such a computation method.

Lemma 15.1 *Assume that A is stable in the transfer matrix $G(s) = (A, B, C, 0)$. Then*

$$\|G\|_2^2 = \mathrm{Tr}(B^T L_o B) = \mathrm{Tr}(C L_c C^T) \tag{15.8}$$

holds. Here, L_c and L_o are, respectively, the controllability Gramian and the observability Gramian which satisfy the following Lyapunov equations:

$$AL_c + L_c A^T + BB^T = 0, \qquad A^T L_o + L_o A + C^T C = 0. \tag{15.9}$$

Proof. The inverse Laplace transform of $G(s)$ is

$$g(t) = \mathcal{L}^{-1}(G) = \begin{cases} Ce^{At}B, & t \geq 0 \\ 0, & t < 0. \end{cases}$$

Based on it, we have

$$\|G\|_2^2 = \int_0^\infty \mathrm{Tr}\ (g(t)g^T(t))\ dt = \int_0^\infty \mathrm{Tr}\ (Ce^{At}BB^Te^{A^Tt}C^T)\ dt.$$

Set

$$L_c = \int_0^\infty e^{At}BB^Te^{A^Tt}\ dt.$$

From the dual of Theorem 4.7, we see that this matrix is the controllability Gramian of (A, B) and satisfies $AL_c + L_cA^T + BB^T = 0$. As such, we have proved the first statement.

Finally, the second statement can be proved based on $\|G\|_2^2 = \int_0^\infty \mathrm{Tr}(g^Tg)dt$ (refer to Exercise 15.3). •

15.1.4 *Condition for $\|G\|_2 < \gamma$*

The next lemma follows immediately from the calculation method of $\mathcal{H}_2$ norm. This lemma can be used to solve the singular $\mathcal{H}_2$ control problem.

Lemma 15.2 *The following statements are equivalent:*

1. *A is stable and $\|C(sI - A)^{-1}B\|_2 < \gamma$.*
2. *There is a matrix $X = X^T > 0$ satisfying*

$$XA + A^T X + C^T C < 0 \tag{15.10}$$

$$\mathrm{Tr}(B^T XB) < \gamma^2. \tag{15.11}$$

3. *There exist matrices $X = X^T$ and $W = W^T$ satisfying*

$$\begin{bmatrix} XA + A^T X & C^T \\ C & -I \end{bmatrix} < 0 \tag{15.12}$$

$$\begin{bmatrix} W & B^T X \\ XB & X \end{bmatrix} > 0 \tag{15.13}$$

$$\mathrm{Tr}(W) < \gamma^2. \tag{15.14}$$

Proof. (1) ⇒ (2) If (1) is true, there must be a sufficiently small $\epsilon > 0$ such that

$$\left\| \begin{bmatrix} C \\ \sqrt{\epsilon} I \end{bmatrix} (sI - A)^{-1} B \right\|_2 < \gamma$$

still holds even if C is augmented to $\left[\begin{smallmatrix} C \\ \sqrt{\epsilon} I \end{smallmatrix}\right]$. Applying Lemma 15.1 to this new system, we have that there is an $X = X^T \geq 0$ meeting

$$XA + A^T X + C^T C = -\epsilon I < 0 \tag{15.15}$$

and

$$\mathrm{Tr}(B^T XB) < \gamma^2.$$

If matrix X is singular, there must be a nonzero vector v satisfying $Xv = 0$. Multiplying both sides of (15.15) by v^T and v, respectively, we get $v^T C^T Cv < 0$. This is a contradiction. Therefore, $X > 0$ must be true.

(2) ⇒ (1) According to Theorem 4.7, A is stable when (15.10) has a solution $X > 0$. Here, set $\hat{C}$ as the matrix defined as follows:

$$0 > XA + A^T X + C^T C = -\hat{C}^T \hat{C}.$$

From Lemma 15.1, we have

$$\left\| \begin{bmatrix} C \\ \hat{C} \end{bmatrix} (sI - A)^{-1} B \right\|_2 < \gamma.$$

So, $\|C(sI - A)^{-1}B\|_2 < \gamma$ holds.

(2) ⇔ (3) According to Schur's lemma, (15.10) and (15.12) are equivalent. Further, we can prove that (15.11) is equivalent to (15.14) and $W - B^T XB > 0$[4]. Finally, $W - B^T XB > 0$ and $X > 0$ are equivalent to (15.13). •

[4] When (15.14) and $W - B^T XB > 0$ hold, (15.11) is obviously true. On the contrary, when (15.11) is true, we have $\delta = \gamma^2 - \mathrm{Tr}(B^T XB) > 0$. Consider $\epsilon < \delta/n$ (n is the dimension of matrix X) and matrix

15.2 $\mathcal{H}_2$ Control Problem

This section describes the $\mathcal{H}_2$ control problem and related assumptions. Let us look at an example first.

Example 15.2 *2-DOF system*

The main purpose of the 2-degree-of-freedom system is to improve the performance of signal tracking. So we hope to be able to exploit some norm to evaluate the transient response properly. $\mathcal{H}_2$ norm is ideally suited for this purpose.

In addition, in order to limit the control input, we also want to take the input as a performance output. However, in reference tracking, the reference input is a persistent signal such as step signal. In order to maintain such an output, the input is generally a persistent signal in the steady state. This means that its $\mathcal{H}_2$ norm is not finite. However, the steady-state value of the input is necessary for achieving reference tracking and should not be restricted. What should be limited is its rate of change in the transient state. In order to remove the steady-state value of the input, we may use W_r^{-1}, the inverse of reference model, to filter the input. For example, for the step signal $r(t) = 1(t)$, its model is $W_r(s) = 1/s$. Since its inverse $W_r^{-1}(s) = s$ is a differentiator, when the input is filtered by $W_r^{-1}(s) = s$, the obtained signal is the first derivative $\dot{u}(t)$ of the input, that is, the rate of change. In this case, the steady-state value of the input is a constant and its derivative is zero. Thus, we achieve the purpose of eliminating steady-state input.

Such treatment is equivalent to Figure 15.4. This figure is the block diagram where W_r is moved behind the tracking error, and W_u is a stable weighting function used to penalize the transient of input. The input–output relationship of the system can be written as

$$\begin{bmatrix} z_1 \\ z_2 \\ \hline w \\ y \end{bmatrix} = \left[\begin{array}{c|c} W_r & -W_r P \\ 0 & W_u \\ \hline I & 0 \\ 0 & P \end{array}\right] \begin{bmatrix} w \\ u \end{bmatrix} = G \begin{bmatrix} w \\ u \end{bmatrix}$$

$$u = K \begin{bmatrix} w \\ y \end{bmatrix}.$$

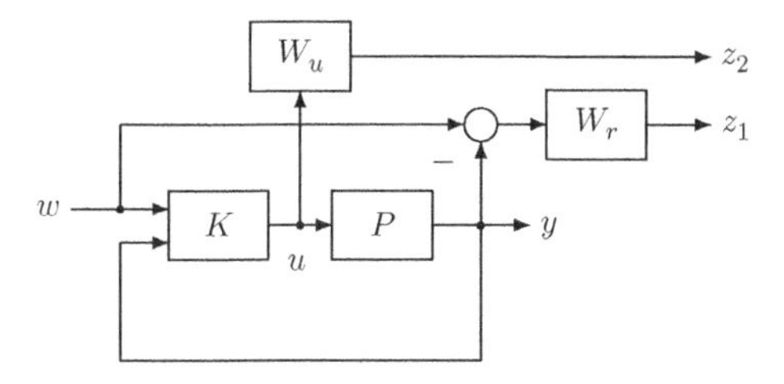

Figure 15.4 $\mathcal{H}_2$ tracking control problem of 2-DOF system

$W = \epsilon I + B^T XB$. First, $W - B^T XB = \epsilon I > 0$. Second, $\mathrm{Tr}(W) = n\epsilon + \mathrm{Tr}(B^T XB) = n\epsilon + \gamma^2 - \delta = \gamma^2 - n(\delta/n - \epsilon) < \gamma^2$ holds. This indicates that they are equivalent.

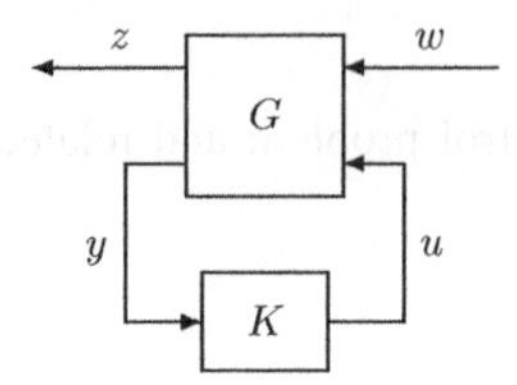

Figure 15.5 Generalized feedback system

Note also that the external input w has been converted to a unit impulse. Hence, by the relationship between the $\mathcal{H}_2$ norm of a transfer matrix and its input–output, we need only minimize the $\mathcal{H}_2$ norm of the closed-loop transfer function from w to $\begin{bmatrix} z_1 \\ z_2 \end{bmatrix}$ in order to achieve the goals. This problem is called an $\mathcal{H}_2$ control problem.

In the following sections, we consider the system shown in Figure 15.5.

Definition 15.1 *The so-called $\mathcal{H}_2$ optimal control problem is to seek a proper controller $K(s)$ such that the closed-loop system is stable and the $\mathcal{H}_2$ norm of the transfer matrix $H_{zw}(s)$ from w to z is minimized. A controller meeting this specification is called an $\mathcal{H}_2$ optimal controller. In addition, for any given $\gamma > 0$, designing a controller satisfying $\|H_{zw}\|_2 < \gamma$ is called γ-optimal $\mathcal{H}_2$ control problem. Such controller is called a γ-optimal $\mathcal{H}_2$ controller.*

Assume that a state-space realization of the generalized plant G is

$$G(s) = \left[\begin{array}{c|cc} A & B_1 & B_2 \\ \hline C_1 & 0 & D_{12} \\ C_2 & D_{21} & 0 \end{array}\right]. \tag{15.16}$$

Note that both D_{11} and D_{22} are assumed to be zero. For the $D_{11} \neq 0$ case, refer to Exercise 15.7 or Ref. [100].

Further, we make the following assumptions:

(A1) (A, B_2) is stabilizable and (C_2, A) is detectable.
(A2) D_{12} has full column rank and $D_{12}^T D_{12} = I$; D_{21} has full row rank and $D_{21} D_{21}^T = I$.
(A3) For all ω, $\begin{bmatrix} A - j\omega I & B_2 \\ C_1 & D_{12} \end{bmatrix}$ has full column rank.
(A4) For all ω, $\begin{bmatrix} A - j\omega I & B_1 \\ C_2 & D_{21} \end{bmatrix}$ has full row rank.

Assumption (A1) is for the stabilization of $G(s)$ by output feedback. Assumptions (A2), (A3), and (A4) are called the nonsingularity condition and are assumed in order to solve the $\mathcal{H}_2$ control problem by using the Riccati equation. Note that the condition $D_{12}^T D_{12} = I$ in (A2) is assumed to simplify the description of solution; it is not a hard constraint[5].

[5] Refer to Exercise 15.6 for the technique on how to convert a general problem equivalently to a problem satisfying this condition.

A problem meeting these assumptions is called a *nonsingular problem*. Its solution is very simple. In contrast, a problem not satisfying the condition (A2)–(A4) is called a *singular problem*. For the γ-optimal problem $\|H_{zw}\|_2 < \gamma$ in singular problems, we will introduce a method based on linear matrix inequality (LMI) in Section 15.5. However, there is still no good method until now for the optimal problem $\min_K \|H_{zw}\|_2$ in the singular case.

15.3 Solution to Nonsingular $\mathcal{H}_2$ Control Problem

This section explains how to solve a $\mathcal{H}_2$ control problem based on the nonsingularity conditions (A1)–(A4). First, according to Corollary 9.1, the following two Hamiltonians

$$H := \begin{bmatrix} A & 0 \\ -C_1^T C_1 & -A^T \end{bmatrix} - \begin{bmatrix} B_2 \\ -C_1^T D_{12} \end{bmatrix} \begin{bmatrix} D_{12}^T C_1 & B_2^T \end{bmatrix} \tag{15.17}$$

$$J := \begin{bmatrix} A^T & 0 \\ -B_1 B_1^T & -A \end{bmatrix} - \begin{bmatrix} C_2^T \\ -B_1 D_{21}^T \end{bmatrix} \begin{bmatrix} D_{21} B_1^T & C_2 \end{bmatrix} \tag{15.18}$$

belong to $\mathrm{dom}(Ric)$. Further, $X := Ric(H) \geq 0$ and $Y := Ric(J) \geq 0$ hold. For simplicity of presentation, we define matrices

$$F_2 := -(B_2^T X + D_{12}^T C_1), \quad L_2 := -(Y C_2^T + B_1 D_{21}^T)$$

and

$$A_{F_2} := A + B_2 F_2, \quad C_{F_2} := C_1 + D_{12} F_2$$
$$A_{L_2} := A + L_2 C_2, \quad B_{L_2} := B_1 + L_2 D_{21}.$$

A_{F_2} and A_{L_2} are stable according to Theorem 9.2.

The solution of $\mathcal{H}_2$ control problem is given by the following theorem.

Theorem 15.1 *Assume (A1)–(A4). Then, the $\mathcal{H}_2$ optimal controller is unique and given by*

$$K_{\mathrm{opt}}(s) := \left[\begin{array}{c|c} A + B_2 F_2 + L_2 C_2 & -L_2 \\ \hline F_2 & 0 \end{array}\right]. \tag{15.19}$$

Moreover, the minimal $\mathcal{H}_2$ norm of the closed-loop transfer matrix is given by

$$\min \|H_{zw}\|_2^2 = \mathrm{Tr}(B_1^T X B_1) + \mathrm{Tr}(F_2 Y F_2^T). \tag{15.20}$$

In addition, the following theorem gives the solution to the γ-optimal control problem.

Theorem 15.2 *Assume (A1)–(A4). Then, for any $\gamma(> \min \|H_{zw}\|_2)$, all γ-optimal $\mathcal{H}_2$ controllers satisfying $\|H_{zw}\|_2 < \gamma$ are characterized by the transfer matrix from y to u in Figure 15.6. Here, the coefficient matrix is*

$$M(s) = \left[\begin{array}{c|cc} A + B_2 F_2 + L_2 C_2 & -L_2 & B_2 \\ \hline F_2 & 0 & I \\ -C_2 & I & 0 \end{array}\right]. \tag{15.21}$$

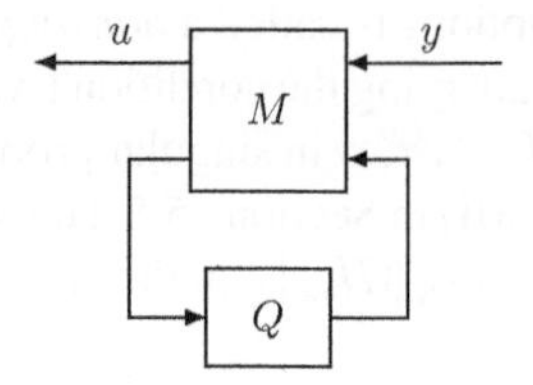

Figure 15.6 Parameterization of γ-optimal $\mathcal{H}_2$ controller

Q is any strictly proper, stable transfer matrix with a compatible dimension and satisfies $\|Q\|_2^2 < \gamma^2 - \min \|H_{zw}\|_2^2$.

γ-Optimal controller is given in the form of linear fractional transformation about the free parameter Q. When $Q = 0$, we get the optimal controller K_{opt}. The proofs of these two theorems are provided in the following section.

15.4 Proof of Nonsingular Solution

15.4.1 Preliminaries

As a preparation, we first state two lemmas about the properties of Riccati equation.

Lemma 15.3 *For $X = Ric(H)$ and $Y = Ric(J)$, the following equations hold:*

1. $A_{F_2}^T X + X A_{F_2} + C_{F_2}^T C_{F_2} = 0$, $X B_2 + C_{F_2}^T D_{12} = 0$.
2. $A_{L_2} Y + Y A_{L_2}^T + B_{L_2} B_{L_2}^T = 0$, $C_2 Y + D_{21} B_{L_2}^T = 0$.

Proof. Only part 1 is proved. Part 2 can be proved in the same manner (refer to Exercise 15.4).

First, the Riccati equation about X is

$$(A - B_2 D_{12}^T C_1)^T X + X(A - B_2 D_{12}^T C_1) - X B_2 B_2^T X + C_1^T (I - D_{12} D_{12}^T) C_1 = 0.$$

It can be arranged into

$$(A + B_2 F_2)^T X + X(A + B_2 F_2) + X B_2 B_2^T X + C_1^T (I - D_{12} D_{12}^T) C_1 = 0.$$

Substituting $B_2^T X = -(F_2 + D_{12}^T C_1)$ into this equation and completing square by using $D_{12}^T D_{12} = I$, we have

$$\begin{aligned} X B_2 B_2^T X + C_1^T (I - D_{12} D_{12}^T) C_1 &= F_2^T F_2 + F_2^T D_{12}^T C_1 + C_1^T D_{12} F_2 + C_1^T C_1 \\ &= (C_1 + D_{12} F_2)^T (C_1 + D_{12} F_2). \end{aligned}$$

So $A_{F_2}^T X + X A_{F_2} + C_{F_2}^T C_{F_2} = 0$ is true. Finally, it can be easily verified that $X B_2 + C_{F_2}^T D_{12} = 0$ via the substitution of C_{F_2} and F_2. •

Lemma 15.4 *Define the following stable transfer matrices:*

$$U := \left[\begin{array}{c|c} A_{F_2} & B_2 \\ \hline C_{F_2} & D_{12} \end{array}\right], \quad V := \left[\begin{array}{c|c} A_{L_2} & B_{L_2} \\ \hline C_2 & D_{21} \end{array}\right]$$

$$G_c(s) := \left[\begin{array}{c|c} A_{F_2} & I \\ \hline C_{F_2} & 0 \end{array}\right], \quad G_f(s) := \left[\begin{array}{c|c} A_{L_2} & B_{L_2} \\ \hline I & 0 \end{array}\right].$$

Then, the following statements hold:

1. *$U^{\sim}U = I$ and $U^{\sim}G_c$ is antistable.*
2. *$VV^{\sim} = I$ and $G_fV^{\sim}$ is antistable.*

Proof. (1) First of all, according to Lemma 15.3 U satisfies the condition of Theorem 9.5, so U is an inner. Further,

$$U^T(-s) = \left[\begin{array}{c|c} -A_{F_2}^T & -C_{F_2}^T \\ \hline B_2^T & D_{12}^T \end{array}\right].$$

Application of the formula of cascade connection yields

$$U^{\sim}(s)G_c(s) = \left[\begin{array}{cc|c} -A_{F_2}^T & -C_{F_2}^T C_{F_2} & 0 \\ 0 & A_{F_2} & I \\ \hline B_2^T & D_{12}^T C_{F_2} & 0 \end{array}\right].$$

Performing a similarity transformation on this transfer matrix with the transformation matrix $T = \left[\begin{smallmatrix} I & X \\ 0 & I \end{smallmatrix}\right]$ and then substituting the equations

$$A_{F_2}^T X + XA_{F_2} + C_{F_2}^T C_{F_2} = 0, \quad B_2^T X + D_{12}^T C_{F_2} = 0$$

into the transformed transfer matrix, we obtain

$$U^{\sim}(s)G_c(s) = \left[\begin{array}{cc|c} -A_{F_2}^T & 0 & -X \\ 0 & A_{F_2} & I \\ \hline B_2^T & 0 & 0 \end{array}\right] = \left[\begin{array}{c|c} -A_{F_2}^T & -X \\ \hline B_2^T & 0 \end{array}\right].$$

Here, since the eigenvalues of A_{F_2} are not observable, we have removed it from the transfer matrix. Obviously, the transfer matrix is antistable.

Finally, statement 2 can be proved based on duality (Exercise 15.5). •

Remark 15.1 *The transformation matrix used in the proof of Lemma 15.4 is in fact obtained from an elementary transformation of matrix. Specifically, to know which pole can be eliminated, it is better to transform the A matrix into a triangular or diagonal matrix. Therefore, we focus on the equation $A_{F_2}^T X + XA_{F_2} + C_{F_2}^T C_{F_2} = 0$ and consider how to eliminate the nondiagonal block $-C_{F_2}^T C_{F_2}$. To this end, it is necessary to postmultiply the first column by X and add it into the second column. The matrix description of this column transformation is exactly the matrix T. The corresponding row transformation has a matrix description T^{-1}, that is, premultiplying the second row with $-X$ and adding it to the first row. Through this similarity transformation, we succeed in eliminating the nondiagonal block.*

15.4.2 Proof of Theorems 15.1 and 15.2

Set a transfer matrix as

$$M(s) = \left[\begin{array}{c|cc} A + B_2F_2 + L_2C_2 & -L_2 & B_2 \\ \hline F_2 & 0 & I \\ -C_2 & I & 0 \end{array}\right].$$

Then any stabilizing controller can be expressed as $K(s) = \mathcal{F}_\ell(M, Q)$ with a stable $Q(s)$ (see Theorem 7.2). Thus, the block diagram of the closed-loop system is as shown in Figure 15.7.

Moreover, $H_{zw}(s)$ can be written as $H_{zw}(s) = \mathcal{F}_\ell(G, \mathcal{F}_\ell(M, Q)) = \mathcal{F}_\ell(N, Q)$ according to Section 7.4.2 in which

$$N(s) = \left[\begin{array}{cc|cc} A_{F_2} & -B_2F_2 & B_1 & B_2 \\ 0 & A_{L_2} & B_{L_2} & 0 \\ \hline C_{F_2} & -D_{12}F_2 & 0 & D_{12} \\ 0 & C_2 & D_{21} & 0 \end{array}\right].$$

It is easy to see that $N_{11} = G_cB_1 - UF_2G_f$, $N_{12} = U$, $N_{21} = V$, and $N_{22} = 0$. Expanding H_{zw} and then substituting into it the transfer matrices U, V, G_c, and G_f defined in Lemma 15.4, we get

$$H_{zw}(s) = G_cB_1 - UF_2G_f + UQV.$$

So,

$$\begin{aligned} \|H_{zw}\|_2^2 =& \langle G_cB_1 - U(F_2G_f - QV), G_cB_1 - U(F_2G_f - QV)\rangle \\ =& \|G_cB_1\|_2^2 - 2\Re[\langle G_cB_1, U(F_2G_f - QV)\rangle] \\ &+ \langle U(F_2G_f - QV), U(F_2G_f - QV)\rangle \\ =& \|G_cB_1\|_2^2 - 2\Re[\langle U^\sim G_cB_1, F_2G_f - QV\rangle] \\ &+ \langle U^\sim U(F_2G_f - QV), F_2G_f - QV\rangle \\ =& \|G_cB_1\|_2^2 + \langle F_2G_f - QV, F_2G_f - QV\rangle \end{aligned}$$

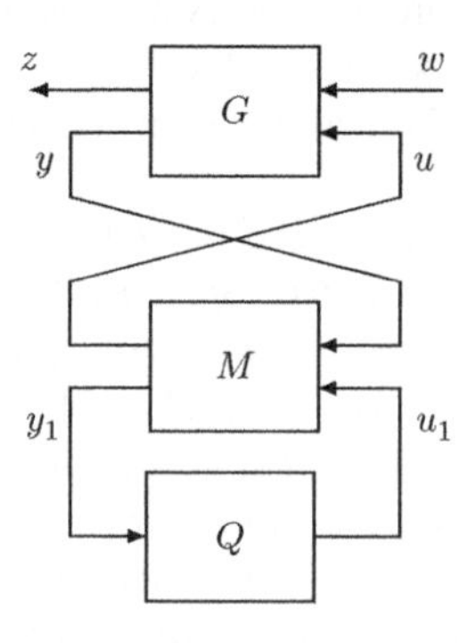

Figure 15.7 Block diagram of the closed-loop system

$$
\begin{aligned}
&=\|G_cB_1\|_2^2+\|F_2G_f\|_2^2-2\Re[\langle F_2G_f,QV\rangle]+\langle QV,QV\rangle\\
&=\|G_cB_1\|_2^2+\|F_2G_f\|_2^2-2\Re[\langle F_2G_fV^\sim,Q\rangle]+\langle QVV^\sim,Q\rangle\\
&=\|G_cB_1\|_2^2+\|F_2G_f\|_2^2+\|Q\|_2^2
\end{aligned}
$$

holds. In the derivation, we have used the fact that stable transfer matrix is orthogonal to antistable transfer matrix (i.e., their inner product is zero; see Lemma 2.6), Lemma 15.4, and the properties $\langle X,Y\rangle=\overline{\langle Y,X\rangle}$, $\langle Z^\sim X,Y\rangle=\langle Y,ZX\rangle$ of inner product. Obviously, $\|H_{zw}\|_2$ is minimum when $Q=0$. According to Lemma 15.1, we have $\|G_cB_1\|_2^2=\mathrm{Tr}(B_1^TXB_1)$, $\|F_2G_f\|_2^2=\mathrm{Tr}(F_2YF_2^T)$. So the minimum of $\|H_{zw}\|_2$ is $\mathrm{Tr}(B_1^TXB_1)+\mathrm{Tr}(F_2YF_2^T)$. Moreover, the unique solution for the $\mathcal{H}_2$ optimal controller is $K_{\mathrm{opt}}=\mathcal{F}_\ell(M,0)=M_{11}$.

In addition, it is easy to know from the preceding equation that the necessary and sufficient condition for $\|H_{zw}\|_2<\gamma$ is $\|Q\|_2^2<\gamma^2-\min\|H_{zw}\|_2^2$. This completes the proof for Theorems 15.1 and 15.2.

15.5 Singular $\mathcal{H}_2$ Control

In this section, we consider the $\mathcal{H}_2$ control problems in which the nonsingularity condition is not satisfied. In this case, it becomes extremely difficult to solve the optimization problem. So we only discuss the γ-optimal $\mathcal{H}_2$ control problem. This problem can be solved by linear matrix inequality (LMI) approach. The LMI solution is given by the following theorem.

Theorem 15.3 *Assume that (A1) holds. Then, the following statements are true:*

1. *The γ-optimal $\mathcal{H}_2$ control problem is solvable iff there exist matrices $X=X^T$, $Y=Y^T$, $\mathbb{A}$, $\mathbb{B}$, $\mathbb{C}$, and $W=W^T$ satisfying the following LMIs:*

$$
\mathrm{He}\begin{bmatrix} AX+B_2\mathbb{C} & A & 0\\ \mathbb{A} & YA+\mathbb{B}C_2 & 0\\ C_1X+D_{12}\mathbb{C} & C_1 & -\frac{1}{2}I\end{bmatrix}<0 \tag{15.22}
$$

$$
\begin{bmatrix} W & B_1^T & B_1^TY\\ B_1 & X & I\\ YB_1 & I & Y\end{bmatrix}>0 \tag{15.23}
$$

$$
\mathrm{Tr}(W)<\gamma^2. \tag{15.24}
$$

2. *When the LMIs (15.22)–(15.24) hold, a γ-optimal $\mathcal{H}_2$ controller*

$$
K(s)=(A_K,B_K,C_K,0) \tag{15.25}
$$

is given by

$$
A_K=N^{-1}(\mathbb{A}-NB_KC_2X-YB_2C_KM^T-YAX)(M^{-1})^T \tag{15.26}
$$

$$
C_K=\mathbb{C}(M^{-1})^T,\quad B_K=N^{-1}\mathbb{B}. \tag{15.27}
$$

M, N are nonsingular matrices satisfying

$$MN^T = I - XY. \tag{15.28}$$

Proof. Note that when $D_{11} = 0$, $D_c = D_{11} + D_{12}D_K D_{21} = 0$ requires $D_K = 0$. Then, the realization of the closed-loop system given in Section 7.4.1 becomes

$$\begin{bmatrix} \dot{x} \\ \dot{x}_K \\ \hline z \end{bmatrix} = \begin{bmatrix} A_c & B_c \\ C_c & D_c \end{bmatrix} \begin{bmatrix} x \\ x_K \\ \hline w \end{bmatrix}, \quad \begin{bmatrix} A_c & B_c \\ C_c & D_c \end{bmatrix} = \left[\begin{array}{cc|c} A & B_2 C_K & B_1 \\ B_K C_2 & A_K & B_K D_{21} \\ \hline C_1 & D_{12} C_K & 0 \end{array}\right].$$

According to Lemma 15.2, the γ-optimal $\mathcal{H}_2$ control problem has a solution if there exist matrices $P = P^T$ and $W = W^T$ satisfying

$$\begin{bmatrix} PA_c + A_c^T P & C_c^T \\ C_c & -I \end{bmatrix} < 0, \quad \begin{bmatrix} W & B_c^T P \\ PB_c & P \end{bmatrix} > 0, \quad \mathrm{Tr}(W) < \gamma^2.$$

According to the variable change method of Subsection 3.2.4, the matrix P can be factorized as

$$P\Pi_1 = \Pi_2, \quad \Pi_1 = \begin{bmatrix} X & I \\ M^T & 0 \end{bmatrix}, \quad \Pi_2 = \begin{bmatrix} I & Y \\ 0 & N^T \end{bmatrix}.$$

Performing congruent transformations on the first two matrix inequalities using $\mathrm{diag}(\Pi_1, I)$ and $\mathrm{diag}(I, \Pi_1)$, respectively, we obtain the following equivalent inequalities:

$$\begin{bmatrix} \Pi_2^T A_c \Pi_1 + (\Pi_2^T A_c \Pi_1)^T & (C_c \Pi_1)^T \\ C_c \Pi_1 & -I \end{bmatrix} < 0, \quad \begin{bmatrix} W & (\Pi_2^T B_c)^T \\ \Pi_2^T B_c & \Pi_2^T \Pi_1 \end{bmatrix} > 0.$$

After the substitution of coefficient matrices of the closed-loop system, we get

$$\Pi_2^T A_c \Pi_1 = \begin{bmatrix} AX + B\mathbb{C} & A + B\mathbb{D}C \\ \mathbb{A} & YA + \mathbb{B}C \end{bmatrix}, \quad \Pi_2^T B_c = \begin{bmatrix} B_1 \\ YB_1 + \mathbb{B}D_{21} \end{bmatrix}$$

$$C_c \Pi_1 = [C_1 X + D_{12}\mathbb{C} \;\; C_1], \quad \Pi_2^T \Pi_1 = \begin{bmatrix} X & I \\ I & Y \end{bmatrix}$$

via detailed computation. Here, the new variables $\mathbb{A}, \mathbb{B}, \mathbb{C}$ are defined by

$$\mathbb{A} = NA_K M^T + NB_K C_2 X + YB_2 C_K M^T + YAX, \quad \mathbb{B} = NB_K, \quad \mathbb{C} = C_K M^T.$$

All inequalities in the theorem are derived via the substitution of these equations. Further, $\left[\begin{smallmatrix} X & I \\ I & Y \end{smallmatrix}\right] > 0$ follows from the inequality (15.23), which in turn is equivalent to $XY - I > 0$. Therefore, the matrices M, N are nonsingular. •

In addition, Ref. [83] provides another form of LMI solution.

15.6 Case Study: $\mathcal{H}_2$ Control of an RTP System

Rapid thermal processing (RTP) system is used for the thermal processing of semiconductor wafers. It is capable of shortening the thermal processing period and has a small size. However, the temperature control is very difficult due to the stringent requirement on temperature uniformity and response speed, the nonlinearity of radiation heat, and the strong coupling between lamps.

Traditional proportional–integral–derivative (PID) compensators cannot achieve such a high performance. For this reason, model-based control designs have been studied since the 1990s. In this section, we introduce the method of Takagi and Liu [85]. Through the case study, we hope that the reader realizes that besides the control design, the locations of the actuator and sensor are also vital in achieving high performance.

The RTP device is illustrated in Figure 15.8(a). The wafer and halogen lamps are placed axisymmetrically like doughnuts. There are three circular lamp zones on top of the wafer and one circumvallating the wafer. Lamps 1–3 are for the thermal processing of wafer surface, while Lamp 4 is used to compensate for the heat leaking from the side of the wafer. Figure 15.8(b) illustrates the cross section of RTP. Further, pyrometers are installed below the bottom of the wafer to measure the temperature. The parameters of RTP are listed in Table 15.1.

15.6.1 *Model of RTP*

The heat conduction equation of wafer is given by

$$k\nabla^2 T = \rho C \frac{\partial T}{\partial t} \tag{15.29}$$

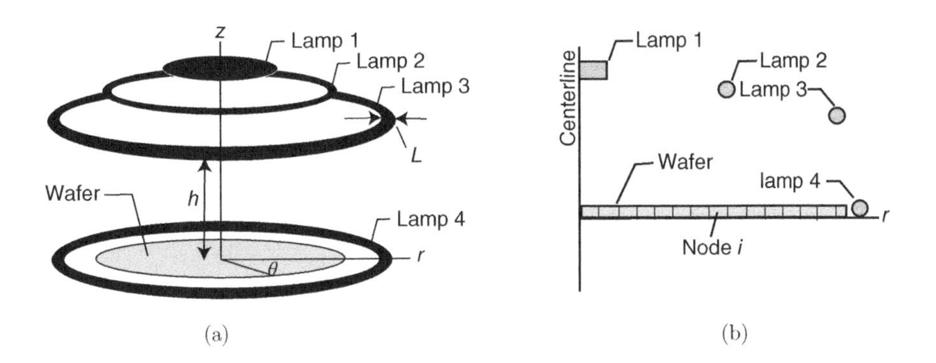

Figure 15.8 RTP system (a) 3D illustration (b) Cross section

Table 15.1 Parameters of RTP

Chamber	Height	100 mm
	Radius	200 mm
Wafer	Radius, R	100 mm
	Thickness, Z	1.0 mm

in which T is the surface temperature of wafer, k the thermal conductivity, ρ the density, and C the specific heat. In the polar coordinate, (15.29) becomes

$$\frac{\partial}{r\partial r}\left(kr\frac{\partial T}{\partial r}\right)+\frac{\partial}{r^2\partial\theta}\left(k\frac{\partial T}{\partial\theta}\right)+\frac{\partial}{\partial z}\left(k\frac{\partial T}{\partial z}\right)=\rho C\frac{\partial T}{\partial t}. \tag{15.30}$$

Here, θ is the angle, r the radius, and z the thickness (Figure 15.8(a)). Since both the wafer and RTP are axisymmetric and the wafer is very thin, we may ignore the thermal gradients in the θ direction and z direction. Then, we divide the wafer as n concentric annular zones (finite elements) with equal areas so as to build a finite dimensional model. When the division is fine enough, it can be assumed that each annular zone has a uniform temperature. Then, the partial differential equation (15.30) reduces to a set of ordinary differential equations:

$$m_iC_i\frac{dT_i}{dt}=q_i^{\text{ab}}+q_i^{\text{em}}+q_i^{\text{conv}}+q_i^{\text{cond}},\quad i=1,\ldots,n \tag{15.31}$$

in which T_i and m_i are the temperature and mass of the ith annular zone. Moreover, the specific heat C_i (J/kg/K) is a function of temperature and described by $C_i=748.249+0.168T_i$.

15.6.1.1 Emission Heat q_i^{em}

According to the Stefan–Boltzmann rule, the heat emitted from the ith annular zone is

$$q_i^{\text{em}}=-\epsilon_i\sigma A_iT_i^4,\quad \epsilon_i=\frac{0.7128}{1+\exp\left(\dfrac{T_i-666.15}{-64.70}\right)} \tag{15.32}$$

in which $\sigma=5.67\times10^{-8}\ \text{W}/m^2K^4$ is the Stefan–Boltzmann constant, A_i the surface area, and ϵ_i the emissivity of the ith annular zone.

15.6.1.2 Convection Heat q_i^{conv}

The heat convected from the ambient gas to the ith annular zone is

$$q_i^{\text{conv}}=-h_iA_i(T_i-T_{\text{gas}}),\quad h_i=14.2+8.6\left(\frac{r_i}{R}\right) \tag{15.33}$$

in which T_{gas} is the gas temperature and h_i $(W/m^2/K)$ the convective heat transfer coefficient.

15.6.1.3 Conduction Heat q_i^{cond}

The heat conducted between adjacent annular zones is described by

$$q_i^{\text{cond}}=-2\pi k_iZ\left(\frac{T_i-T_{i-1}}{r_i^{\text{cen}}-r_{i-1}^{\text{cen}}}r_i^{\text{in}}+\frac{T_i-T_{i+1}}{r_i^{\text{cen}}-r_{i+1}^{\text{cen}}}r_i^{\text{out}}\right) \tag{15.34}$$

in which $r_i^{\text{in}},r_i^{\text{cen}},r_i^{\text{out}}$ are, respectively, the inner radius, central radius, and outer radius of the ith annular zone. The thermal conductivity k_i (W/m/K) is a function of temperature:

$$k_i=0.575872-\frac{7.84727\times10^8}{T_i^3}+\frac{4.78671\times10^6}{T_i^2}+\frac{3.10482\times10^4}{T_i}. \tag{15.35}$$

15.6.1.4 Radiation Heat q_i^{ab} of Lamp

q_i^{ab} denotes the heat irradiated from the lamps to the ith annular zone. Let the number of lamps be J and the heat of lamp j be P_j, then the heat absorbed by the ith annular zone is

$$q_i^{\text{ab}} = \alpha_i \sum_{j=1}^{J} L_{i,j} P_j. \tag{15.36}$$

α_i is the radiant energy absorptivity of the wafer and equal to the emissivity when the wafer surface is gray. $L_{i,j}$ is the radiation view factor from the ith annular zone to the jth circular lamp zone.

In this study, the wafer is divided into 20 annular zones.

15.6.2 *Optimal Configuration of Lamps*

In order to realize a high uniformity in the surface temperature of wafer, the heat irradiated from the lamps to the wafer surface must be uniform. So, the configuration of lamps needs to be optimized. The configuration designed according to the method of Yaw and Lin [96] is shown in Table 15.2. The view factor relation between the annular zones and lamps is illustrated in Figure 15.9(a). Figure 15.9(b) shows the heat absorbed by the wafer surface. Obviously, the heat fluxes are almost uniform except the outmost annular zone (Figure 15.10).

Table 15.2 Optimal configuration of lamps

Lamp no.	Inner radius (mm)	Height (mm)	Width (mm)
Lamp 1	0	100	25
Lamp 2	84.2	87.9	10
Lamp 3	100	26	10
Lamp 4	181	0	10

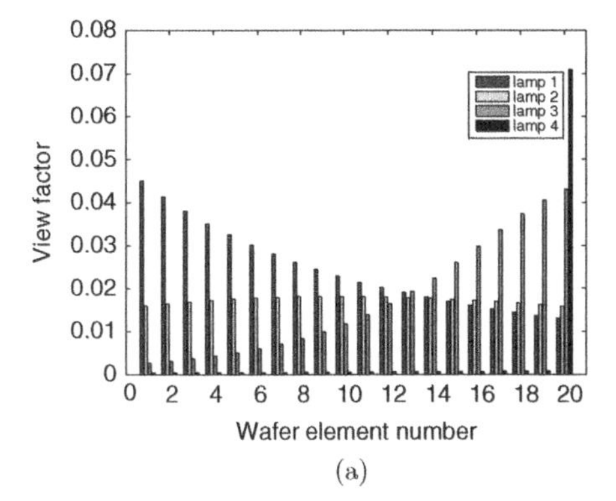

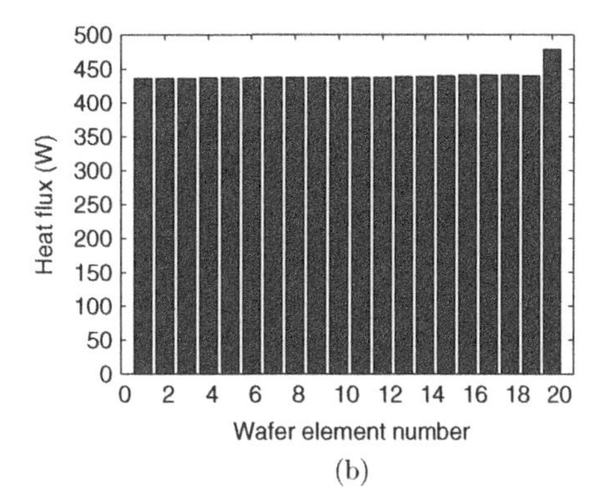

Figure 15.9 View factors and radiation heat on wafer surface (a) View factors between annular zones and lamps (b) Radiation heat on wafer surface

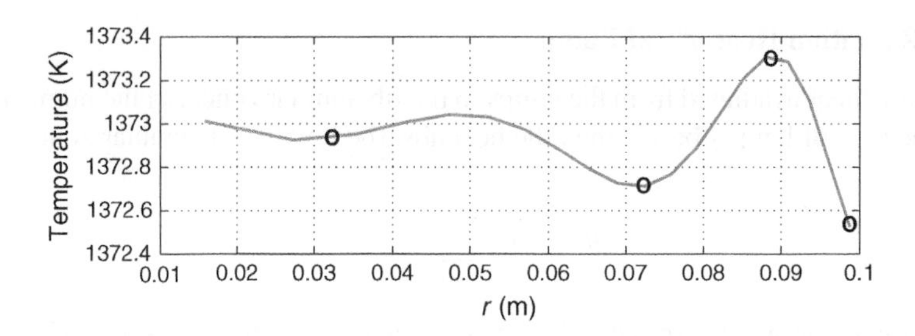

Figure 15.10 Steady-state temperature distribution of wafer surface

15.6.3 Location of Sensors

In feedback control, what is important is not only the location of the actuator but also that of the sensor. The temperature information on the whole wafer surface is not accessible. We can only measure finite points. Therefore, it is important to determine the number of sensors and locate them suitably. In correspondence with the number of circular lamp zones, four pyrometers are used. Then, we investigate the (radial) temperature distribution when the steady-state input u_0 is applied. The sensors are located at the annular zones numbered 3, 11, 18, and 20, where the peaks of temperature fluctuation occur.

15.6.4 $\mathcal{H}_2$ Control Design

15.6.4.1 Specification on RTP Temperature Control

As shown in Figure 15.11, the process is to start from the initial temperature 873 K and raise the temperature to 1373 K at a rate of 100 K/s and then keep the state for 5 seconds and carry out the thermal processing. After that, lower the temperature at a rate of −40 K/s from 10 s to 17.5 s. In the whole process, the maximal temperature fluctuation must be less than ±1 K.

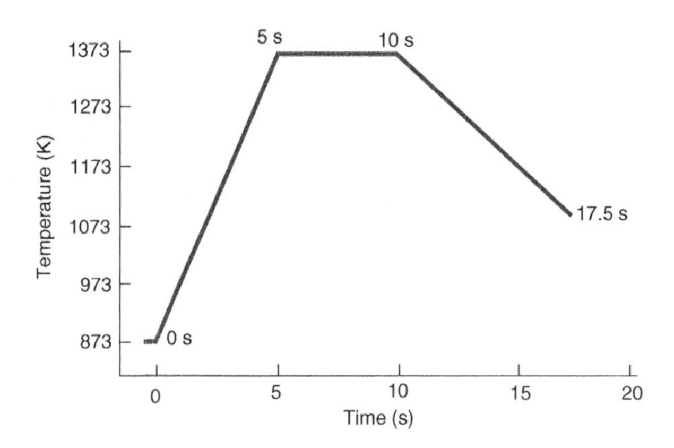

Figure 15.11 Reference trajectory of the thermal processing

Let $T = [T_1, \ldots, T_{20}]^T$ denote the temperature vector of wafer surface and $u = [u_1, \ldots, u_4]^T$ denote the lamp power vector. Linear approximation about the target temperature 1373 K is obtained as

$$\Delta\dot{T} = A\Delta T + B\Delta u, \quad y = C\Delta T \tag{15.37}$$

in which $(\Delta T, \Delta u)$ are the error vectors w.r.t. the equilibrium (T_0, u_0):

$$\Delta T = T - T_0, \quad \Delta u = u - u_0. \tag{15.38}$$

Further, the coefficient matrices A, B are as follows, in which B_{ij} denotes the (i, j) element of matrix B:

$$A = \left(\frac{\partial q_{em}}{\partial T} + \frac{\partial q_{conv}}{\partial T} + \frac{\partial q_{cond}}{\partial T} \right)\Bigg|_{T=T_0}$$

$$B_{ij} = \alpha_i L_{i,j}, \quad j = 1, \ldots, 4, \quad i = 1, \ldots, 20.$$

In the RTP system, the coupling between the radiation heat of lamps is so strong that multiple-input multiple-output (MIMO) control has to be adopted. $\mathcal{H}_2$ control is applied since it is capable of optimizing the time response directly. The generalized plant is given in Figure 15.12. The performance outputs are the tracking error z_1 and the control input z_2. The disturbance is the impulse input of the reference model. Note that the performance output includes only the errors between the measured outputs and the reference, not the tracking errors of all annular zones. This is because the number of weighting functions increases with that of performance outputs, which results in a high controller degree. Further, it will be shown later that with a suitable location of sensors, this design can control indirectly the temperature of the whole wafer surface.

In this figure, $W_R(s)$ is the reference model and $W_u(s)$ is used to tune the input. For simplicity of tuning, they are all set as diagonal. $W_R(s)$ is chosen as an integrator in order to track step reference. $W_u(s)$ is chosen as a high-pass filter so as to suppress the amplitude and the high-frequency components of the input. The rule of tuning is to raise the gain of $W_R(s)$

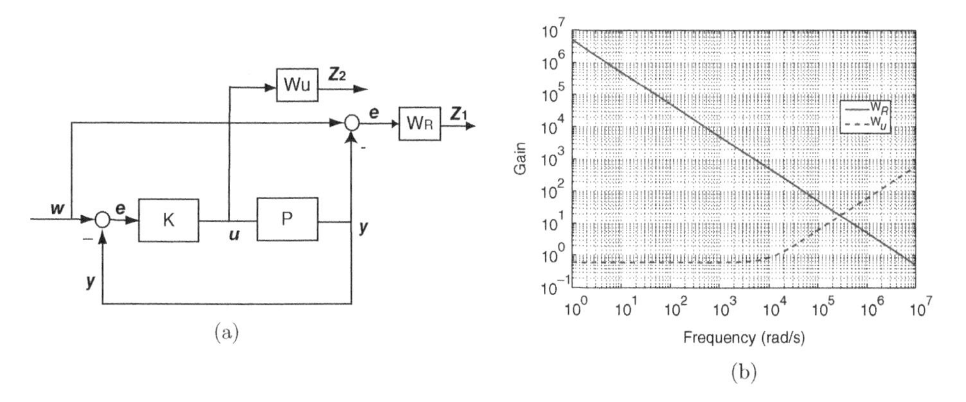

Figure 15.12 Generalized plant and weighting functions (a) Generalized plant (b) Frequency response of weighting functions W_R, W_u

first and then tune the gain of $W_u(s)$ so as to avoid saturation of lamp power. Moreover, the high-frequency gain of controller can be limited by lowering the first corner frequency of $W_u(s)$. Via trial and error, the weighting functions are determined as shown in Figure 15.12(b).

15.6.5 Simulation Results

$\mathcal{H}_2$ controller is designed based on the preceding discussions. Simulations are done w.r.t. the nonlinear model (15.31). Shown in Figure 15.13 are the lamp power u and the maximal absolute error $\max_{1\leq i\leq 20}|T_i(t) - T_{ref}(t)|$ among all annular zones. Further, to check the effect of sensor location, the system is redesigned w.r.t. traditional equal distance location. The result is illustrated in Figure 15.14. Apparently, the error is much smaller in the case of optimized sensor location.

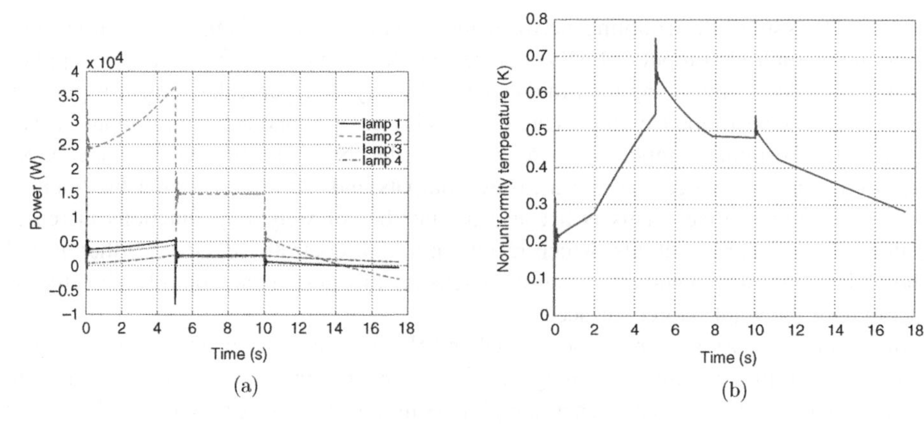

Figure 15.13 Result of optimal sensor location (a) Lamp powers (b) Nonuniformity of temperature

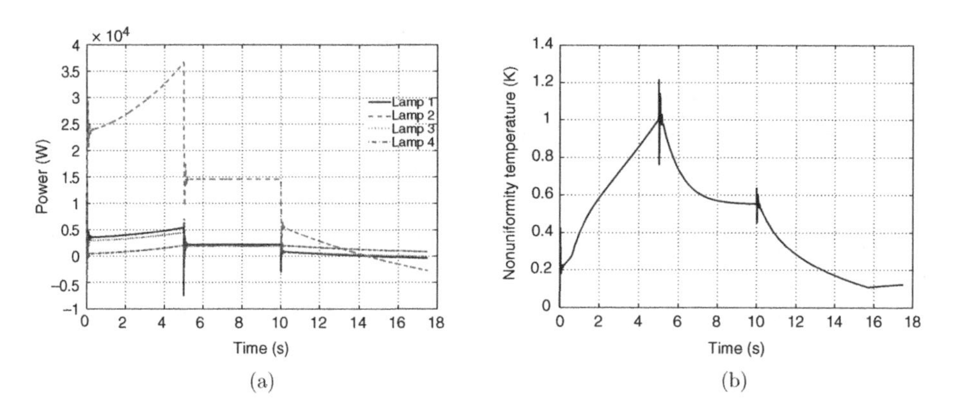

Figure 15.14 Result of equal distance sensor location (a) Lamp powers (b) Nonuniformity of temperature

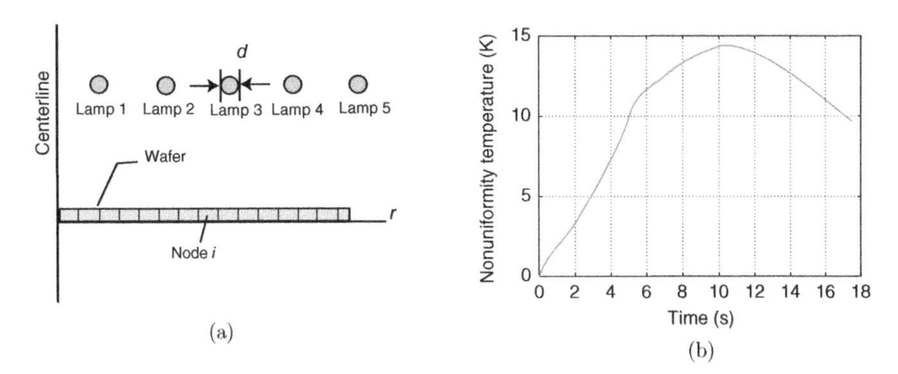

Figure 15.15 RTP with lamps located at the same height and equal distance (a) Lamp configuration (b) Nonuniformity of temperature

Finally, to validate the effect of lamp configuration, a state feedback $\mathcal{H}_2$ design is carried out w.r.t. the traditional configuration which places the lamps at the same height and equal distance (Figure 15.15(a)). A very large error is observed in the simulation (Figure 15.15(b)), which indicates the effectivity and importance of lamp configuration.

Exercises

15.1 For a stable transfer function

$$G(s) = \frac{s+5}{(s+1)(s+10)},$$

calculate its $\mathcal{H}_2$ norm by using the methods of residue, impulse response, and Lyapunov equation. Further, solve the following problems:

(a) When the input is $u(t) = 2\delta(t)$, compute the $\mathcal{H}_2$ norm of the output.

(b) Let $w_n(t)$ be a unit white noise. When the input is $u(t) = 3w_n(t)$, compute the steady-state variance $E[y(\infty)y^T(\infty)]$ of output.

15.2 In the closed-loop system of Figure 15.16, the plant is $P(s) = 1/s$ and controller is $K(s) = k$. When $r(t) = 1$ $(t \geq 0)$, find a gain k such that the tracking error $e(t)$ satisfies the specification $\|e\|_2 \leq 0.1$.

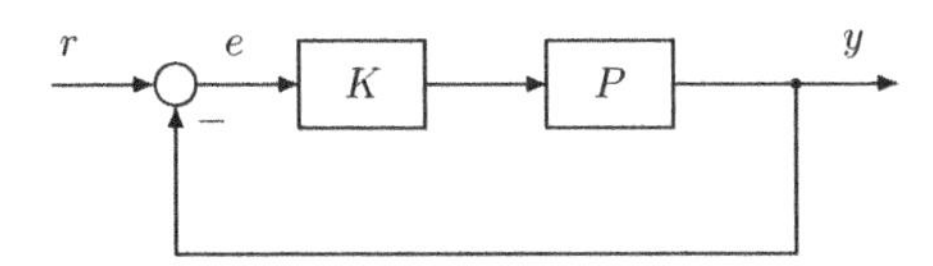

Figure 15.16 Response improvement

15.3 Prove that the $\mathcal{H}_2$ norm of a stable matrix $G(s) = (A, B, C, 0)$ can be computed as follows:

$$\|G\|_2^2 = \text{Tr}(B^T L_o B)$$

in which L_o is the observability Gramian satisfying $A^T L_o + L_o A + C^T C = 0$.

15.4 Prove statement 2 of Lemma 15.3.

15.5 Prove statement 2 of Lemma 15.4.

15.6 When both D_{12} and D_{21} satisfy the full rank condition in (A2), but not normalized, find a transformation on the input and output signals such that the new generalized plant satisfies (A2). (Hint: use the singular value decomposition.)

15.7 Assume (A2). When $D_{11} \neq 0$, a necessary condition for the existence of the $\mathcal{H}_2$ norm of the closed-loop system is

$$D_c = D_{11} + D_{12} D_K D_{21} = 0.$$

Answer the question that under what condition this equation has a solution D_K and solve for it.

Next, we transform the system as shown in Figure 15.17. Prove that $\hat{D}_{11} = 0$ in the new generalized plant $\hat{G}$. This means that we can solve the $\mathcal{H}_2$ control problem w.r.t. $\hat{G}$ first and obtain the corresponding optimal $\mathcal{H}_2$ controller $\hat{K}$. After that, we may reverse $\hat{K}$ back to K. Derive the corresponding formula.

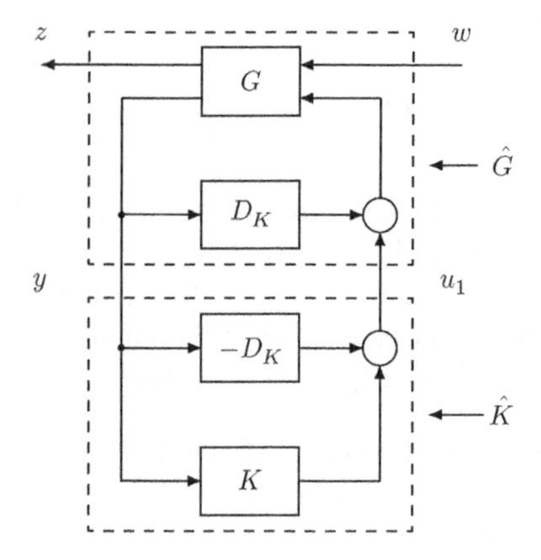

Figure 15.17 Eliminating the D_{11} term

15.8 In the case of state feedback $y = x$, prove that the $\mathcal{H}_2$ optimal controller is $u = F_2 x$. Further, prove that the minimum norm is $\min \|H_{zw}\|_2 = \text{Tr}(B_1^T X B_1)$.

15.9 Derive a state feedback solution for the γ-optimal $\mathcal{H}_2$ control by using the LMI approach.

Notes and References

$\mathcal{H}_2$ control was also known as LQ control (in state feedback) or LQG control (in output feedback) in the past [2, 92]. Safanov and Athans [82] showed that the LQ control has a certain degree of stability margin. Sato and Liu [83] gave an LMI solution to the $\mathcal{H}_2$ control problem and Ref. [65] solved the $\mathcal{H}_2$ control problem with unstable weighting functions. Takagi and Liu [85] applied $\mathcal{H}_2$ control to the temperature processing of semiconductor wafers and validated the importance of actuator/sensor locations.

16

$\mathcal{H}_\infty$ Control

As learnt in Chapter 11, many robustness conditions are given as inequalities about the $\mathcal{H}_\infty$ norms of transfer functions. So it is necessary to establish a method for designing controllers satisfying such $\mathcal{H}_\infty$ norm inequality. The main objective of this chapter is to introduce such design approaches. Moreover, details will be provided on how to design a generalized plant and how to determine weighting functions. These knowledge are very important in practical application of $\mathcal{H}_\infty$ control. Finally, a case study will be described in some detail.

16.1 Control Problem and $\mathcal{H}_\infty$ Norm

16.1.1 Input–Output Relation of Transfer Matrix's $\mathcal{H}_\infty$ Norm

Let $u(t)$ be the input of stable system $G(s)$ and $y(t)$ the output. Then, the following relationship holds:

$$\|G\|_\infty = \sup_{\|u\|_2 \neq 0} \frac{\|y\|_2}{\|u\|_2}. \tag{16.1}$$

Since $\|y\|_2/\|u\|_2$ is the ratio of the square roots of input energy and output energy, this equation shows that the supremum of the ratios for all energy-bounded input $u(t)$ is equal to the $\mathcal{H}_\infty$ norm $\|G\|_\infty$ of the transfer function. Therefore, if we want to lower the output response $y(t)$ to an energy-bounded disturbance $u(t)$ (i.e., $\|u\|_2 < \infty$), we need only ensure that

$$\|G\|_\infty \to 0.$$

Further, to make the input–output ratio less than a given value $\gamma > 0$, it is sufficient to guarantee that

$$\|G\|_\infty < \gamma.$$

However, this relationship is purely mathematical and having no engineering meaning. Why? This is because a disturbance in practice is not bounded in energy. Instead, it is a persistent signal like the step signal whose energy is unbounded. In addition, the essence of weighting function cannot be explained from this angle of viewpoint.

Robust Control: Theory and Applications, First Edition. Kang-Zhi Liu and Yu Yao.

16.1.2 Disturbance Control and Weighting Function

Now, we investigate the relationship between the $\mathcal{H}_\infty$ norm of a transfer matrix and its input–output from another angle. For single-input single-output (SISO) system, the $\mathcal{H}_\infty$ norm

$$\|G\|_\infty = \sup_\omega |G(j\omega)|$$

can be interpreted as the maximum amplitude of the frequency response of the system to a unit impulse input. In addition, for multiple-input multiple-output (MIMO) systems, there holds the following relationship [62]:

$$\|G\|_\infty = \sup_{\substack{u\in\mathbb{C}^m \\ \|u\|\leq 1}} \|Gu\|_\infty, \quad \|Gu\|_\infty = \sup_\omega \|G(j\omega)u\|_2. \tag{16.2}$$

Here the complex space $\mathbb{C}^m$ can be regarded as a space of impulse vector signals containing time delay. So, equation (16.2) implies that the $\mathcal{H}_\infty$ norm $\|G\|_\infty$ is the maximum amplitude of all the frequency responses w.r.t. unit impulse vectors whose elements are imposed at arbitrary instants.

Hence, if the frequency characteristic of disturbance $d(t)$ is $W(s)$, $y(t)$ can be regarded as the impulse response of the transfer function GW with weighting function by Figure 16.1. In order to suppress the output response to this disturbance, we need only guarantee that

$$\|\hat{y}\|_\infty \leq \|GW\|_\infty < \gamma \tag{16.3}$$

for a specified value $\gamma > 0$.

Such problem is known as a problem with weighting function, and $W(s)$ is called the weighting function. Even if we cannot know exactly the frequency characteristics of a disturbance, its upper bound can be used as a weighting function if the upper bound can be estimated. In fact, for any frequency ω, when the amplitude of the frequency characteristic of disturbance $d(t)$ is less than $|W(j\omega)|$, that is,

$$|\hat{d}(j\omega)| \leq |W(j\omega)| \quad \forall\omega, \tag{16.4}$$

(refer to Figure 16.2), then the maximum amplitude of the frequency response $\hat{y}(j\omega)$ is suppressed below γ if (16.3) holds.

Example 16.1 *In the closed-loop system of Exercise 15.2, assume that the plant is $P(s) = 1/s$ and the controller is $K(s) = k$. When the reference input is $r(t) = 1(t)$, seek a gain k such that the tracking error $e(t)$ satisfies the performance specification* $\sup_\omega |\hat{e}(j\omega)| \leq 0.1$.

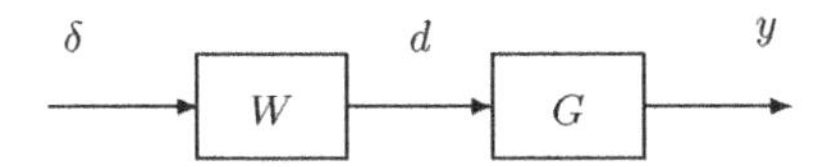

Figure 16.1 Response to disturbance with frequency characteristic $W(s)$

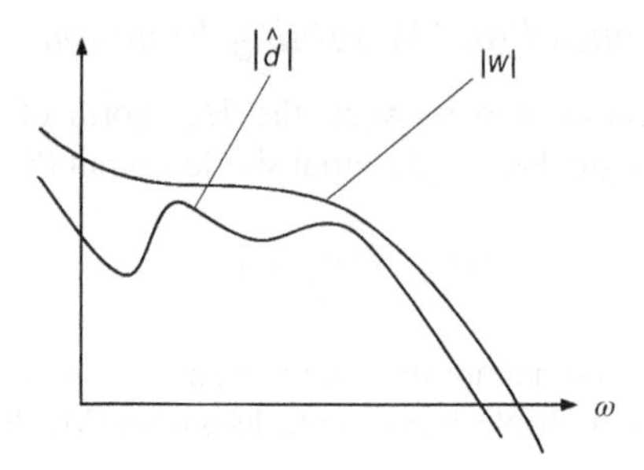

Figure 16.2 Weighting function and frequency characteristic of disturbance

Solution Reference input $r(t)$ is the unit impulse response of $W(s) = 1/s$. Hence, $e(t)$ can be regarded as the unit impulse response of the weighted transfer function WS. Therefore, we have $\sup_\omega |\hat{e}(j\omega)| = \|WS\|_\infty$ in which

$$S(s) = \frac{1}{1+PK} = \frac{s}{s+k}$$

is the sensitivity function. To ensure the internal stability, $k > 0$ is necessary. Then, from the tracking performance specification, we finally obtain

$$\|WS\|_\infty = \left\|\frac{1}{s+k}\right\|_\infty = \frac{1}{k} \le 0.1 \quad \Rightarrow \quad k \ge 10. \qquad \bigtriangledown$$

This control problem is called an $\mathcal{H}_\infty$ control problem. The general $\mathcal{H}_\infty$ control problem will be elaborated in the next section.

16.2 $\mathcal{H}_\infty$ Control Problem

For simplicity of description, the state-space realization of the generalized plant $G(s)$ in Figure 16.3 is decomposed as

$$G(s) = \left[\begin{array}{c|c} A & B \\ \hline C & D \end{array}\right] = \left[\begin{array}{c|cc} A & B_1 & B_2 \\ \hline C_1 & D_{11} & D_{12} \\ C_2 & D_{21} & 0 \end{array}\right] \tag{16.5}$$

in accordance with the dimensions of input and output. Here, the realization should be minimal.

The so-called $\mathcal{H}_\infty$ control problem is to design a controller such that the generalized feedback system in Figure 16.3 is stable and the $\mathcal{H}_\infty$ norm of the closed-loop transfer matrix from disturbance w to performance output z is less than a given positive number γ. The $\mathcal{H}_\infty$ control problem has two kinds of solutions: one is based on the algebraic Riccati equation (ARE) and the other based on the linear matrix inequality (LMI).

Historically, ARE solution was proposed by Doyle, Glover, Khargonekar, and Francis in the late 1980s which has had a profound impact on the research of robust control since then. However, this approach requires a so-called nonsingularity condition on the generalized plant

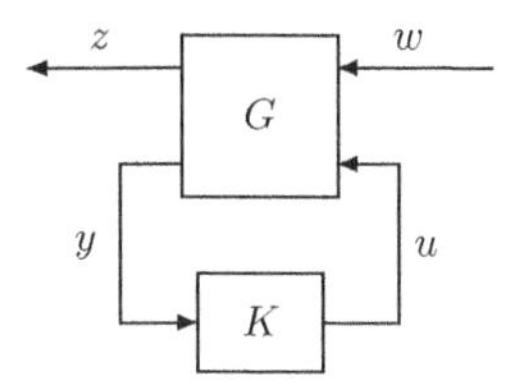

Figure 16.3 Generalized feedback system

which is not always satisfied in practical applications. Moreover, this approach can handle only a single objective, that is, $\mathcal{H}_\infty$ norm minimization. So in this book, we focus on the LMI solution. As for the ARE solution, interested readers may consult [100, 26, 36].

Obviously, the $\mathcal{H}_\infty$ norm specification can be characterized by bounded-real lemma. So, the central idea of LMI approach is to transform the $\mathcal{H}_\infty$ norm specification into a matrix inequality by using bounded-real lemma and derive the existence condition and the solution based on LMI techniques.

The following assumption is made, which is necessary for the stabilization of the feedback system. It basically requires that all the weighting functions are stable.

(A1) (A, B_2) is stabilizable and (C_2, A) is detectable.

16.3 LMI Solution 1: Variable Elimination

First, we define the following two matrices:

$$N_Y = [C_2 \quad D_{21}]_\perp, N_X = [B_2^T \quad D_{12}^T]_\perp.$$

For the generalized plant $G(s)$ given in (16.5), the solvability condition for the $\mathcal{H}_\infty$ control problem is given by the following theorem.

Theorem 16.1 *Suppose that (A1) is satisfied. There is a controller $K(s)$ such that the closed-loop system in Figure 16.3 is stable and satisfies $\|H_{zw}\|_\infty < \gamma$ iff the following LMIs have positive definite solutions X, Y:*

$$\begin{bmatrix} N_X^T & 0 \\ 0 & I_{n_w} \end{bmatrix} \begin{bmatrix} AX + XA^T & XC_1^T & B_1 \\ C_1X & -\gamma I & D_{11} \\ B_1^T & D_{11}^T & -\gamma I \end{bmatrix} \begin{bmatrix} N_X & 0 \\ 0 & I_{n_w} \end{bmatrix} < 0 \tag{16.6}$$

$$\begin{bmatrix} N_Y^T & 0 \\ 0 & I_{n_z} \end{bmatrix} \begin{bmatrix} YA + A^TY & YB_1 & C_1^T \\ B_1^TY & -\gamma I & D_{11}^T \\ C_1 & D_{11} & -\gamma I \end{bmatrix} \begin{bmatrix} N_Y & 0 \\ 0 & I_{n_z} \end{bmatrix} < 0 \tag{16.7}$$

$$\begin{bmatrix} X & I \\ I & Y \end{bmatrix} \geq 0, \ \mathrm{rank} \begin{bmatrix} X & I \\ I & Y \end{bmatrix} \leq n + n_K. \tag{16.8}$$

Here, n, n_K are the degrees of $G(s), K(s)$, respectively.

16.3.1 Proof of Theorem 16.1

The proof is quite elementary and only requires some knowledge of algebra.

Let a realization of the closed-loop transfer matrix $H_{zw}(s)$ be $H_{zw}(s) = (A_c, B_c, C_c, D_c)$. According to the bounded-real lemma, the $\mathcal{H}_\infty$ control problem is solvable iff there exists a positive definite matrix P satisfying

$$\begin{bmatrix} A_c^T P + PA_c & PB_c & C_c^T \\ B_c^T P & -\gamma I & D_c^T \\ C_c & D_c & -\gamma I \end{bmatrix} < 0. \tag{16.9}$$

Further, set the controller as $K(s) = (A_K, B_K, C_K, D_K)$. According to Section 7.4.1 (A_c, B_c, C_c, D_c) is given by (7.43). Substitution of these matrices shows that (16.9) is equivalent to

$$Q + E^T \mathcal{K} F + F^T \mathcal{K}^T E < 0 \tag{16.10}$$

in which

$$\begin{bmatrix} Q & E^T \\ F & \end{bmatrix} = \left[\begin{array}{ccc|c} \overline{A}^T P + P\overline{A} & P\overline{B}_1 & \overline{C}_1^T & P\overline{B}_2 \\ \overline{B}_1^T P & -\gamma I & \overline{D}_{11}^T & 0 \\ \overline{C}_1 & \overline{D}_{11} & -\gamma I & \overline{D}_{12} \\ \hline \overline{C}_2 & \overline{D}_{21} & 0 & \end{array}\right], \quad \mathcal{K} = \begin{bmatrix} D_K & C_K \\ B_K & A_K \end{bmatrix}.$$

Here, all matrices marked by bar are defined in (7.44). Owing to Lemma 3.1, (16.10) is equivalent to the following inequalities:

$$E_\perp^T Q E_\perp < 0, \quad F_\perp^T Q F_\perp < 0. \tag{16.11}$$

By direct calculation, the orthogonal matrices of E, F are obtained as

$$E_\perp = \begin{bmatrix} P^{-1} & & \\ & I & 0 \\ & 0 & I \end{bmatrix} \begin{bmatrix} I & & & \\ & 0 & 0 & I \\ & 0 & I & 0 \\ & I & 0 & 0 \end{bmatrix} \begin{bmatrix} N_X & 0 \\ 0 & I \\ 0 & 0 \end{bmatrix}$$

$$F_\perp = \begin{bmatrix} I & & & \\ & 0 & I & 0 \\ & I & 0 & 0 \\ & 0 & 0 & I \end{bmatrix} \begin{bmatrix} N_Y & 0 \\ 0 & 0 \\ 0 & I \end{bmatrix}.$$

Further, we decompose the positive definite matrix P as

$$P = \begin{bmatrix} Y & * \\ * & * \end{bmatrix}, \quad P^{-1} = \begin{bmatrix} X & * \\ * & * \end{bmatrix}.$$

Then, conditions (16.6) and (16.7) are derived via detailed calculation based on the substitution of $E_\perp, F_\perp$ and the decomposed matrices P, P^{-1} into (16.11). Finally, the condition (16.8) is obtained from the positive definiteness of matrix P (refer to Lemma 3.1).

16.3.2 Computation of Controller

When the $\mathcal{H}_\infty$ control problem is solvable, we obtain positive definite solutions X, Y of the LMIs. In calculating the corresponding controller, we first compute a matrix M satisfying

$$MM^T = Y - X^{-1}. \tag{16.12}$$

After that, we set

$$P = \begin{bmatrix} Y & M \\ M^T & I \end{bmatrix} \tag{16.13}$$

and substitute it back into (16.10); then we get the coefficient matrix $\mathcal{K}$ of the controller directly by numerical calculation.

In addition, we can also use the solutions of X, Y to calculate $\mathcal{K}$ directly. Refer to references [43, 32] for the formula.

16.4 LMI Solution 2: Variable Change

The variable change method can also be used to solve the $\mathcal{H}_\infty$ control problem. We have shown that the $\mathcal{H}_\infty$ control problem is solvable iff there exists a positive definite matrix P satisfying the inequality (16.9). Now, we substitute the realization (7.41) of the closed-loop system into this inequality and note that matrix P has the following structure:

$$P\Pi_1 = \Pi_2, \quad \Pi_1 = \begin{bmatrix} X & I \\ M^T & 0 \end{bmatrix}, \quad \Pi_2 = \begin{bmatrix} I & Y \\ 0 & N^T \end{bmatrix}.$$

Similar to Subsection 3.2.4, after the following variable change

$$\begin{aligned} \mathbb{A} &= NA_K M^T + NB_K CX + YBC_K M^T + Y(A + BD_K C)X \\ \mathbb{B} &= NB_K + YBD_K, \quad \mathbb{C} = C_K M^T + D_K CX, \quad \mathbb{D} = D_K, \end{aligned} \tag{16.14}$$

we obtain the solvability conditions that follow:

$$\text{He}\begin{bmatrix} AX + B_2\mathbb{C} & A + B_2\mathbb{D}C_2 & B_1 + B_2\mathbb{D}D_{21} & 0 \\ \mathbb{A} & YA + \mathbb{B}C_2 & YB_1 + \mathbb{B}D_{21} & 0 \\ 0 & 0 & -\frac{\gamma}{2}I & 0 \\ C_1X + D_{12}\mathbb{C} & C_1 + D_{12}\mathbb{D}C_2 & D_{11} + D_{12}\mathbb{D}D_{21} & -\frac{\gamma}{2}I \end{bmatrix} < 0 \tag{16.15}$$

$$\begin{bmatrix} X & I \\ I & Y \end{bmatrix} > 0. \tag{16.16}$$

Here, to simplify the description of large matrix, we have used the notation $\text{He}(A) = A + A^T$. Finally, coefficient matrices of the controller $K(s) = (A_K, B_K, C_K, D_K)$ are given by

$$\begin{aligned} D_K &= \mathbb{D}, \quad C_K = (\mathbb{C} - D_K CX)(M^{-1})^T, \quad B_K = N^{-1}(\mathbb{B} - YBD_K) \\ A_K &= N^{-1}(\mathbb{A} - NB_K CX - YBC_K M^T - Y(A + BD_K C)X)(M^{-1})^T. \end{aligned} \tag{16.17}$$

The detailed derivation is left for the reader to complete (Exercise 16.5).

16.5 Design of Generalized Plant and Weighting Function

In the application of control theory such as $\mathcal{H}_\infty$, $\mathcal{H}_2$ to real systems, the most important is how to determine the generalized plant and the weighting functions. This is because all the required performance specifications have to be reflected on the generalized plant and the weighting functions. In principle, selection of the generalized plant and weighting functions must be the symptomatic treatment, that is, they should be determined in accordance with the feature of the problem. In general, trial and error is needed in order to achieve the best performance. It may be said that the selection of generalized plant and weighting function reflects the ability of an engineer and tests the strength of his/her engineering sense.

16.5.1 Principle for Selection of Generalized Plant

16.5.1.1 Consideration of Disturbance Control

$\mathcal{H}_\infty$ control is mainly useful in disturbance attenuation and robustness to model uncertainty. Therefore, before the design we need to find out all possible disturbances and single out the major one. Then, we put the response to this disturbance into the performance output. Furthermore, we need to examine the frequency characteristics of the disturbance and use its estimate as the weighting function. In the case of reference tracking, if the tracking error is treated as a performance output, then the reference input can be regarded as a disturbance.

16.5.1.2 Consideration of Model Uncertainty

First of all, we need to estimate the scope of the uncertain parameters and the unmodeled high-frequency dynamics. In order to ensure the robustness to uncertainty, we may apply the small-gain theorem introduced in Chapter 11. Namely, remove the uncertainty Δ from the loop, and penalize the $\mathcal{H}_\infty$ norm of the transfer function from its output to its input.

Particularly, in the case of multiple uncertain parameters, in order to figure out which should be considered in the design, we may change the parameters one by one and analyze the corresponding Bode plots of the plant. Since the key factors affecting the performance of a control system are the low-frequency gain of the plant, the resonance frequency, and the resonance peak, we must ensure the robustness to those uncertain parameters that have big impact on these factors. There are two ways: one is to transform the uncertain parameters into additive or multiplicative dynamic uncertainty, and another is to use the polytopic model of Section 18.3 and the common Lyapunov function method. As small-gain theorem is very conservative for parameter uncertainty, the latter is recommended.

16.5.1.3 Consideration of Input Constraint

In order to avoid input command that causes saturation of the actuator or impulsive input which has an adverse effect on the system, we must put the input into the performance output.

16.5.2 *Selection of Weighting Function*

The weighting functions should be determined based on the following principles.

16.5.2.1 Weighting Function of Dynamic Uncertainty

Draw the estimated frequency responses of uncertainty on the Bode plot, and then find a low-order transfer function whose Bode plot covers these frequency responses. In determining a weighting function, we should better use a line segment approximation to determine a transfer function roughly and then draw the Bode gain plots of the weighting function and uncertainty on the same plot using MATLAB so that we can make sure that the gain of weighting function is higher than that of uncertainty. If necessary, we make a fine-tuning on the weighting function. Further, since control is not required in the high-frequency band beyond the control bandwidth, the gain of the uncertainty weighting function should be lifted high enough in this band in order to suppress the high-frequency oscillatory part of the control input.

16.5.2.2 Weighting Function of Input

The input weighting function is mainly used to reduce the high-frequency components of the input. So it should be selected as a high-pass transfer function. The control bandwidth can be determined based on the fast response specification. The gain of input weighting function should be sufficiently low within the control bandwidth, but it should have a high gain beyond the control bandwidth.

16.5.2.3 Weighting Function of Performance

Usually, a disturbance contains a lot of low-frequency components, which is characterized by its high low-frequency gain. In order to suppress the influence of disturbance effectively, we should set its weighting function as a low-pass transfer function. If we have some a priori information on the disturbance, we can estimate its frequency response based on it. For example, a disturbance taking roughly a constant value over a long time can be treated as a step signal and described by an integrator. The gain of weighting function can be tuned via repeated design and simulation. The higher the gain of performance weighting function is, the better the performance will be. In addition, an effective way for the adjustment of weighting functions is to place the dynamics of a weighting function at the performance output port and place a tuning gain at the disturbance input port.

Among these three kinds of weighting functions, the uncertainty weighting function is the most easy to determine and should be determined first. Then, in choosing the weighting functions for performance and input, we should try to achieve the best control performance by tuning them through repeated design and simulation. In general, a high gain of performance weighting function in the Low- and middle-frequency domain leads to a better disturbance attenuation performance and a faster response. Meanwhile, lifting the high-frequency gain of input weighting function yields an efficient suppression of the high-frequency oscillation of input. The most difficult part is how to ensure that these two kinds of weighting functions do not conflict in the middle-frequency band. Otherwise, no solution can be obtained. So trial and error is inevitable.

In the tuning of weighting function, the following transfer function may be used:

$$W(s) = k\frac{s+a}{s+b}, \quad k = \sqrt{\frac{\omega_c^2 + b^2}{\omega_c^2 + a^2}} \tag{16.18}$$

if we simply want to tune the high-frequency gain and low-frequency gain without changing the crossover frequency ω_c.

16.6 Case Study

The procedure of $\mathcal{H}_\infty$ control design is illustrated in this section through a case study.

In the head positioning control of hard disk drive (HDD) of Example 11.3, the plant has a multiplicative uncertainty, and the head is subject to a wind disturbance brought by the high-speed rotation of the disk. We hope to suppress the wind disturbance so that the head can be accurately positioned on the designated track. In order to achieve this goal, we choose the generalized plant shown in Figure 16.4.

This generalized plant is determined based on the following considerations. Firstly, the HDD is sealed and the disk rotates at a constant speed, so the wind can be regarded as a constant torque disturbance (i.e., a step signal) which is applied at the input port. Since the objective is positioning the head, the head position should be chosen as a performance output. As for the multiplicative uncertainty, it can be treated by the small-gain theorem. In Figure 16.4, w_2 and z_2 are the input and output used to ensure robustness to the multiplicative uncertainty (between them is the uncertainty); z_3 is a performance output used to penalize the control input u; w_1 and z_1 are the input and output used to penalize the disturbance response. W_2 expresses the gain of the multiplicative uncertainty, and W_1 represents the dynamics of the disturbance. W_3 is a parameter mainly used to tune the response speed. W_4 is a weighting function used to adjust the control input. If the $\mathcal{H}_\infty$ norm of the closed-loop transfer matrix about the generalized plant is less than 1, the robust disturbance attenuation performance is guaranteed.

As shown in Figure 16.5(a), the gain of uncertainty weighting function W_2 starts rising sharply around $\omega = 2 \times 10^4$ rad/s. So it is impossible to achieve effective control in the frequency domain above it. Therefore, the disturbance weighting function W_1 and the input weighting function W_4 should intersect in the vicinity of this frequency. The wind disturbance is set as a step signal whose transfer function is W_1. The gain of W_1 should be chosen as high

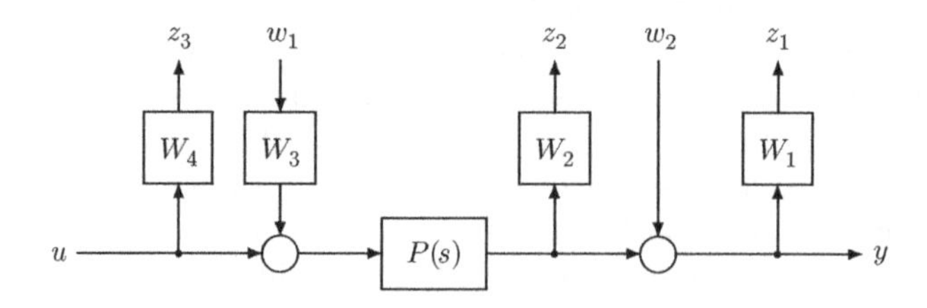

Figure 16.4 Generalized plant for HDD

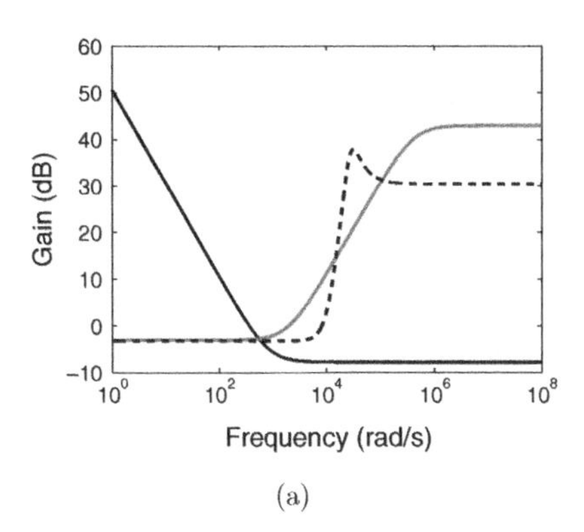

(a)

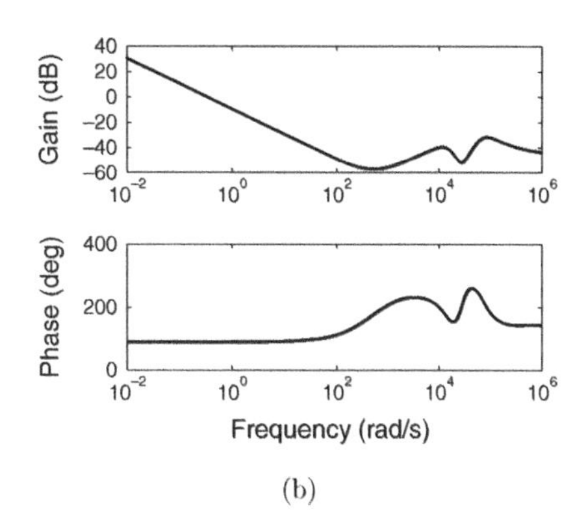

(b)

Figure 16.5 Weighting functions and $\mathcal{H}_\infty$ controller (a) Weighting functions (b) $\mathcal{H}_\infty$ controller

as possible. Through trial and error, the following weighting functions are determined finally:

$$W_1(s) = \frac{s + 8.1 \times 10^2}{s + 1.0 \times 10^{-6}} \times 4.1 \times 10^{-4}$$

$$W_2(s) = \left(\frac{s^2 + 1.4 \times 10^4 s + 1.1 \times 10^8}{s^2 + 1.9 \times 10^4 s + 7.6 \times 10^8}\right)^2 \times 33$$

$$W_3 = 1.0 \times 10^{-3}, W_4(s) = \frac{s + 2.0 \times 10^3}{s + 4.0 \times 10^5} \times 1.4 \times 10^2.$$

Figure 16.5(a) shows, respectively, the weighting function $W_2(s)$ of the multiplicative uncertainty (dashed line, high pass), the disturbance weighting function $W_1(s)$ (solid line, low pass), and the input weighting function $W_4(s)$ (solid line, high pass). The Bode plot of the designed $\mathcal{H}_\infty$ controller is shown in Figure 16.5(b) whose upper side is the gain and lower side is the phase. It has an integrator characteristic in the low frequency, contains a notch and a phase lead compensation in the middle-frequency band, and rolls off in the high frequency. An effective and frequently used method in vibration control is to insert a notch filter in the band of resonant modes so as to lower the open-loop gain in this band. In robust control design, such notch characteristic of controller is automatically obtained by penalizing the uncertainty.

The output response to a step disturbance at the input port is shown in Figure 16.6. Meanwhile, the input response is shown in Figure 16.7. There is no noticeable difference in the outputs, which shows that almost the same output response has been achieved. Meanwhile, input of the actual system is much more oscillatory.

16.7 Scaled $\mathcal{H}_\infty$ Control

In the robust performance design to be illustrated in the subsequent chapters, we will often encounter $\mathcal{H}_\infty$ control problems with a constant scaling matrix L, that is,

$$\|L^{1/2} H_{zw} L^{-1/2}\|_\infty < \gamma. \tag{16.19}$$

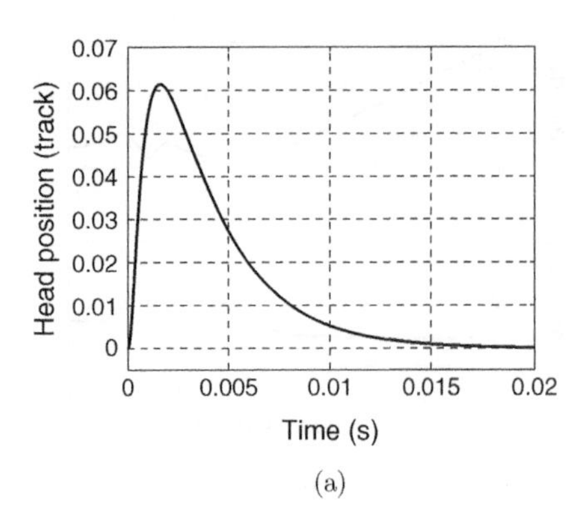

(a)

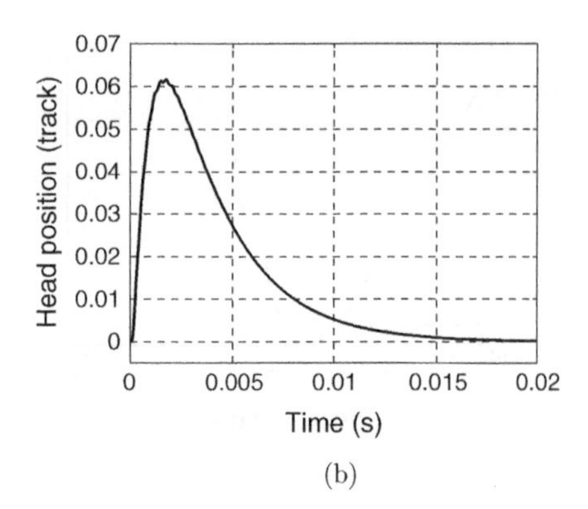

(b)

Figure 16.6 Step disturbance response (output) (a) Nominal (b) Actual

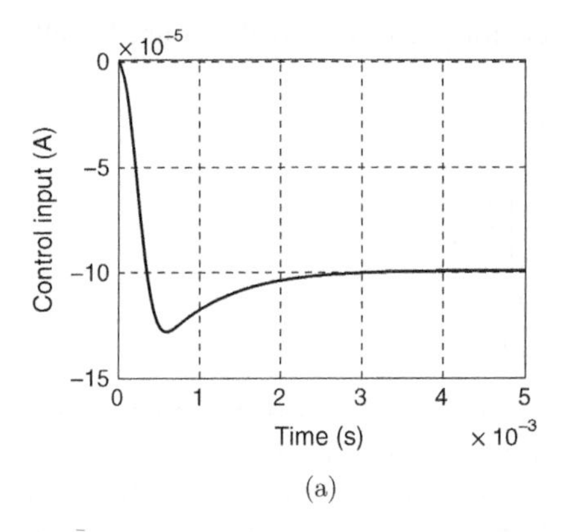

(a)

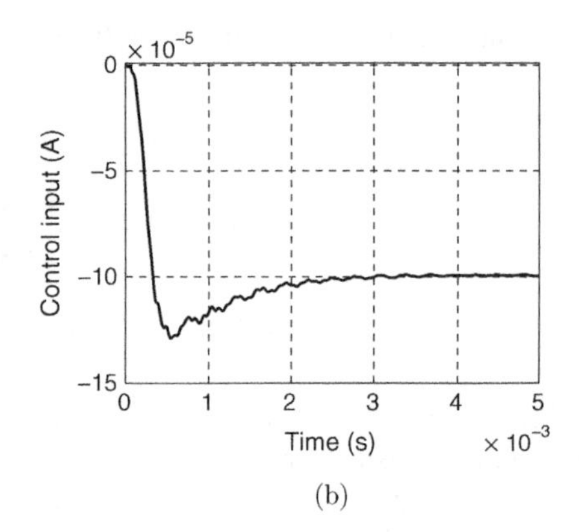

(b)

Figure 16.7 Step disturbance response (input) (a) Nominal (b) Actual

Therefore, this section describes the solution for such problems.

Without loss of generality, we may limit the scaling matrix L to positive definite matrix. According to Exercise 8.2, for a matrix $L > 0$ with a suitable dimension, the closed-loop system $H_{zw}(s) = (A_c, B_c, C_c, D_c)$ satisfies the $\mathcal{H}_\infty$ norm specification (16.19) iff there exist positive definite matrices P, L satisfying

$$\begin{bmatrix} A_c^T P + PA_c & PB_c & C_c^T \\ B_c^T P & -\gamma L & D_c^T \\ C_c & D_c & -\gamma L^{-1} \end{bmatrix} < 0. \tag{16.20}$$

Starting from this inequality, it is easy to derive the solvability condition for this scaled $\mathcal{H}_\infty$ control problem by following the variable elimination method of Section 16.3.

Theorem 16.2 *Assume that $G(s)$ satisfies the condition (A1). There are a controller $K(s)$ and a scaling matrix $L > 0$ such that the closed-loop system is stable and $\|L^{1/2} H_{zw} L^{-1/2}\|_\infty < \gamma$*

iff there exist matrices $X > 0, Y > 0$ and L, J satisfying the following conditions:

$$\begin{bmatrix} N_X^T & 0 \\ 0 & I_{n_w} \end{bmatrix} \begin{bmatrix} AX + XA^T & XC_1^T & B_1 \\ C_1X & -\gamma J & D_{11} \\ B_1^T & D_{11}^T & -\gamma L \end{bmatrix} \begin{bmatrix} N_X & 0 \\ 0 & I_{n_w} \end{bmatrix} < 0 \tag{16.21}$$

$$\begin{bmatrix} N_Y^T & 0 \\ 0 & I_{n_z} \end{bmatrix} \begin{bmatrix} YA + A^TY & YB_1 & C_1^T \\ B_1^TY & -\gamma L & D_{11}^T \\ C_1 & D_{11} & -\gamma J \end{bmatrix} \begin{bmatrix} N_Y & 0 \\ 0 & I_{n_z} \end{bmatrix} < 0 \tag{16.22}$$

$$\begin{bmatrix} X & I \\ I & Y \end{bmatrix} \geq 0 \tag{16.23}$$

$$LJ = I. \tag{16.24}$$

Here, all notations are as defined as Section 16.3.

Unfortunately, the condition $LJ = I$ in this theorem is not convex. So the solvability condition cannot be solved by using LMI approach directly. We need to use the *K–L iteration method* that follows:

Step 1 Let $L = I$.
Step 2 Compute a controller $K(s)$ such that $\|L^{1/2}H_{zw}L^{-1/2}\|_\infty$ is minimized, and denote the minimal norm by γ_K.
Step 3 Fixing the controller $K(s)$, find scaling matrix $L > 0$ such that $\|L^{1/2}H_{zw}L^{-1/2}\|_\infty$ is minimized and denote the minimal norm by γ_L.
Step 4 If $\gamma_K - \gamma_L$ is less than a specified value, end the design and output the controller $K(s)$ obtained in Step 2; otherwise, return to Step 2.

L is known in Step 2, so we can compute γ_K and $P > 0$ by solving a GEVP based on solvability condition of Theorem 16.2, that is,

$$\min\ \gamma \text{ subject to (16.21), (16.22), (16.23)}$$

Then, the controller $K(s)$ is obtained by solving the following LMI:

$$Q + E^T\mathcal{K}F + F^T\mathcal{K}^TE < 0 \tag{16.25}$$

$$\begin{bmatrix} Q & E^T \\ F & \end{bmatrix} = \left[\begin{array}{ccc|c} \overline{A}^TP + P\overline{A} & P\overline{B}_1 & \overline{C}_1^T & P\overline{B}_2 \\ \overline{B}_1^TP & -\gamma_K L & \overline{D}_{11}^T & 0 \\ \overline{C}_1 & \overline{D}_{11} & -\gamma_K J & \overline{D}_{12} \\ \hline \overline{C}_2 & \overline{D}_{21} & 0 & \end{array}\right]. \tag{16.26}$$

Meanwhile, the optimization problem in Step 3 can be solved by solving the GEVP:

$$\min\ \gamma \text{ subject to}$$
$$\begin{bmatrix} A_c^T P + P A_c & P B_c & C_c^T L \\ B_c^T P & -\gamma L & D_c^T L \\ L C_c & L D_c & -\gamma L \end{bmatrix} < 0, \quad P > 0, \quad L > 0. \tag{16.27}$$

•

Exercise

16.1 Let the input and output of stable transfer function $P(s) = \frac{s+1}{s+10}$ be $u(t)$ and $y(t)$, respectively.
(a) Draw the asymptotes of gain and phase of $P(s)$ in the Bode plot.
(b) Calculate $\|P\|_\infty$.
(c) Compute the maximal output energy $\sup \|y\|_2$ w.r.t. all inputs $u(t)$ satisfying $\|u\|_2 \leq 1/2$.

16.2 Let us revisit the cruise control problem discussed in Exercise 12.6. Assume that $M \in [M_1, M_2], \mu \in [\mu_1, \mu_2]$. Starting from the state equation of $\tilde{P}(s)$

$$\dot{x} = -\frac{\mu}{M}x + \frac{1}{M}F, \quad v = x,$$

solve the same problem again by the polytopic method.

16.3 In Figure 16.8, $P(s)$, $W(s)$ are, respectively,

$$P(s) = \frac{1}{s+1}, \quad W(s) = \frac{1}{s}, \quad w(t) = \delta(t).$$

In order to suppress the disturbance d, we hope to ensure $\|H_{yw}\|_\infty < 1$, in which $H_{yw}(s)$ is the closed-loop transfer function from w to y. Examine whether this goal can be achieved by the following controller:

$$K = \frac{Q}{1 - PQ}, \quad Q = \frac{s+1}{as+b}, \quad a > 0, \ b > 0.$$

If possible, find the condition on parameters (a, b). Further, discuss the strategy of disturbance attenuation, i.e., how to tune these two parameters.

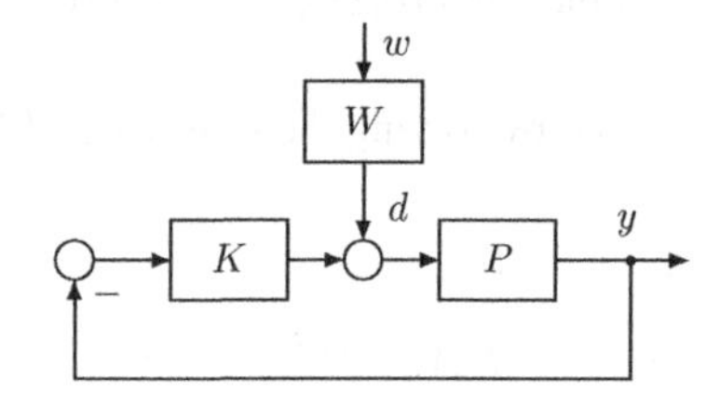

Figure 16.8 Disturbance attenuation

16.4 Derive the LMI solvability condition for state feedback $\mathcal{H}_\infty$ control with $u = Fx$ using the methods of variable elimination and variable change.

16.5 Multiply the first row block and the first column block of (16.9) by Π_1^T and Π_1, respectively; then derive the LMI solutions of Section 16.4 by using the variable change method.

16.6 Follow the variable elimination method of Section 16.3 to derive the solution for the scaled $\mathcal{H}_\infty$ control problem in Section 16.7, based on Exercise 8.2.

16.7 Try to use the variable change method and K–L iteration to solve the scaled $\mathcal{H}_\infty$ control problem in Section 16.7.

Notes and References

Among the approaches to $\mathcal{H}_\infty$ control problem, the interpolation method of the early stage can be found in Refs [48, 25, 29]; ARE solution is in Refs [36, 26, 100]; refer to Refs [33, 42, 32] for the LMI solution. A simple proof for the ARE solution is given in Ref. [54]. Liu and Mita [58, 55] discussed the discrete $\mathcal{H}_\infty$ control problem and the parametrization of state feedback $\mathcal{H}_\infty$ controllers. Some industrial applications of $\mathcal{H}_\infty$ control may be found in Refs [41, 39] (hard disk drive), [74] (mine train), [73] (engine clutch), [70] (multi-area power system), and [75] (paper machine). For the benchmark on HDD head positioning, refer to Ref. [87].

17

μ Synthesis

This chapter discusses the robust stability and robust performance for plants having multiple uncertainties. Based on the small-gain principle, we have succeeded in deriving conditions that can guarantee the robust stability or robust performance for plants with a single norm-bounded uncertainty in Chapter 11. However, for the problems considered in this chapter, it will be very conservative to apply the small-gain principle directly. New concepts and mathematical tools are needed. Here, the key is that we can always aggregate the multiple uncertainties into a block diagonal matrix by appropriate transformation of the block diagram. So, we can focus on the uncertainty in a form of block diagonal matrix. Such uncertainty is called *structured uncertainty*.

Just as the robust stability conditions for uncertainties in the form of single matrix are described by the largest singular value, the robust stability conditions for uncertainties in the form of block diagonal matrix can be expressed by the so-called *structured singular value*, with an alias μ.

More importantly, by using μ, we are able to derive the necessary and sufficient conditions for a class of robust performance, which cannot be obtained from the small-gain theorem. This has a very important practical significance. By using the tool of μ synthesis, it is possible to maximize the potential of feedback control in control design.

This chapter gives a brief description on the concept of μ, the necessary and sufficient condition for robust performance, and the μ synthesis. A design example will also be illustrated in some detail.

17.1 Introduction to μ

17.1.1 Robust Problems with Multiple Uncertainties

In Chapter 11, we have discussed the robust control problem for plant sets with uncertainty. The problem handled that there are about systems containing a single uncertainty. However, it is rare that all uncertainties of the plant gather in the same place; they are usually dispersed

Robust Control: Theory and Applications, First Edition. Kang-Zhi Liu and Yu Yao.

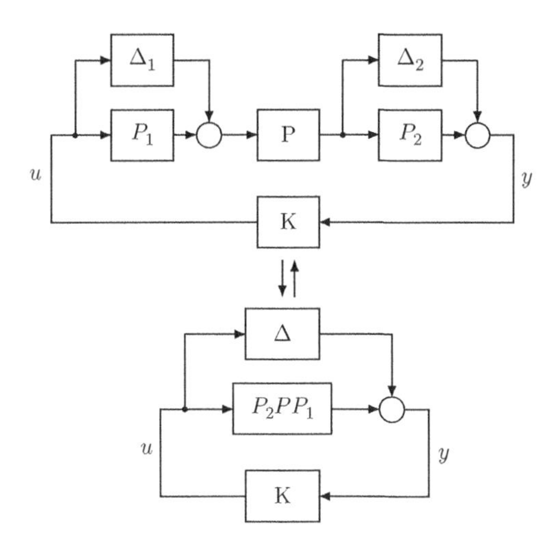

Figure 17.1 Merging two uncertainties into one

in different parts of the system. If these uncertainties are compulsorily put into an uncertainty and handled with the techniques of Chapter 11, usually the result will be very conservative. This is observed in the following example.

Example 17.1 *The plant shown in Figure 17.1 has two uncertainties. When they are put together as in Figure 17.1, by the input–output relation*

$$\hat{y} = (\Delta_2 + P_2)P(\Delta_1 + P_1)\hat{u} = P_2PP_1\hat{u} + \Delta\hat{u}, \tag{17.1}$$

we get

$$\Delta = \Delta_2P\Delta_1 + \Delta_2PP_1 + P_2P\Delta_1. \tag{17.2}$$

In treating this Δ as a norm-bounded uncertainty, we need to estimate an upper bound based on the upper bounds of Δ_1, Δ_2. This estimation cannot be done accurately, and the bound is often enlarged. Further, in the estimated scope of Δ, there are many other uncertainties which are not in the class of (17.2). Thereby, the scope of plant set is expanded. As a consequence, the plant set under consideration is far greater than the actual plant set. Then, the constraint on the controller gets more stringent so that the obtained result is often very conservative. Most typically, the controller gain has to be lowered in the low-frequency band which is crucial for system performance, thus making it impossible to realize good disturbance attenuation and fast response.

However, the scattered uncertainties can always be collected into a block diagonal matrix. In the present case, via the transformation of block diagram shown in Figure 17.2, these two uncertainties can be rewritten as a diagonal matrix $\begin{bmatrix} \Delta_1 & 0 \\ 0 & \Delta_2 \end{bmatrix}$ where

$$\begin{bmatrix} \hat{z}_1 \\ \hat{z}_2 \end{bmatrix} = M \begin{bmatrix} \hat{w}_1 \\ \hat{w}_2 \end{bmatrix}, \tag{17.3}$$

$$M = \begin{bmatrix} KP_2(I - PP_1KP_2)^{-1}P & (I - KP_2PP_1)^{-1}K \\ (I - PP_1KP_2)^{-1}P & PP_1(I - KP_2PP_1)^{-1}K \end{bmatrix}. \tag{17.4}$$

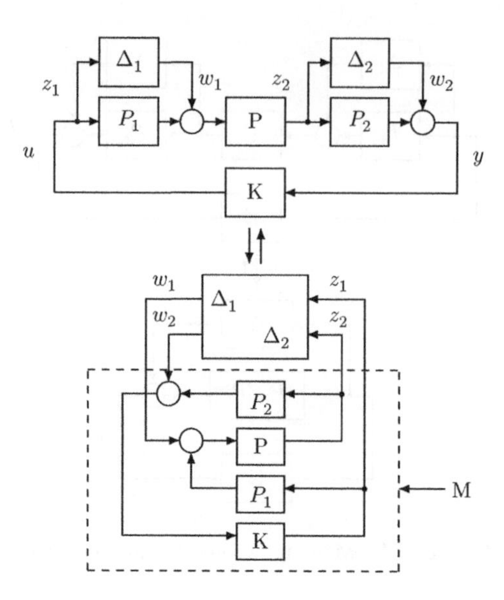

Figure 17.2 Collecting two uncertainties as a diagonal matrix

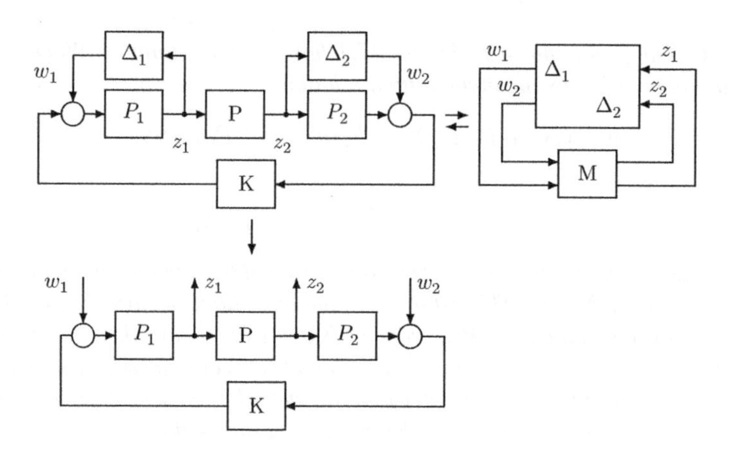

Figure 17.3 How to calculate the M matrix

This transformation does not change the uncertainties. Therefore, it is possible to achieve a less conservative control design.

Example 17.2 *The system shown in Figure 17.3, which contains an additive uncertainty and a feedback uncertainty, can also be transformed into a system with a diagonal uncertainty. Here*

$$\begin{bmatrix} \hat{z}_1 \\ \hat{z}_2 \end{bmatrix} = M \begin{bmatrix} \hat{w}_1 \\ \hat{w}_2 \end{bmatrix}, \tag{17.5}$$

$$M = \begin{bmatrix} (I - P_1KP_2P)^{-1}P_1 & (I - P_1KP_2P)^{-1}P_1K \\ P(I - P_1KP_2P)^{-1}P_1 & P(I - P_1KP_2P)^{-1}P_1K \end{bmatrix}. \tag{17.6}$$

In such transformations, the calculation of M matrix is in fact very easy. As shown in Figure 17.3, the method is to name the inputs and outputs of uncertainties first and then remove the uncertainties. After that, the matrix M is obtained by computing the transfer matrix from the output vector of uncertainties to their input vector.

In general, when there are r uncertainties Δ_i $(i = 1, \ldots, r)$, the closed-loop system can always be rewritten as the one in Figure 17.4. The block diagonal uncertainty Δ is called *structured uncertainty*. Here, both M and Δ_i are transfer matrices. Such transformation does not change the stability of system.

Next, we discuss the condition for the robust stability of closed-loop system. For simplicity, it is assumed that the uncertainty $\Delta = \text{diag}(\Delta_1 \cdots, \Delta_r)$ is stable. Note that the roots of $\det[I - M(s)\Delta(s)] = 0$ are the poles of closed-loop system. The closed-loop system must be stable even when $\Delta = 0$. In this case, only $M(s)$ is left in the loop. This implies that $M(s)$ must be stable, which corresponds to the nominal stability. Moreover, $\det[I - M(s)\Delta] = 1 \neq 0$ holds in the closed right half-plane. Next, we fix the dynamics of uncertainty Δ and increase its gain gradually until the closed-loop system becomes unstable. Since the closed-loop poles vary continuously with the uncertainty, they must cross the imaginary axis before getting unstable (see Figure 17.5). Therefore, the uncertainty that destabilizes the closed-loop system for the first time must be the one with the smallest norm in all Δ's satisfying

$$\det(I - M(j\omega)\Delta(j\omega)) = 0, \quad \exists\omega \in [0, \infty). \tag{17.7}$$

Its norm becomes the supremum of the norms of all uncertainties that guarantee the stability of closed-loop system and is called *stability margin*. It is easy to see that the stability margin depends on the diagonal structure of the uncertainty Δ and the matrix M. The reciprocal of the stability margin is exactly the structured singular value $\mu_\Delta(M(j\omega))$ to be touched in the next section.

17.1.2 Robust Performance Problem

Theorem 12.2 of Section 12.4 shows that, via the introduction of a virtual uncertainty representing the performance, a problem of robust $\mathcal{H}_\infty$ performance can be equivalently converted

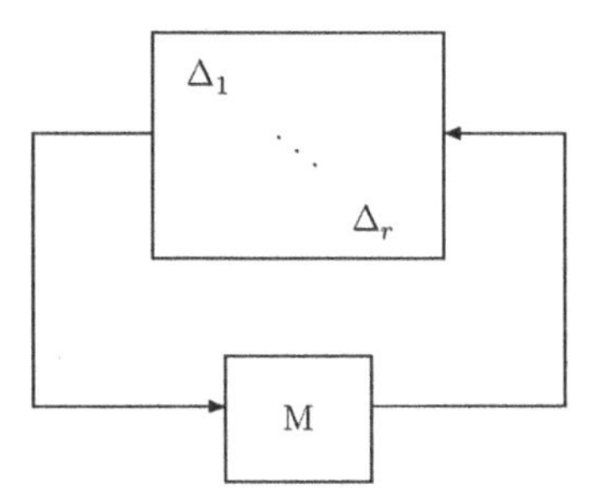

Figure 17.4 System with structured uncertainty

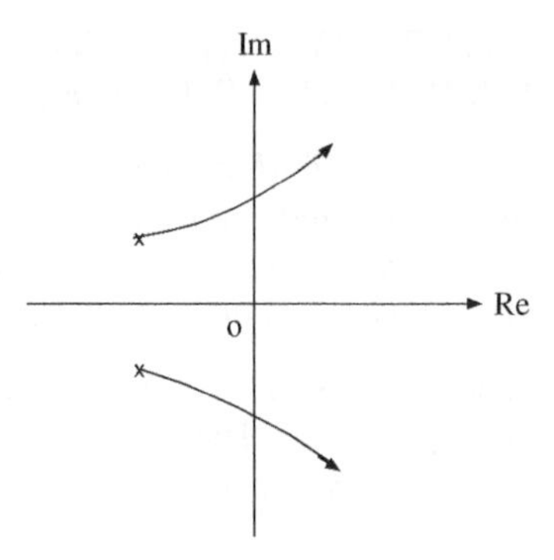

Figure 17.5 Continuity of poles

into a robust stabilization problem of systems with structured uncertainty. This is an even more important motivation for considering the robust stabilization of systems with structured uncertainty in this chapter.

17.2 Definition of μ and Its Implication

When considering a transfer matrix at a frequency, it becomes a complex matrix. Therefore, we will fix the frequency and only consider the set of block diagonal matrices $\mathbf{\Delta} \subset \mathbb{C}^{m\times n}$ and matrix $M \in \mathbb{C}^{n\times m}$. Assume that the set $\mathbf{\Delta}$ has the following structure:

$$\mathbf{\Delta} = \{\Delta \mid \Delta = \text{diag}(\delta_1 I_{r_1}, \ldots, \delta_S I_{r_S}, \Delta_1, \ldots, \Delta_F)\}, \quad \delta_i \in \mathbb{C}, \quad \Delta_j \in \mathbb{C}^{m_j\times n_j} \tag{17.8}$$

in which

$$\sum_{i=1}^{S} r_i + \sum_{j=1}^{F} m_j = m, \quad \sum_{i=1}^{S} r_i + \sum_{j=1}^{F} n_j = n. \tag{17.9}$$

δ_i is called a *scalar uncertainty*, while Δ_j is called a *full-block uncertainty*. Note that a full-block uncertainty is a matrix uncertainty and all of its elements are uncertain.

Definition 17.1 *For any given matrix $M \in \mathbb{C}^{n\times n}$, the structured singular value $\mu_{\mathbf{\Delta}}(M)$ is defined as*

$$\mu_{\mathbf{\Delta}}(M) = \frac{1}{\min\{\sigma_{\max}(\Delta) \mid \Delta \in \mathbf{\Delta},\ \det(I - M\Delta) = 0\}}. \tag{17.10}$$

$\mu_{\mathbf{\Delta}}(M) = 0$ when there is no $\Delta \in \mathbf{\Delta}$ satisfying $\det(I - M\Delta) = 0$.

According to this definition, $\mu_{\mathbf{\Delta}}(M)$ is the reciprocal of the gain of the smallest uncertainty among all $\Delta \in \mathbf{\Delta}$ satisfying $\det(I - M\Delta) = 0$. Therefore, $\det(I - M\Delta) \neq 0$ holds for all $\Delta \in \mathbf{\Delta}$ satisfying $\sigma_{\max}(\Delta) < 1/\mu_{\mathbf{\Delta}}(M)$. Conversely, as long as there is one $\Delta_1 \in \mathbf{\Delta}$ such that $\sigma_{\max}(\Delta_1) \geq 1/\mu_{\mathbf{\Delta}}(M)$, there must be a $\Delta \in \mathbf{\Delta}$ satisfying $\det(I - M\Delta) = 0$.

Remark 17.1 *$\mu_{\mathbf{\Delta}}(M)$ may be interpreted as the reciprocal of the supremum of stability margin for the feedback system in Figure 17.4 when the frequency is fixed. In this regard, we will* *make a more detailed analysis. Here, assume that $M(s)$ and $\Delta(s)$ are stable systems and $\Delta(s)$ belongs to a set of stable uncertainties with the block diagonal structure of $\mathbf{\Delta}$.*

According to the discussion in the previous section, the closed-loop system is robustly stable iff

$$\det[I - M(j\omega)\Delta(j\omega)] \neq 0 \quad \forall\omega$$

holds for all $\Delta(s) \in \mathbf{\Delta}$. Therefore, as long as

$$\sigma_{\max}(\Delta(j\omega)) < \frac{1}{\mu_{\mathbf{\Delta}}(M(j\omega))}$$

on the entire imaginary axis, the stability is guaranteed. Otherwise, there must be an uncertainty destabilizing the closed-loop system (see Exercise 17.2). Roughly speaking, the smaller $\mu_{\mathbf{\Delta}}(M)$ is, the bigger the permissible uncertainty will be. That is, the problem of maximizing the range of uncertainty while guaranteeing the stability of control system has been converted into the problem of minimizing $\mu_{\mathbf{\Delta}}(M)$.

Summarizing these discussions, we obtain the robust stability condition for systems with structured uncertainty.

Theorem 17.1 *In Figure 17.4, assume that the nominal system $M(s)$ and the structured uncertainty $\Delta(s) \in \mathbf{\Delta}$ are stable and $\|\Delta\|_\infty < \gamma$. Then, the closed-loop system is robustly stable iff*

$$\sup_\omega \mu_{\mathbf{\Delta}}(M(j\omega)) \leq \frac{1}{\gamma}. \tag{17.11}$$

17.3 Properties of μ

It is worth noting that the value of $\mu_{\mathbf{\Delta}}(M)$ depends on the structure of Δ, that is, the number of blocks in the block diagonal matrix and the property and dimension of each diagonal block, as well as the nominal system M. It is not determined solely by the matrix M. This makes the calculation of $\mu_{\mathbf{\Delta}}(M)$ very hard. However, when Δ has some special structures, we are able to calculate $\mu_{\mathbf{\Delta}}(M)$ purely based on M. The following presents these special cases.

For simplicity, only square uncertainty Δ is treated hereafter.

17.3.1 Special Cases

17.3.1.1 Single Scalar Block Uncertainty $\Delta = \{\delta I \,|\, \delta \in \mathbb{C}\}$

In this case, $\mu_{\mathbf{\Delta}}(M) = \rho(M)$ holds. Here, $\rho(M)$ denotes the spectral radius of matrix M, that is, the maximum absolute value of all eigenvalues of M.

Proof. First, $\det(I - M\delta) = \det(\delta^{-1}I - M)\det(\delta I) = 0$ holds. So any nonzero δ^{-1} satisfying this equation is an eigenvalue of M. Then, it is easy to see that the reciprocal of the minimum size of uncertainty δ satisfying this equation is the spectral radius $\rho(M)$ of matrix M, that is,

$$\mu_{\mathbf{\Delta}}(M) = \frac{1}{\min(|\delta| : \det(I - M\delta) = 0)}$$

$$= \max(|\delta^{-1}| : \det(\delta^{-1}I - M) = 0)$$
$$= \rho(M).$$

•

17.3.1.2 Full-Block Uncertainty $\boldsymbol{\Delta} = \mathbb{C}^{n \times n}$

In this case, all elements are uncertain and $\mu_{\boldsymbol{\Delta}}(M) = \sigma_{\max}(M)$ holds.

Proof. $\det(I - M\Delta) \neq 0$ holds when $1/\sigma_{\max}(\Delta) > \sigma_{\max}(M)$. Therefore, there must be $\mu_{\boldsymbol{\Delta}}(M) \leq \sigma_{\max}(M)$.

On the other hand, by singular value decomposition, we know that

$$M = V \operatorname{diag}(\sigma_1, \ldots, \sigma_n)U, \quad UU^* = I, \ VV^* = I$$

hold in which $\sigma_{\max}(M) = \sigma_1 \geq \sigma_2 \geq \cdots \geq \sigma_n \geq 0$. Since

$$u = U^*[1, 0, \ldots, 0]^T \Rightarrow Mu = \sigma_1 V[1, 0, \ldots, 0]^T = \sigma_1 v$$

where $v = V[1, 0, \ldots, 0]^T$. Set $\Delta = \sigma_1^{-1}uv^*$. This Δ satisfies $\sigma_{\max}(\Delta) = \sigma_1^{-1}$ and

$$(I - M\Delta)v = v - \sigma^{-1}Muv^*v = 0.$$

So, $\mu_{\boldsymbol{\Delta}}(M) \geq \sigma_{\max}(M)$ holds. Therefore, $\mu_{\boldsymbol{\Delta}}(M) = \sigma_{\max}(M)$. •

17.3.2 Bounds of $\mu_{\boldsymbol{\Delta}}(M)$

The following inclusion relation of sets is true for any set of square uncertainties:

$$\{\delta I_n \mid \delta \in \mathbb{C}\} \subset \boldsymbol{\Delta} \subset \mathbb{C}^{n \times n}. \tag{17.12}$$

When the uncertainty is restricted to its subset, a greater uncertainty magnitude is allowed so that the corresponding μ gets smaller. Therefore, from the aforementioned special cases, we have the following inequalities:

$$\rho(M) \leq \mu_{\boldsymbol{\Delta}}(M) \leq \sigma_{\max}(M). \tag{17.13}$$

Therefore, $\sigma_{\max}(M), \rho(M)$ are, respectively, upper and lower bounds of $\mu_{\boldsymbol{\Delta}}(M)$. Since the structure of $\boldsymbol{\Delta}$ is not taken into account, in general the discrepancy between the upper and lower bounds may be very big. In order to obtain a tighter estimate of $\mu_{\boldsymbol{\Delta}}(M)$, we should make full use of the structure of $\boldsymbol{\Delta}$.

Next, we try to use the diagonal structure of $\boldsymbol{\Delta}$ to calculate upper and lower bounds that are closer to each other. Considered here is the case where each diagonal block of $\boldsymbol{\Delta}$ is a square matrix, that is, $m_i = n_i$. The following set of scaling matrices is introduced:

$$\mathbb{D} = \{D \mid D = \operatorname{diag}(D_1, \ldots, D_S, d_1 I_{m_1}, \ldots, d_{F-1} I_{m_{F-1}} I_{m_F})\} \tag{17.14}$$

in which

$$D_i \in \mathbb{C}^{r_i \times r_i}, \quad D_i = D_i^* > 0, \quad d_j \in \mathbb{R}, d_j > 0$$

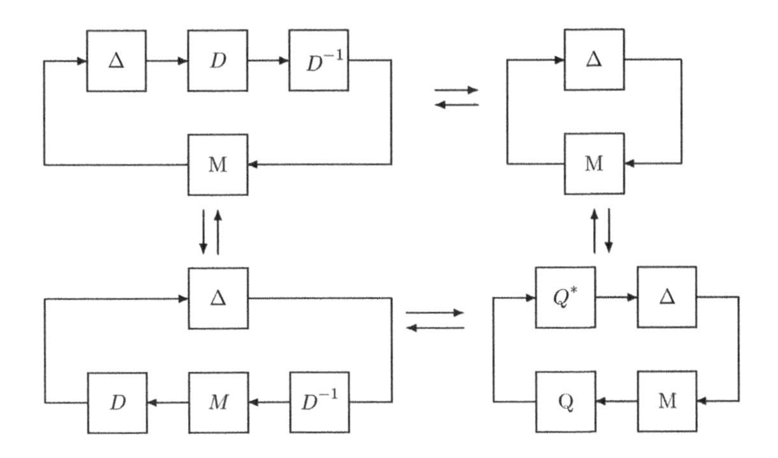

Figure 17.6 Introduction of scaling and calculation of μ

and

$$\mathbb{Q} = \{Q \in \mathbf{\Delta} \,|\, Q^*Q = I_n\}. \tag{17.15}$$

Note that $d_F = 1$. This is because D and D^{-1} will appear always in pairs, and $\mu_{\mathbf{\Delta}}(\alpha DM(\alpha D)^{-1}) = \mu_{\mathbf{\Delta}}(DMD^{-1})$ apparently holds. Therefore, for simplicity, we have normalized D in advance. It is easy to verify the following relations (see Figure 17.6):

$$Q^* \in \mathbb{Q}, \quad Q\Delta \in \mathbf{\Delta}, \quad \Delta Q \in \mathbf{\Delta}$$

$$\sigma_{\max}(Q\Delta) = \sigma_{\max}(\Delta Q) = \sigma_{\max}(\Delta) \tag{17.16}$$

$$D\Delta = \Delta D. \tag{17.17}$$

Then, the following theorem holds.

Theorem 17.2 *For any $Q \in \mathbb{Q}$ and $D \in \mathbb{D}$, there hold*

$$\mu_{\mathbf{\Delta}}(M) = \mu_{\mathbf{\Delta}}(DMD^{-1}) = \mu_{\mathbf{\Delta}}(QM) = \mu_{\mathbf{\Delta}}(MQ). \tag{17.18}$$

Proof. First of all, we obtain $\mu_{\mathbf{\Delta}}(M) = \mu_{\mathbf{\Delta}}(DMD^{-1})$ from $\det(I - M\Delta) = \det(I - MD^{-1}D\Delta) = \det(I - MD^{-1}\Delta D) = \det(I - DMD^{-1}\Delta)$.

Secondly, $\mu_{\mathbf{\Delta}}(M) = \mu_{\mathbf{\Delta}}(MQ)$ holds because $\det(I - M\Delta) = 0 \Leftrightarrow \det(I - MQQ^*\Delta) = 0$, $Q^*\Delta \in \mathbf{\Delta}$ and $\sigma_{\max}(Q^*\Delta) = \sigma_{\max}(\Delta)$. Similarly, we can prove that $\mu_{\mathbf{\Delta}}(M) = \mu_{\mathbf{\Delta}}(QM)$. •

From this theorem, we get the following new bounds for $\mu_{\mathbf{\Delta}}(M)$:

$$\max_{Q \in \mathbb{Q}} \rho(QM) \leq \mu_{\mathbf{\Delta}}(M) \leq \inf_{D \in \mathbb{D}} \sigma_{\max}(DMD^{-1}). \tag{17.19}$$

This indicates that we may approach $\mu_{\mathbf{\Delta}}(M)$ by solving optimization problems about the spectral radius and the largest singular value. When the difference between the upper and lower bounds is small enough, any of them can be used to approximate $\mu_{\mathbf{\Delta}}(M)$.

17.4 Condition for Robust $\mathcal{H}_\infty$ Performance

Recall the robust tracking problem in Section 12.4. For stable uncertainty Δ satisfying $\|\Delta\|_\infty < 1$, the robust tracking condition is given by

$$\|WKS\|_\infty \leq 1, \quad \|W_S S(1+\Delta WKS)^{-1}\|_\infty \leq 1. \tag{17.20}$$

Further, based on the matrix inversion formula and the definition of linear fractional transformation (LFT), we can conduct the following transformation:

$$\begin{aligned} W_S S(1+\Delta WKS)^{-1} &= W_S S - W_S S(1+\Delta WKS)^{-1}\Delta WKS \\ &= \mathcal{F}_u\left(\begin{bmatrix} -WKS & WKS \\ -W_S S & W_S S \end{bmatrix}, \ \Delta\right) \\ &= \mathcal{F}_u(M, \Delta). \end{aligned}$$

Here

$$M(s) = \begin{bmatrix} -WKS & WKS \\ -W_S S & W_S S \end{bmatrix}.$$

Then, the robust tracking condition becomes

$$\|WKS\|_\infty \leq 1, \quad \|\mathcal{F}_u(M, \Delta)\|_\infty \leq 1.$$

Block diagram of the corresponding closed-loop system is shown in Figure 17.7(a). The first norm condition ensures the robust stability and the second ensures the robust reference tracking. According to Theorem 12.2, the performance condition $\|\mathcal{F}_u(M, \Delta)\|_\infty \leq 1$ is equivalent to the robust stability of the closed-loop system after inserting a virtual uncertainty Δ_f with $\mathcal{H}_\infty$ norm less than 1 between the disturbance w and the performance output z, as shown in Figure 17.7(b). As a result, the necessary and sufficient condition for the robust tracking is given by

$$\mu_{\mathbf{\Delta}_P}\left(\begin{bmatrix} -WKS & WKS \\ -W_S S & W_S S \end{bmatrix}(j\omega)\right) \leq 1 \quad \forall \omega \tag{17.21}$$

in which $\mathbf{\Delta}_P$ is a dilated structured uncertainty:

$$\mathbf{\Delta}_P := \left\{\begin{bmatrix} \Delta & 0 \\ 0 & \Delta_f \end{bmatrix} | \Delta \in \mathbf{\Delta}, \Delta_f \in \mathbb{C}^{q\times p}\right\}. \tag{17.22}$$

The $\mathcal{H}_\infty$ norm condition with scaling obtained in Section 12.4 happens to be an upper bound of the μ condition.

This example can be extended to the general case. Specifically, Theorem 17.3 holds.

Theorem 17.3 *Consider the closed-loop system shown in Figure 17.7(a), where M is a stable transfer matrix and the uncertainty $\Delta(s) \in \mathbf{\Delta}$ is stable and satisfies $\|\Delta\|_\infty < 1$. The closed-loop system satisfies $\|\mathcal{F}_u(M, \Delta)\|_\infty \leq 1$ for all uncertainties iff*

$$\sup_{\omega\in\mathbb{R}} \mu_{\mathbf{\Delta}_P}(M(j\omega)) \leq 1. \tag{17.23}$$

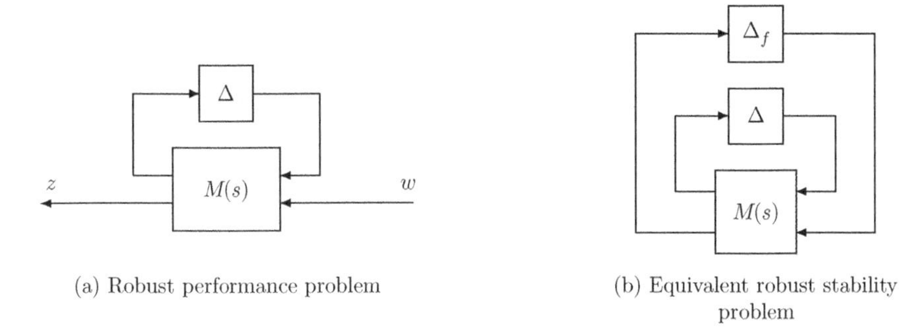

(a) Robust performance problem

(b) Equivalent robust stability problem

Figure 17.7 Robust performance problem and equivalent robust stability problem (a) Robust performance problem, (b) Equivalent robust stability problem

Proof. Firstly, we fix $\Delta(s)$. According to Theorem 12.2, the performance condition $\|\mathcal{F}_u(M,\Delta)\|_\infty \leq 1$ is equivalent to the robust stability condition with an uncertainty $\Delta_f(s)$ whose $\mathcal{H}_\infty$ norm is less than 1 being inserted between the input and output of the closed-loop system, as shown in Figure 17.7(b). The equivalence holds for all the uncertainties satisfying $\|\Delta\|_\infty < 1$. So, we get the conclusion. •

Remark 17.2 *In the previous theorem, $\Delta(s)$ is the real uncertainty. However, another uncertainty $\Delta_f(s)$ is not a true uncertainty but a virtual uncertainty introduced to represent the control performance. The property of $\Delta_f(s)$ depends on the control performance specification. Here, the reason for assuming that the virtual uncertainty is a full-block uncertainty $\Delta_f(s) \in \mathbb{C}^{p\times q}$ is that the $\mathcal{H}_\infty$ norm is used to specify the performance.*

17.5 D–K Iteration Design

So far, we have shown that a robust performance condition can be converted equivalently into a μ condition. In this section, we consider how to design a controller to meet this condition, that is, the μ synthesis problem. Conceptually, μ is a powerful tool for explaining succinctly the robust performance problem as well as the robust stability problem with multiple uncertainties. However, from the point of practical applications, it is already very difficult to compute the value of μ for a given system, not to mention the μ synthesis problem. Therefore, a reasonable approach is to find an approximate solution based on the upper and lower bounds of μ. So, we consider making full use of the relationships

$$\max_{Q\in\mathbb{Q}} \rho(QM) \leq \mu_{\boldsymbol{\Delta}}(M) \leq \inf_{D\in\mathbb{D}} \sigma_{\max}(DMD^{-1}). \tag{17.24}$$

This lower bound can be proved to be exactly equal to μ[24]. Unfortunately, the maximization problem of spectral radius is not convex and contains more than one local maxima. So, it is difficult to calculate μ using the lower bound. On the contrary, the upper bound is the largest singular value, and we can prove that its minimization problem is convex. It is relatively easy to calculate the global minimum. For this reason, the μ synthesis basically uses this upper bound to find an approximate solution.

17.5.1 Convexity of the Minimization of the Largest Singular Value

As $D(s)$ and $M(s)$ in (17.24) are both transfer matrices, the minimization of $\sigma_{\max}(DMD^{-1})$ should be done pointwise in the frequency domain. So, we fix the frequency ω in the analysis. In this case, D and M can be treated as complex matrices.

According to the property of the largest singular value, minimizing $\sigma_{\max}(DMD^{-1})$ is equivalent to minimizing $\gamma > 0$ satisfying

$$(DMD^{-1})^*(DMD^{-1}) \leq \gamma^2 I.$$

Set an Hermitian matrix $X = D^*D$. When M is known, the aforementioned inequality can be rewritten as

$$M^*XM \leq \gamma^2 X. \tag{17.25}$$

This inequality is an LMI about the variable matrix X. Minimizing γ subject to this inequality is a GEVP and convex. So, its solution converges globally. Therefore, we can solve the optimization problem

$$\min \gamma$$
$$\text{subject to (17.25)}$$

to obtain the optimal X. Then, D is computed by using the singular value decomposition method.

17.5.2 Procedure of D–K Iteration Design

In general, any closed-loop transfer matrix M can be expressed as an LFT about the controller $K(s)$:

$$M(s) = \mathcal{F}_\ell(G, K). \tag{17.26}$$

So, the upper bound given in (17.24) is the minimum about the controller $K(s)$ and scaling matrix $D(s)$. When we take the maximum of $\sigma_{\max}(DMD^{-1})$ w.r.t. all frequencies, the maximum singular value $\sigma_{\max}$ becomes the $\mathcal{H}_\infty$ norm. Then, we obtain

$$\sup_\omega \mu_\Delta(M) \leq \inf_{D\in\mathcal{D}} \|DMD^{-1}\|_\infty \tag{17.27}$$

in which $\mathcal{D}$ denotes the set

$$\mathcal{D} := \{D(s) \mid D(s),\ D^{-1}(s) \text{ stable and } D(j\omega) \in \mathbb{D}\}.$$

In designing the controller, there is no way to solve for the scaling matrix D and controller $K(s)$ simultaneously. The so-called D–K iteration method is usually used. The idea is that when the controller $K(s)$ is known, the closed-loop transfer matrix $M(s)$ is also fixed. So, the scaling matrix $D(s)$ can be calculated pointwise. Meanwhile, when the scaling matrix $D(s)$ is given, the controller $K(s)$ can be obtained by solving an $\mathcal{H}_\infty$ control problem.

Specifically, the D–K iteration uses successively these two parameters to do the minimization. That is, fix $D(s)$ first and use $K(s)$ to do the optimization. Next, fix the obtained $K(s)$

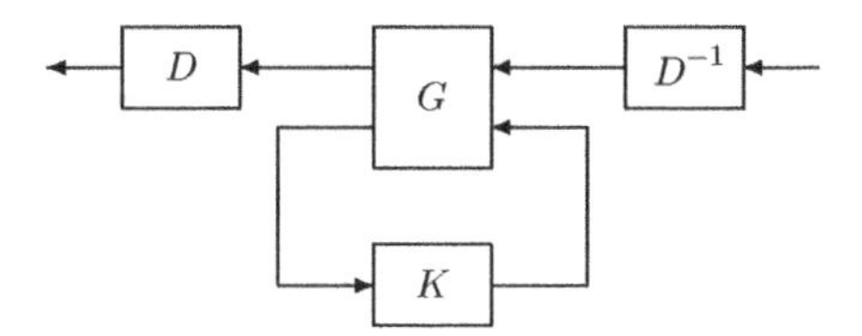

Figure 17.8 μ synthesis using scaling

and use $D(s)$ to do the optimization. Then, repeat this process. In the case without any scalar block uncertainty, the detail is as follows:

1. Determine an initial scaling matrix $D_\omega = \text{diag}(d_1^\omega I, \ldots, d_{F-1}^\omega I, I)$ in the whole frequency domain pointwise. Usually, the initial value is taken as the identity matrix.
2. For $i = 1, 2, \ldots, (F-1)$, find a stable minimum-phase transfer function $d_i(s)$ satisfying $|d_i(j\omega)| \approx d_i^\omega$.
3. Set $D(s) = \text{diag}(d_1(s)I, \ldots, d_{F-1}(s)I, I)$ and calculate the generalized plant shown in Figure 17.8:
$$\hat{G}(s) = \begin{bmatrix} D(s) & \\ & I \end{bmatrix} G(s) \begin{bmatrix} D^{-1}(s) & \\ & I \end{bmatrix}.$$
4. Solve the $\mathcal{H}_\infty$ control problem to minimize the norm $\|\mathcal{F}_\ell(\hat{G}, K)\|_\infty$. Denote the optimal controller as $\hat{K}$. Note that the generalized plant $\hat{G}$ contains the scaling matrix.
5. Seek a D_ω that minimizes $\sigma_{\max}(D_\omega \mathcal{F}_\ell(G, \hat{K}) D_\omega^{-1})$ at each frequency over the whole domain and denote this new function as $\hat{D}_\omega$. Note that in this optimization, the optimal $\hat{K}$ obtained from previous step is used.
6. Compare this $\hat{D}_\omega$ with the last D_ω. If they are close enough, end the calculation. Otherwise, replace G and D_ω by $\hat{G}$ and $\hat{D}_\omega$, respectively; then return to Step 2.

As long as any of K and D is fixed, the optimal problem has a global optimum w.r.t. the remaining variable. The optimal solution can be obtained either by optimization methods or by $\mathcal{H}_\infty$ control. What needs to be emphasized is that the original problem is a simultaneous optimization problem about two variables K and D, which is not convex. So, D–K iteration cannot guarantee the global convergence of the solution. However, numerous applications have proven that the D–K iteration method is quite effective.

Moreover, the D iteration part in D–K iteration may be viewed as an automatic tuning of weighting functions in $\mathcal{H}_\infty$ design. Another feature of μ synthesis is that, after each iteration, the scaling function is added to the generalized plant. So the order of generalized plant gets higher and higher, which leads to a very high order of the final controller. Usually, model reduction is necessary before the controller is implemented.

17.6 Case Study

In Section 16.6, we designed an $\mathcal{H}_\infty$ controller for the head positioning control of a hard disk drive. The design is based on a sufficient condition and may not bring the potential of feedback

Table 17.1 Transitions of μ and $\mathcal{H}_\infty$ norm

Number of D-K iterations	1	2
μ	1.378	0.998
$\mathcal{H}_\infty$ norm	1.432	0.999

control into full play. So here we try to use the D–K iteration to conduct the μ synthesis, aiming at further improving the performance of the control system. The generalized plant is identical to Figure 16.4. But the disturbance specification is more stringent than that in Example 16.6 (the gain of weighting function $W_1(s)$ is 60% higher).

The weighting functions are as follows:

$$W_1(s) = 1.6 \times \frac{s + 1.1 \times 8.1 \times 10^2}{s + 1.0 \times 10^{-6}} \times 4.1 \times 10^{-4}$$

$$W_2(s) = \left(\frac{s^2 + 1.4 \times 10^4 s + 1.1 \times 10^8}{s^2 + 1.9 \times 10^4 s + 7.6 \times 10^8}\right)^2 \times 33$$

$$W_3 = 1.0 \times 10^{-3}, \quad W_4(s) = \frac{s + 2.0 \times 10^3}{s + 4.0 \times 10^5} \times 1.4 \times 10^2.$$

The transitions of μ and $\mathcal{H}_\infty$ norms are shown as in Table 17.1. A value of μ less than 1 is obtained in the second D–K iteration. The corresponding μ controller is shown in Figure 17.9 (solid line). Compared with the $\mathcal{H}_\infty$ controller (dashed line), its low-frequency gain is higher. The disturbance response and control input are shown in Figure 17.10 and Figure 17.11, respectively, where the solid line is for μ and the dashed line for $\mathcal{H}_\infty$. First of all, as in the $\mathcal{H}_\infty$ control case, there is almost no difference in the output response. Secondly, compared with $\mathcal{H}_\infty$ control although the magnitudes of the two inputs are roughly the same, there is a significant

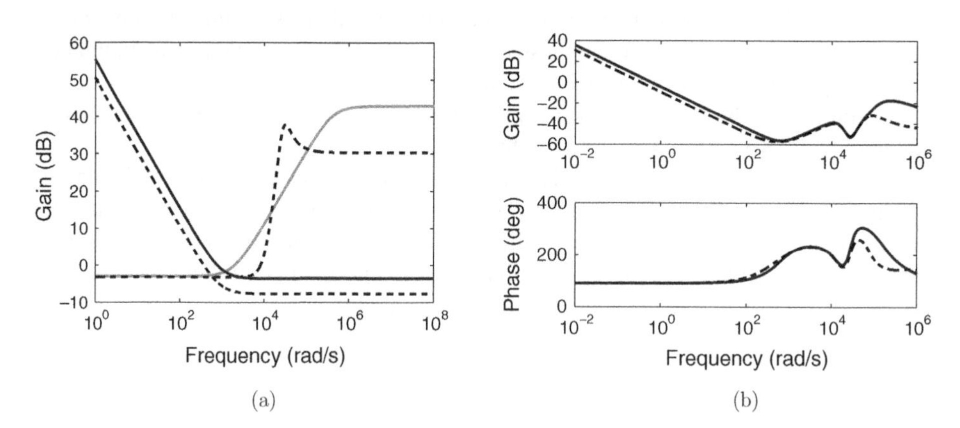

Figure 17.9 Comparison of μ and $\mathcal{H}_\infty$ controllers (solid: μ, dashed: $\mathcal{H}_\infty$) (a) Weighting functions (b) Controllers

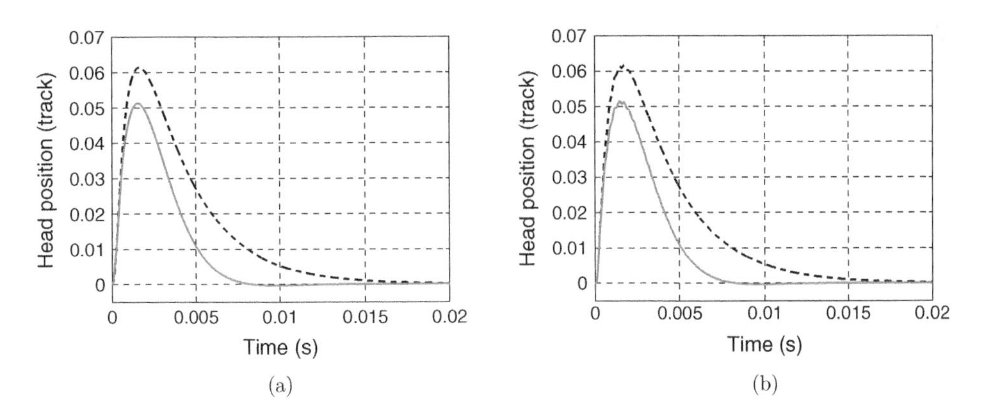

Figure 17.10 Step disturbance response (output; solid: μ, dashed: $\mathcal{H}_\infty$) (a) Norminal output response (b) Actual output response

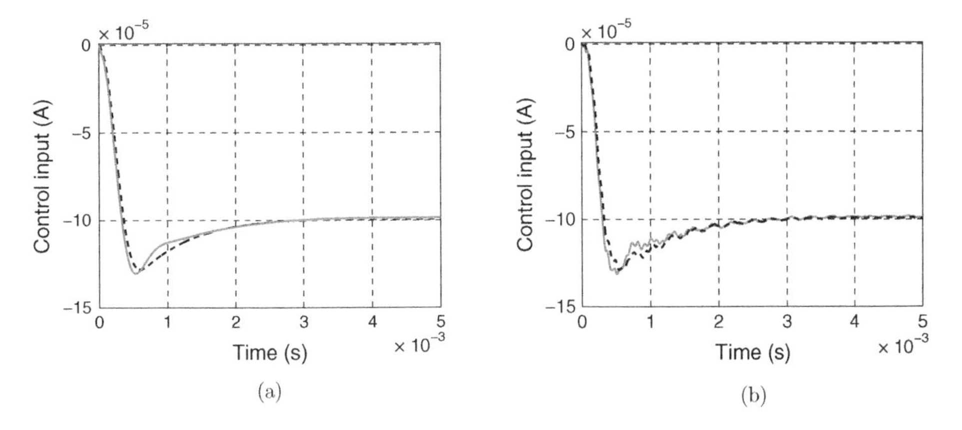

Figure 17.11 Step disturbance response (input; solid: μ, dashed: $\mathcal{H}_\infty$) (a) Nominal response (b) Actual response

gap in the disturbance responses. The μ controller has greatly improved the response. Looking at the input, the μ controller acts slightly faster.

Exercises

17.1 In the system with structured uncertainty shown in Figure 17.4, assume that the nominal system $M(s)$ and uncertainty $\Delta(s)$ are stable. Prove that the closed-loop system is robustly stable iff

$$\det[I - M(j\omega)\Delta(j\omega)] \neq 0 \quad \forall\omega$$

holds for all $\Delta(s)$, based on Nyquist stability criterion.

17.2 Prove the necessity of Theorem 17.1 following the steps below:

(a) When there is a frequency ω_0 at which

$$\mu_\Delta(M(j\omega_0)) > \frac{1}{\gamma}$$

holds, there must be a $\gamma_0 < \gamma$ and an uncertainty

$$\Delta_0 = \mathrm{diag}(\delta_1^0 I_{r_1}, \ldots, \delta_S^0 I_{r_S}, \ \Delta_1^0, \ldots, \Delta_F^0), \quad \|\Delta_0\| = \gamma_0$$

as well as a nonzero complex vector u such that $[I - M(j\omega_0)\Delta_0]u = 0$.

(b) Define $v = \Delta_0 u$ and partition (u, v) in accordance with the structure of Δ_0 so that $v_i = \delta_i^0 u_i, v_k = \Delta_k^0 u_k$ (the subscripts i, k indicate scalar block and full-block uncertainties, respectively).

(c) Following the proof of small-gain theorem to construct a stable transfer function $\delta_i(s)$ and a stable transfer matrix $\Delta_k(s)$ such that

$$\delta_i(j\omega_0) = \delta_i^0, \quad \|\delta_i\|_\infty = |\delta_i^0| \le \gamma_0$$

$$\Delta_k(j\omega_0) = v_k \frac{u_k^*}{\|u_k\|^2}, \quad \|\Delta_i\|_\infty = \|\Delta_i^0\| \le \gamma_0.$$

(d) Construct $\Delta(s) \in \mathbf{\Delta}$ using $\delta_i(s)$ and $\Delta_k(s)$ then prove that it satisfies $\Delta(j\omega_0)u = v$ and $\|\Delta\|_\infty < \gamma$.

(e) Prove that $j\omega_0$ becomes a pole of the closed-loop system for this uncertainty $\Delta(s)$.

Notes and References

μ theory was proposed by Doyle [24]. Refer to Refs [77, 100, 5, 27] for more details as well as tutorials. Hirata *et al.* [40] gave a method to transform parametric systems so that they can be solved by μ.

18

Robust Control of Parametric Systems

In Chapter 13, we analyzed the quadratic stability of parametric systems. Based on these analyses, this chapter deals with the robust stabilization of parametric systems. Further, we illustrate in detail the $\mathcal{H}_\infty$ control design methods for this class of systems. This chapter focuses on polytopic systems and systems with norm-bounded parameter uncertainty.

18.1 Quadratic Stabilization of Polytopic Systems

18.1.1 State Feedback

To quadratically stabilize the polytopic system

$$\dot{x} = \left(\sum_{i=1}^{N} \lambda_i A_i\right) x + \left(\sum_{i=1}^{N} \lambda_i B_i\right) u, \quad \lambda_i \geq 0, \quad \sum_{i=1}^{N} \lambda_i = 1, \tag{18.1}$$

how should we design the feedback control?

Consider the state feedback first

$$u = Fx. \tag{18.2}$$

Here, the closed-loop system is

$$\begin{aligned} \dot{x} &= \left(\sum_{i=1}^{N} \lambda_i A_i + \sum_{i=1}^{N} \lambda_i B_i F\right) x = \sum_{i=1}^{N} \lambda_i \left(A_i + B_i F\right) x \\ &\lambda_i \geq 0, \quad \sum_{i=1}^{N} \lambda_i = 1. \end{aligned} \tag{18.3}$$

From Chapter 13, we know that the condition for the quadratic stabilization of a polytopic system boils down to the existence of a gain matrix F and a positive definite matrix P satisfying the inequality

$$(A_i + B_i F)^T P + P(A_i + B_i F) < 0 \quad \forall i.$$

That is,

$$A_i^T P + PA_i + (B_i F)^T P + PB_i F < 0 \quad \forall i. \tag{18.4}$$

However, the product terms $(B_i F)^T P, PB_i F$ of unknown matrices appear in the inequality. So this inequality is a BMI rather than an LMI, which makes numerical computation very difficult.

To get over this difficulty, variable change is effective. In particular, we focus on the positive definiteness of matrix P. Here, let us try setting $M = PB_i F$. The idea is to calculate (P, M) first and then reverse to F. Unfortunately, B_i is singular in general. So F cannot be calculated from M. Hence, it is necessary to change the matrix product into a direct product of P and F first. This is done by premultiplying and postmultiplying (18.4) by $Q = P^{-1}$. Then, (18.4) turns into

$$QA_i^T + A_i Q + (FQ)^T B_i^T + B_i FQ < 0.$$

So, a change of variable

$$Q = P^{-1}, \;\; X = FQ \Longleftrightarrow F = XQ^{-1} \tag{18.5}$$

transforms the condition (18.4) for quadratic stability into

$$QA_i^T + A_i Q + X^T B_i^T + B_i X < 0 \quad \forall i. \tag{18.6}$$

(18.6) is an LMI about the new variable matrices (Q, X) and can be solved numerically.

18.1.2 Output Feedback

In the variable change method introduced in Section 3.2.4, the new variables contain the coefficient matrices of the plant. For this reason, in general the variable change method cannot be applied to solve the robust control problem for systems with uncertain parameters. But we may try the variable elimination method.

18.1.2.1 Quadratic Stabilization of Nominal System

Consider the quadratic stabilization of the nominal system first. Let the realizations of the plant and the (full-order) controller be

$$P(s) = (A,\ B,\ C,\ 0), \quad K(s) = (A_K,\ B_K,\ C_K,\ D_K), \tag{18.7}$$

respectively. The A matrix of the closed-loop system formed by a positive feedback interconnection of them is

$$A_c = \overline{A} + \overline{B}\mathcal{K}\overline{C}$$

$$\mathcal{K} = \begin{bmatrix} D_K & C_K \\ B_K & A_K \end{bmatrix}, \; \overline{A} = \begin{bmatrix} A & 0 \\ 0 & 0 \end{bmatrix}, \; \overline{B} = \begin{bmatrix} B & 0 \\ 0 & I \end{bmatrix}, \; \overline{C} = \begin{bmatrix} C & 0 \\ 0 & I \end{bmatrix}. \tag{18.8}$$

Therefore, the quadratic stability condition reduces to the existence of a positive definite matrix P satisfying

$$A_c^T P + PA_c = \overline{A}^T P + P\overline{A} + \overline{C}^T \mathcal{K}^T \overline{B}^T P + P\overline{B}\mathcal{K}\overline{C} < 0. \tag{18.9}$$

According to Lemma 3.1, this inequality is solvable iff

$$\overline{C}_\perp^T(\overline{A}^T P + P\overline{A})\overline{C}_\perp < 0, \quad \overline{B}_\perp^T P^{-1}(\overline{A}^T P + P\overline{A})P^{-1}\overline{B}_\perp^T < 0$$

holds simultaneously. It is easy to know that

$$\overline{C}_\perp = \begin{bmatrix} C_\perp \\ 0 \end{bmatrix}, \quad \overline{B}_\perp = [B_\perp \; 0].$$

Next, we decompose the positive definite matrix P as

$$P = \begin{bmatrix} Y & * \\ * & * \end{bmatrix}, \quad P^{-1} = \begin{bmatrix} X & * \\ * & * \end{bmatrix}.$$

Substituting this decomposition into the previous conditions, we arrive at the solvable condition

$$\begin{aligned} & C_\perp^T(A^T Y + YA)C_\perp < 0 \\ & B_\perp(A^T X + XA)B_\perp^T < 0 \\ & \begin{bmatrix} X & I \\ I & Y \end{bmatrix} > 0 \end{aligned} \tag{18.10}$$

for the output feedback quadratic stabilization problem after some simple calculation.

18.1.2.2 Quadratic Stabilization of Parametric Systems

From the proof of the sufficiency of Theorem 3.1, we see that the constructed solution depends on all matrices in the inequality. When some matrix has uncertainty, the obtained solution is not realizable. Therefore, for parametric systems, (18.10) is only a necessary condition, not sufficient. In addition, it is seen from the foregoing results that this method needs the orthogonal matrices of B and C. For an uncertain matrix, the relationship between its orthogonal matrix and the uncertain parameters is very complicated. Hence, it is very difficult to reduce the preceding solvability condition to finite number of LMIs about the vertices of the parameter polytope. So the following assumptions are made.

Assumption Matrices B and C are independent of the uncertain parameters, while A is a matrix polytope of the uncertain parameter vector λ, that is,

$$A = \sum_{i=1}^{N} \lambda_i A_i, \quad \lambda_i \geq 0, \; \sum_{i=1}^{N} \lambda_i = 1. \tag{18.11}$$

Under this assumption, the necessary condition (18.10) for the solvability of the output feedback quadratic stabilization problem can be reduced to the vertex conditions about the parameter vector polytope. Specifically, the solvability of the following LMIs is necessary for the existence of a fixed controller such that the feedback system consisting of the plant of (18.1) and the controller is quadratic stable ($i = 1, \ldots, N$):

$$\begin{aligned} B_\perp(A_i^T X + X A_i)B_\perp^T &< 0 \\ C_\perp^T(A_i^T Y + Y A_i)C_\perp &< 0 \\ \begin{bmatrix} X & I \\ I & Y \end{bmatrix} &> 0. \end{aligned} \tag{18.12}$$

Calculating P using (X, Y) according to Lemma 3.1 and then substituting it back into LMIs

$$\overline{A_i^T} P + P\overline{A}_i + \overline{C}^T \mathcal{K}^T \overline{B}^T P + P\overline{B}\mathcal{K}\overline{C} < 0, \ i = 1, \ldots, N, \tag{18.13}$$

we may get a controller. Here, $\overline{A}_i$ is a matrix obtained by replacing the matrix A in $\overline{A}$ by A_i. The output feedback quadratic stabilization can be realized if these LMIs have a solution $\mathcal{K}$.

The next question is how to design a quadratically stable feedback system when B or C contains parameter uncertainty. Let us seek the answer from engineering practice. In actual systems, in order to avoid being sensitive to sensor noise, the controller often needs to be rolled off in the high-frequency domain. That is, only a strictly proper controller is practical. If we hope to roll off the controller in the frequency band higher than ω_c, we can insert the following low-pass filter to the controller in advance:

$$F(s) = \frac{\omega_c}{s + \omega_c} I = \left[\begin{array}{c|c} -\omega_c I & \omega_c I \\ \hline I & 0 \end{array}\right]. \tag{18.14}$$

That is, to use the following controller in the feedback control

$$C(s) = F(s)K(s). \tag{18.15}$$

In the design, we may absorb the filter $F(s)$ into the plant and design a controller $K(s)$ for the augmented plant:

$$G(s) = P(s)F(s) = \left[\begin{array}{cc|c} A & B & 0 \\ 0 & -\omega_c I & \omega_c I \\ \hline C & 0 & 0 \end{array}\right]. \tag{18.16}$$

Thus, even if the matrix B contains uncertain parameters, after augmentation the new B does not depend on the uncertain parameters. If the matrix C is also independent of parameter uncertainty, then in the augmented plant, only the matrix A contains uncertain parameters. Hence we can apply the previous results. For example, for the mass–spring–damper system

$$\dot{x} = \begin{bmatrix} 0 & 1 \\ -\dfrac{k}{m} & -\dfrac{b}{m} \end{bmatrix} x + \begin{bmatrix} 0 \\ \dfrac{1}{m} \end{bmatrix} u, \ \ y = [1 \ 0]x,$$

this method is applicable. Moreover, if integral control is desired, the filter can be set as an integrator $F(s) = 1/s$.

As for the case where the matrix C contains uncertain parameters, the output can be filtered by a low-pass filter before being applied to the controller. Then, the new plant becomes

$$G(s) = F(s)P(s) = \left[\begin{array}{cc|c} A & 0 & B \\ \omega_c C & -\omega_c I & 0 \\ \hline 0 & I & 0 \end{array}\right]. \tag{18.17}$$

The new C matrix is independent of the uncertain parameters. It is a rare case that both B and C contain uncertain parameters. In fact, when the states are all selected as physical variables, the matrix C will not contain any physical parameter since the measured output must be physical variables.

18.2 Quadratic Stabilization of Norm-Bounded Parametric Systems

The discussion in the previous section shows that the output feedback design of polytopic systems is very difficult. So, we discuss the output feedback control problem of norm-bounded parametric systems in this section. Based on Section 11.3.3, we know that a general model for stabilization is given by

$$G\begin{cases}\dot{x} = Ax + B_1w + B_2u \\ z = C_1x + D_{11}w + D_{12}u \\ y = C_2x + D_{21}w\end{cases} \tag{18.18}$$

$$w = \Delta z, \quad \|\Delta\|_2 \leq 1. \tag{18.19}$$

We design a dynamic output feedback controller K

$$\begin{aligned}\dot{x}_K &= A_Kx_K + B_Ky \\ u &= C_Kx_K + D_Ky\end{aligned} \tag{18.20}$$

to quadratically stabilize the closed-loop system. The part beside Δ is an LFT interconnection between the generalized plant G and the controller K:

$$M(s) = \mathcal{F}_\ell(G, K). \tag{18.21}$$

According to the analysis in Section 13.2.3, the quadratic stability condition is that the nominal closed-loop system $M(s)$ is stable and satisfies the $\mathcal{H}_\infty$ norm condition:

$$\|M\|_\infty < 1. \tag{18.22}$$

This is an $\mathcal{H}_\infty$ control problem whose solution is known. Refer to Chapter 16 for the details.

Compared with the polytopic model, the matching of norm-bounded parametric model and the physical system is not good enough. Strong conservatism may be introduced in such modeling. However, its advantage lies in the simplicity of controller design.

18.3 Robust $\mathcal{H}_\infty$ Control Design of Polytopic Systems

This section discusses the robust $\mathcal{H}_\infty$ control design problem for the polytopic system

$$G(s,\theta) = \left[\begin{array}{c|cc} A(\theta) & B_1(\theta) & B_2 \\ \hline C_1(\theta) & D_{11}(\theta) & D_{12} \\ C_2 & D_{21} & 0\end{array}\right]. \tag{18.23}$$

Here

$$A(\theta) = \sum_{i=1}^{N} \lambda_i A(\theta_i), \quad B_1(\theta) = \sum_{i=1}^{N} \lambda_i B(\theta_i),$$

$$C_1(\theta) = \sum_{i=1}^{N} \lambda_i C(\theta_i), \quad D_{11}(\theta) = \sum_{i=1}^{N} \lambda_i D_{11}(\theta_i)$$

and $\lambda_i \geq 0, \sum_{i=1}^{N} \lambda_i = 1$. θ is an uncertain parameter vector belonging to a polytope with N vertices θ_i. For all uncertain parameters θ, it is assumed that $(A(\theta), B_2)$ is controllable and $(C_2, A(\theta))$ is detectable.

When the uncertain generalized plant is controlled by a fixed controller $K(s) = (A_K, B_K, C_K, D_K)$, the closed-loop system in Figure 16.3 becomes

$$H_{zw}(s,\theta) = (A_c(\theta),\ B_c(\theta),\ C_c(\theta),\ D_c(\theta)) \tag{18.24}$$

in which

$$\begin{bmatrix} A_c(\theta) & B_c(\theta) \\ C_c(\theta) & D_c(\theta) \end{bmatrix} = \begin{bmatrix} \overline{A}(\theta) & \overline{B}_1(\theta) \\ \overline{C}_1(\theta) & \overline{D}_{11}(\theta) \end{bmatrix} + \begin{bmatrix} \overline{B}_2 \\ \overline{D}_{12} \end{bmatrix} \mathcal{K}[\overline{C}_2,\ \overline{D}_{21}] \tag{18.25}$$

$$\begin{bmatrix} \overline{A}(\theta) & \overline{B}_1(\theta) & \overline{B}_2 \\ \overline{C}_1(\theta) & \overline{D}_{11}(\theta) & \overline{D}_{12} \\ \overline{C}_2 & \overline{D}_{21} & \end{bmatrix} = \left[\begin{array}{cc|c|cc} A(\theta) & 0 & B_1(\theta) & B_2 & 0 \\ 0 & 0 & 0 & 0 & I \\ \hline C_1(\theta) & 0 & D_{11}(\theta) & D_{12} & 0 \\ \hline C_2 & 0 & D_{21} & & \\ 0 & I & 0 & & \end{array}\right]. \tag{18.26}$$

According to the bounded-real lemma, the closed-loop system is robustly stable and the norm condition $\|H_{zw}\|_\infty < \gamma$ holds iff there is a positive definite matrix P satisfying the following inequality

$$\begin{bmatrix} A_c(\theta)^T P + PA_c(\theta) & PB_c(\theta) & C_c(\theta)^T \\ B_c(\theta)^T P & -\gamma I & D_c(\theta)^T \\ C_c(\theta) & D_c(\theta) & -\gamma I \end{bmatrix} < 0. \tag{18.27}$$

Since the system contains an uncertain parameter vector θ, as a necessary and sufficient condition, the matrix P should be a function of θ. However, in order to simplify the design, a constant P is used. In this case, the condition (18.27) is just sufficient, not necessary. This is equivalent to using a common Lyapunov function to ensure that the $\mathcal{H}_\infty$ norm of the closed-loop system is less than γ. By Theorem 16.1, the robust $\mathcal{H}_\infty$ control problem has a solution only if the following LMIs

$$\begin{bmatrix} N_X^T & 0 \\ 0 & I_{n_w} \end{bmatrix} \begin{bmatrix} A(\theta)X + XA(\theta)^T & XC_1(\theta)^T & B_1(\theta) \\ C_1(\theta)X & -\gamma I & D_{11}(\theta) \\ B_1(\theta)^T & D_{11}(\theta)^T & -\gamma I \end{bmatrix} \begin{bmatrix} N_X & 0 \\ 0 & I_{n_w} \end{bmatrix} < 0$$

$$\begin{bmatrix} N_Y^T & 0 \\ 0 & I_{n_z} \end{bmatrix} \begin{bmatrix} YA(\theta) + A(\theta)^T Y & YB_1(\theta) & C_1(\theta)^T \\ B_1(\theta)^T Y & -\gamma I & D_{11}(\theta)^T \\ C_1(\theta) & D_{11}(\theta) & -\gamma I \end{bmatrix} \begin{bmatrix} N_Y & 0 \\ 0 & I_{n_z} \end{bmatrix} < 0$$

$$\begin{bmatrix} X & I \\ I & Y \end{bmatrix} \geq 0, \quad \text{rank} \begin{bmatrix} X & I \\ I & Y \end{bmatrix} \leq n + n_K$$

have positive definite solution X, Y.

These conditions rely on the uncertain parameter θ, so the number of LMIs is not finite. In order to find a finite number LMI condition, we note the assumption that $(B_2, C_2, D_{12}, D_{21})$ is independent of the uncertain parameter θ. Then, the matrix N_X, N_Y are also constant. So, the aforementioned LMIs become affine functions of the uncertain parameter vector θ and can be reduced to the finite vertex conditions $(i = 1, \dots, N)$ that follow:

$$\begin{bmatrix} N_X^T & 0 \\ 0 & I_{n_w} \end{bmatrix} \begin{bmatrix} A(\theta_i)X + XA(\theta_i)^T & XC_1(\theta_i)^T & B_1(\theta_i) \\ C_1(\theta_i)X & -\gamma I & D_{11}(\theta_i) \\ B_1(\theta_i)^T & D_{11}(\theta_i)^T & -\gamma I \end{bmatrix} \begin{bmatrix} N_X & 0 \\ 0 & I_{n_w} \end{bmatrix} < 0 \tag{18.28}$$

$$\begin{bmatrix} N_Y^T & 0 \\ 0 & I_{n_z} \end{bmatrix} \begin{bmatrix} YA(\theta_i) + A(\theta_i)^T Y & YB_1(\theta_i) & C_1(\theta_i)^T \\ B_1(\theta_i)^T Y & -\gamma I & D_{11}(\theta_i)^T \\ C_1(\theta_i) & D_{11}(\theta_i) & -\gamma I \end{bmatrix} \begin{bmatrix} N_Y & 0 \\ 0 & I_{n_z} \end{bmatrix} < 0 \tag{18.29}$$

$$\begin{bmatrix} X & I \\ I & Y \end{bmatrix} > 0. \tag{18.30}$$

Here, we have strengthened the third inequality to a strict one. The designed controller has full order, that is, its order is the same as that of the generalized plant.

To compute the controller, after expanding the bounded-real condition (18.27), we rearrange it with respect to $\mathcal{K}$ as

$$Q(\theta) + E^T \mathcal{K} F + F^T \mathcal{K}^T E < 0$$

$$\begin{bmatrix} Q(\theta) & E^T \\ F & \end{bmatrix} = \left[\begin{array}{ccc|c} \overline{A}(\theta)^T P + P\overline{A}(\theta) & P\overline{B}_1(\theta) & \overline{C}_1(\theta)^T & P\overline{B}_2 \\ \overline{B}_1(\theta)^T P & -\gamma I & \overline{D}_{11}(\theta)^T & 0 \\ \overline{C}_1(\theta) & \overline{D}_{11}(\theta) & -\gamma I & \overline{D}_{12} \\ \hline \overline{C}_2 & \overline{D}_{21} & 0 & \end{array}\right].$$

The matrix P can be calculated by the method of Section 16.3.2. Further, only $Q(\theta)$ is an affine function of the uncertain parameter θ in the previous LMI. So it can be reduced to the following vertex conditions:

$$Q(\theta_i) + E^T \mathcal{K} F + F^T \mathcal{K}^T E < 0 \quad \forall \quad i = 1, \dots, N. \tag{18.31}$$

Solving these coupled LMIs leads to the coefficient matrix $\mathcal{K}$ of the controller. It is noted that in general the feasibility of this LMI is not guaranteed.

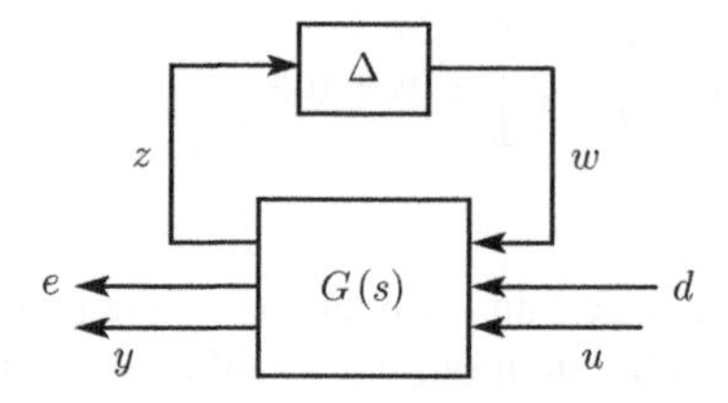

Figure 18.1 Norm-bounded parametric system

18.4 Robust $\mathcal{H}_\infty$ Control Design of Norm-Bounded Parametric Systems

The general model of norm-bounded parametric systems is shown in Figure 18.1 in which

$$G(s) = \left[\begin{array}{c|cc:c} A & B_1 & B_2 & B_3 \\ \hline C_1 & D_{11} & D_{12} & D_{13} \\ C_2 & D_{21} & D_{22} & D_{23} \\ \hdashline C_3 & D_{31} & D_{32} & 0 \end{array}\right], \quad \|\Delta\|_2 \leq 1. \tag{18.32}$$

d is the disturbance signal and e is the performance output, while (z, w) are the input and output of the uncertain parameter matrix Δ.

Now, we discuss the $\mathcal{H}_\infty$ robust disturbance control problem

$$\|H_{ed}\|_\infty < 1 \quad \forall \|\Delta\|_2 \leq 1. \tag{18.33}$$

According to the equivalence between the $\mathcal{H}_\infty$ norm performance and the robust stability (Section 12.3), this performance specification is equivalent to the robust stability condition of the closed-loop system with a virtual norm-bounded uncertainty $\Delta_P(s)$ ($\|\Delta_P\|_\infty \leq 1$) being inserted between the signals d and e. Then, the overall uncertainty becomes $\text{diag}(\Delta, \Delta_P)$. In order to reduce the conservatism of the small-gain method, we introduce a constant scaling matrix following the method of Section 12.4.2. Since these two uncertainties are full-block matrices, the scaling matrix permutable with $\text{diag}(\Delta, \Delta_P)$ is

$$L = \text{diag}(I, \sqrt{\ell} I).$$

This robust $\mathcal{H}_\infty$ control problem can be solved by the scaled robust $\mathcal{H}_\infty$ control method of Section 16.7.

Exercises

18.1 Suppose that $(B_2, D_{12}, C_2, D_{21})$ are also affine functions of the uncertain parameter vector θ in the polytopic system discussed in Section 18.3. Try to propose a method to solve the robust $\mathcal{H}_\infty$ control design problem.

18.2 Consider an uncertain system with a norm-bounded time-varying uncertainty:

$$\dot{x}(t) = (A + B\Delta(t)C)x(t), \quad \|\Delta(t)\|_2 \leq 1 \quad \forall t \geq 0. \tag{18.34}$$

Use the following two methods to prove that this system is quadratically stable if there is a matrix $P > 0$ satisfying

$$\begin{bmatrix} PA + A^T P & PB & C^T \\ B^T P & -I & 0 \\ C & 0 & -I \end{bmatrix} < 0. \tag{18.35}$$

(a) Show that the system is equivalent to a positively feedback-connected system of $\Delta(t)$ and $M(s) = (A, B, C, 0)$; then apply the bounded-real lemma.

(b) Start from the definition of quadratic stability, that is, there is a matrix $P > 0$ such that

$$(A + B\Delta C)^T P + P(A + B\Delta C) < 0 \quad \forall \|\Delta(t)\|_2 \leq 1.$$

Show via the completion of square w.r.t. $\Delta(t)$ that this inequality holds subject to the given LMI (18.35).

19

Regional Pole Placement

In conventional modern control theory, as the model uncertainty is not taken into consideration, the so-called pole placement is to place the poles to fixed points in the complex plane. However, it is impossible to fix the closed-loop poles to specific points when the system has uncertainty. This is because the closed-loop poles move with the variation of plant. Nevertheless, it is still possible to place the closed-loop poles in a region. In addition, from the viewpoint of robust performance, the response quality of the closed-loop system is guaranteed if the closed-loop poles can be locked in a prescribed region. This is what we are yearning for. For this reason, we will first discuss this issue for the nominal system and then extend to the regional pole placement of uncertain systems in this chapter.

19.1 Convex Region and Its Characterization

19.1.1 Relationship between Control Performance and Pole Location

First of all, we consider how to place the system poles so as to achieve a satisfactory transient response. We look at the prototype second-order system

$$G(s) = \frac{\omega_n^2}{s^2 + 2\zeta\omega_n s + \omega_n^2}, \quad 0 < \zeta < 1 \tag{19.1}$$

first. The poles of $G(s)$ are $p = -\zeta\omega_n \pm j\omega_n\sqrt{1-\zeta^2}$. Since the rise time is inversely proportional to the natural frequency ω_n, ω_n must be greater than a certain number $r > 0$ in order to guarantee a short rise time. Meanwhile, to avoid a too large control input, it is necessary to let ω_n be smaller than a certain value $R > 0$. Namely, the size of pole $|p| = \omega_n$ should be in the following range:

$$r \leq |p| = \omega_n \leq R.$$

Moreover, if the damping ratio corresponding to the greatest allowable overshoot is ζ_p, the damping ratio must satisfy

$$\zeta \geq \zeta_p.$$

Robust Control: Theory and Applications, First Edition. Kang-Zhi Liu and Yu Yao.

Hence, the angle between the poles and the real axis must satisfy

$$\tan\theta = \frac{\sqrt{1-\zeta^2}}{\zeta} = \sqrt{\frac{1}{\zeta^2} - 1} \le \sqrt{\frac{1}{\zeta_p^2} - 1}$$

$$\Rightarrow \theta \le \theta_p := \arctan\sqrt{\frac{1}{\zeta_p^2} - 1}.$$

Further, the real parts of the poles are $-\zeta\omega_n$. So, to shorten the settling time,

$$-\zeta\omega_n \le -\sigma$$

should be satisfied w.r.t. the required convergence rate σ. Basically this can be ensured by adjusting r.

Drawing the region satisfying these conditions, we obtain the shaded part in Figure 19.1. Obviously, this region is not convex. Generally speaking, the region of desirable poles is not necessarily convex. However, it is extremely difficult to design a feedback system to place the poles in a nonconvex region. For simplicity of design, the convex region has more advantage. In the prototype second-order system, we can get a convex region by adding to the shaded portion in Figure 19.1 an area surrounded by the dashed vertical line $\Re[z] = -\zeta\omega_n$ and the arc with radius r.

19.1.2 *LMI Region and Its Characterization*

Among convex regions, the *LMI region* described by LMI is particularly important. Figure 19.2 presents four typical LMI regions. We investigate the characterizations of these regions in this subsection.

The point $z = x + jy$ located in the left of straight line $x = -\sigma$ in Figure 19.2(a) can be expressed as

$$x < -\sigma \Leftrightarrow z + \bar{z} < -2\sigma. \tag{19.2}$$

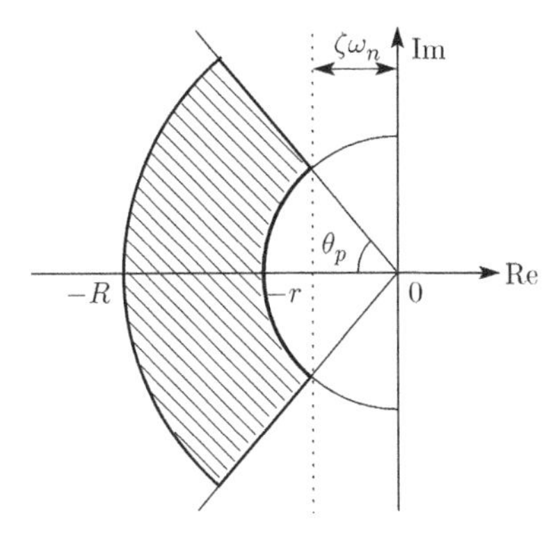

Figure 19.1 Desirable pole region for second-order systems

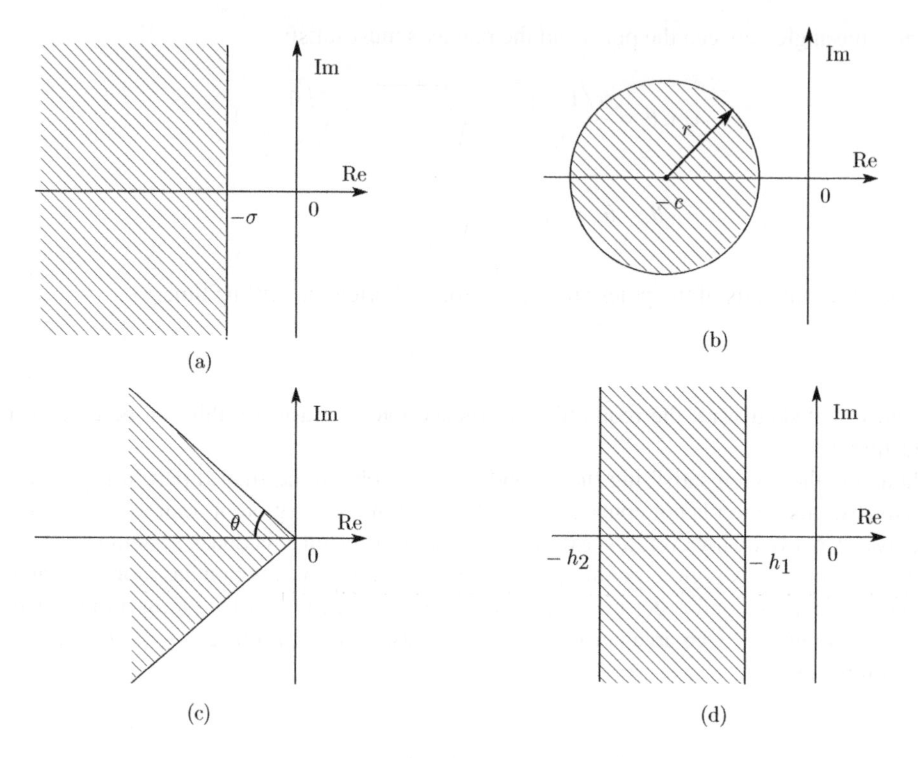

Figure 19.2 Typical examples of LMI region (a) σ region (b) Disk (c) Sector (d) Strip

Secondly, the point $z = x + jy$ located in the disk centered at $(-c, 0)$ and with a radius r in Figure 19.2(b) can be expressed as follows using z and $\overline{z}$:

$$\overline{(z+c)}(z+c) < r^2 \Leftrightarrow -r - (\overline{z}+c) \cdot \frac{1}{-r} \cdot (z+c) < 0 \Leftrightarrow \begin{bmatrix} -r & z+c \\ \overline{z}+c & -r \end{bmatrix} < 0.$$

Here, we have applied Schur's lemma. Splitting the inequality further into terms about z and $\overline{z}$, it can be rewritten as

$$\begin{bmatrix} -r & c \\ c & -r \end{bmatrix} + z \begin{bmatrix} 0 & 1 \\ 0 & 0 \end{bmatrix} + \overline{z} \begin{bmatrix} 0 & 0 \\ 1 & 0 \end{bmatrix} < 0. \tag{19.3}$$

As for the point $z = x + jy$ in the sector $|\arg z - \pi| < \theta$ of the left half-plane in Figure 19.2(c), it satisfies

$$\frac{|y|}{-x} < \tan\theta \Leftrightarrow x\sin\theta < -|y|\cos\theta < 0$$
$$\Leftrightarrow (x\sin\theta)^2 > (y\cos\theta)^2, \quad x\sin\theta < 0.$$

Since $x\sin\theta < 0$, we get

$$x\sin\theta < \frac{(y\cos\theta)^2}{x\sin\theta} \Leftrightarrow \begin{bmatrix} x\sin\theta & jy\cos\theta \\ -jy\cos\theta & x\sin\theta \end{bmatrix} < 0.$$

Substituting $x = (z + \overline{z})/2$, $jy = (z - \overline{z})/2$, and then eliminating $1/2$, we finally obtain

$$\begin{bmatrix} (z+\overline{z})\sin\theta & (z-\overline{z})\cos\theta \\ -(z-\overline{z})\cos\theta & (z+\overline{z})\sin\theta \end{bmatrix} < 0 \Leftrightarrow$$

$$z\begin{bmatrix} \sin\theta & \cos\theta \\ -\cos\theta & \sin\theta \end{bmatrix} + \overline{z}\begin{bmatrix} \sin\theta & -\cos\theta \\ \cos\theta & \sin\theta \end{bmatrix} < 0. \tag{19.4}$$

Note that these characterizations of regions are described by linear matrix inequalities about variables z and $\overline{z}$.

Generalizing these examples, we call the region D of complex number z characterized by

$$D = \{z \in \mathbb{C} \mid f_D(z) < 0\} \tag{19.5}$$

an LMI region, in which

$$f_D(z) = L + zM + \overline{z}M^T. \tag{19.6}$$

Matrix $f_D(z)$ is called the *characteristic function*. Here, L and M are both square matrices.

19.2 Condition for Regional Pole Placement

Consider the following system:

$$\dot{x} = Ax. \tag{19.7}$$

Let us analyze the required condition for its eigenvalues being located in an LMI region. As an example, we first consider a second-order system with distinct eigenvalues. In this case, we may use a (complex in general) matrix T to transform the matrix A into a diagonal one:

$$T^{-1}AT = \begin{bmatrix} z_1 & \\ & z_2 \end{bmatrix}. \tag{19.8}$$

For all eigenvalues of A to be located in the LMI region D of Eq. (19.5),

$$f_D(z_1) = L + z_1 M + \overline{z}_1 M^T < 0, \quad f_D(z_2) = L + z_2 M + \overline{z}_2 M^T < 0$$

must be satisfied. These two inequalities can be summarized as a matrix inequality:

$$\begin{bmatrix} f_D(z_1) & \\ & f_D(z_2) \end{bmatrix} = \begin{bmatrix} L & \\ & L \end{bmatrix} + \begin{bmatrix} z_1 M & \\ & z_2 M \end{bmatrix} + \begin{bmatrix} \overline{z}_1 M^T & \\ & \overline{z}_2 M^T \end{bmatrix} < 0.$$

Set $n_f = \dim(f_D)$. The (i, j) and $(n_f + i, n_f + j)$ elements of the matrix in the left-hand side are, respectively,

$$l_{ij} + z_1 m_{ij} + \overline{z}_1 m_{ji}, \quad l_{ij} + z_2 m_{ij} + \overline{z}_2 m_{ji}$$

in which l_{ij}, m_{ij} denote the (i, j) elements of L, M. They can be collected into a block ((i, j) block):

$$\begin{aligned} & l_{ij}\begin{bmatrix} 1 & \\ & 1 \end{bmatrix} + m_{ij}\begin{bmatrix} z_1 & \\ & z_2 \end{bmatrix} + m_{ji}\begin{bmatrix} \overline{z}_1 & \\ & \overline{z}_2 \end{bmatrix} \\ & = l_{ij} I_2 + m_{ij}(T^{-1}AT) + m_{ji}(T^{-1}AT)^* \end{aligned}$$

via exchange of row and column. The matrix obtained after performing this transformation on all elements can be written compactly as

$$L \otimes I_2 + M \otimes (T^{-1}AT) + M^T \otimes (T^{-1}AT)^*$$

by using the Kronecker product. Since the transformation is congruent, it does not alter the negative definiteness of the matrix. So we have

$$L \otimes I_2 + M \otimes (T^{-1}AT) + M^T \otimes (T^{-1}AT)^* < 0. \tag{19.9}$$

Moreover, to describe this inequality using A directly, we apply the property of the Kronecker product

$$(AC) \otimes (BD) = (A \otimes B)(C \otimes D)$$

to $M \otimes (T^{-1}AT)$:

$$\begin{aligned} M \otimes (T^{-1}AT) &= (I_2 \cdot M) \otimes (T^{-1} \cdot AT) \\ &= (I_2 \otimes T^{-1})[M \otimes (AT)] \\ &= (I_2 \otimes T^{-1})[(M \cdot I_2) \otimes (ATT^* \cdot T^{-*})] \\ &= (I_2 \otimes T^{-1})[M \otimes (ATT^*)](I \otimes T^{-*}). \end{aligned}$$

Similarly, the following equations are also true:

$$\begin{aligned} M^T \otimes (T^{-1}AT)^* = M^T \otimes (T^* A^T T^{-*}) &= (I_2 \otimes T^{-1})[M^T \otimes (ATT^*)^*](I_2 \otimes T^{-*}), \\ L \otimes I_2 = L \otimes (T^* T^{-*}) &= (I_2 \otimes T^{-1})[L \otimes (TT^*)](I_2 \otimes T^{-*}). \end{aligned}$$

Note that

$$I_2 \otimes T^{-1} = \begin{bmatrix} T^{-1} & \\ & T^{-1} \end{bmatrix}, \quad I_2 \otimes T = \begin{bmatrix} T & \\ & T \end{bmatrix}.$$

Multiplying (19.9) by $I_2 \otimes T$ and $I_2 \otimes T^*$, respectively, from the left and right, we obtain

$$L \otimes (TT^*) + M \otimes (ATT^*) + M^T \otimes (ATT^*)^* < 0. \tag{19.10}$$

This is a necessary condition for all eigenvalues of A being located in the LMI region. For convenience, set $X = TT^*$. Then, the inequality (19.10) can be further rewritten as

$$L \otimes X + M \otimes (AX) + M^T \otimes (AX)^* < 0. \tag{19.11}$$

Obviously, the matrix X is Hermitian and positive definite. But in general, X is complex.

The condition (19.11) is an LMI about a complex matrix X. But the calculation of complex matrix is not easy. So we study whether we can replace the complex matrix X by a real one. For this purpose, we consider an arbitrary Hermitian matrix Z first. Denote the real and imaginary parts of Z by Z_{re} and Z_{im}. Due to the property of Hermitian,

$$Z_{\mathrm{re}} + jZ_{\mathrm{im}} = (Z_{\mathrm{re}} + jZ_{\mathrm{im}})^* = Z_{\mathrm{re}}^T - jZ_{\mathrm{im}}^T$$

holds. The real parts and imaginary parts of both sides are equal, respectively. So, we have

$$Z_{\mathrm{re}} = Z_{\mathrm{re}}^T, \ Z_{\mathrm{im}} + Z_{\mathrm{im}}^T = 0.$$

Then, owing to the skew symmetry of the imaginary part $Z_{\rm im}$ and the property $(v^T Z_{\rm im} v)^T = v^T Z_{\rm im}^T v$ of scalar $v^T Z_{\rm im} v$, we see that for any compatible real vector v,

$$v^T Z v = v^T Z_{\rm re} v + j v^T Z_{\rm im} v = v^T Z_{\rm re} v + \frac{j}{2} v^T (Z_{\rm im} + Z_{\rm im}^T) v = v^* Z_{\rm re} v$$

holds. That is, $Z < 0$ and $Z_{\rm re} < 0$ are equivalent.

Let us look back at the condition of (19.11), noting that all matrices are real except the Hermitian matrix X. Splitting X as $X = X_{\rm re} + jX_{\rm im}$, the real part of the left-hand side of (19.11) is equal to the matrix obtained with X replaced by $X_{\rm re}$. Then, as $X_{\rm re}^* = X_{\rm re}^T$, a condition equivalent to (19.11) is obtained via the preceding discussion:

$$L \otimes X_{\rm re} + M \otimes (AX_{\rm re}) + M^T \otimes (AX_{\rm re})^T < 0.$$

Finally, we still use X to denote $X_{\rm re}$ for simplicity. Then, for all eigenvalues of A to be located in LMI region D, there must be a real symmetrical positive definite matrix X satisfying

$$L \otimes X + M \otimes (AX) + M^T \otimes (AX)^T < 0. \tag{19.12}$$

Then, what about the case of multiple poles? When the geometric multiplicity is 2, the condition is the same. In the other case, we can convert A into the Jordan canonical form via similarity transformation:

$$T^{-1}AT = \begin{bmatrix} z & 1 \\ & z \end{bmatrix}.$$

Multiplying both sides by

$$T_k^{-1} = \begin{bmatrix} \frac{1}{k} & \\ & 1 \end{bmatrix}, \quad T_k = \begin{bmatrix} k & \\ & 1 \end{bmatrix}$$

from left and right and then taking the limit $k \to \infty$, we have

$$\begin{aligned} T_k^{-1}T^{-1}ATT_k &= \begin{bmatrix} \frac{1}{k} & \\ & 1 \end{bmatrix} \begin{bmatrix} z & 1 \\ & z \end{bmatrix} \begin{bmatrix} k & \\ & 1 \end{bmatrix} = \begin{bmatrix} z & \frac{1}{k} \\ & z \end{bmatrix} \\ &\to \begin{bmatrix} z & \\ & z \end{bmatrix}. \end{aligned}$$

Now, set $S = TT_k$. We have from (19.9) that for sufficiently large k,

$$L \otimes I_2 + M \otimes (S^{-1}AS) + M^T \otimes (S^* A^T S^{-*}) < 0 \tag{19.13}$$

holds. Therefore, like the case of distinct poles, Eq. (19.12) holds for $X = (SS^*)_{\rm re} > 0$.

Hereafter, we set the matrix in the inequality (19.12) as

$$M_D(A, X) = L \otimes X + M \otimes (AX) + M^T \otimes (AX)^T. \tag{19.14}$$

This discussion applies to matrix A with any dimension. So we get the following theorem.

Theorem 19.1 *All eigenvalues of matrix A are located in an LMI region D iff there exists a real symmetrical positive definite matrix X such that*

$$M_D(A,X) < 0. \tag{19.15}$$

Proof. We have shown that when the eigenvalues of A are all located in an LMI region D, such matrix $X > 0$ exists.

Now, we prove its converse, that is, when a real symmetrical matrix $X > 0$ satisfying the inequality (19.15) exists, the eigenvalues of the matrix A are all located in the LMI region D. To this end, suppose that z is an eigenvalue of A and $v \neq 0$ is the corresponding eigenvector satisfying $v^*A = zv^*$. For the (i,j) block of $M_D(A,X)$, there holds

$$\begin{aligned} v^*(l_{ij}X + m_{ij}AX + m_{ji}XA^T)v &= l_{ij}v^*Xv + m_{ij}v^*AXv + m_{ji}v^*XA^Tv \\ &= l_{ij}v^*Xv + m_{ij}zv^*Xv + m_{ji}\overline{z}v^*Xv \\ &= v^*Xv(l_{ij} + m_{ij}z + m_{ji}\overline{z}) \\ &= v^*Xv \cdot (f_D(z))_{ij} \end{aligned}$$

in which $(f_D(z))_{ij}$ denotes the (i,j) element of the characteristic function $f_D(z)$. So we have

$$(I \otimes v^*)M_D(A,X)(I \otimes v) = (v^*Xv) \cdot f_D(z).$$

Owing to the condition $M_D(A,X) < 0$ and $v^*Xv > 0$, we see that $f_D(z) < 0$ and $z \in D$. •

Next, we consider the examples of LMI region examined in Section 19.1.2.

Example 19.1 *Let us write down the detailed condition for the eigenvalues of A to be located in a disk centered at $(-c, 0)$ and with radius r. The characteristic function of this LMI region is*

$$f_D(z) = \begin{bmatrix} -r & c \\ c & -r \end{bmatrix} + z\begin{bmatrix} 0 & 1 \\ 0 & 0 \end{bmatrix} + \overline{z}\begin{bmatrix} 0 & 0 \\ 1 & 0 \end{bmatrix},$$

that is,

$$L = \begin{bmatrix} -r & c \\ c & -r \end{bmatrix}, \quad M = \begin{bmatrix} 0 & 1 \\ 0 & 0 \end{bmatrix}.$$

According to Theorem 19.1, the required condition is

$$\begin{aligned} M_D(A,X) &= L \otimes X + M \otimes (AX) + M^T \otimes (AX)^T \\ &= \begin{bmatrix} -rX & cX \\ cX & -rX \end{bmatrix} + \begin{bmatrix} 0 & AX \\ 0 & 0 \end{bmatrix} + \begin{bmatrix} 0 & 0 \\ (AX)^T & 0 \end{bmatrix} \\ &= \begin{bmatrix} -rX & cX + AX \\ cX + (AX)^T & -rX \end{bmatrix} < 0. \end{aligned} \tag{19.16}$$

Observing $M_D(A,X)$ and $f_D(z)$ in this example carefully, we notice that the constant 1 in $f_D(z)$ corresponds to the matrix X in $M_D(A,X)$; variables $z, \overline{z}$ correspond to matrices $AX, (AX)^T$, respectively. That is,

$$(1,\ z,\ \overline{z}) \Leftrightarrow (X,\ AX,\ (AX)^T). \tag{19.17}$$

This correspondence is true for all cases.

Look at another example. The characteristic function for the sector in the left half-plane and with an angle θ is

$$f_D(z) = z\begin{bmatrix} \sin\theta & \cos\theta \\ -\cos\theta & x\sin\theta \end{bmatrix} + \overline{z}\begin{bmatrix} \sin\theta & -\cos\theta \\ \cos\theta & \sin\theta \end{bmatrix}.$$

That is,

$$L = 0, \quad M = \begin{bmatrix} \sin\theta & \cos\theta \\ -\cos\theta & \sin\theta \end{bmatrix}.$$

Substitution of the correspondence relationship in (19.17), we obtain the condition

$$M_D(A,X) = \begin{bmatrix} (AX + XA^T)\sin\theta & (AX - XA^T)\cos\theta \\ -(AX - XA^T)\cos\theta & (AX + XA^T)\sin\theta \end{bmatrix} < 0. \tag{19.18}$$

On the other hand, we can get the same result by calculating $M_D(A,X)$ according to its definition:

$$\begin{aligned} M_D(A,X) &= \begin{bmatrix} \sin\theta & \cos\theta \\ -\cos\theta & \sin\theta \end{bmatrix} \otimes (AX) + \begin{bmatrix} \sin\theta & \cos\theta \\ -\cos\theta & \sin\theta \end{bmatrix}^T \otimes (AX)^T \\ &= \begin{bmatrix} (AX + XA^T)\sin\theta & (AX - XA^T)\cos\theta \\ -(AX - XA^T)\cos\theta & (AX + XA^T)\sin\theta \end{bmatrix}. \end{aligned}$$

Moreover, the characteristic function for the half-plane D in the left of straight line $\Re[z] = -\sigma$ is $f_D(z) = z + \overline{z} + 2\sigma$. So, the eigenvalues of matrix A are all located in this region iff

$$M_D(A,X) = AX + (AX)^T + 2\sigma X < 0, \ \ X > 0. \tag{19.19}$$

This is identical to the condition for the state to exponentially converge at a rate of σ, which was shown in Section 13.1.2.

Example 19.2 *Let the system matrix A be*

$$A = \begin{bmatrix} 0 & 1 \\ -10 & -6 \end{bmatrix}.$$

It is easy to know that the eigenvalues of A are $p = -3 \pm j$. Set the parameters of the aforementioned regions as

$$(1)\ disk: c = 0,\ \ r = 5;\ \ (2)\ half\text{-}plane: \sigma = 2;\ \ (3) sector: \theta = \pi/4.$$

Obviously, the eigenvalues of matrix A are contained in all of these LMI regions. Denote the solutions of (19.16), (19.19), and (19.18), respectively, by X_1, X_2, and X_3. Numerical computation gives

$$X_1 = \begin{bmatrix} 0.1643 & -0.2756 \\ -0.2756 & 0.8687 \end{bmatrix}, \quad X_2 = \begin{bmatrix} 0.1567 & -0.3532 \\ -0.3532 & 1.0106 \end{bmatrix}$$

$$X_3 = \begin{bmatrix} 0.1335 & -0.2187 \\ -0.2187 & 0.4799 \end{bmatrix}.$$

They are all positive definite matrices. So, according to Theorem 19.1, it is confirmed that the eigenvalues of matrix A are located in these three regions.

But, when σ increases to $\sigma = 4$, the system poles move to the right of half-plane $\Re[z] < -4$. In this case, computation of (19.19) shows that it has no positive definite solution.

19.3 Composite LMI Region

Now, what condition is required for the eigenvalues of matrix A to be located in the intersection of several LMI regions? Some good examples are the intersection of a disk and a half-plane or the intersection of a sector and a half-plane.

To find this condition, we study the case of two regions first. Let the characteristic functions of LMI regions D_1 and D_2 be

$$f_{D_1}(z) = L_1 + zM_1 + \overline{z}M_1^T, \ f_{D_2}(z) = L_2 + zM_2 + \overline{z}M_2^T, \tag{19.20}$$

respectively. Then, the condition for the eigenvalues of A to be located in LMI region D_1 is that there is a positive definite matrix X_1 such that

$$M_{D_1}(A, X_1) = L_1 \otimes X_1 + M_1 \otimes (AX_1) + M_1^T \otimes (AX_1)^T < 0 \tag{19.21}$$

holds. Meanwhile, the condition for the eigenvalues of A to be located in LMI region D_2 is that there is a positive definite matrix X_2 satisfying

$$M_{D_2}(A, X_2) = L_2 \otimes X_2 + M_2 \otimes (AX_2) + M_2^T \otimes (AX_2)^T < 0. \tag{19.22}$$

So, if these two matrix inequalities have solutions X_1, X_2 simultaneously, the eigenvalues of A are located in the intersection $D_1 \cap D_2$ of LMI regions D_1, D_2.

However, in control design it is rather inconvenient to use different matrices X_1, X_2 to describe this property. So, let us investigate whether we can make $X_1 = X_2$. To find a clue, we go back to (19.11) first. The unknown matrix X and the matrix T used in the diagonal transformation of A has a relationship of $X = TT^*$. The transforming matrix T has nothing to do with the LMI region. So, X should be independent of the LMI region. That is to say, when considering the intersection $D_1 \cap D_2$ of LMI regions, these two matrices should have a common solution.

The detailed proof is as follows. First of all, the characteristic function $f_{D_1 \cap D_2}$ for the complex region $D_1 \cap D_2$ is obviously

$$f_{D_1 \cap D_2}(z) = \begin{bmatrix} f_{D_1}(z) & 0 \\ 0 & f_{D_2}(z) \end{bmatrix}$$

$$= \begin{bmatrix} L_1 + zM_1 + \overline{z}M_1^T & 0 \\ 0 & L_2 + zM_2 + \overline{z}M_2^T \end{bmatrix}$$

$$= \begin{bmatrix} L_1 & 0 \\ 0 & L_2 \end{bmatrix} + z \begin{bmatrix} M_1 & 0 \\ 0 & M_2 \end{bmatrix} + \overline{z} \begin{bmatrix} M_1 & 0 \\ 0 & M_2 \end{bmatrix}^T. \tag{19.23}$$

From it, we see that $D_1 \cap D_2$ is also an LMI region. So, we can use Theorem 19.1 to prove the following corollary.

Corollary 19.1 *Given two LMI regions D_1 and D_2, all eigenvalues of matrix A are in the composite region $D_1 \cap D_2$ iff there exists a positive definite matrix X satisfying $M_{D_1}(A,X) < 0$ and $M_{D_2}(A,X) < 0$.*

Proof. According to Theorem 19.1, the condition for the pole placement in composite region $D_1 \cap D_2$ is that there is a matrix $X > 0$ satisfying the following inequality:

$$\begin{aligned} & M_{D_1 \cap D_2}(A,X) \\ &= \begin{bmatrix} L_1 & 0 \\ 0 & L_2 \end{bmatrix} \otimes X + \begin{bmatrix} M_1 & 0 \\ 0 & M_2 \end{bmatrix} \otimes (AX) + \begin{bmatrix} M_1^T & 0 \\ 0 & M_2^T \end{bmatrix} \otimes (AX)^T \\ &= \begin{bmatrix} L_1 \otimes X & 0 \\ 0 & L_2 \otimes X \end{bmatrix} + \begin{bmatrix} M_1 \otimes (AX) & 0 \\ 0 & M_2 \otimes (AX) \end{bmatrix} \\ &\quad + \begin{bmatrix} M_1^T \otimes (AX)^T & 0 \\ 0 & M_2^T \otimes (AX)^T \end{bmatrix} \\ &= \begin{bmatrix} M_{D_1}(A,X) & 0 \\ 0 & M_{D_2}(A,X) \end{bmatrix} < 0. \end{aligned} \tag{19.24}$$

Obviously, this condition is equivalent to $M_{D_1}(A,X) < 0$ and $M_{D_2}(A,X) < 0$. •

This conclusion applies to the case of more than two regions. What we need to do is to use a common solution X.

Let us look at an example: find a condition such that all eigenvalues of A are located in the composite region shown in Figure 19.3. This region is the intersection of a disk, a sector, and a half-plane. So the condition is that there is a common solution $X > 0$ for the conditions about all the regions. That is, there is an $X > 0$ satisfying all of the following matrix inequalities:

$$\begin{bmatrix} -rX & cX + AX \\ cX + (AX)^T & -rX \end{bmatrix} < 0$$

$$\begin{bmatrix} (AX + XA^T)\sin\theta & (AX - XA^T)\cos\theta \\ -(AX - XA^T)\cos\theta & (AX + XA^T)\sin\theta \end{bmatrix} < 0$$

$$AX + (AX)^T + 2\sigma X < 0.$$

Example 19.3 *In Example 19.2, we verified that the eigenvalues of system matrix*

$$A = \begin{bmatrix} 0 & 1 \\ -10 & -6 \end{bmatrix}$$

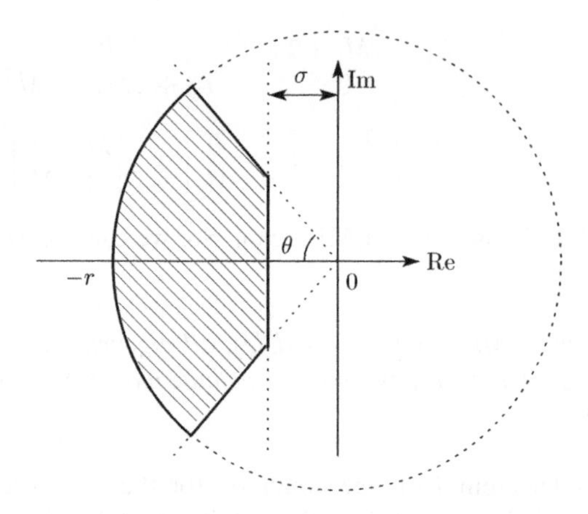

Figure 19.3 Composite region of disk, half-plane, and sector

are located in the following regions, respectively:

$$(1)\ disk : c = 0,\ \ r = 5; (2)\ \ half\text{-}plane : \sigma = 2; (3)\ \ sector : \theta = \pi/4.$$

The intersection of these three regions is the shaded part of Figure 19.3. Here, let us confirm that all eigenvalues of A are located in this composite region by using Corollary 19.1. Solving for the common solution of those three LMIs preceding the example, we obtain

$$X = \begin{bmatrix} 18.357 & -38.4586 \\ -38.4586 & 120.2923 \end{bmatrix}.$$

This matrix is positive definite. Hence, the same conclusion is obtained.

19.4 Feedback Controller Design

In this section, we discuss how to design a feedback controller to place the poles of closed-loop system in a specified LMI region and provide a design example.

19.4.1 Design Method

Let the plant be

$$\dot{x} = Ax + Bu, \quad y = Cx \tag{19.25}$$

and assume that the state equation of controller is

$$\dot{x}_K = A_K x_K + B_K y, \quad u = C_K x_K + D_K y. \tag{19.26}$$

Then, the state equation of the closed-loop system is given by (refer to Subsection 7.4.1)

$$\begin{bmatrix} \dot{x} \\ \dot{x}_K \\ z \end{bmatrix} = \begin{bmatrix} A_c & B_c \\ C_c & D_c \end{bmatrix} \begin{bmatrix} x \\ x_K \\ w \end{bmatrix} \tag{19.27}$$

in which

$$A_c = \begin{bmatrix} A + B_2 D_K C_2 & B_2 C_K \\ B_K C_2 & A_K \end{bmatrix}.$$

Although A_c is an affine function of the coefficient matrix $\mathcal{K} = \left[\begin{smallmatrix} D_K & C_K \\ B_K & A_K \end{smallmatrix}\right]$ of the controller, in the regional pole placement condition

$$M_D(A_c, P) = L \otimes P + M \otimes (A_c P) + M^T \otimes (A_c P)^T < 0, \tag{19.28}$$

A_c appears in almost all the blocks, so $M_D(A_c, P)$ cannot be described in the form of $Q + E^T \mathcal{K} F + F^T \mathcal{K}^T E$ by using a single $\mathcal{K}$. This means that the variable elimination method does not apply. In the sequel, we use the variable change method of Section 3.2.4. Note that the matrix P can be factorized as $P = \Pi_2 \Pi_1^{-1}$. Multiplying Π_1^T, Π_1 to all blocks of the matrices in the left of this inequality from both sides, we obtain an equivalent condition:

$$L \otimes (\Pi_1^T P \Pi_1) + M \otimes (\Pi_1^T A_c P \Pi_1) + M^T \otimes (\Pi_1^T A_c P \Pi_1)^T < 0. \tag{19.29}$$

Next, we need only to calculate $\Pi_1^T P \Pi_1$, $\Pi_1^T A_c P \Pi_1$ and substitute them into this inequality. In Section 3.2.4, we have shown that

$$\Pi_1^T P \Pi_1 = \Pi_1^T \Pi_2 = \begin{bmatrix} X & I \\ I & Y \end{bmatrix}, \quad \Pi_1^T A_c P \Pi_1 = \begin{bmatrix} AX + B\mathbb{C} & A + B\mathbb{D}C \\ \mathbb{A} & YA + \mathbb{B}C \end{bmatrix}$$

in which the new unknown matrices $\mathbb{A}, \mathbb{B}, \mathbb{C}, \mathbb{D}$ are

$$\begin{aligned} \mathbb{A} &= NA_K M^T + NB_K CX + YBC_K M^T + Y(A + BD_K C)X \\ \mathbb{B} &= NB_K + YBD_K, \ \ \mathbb{C} = C_K M^T + D_K CX, \ \ \mathbb{D} = D_K, \end{aligned} \tag{19.30}$$

respectively. Substituting them into Eq. (19.29), we get the design condition:

$$\begin{aligned} L \otimes \begin{bmatrix} X & I \\ I & Y \end{bmatrix} + M \otimes \begin{bmatrix} AX + B\mathbb{C} & A + B\mathbb{D}C \\ \mathbb{A} & YA + \mathbb{B}C \end{bmatrix} \\ + M^T \otimes \begin{bmatrix} AX + B\mathbb{C} & A + B\mathbb{D}C \\ \mathbb{A} & YA + \mathbb{B}C \end{bmatrix}^T < 0. \end{aligned} \tag{19.31}$$

In addition, when we use a full-order controller, by the positive definiteness of matrix P, we have

$$\begin{bmatrix} X & I \\ I & Y \end{bmatrix} > 0. \tag{19.32}$$

In this case, coefficient matrices of the controller can be computed from the matrices $\mathbb{A}, \mathbb{B}, \mathbb{C}, \mathbb{D}$ uniquely. Refer to Section 3.2.4 for the details.

19.4.2 Design Example: Mass–Spring–Damper System

In the mass–spring–damper system, assume that $m = 1$ kg, the damping ratio $b = 0$, and the spring constant $k = 100$ N/m. Then, the state equation becomes

$$\dot{x} = \begin{bmatrix} 0 & 1 \\ -100 & 0 \end{bmatrix} x + \begin{bmatrix} 0 \\ 1 \end{bmatrix} u.$$

Now, we design an output feedback controller to place the poles of closed-loop system in the following composite region:

$$(1)\ \text{disk}: c = 0,\quad r = 5;\ (2)\ \text{half-plane}: \sigma = 2;\ (3)\ \text{sector}: \theta = \pi/4.$$

Substitution of the coefficient matrices (L, M) of these three LMI regions into Eq. (19.31), respectively, we get three LMIs. With (19.32) added, it is required to solve four simultaneous LMIs in total. The obtained matrices are

$$X = \begin{bmatrix} 85.6357 & -189.7711 \\ -189.7711 & 629.9055 \end{bmatrix},\quad Y = \begin{bmatrix} 629.9055 & -189.7711 \\ -189.7711 & 85.6357, \end{bmatrix}$$

respectively. And the controller is

$$K(s) = \frac{90s^2 + 1087s + 13023}{s^2 + 12.2859s - 128.9044}.$$

Simple calculation shows that poles of the closed-loop system are

$$-2.8568 \pm j1.1716,\quad -3.2861 \pm j1.7570$$

and they are located in the three specified LMI regions.

19.5 Analysis of Robust Pole Placement

In this section, we analyze the condition for all poles of a parametric system being placed in a designated LMI region. Two cases are discussed.

19.5.1 Polytopic System

First of all, we inspect the example of mass–spring–damper system

$$\dot{x} = \begin{bmatrix} 0 & 1 \\ -\dfrac{k}{m} & -\dfrac{b}{m} \end{bmatrix} x + \begin{bmatrix} 0 \\ \dfrac{1}{m} \end{bmatrix} u.$$

Let $p_1 = k/m,\ p_2 = b/m$. When the input of this system is $u = 0$, the state equation can be written as

$$\dot{x} = \left\{ \begin{bmatrix} 0 & 1 \\ 0 & 0 \end{bmatrix} + p_1 \begin{bmatrix} 0 & 0 \\ -1 & 0 \end{bmatrix} + p_2 \begin{bmatrix} 0 & 0 \\ 0 & -1 \end{bmatrix} \right\} x.$$

That is, the system matrix A is an affine function of the parameter vector $p = [p_1 \quad p_2]^T$. Of course, the eigenvalues of A change with p.

Generally, when the vector of uncertain parameter is $p \in \mathbb{R}^N$, the state equation of a system is described by

$$\dot{x} = A(p)x, \quad x(0) \neq 0. \tag{19.33}$$

The relationship between system matrix $A(p)$ and parameter vector p may take many forms such as polytope, LFT, and so on, besides the affine form shown in the preceding example.

So far, we have derived the condition for the poles of a system being located in an LMI region when p is given. This condition requires that there is a positive definite matrix X satisfying the matrix inequality $M_D(A, X) < 0$. It should be noted that the matrix X depends on the system matrix A. Reasoning from this fact, X should change when A changes with p. That is, X is a function of the parameter vector p, denoted as $X(p)$. Hence, we immediately get the following corollary from Theorem 19.1.

Corollary 19.2 *All eigenvalues of matrix $A(p)$ are located in LMI region D iff there is a real positive definite matrix $X(p)$ satisfying*

$$M_D(A(p), X(p)) < 0. \tag{19.34}$$

However, it is very difficult to figure out the relationship between $X(p)$ and p, and it remains an open problem so far. Here, although just a sufficient condition, we follow the philosophy of quadratic stability and fix $X(p)$ as a constant matrix so as to simplify the problem. Then, obviously the eigenvalues of matrix $A(p)$ are all located in LMI region D so long as the following matrix inequality

$$M_D(A(p), X) < 0 \tag{19.35}$$

has a real positive definite solution.

But, this condition still depends on the uncertain parameter vector p. So it cannot be checked. In the sequel, we will derive a condition independent of the parameter vector p. For this, consider the polytopic system:

$$A(p) = \sum_{i=1}^{N} p_i A_i, \quad \sum_{i=1}^{N} p_i = 1, \ p_i \geq 0. \tag{19.36}$$

This matrix polytope has N vertices. To ensure the condition (19.35),

$$M_D(A_i, X) < 0 \quad \forall i = 1, \ldots, N \tag{19.37}$$

must be true at each vertex A_i. Conversely, any constant matrix L can be described as $L = \sum_{i=1}^{N} p_i L$ since $\sum_{i=1}^{N} p_i = 1$. Further, noting that $M_D(A(p), X)$ is an affine function of $A(p)$ and $p_i \geq 0$, when the inequality (19.37) holds, there also holds

$$M_D(A(p), X) = \sum_{i=1}^{N} p_i M_D(A_i, X) < 0,$$

that is, (19.37) is equivalent to (19.35). So we obtain the following result.

Corollary 19.3 *For a matrix polytope $A(p)$, if there is a real positive definite matrix X satisfying the matrix inequality*

$$M_D(A_i, X) < 0 \quad \forall i = 1, \ldots, N, \tag{19.38}$$

then the eigenvalues of $A(p)$ are all located in the LMI region D.

Example 19.4 *Consider the system matrix:*

$$A = \begin{bmatrix} 0 & 1 \\ -3 & -b \end{bmatrix}, \quad b > 0.$$

It corresponds to the case where the mass is 1, spring constant is 3, and frictional coefficient is b in the mass–spring–damper system. Let the LMI region be a disk centered at $(-c, 0) = (-2.5, 0)$ and with a radius $r = 2$. The eigenvalues of matrix A are $\frac{-b \pm \sqrt{b^2 - 12}}{2}$. In the case of complex roots ($b < \sqrt{12}$), the condition for the eigenvalues being located inside the disk is

$$r^2 > \left(c - \frac{b}{2}\right)^2 + \left(\frac{\sqrt{12 - b^2}}{2}\right)^2 \quad \Rightarrow \quad b > \frac{c^2 - r^2 + 3}{c}.$$

Meanwhile, in the case of real roots, the condition becomes

$$c + r > \frac{d + \sqrt{b^2 - 12}}{2} \quad \Rightarrow \quad b < c + r + \frac{3}{c + r}.$$

After the substitution of $-c = -2.5$ and $r = 2$, we see that the condition for all eigenvalues being located in the given disk is

$$2.1 < b < 5.167.$$

For this range of b, applying the condition of Corollary 19.3, we fail to obtain a positive definite solution. The condition of Corollary 19.3 is not satisfied until the range of friction coefficient is reduced to the interval $[2.3,\ 4]$.

From this example, we realize that it is rather conservative using a common solution to guarantee the robust pole placement.

19.5.2 Norm-Bounded Parametric System

Consider the following parametric system:

$$\dot{x} = A_\Delta x = (A + B\Delta(I - D\Delta)^{-1}C)x, \quad \|\Delta(t)\|_2 \le 1. \tag{19.39}$$

As shown in Figure 19.4, the system is equivalent to the closed-loop system made up by the nominal system $M(s)$

$$M \begin{cases} \dot{x} = Ax + Bw \\ z = Cx + Dw \end{cases} \tag{19.40}$$

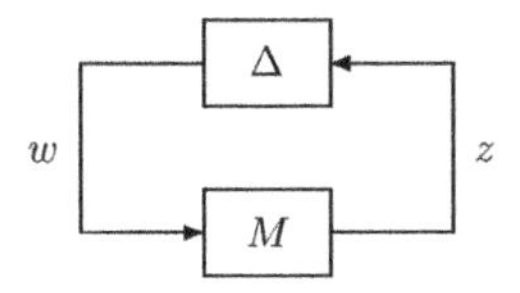

Figure 19.4 Parametric system

and the norm-bounded parameter uncertainty $\Delta(t)$

$$w = \Delta z, \quad \|\Delta(t)\|_2 \leq 1.$$

In addition, as $\Delta(t)$ varies arbitrarily in the range of $\|\Delta(t)\|_2 \leq 1$, the matrix $I - D\Delta$ is invertible iff $\|D\|_2 < 1$.

In system (19.39), denote the signals corresponding to the initial state $x_i(0)$ by $x_i(t),\ w_i(t),\ z_i(t)$. That is,

$$\begin{aligned} \dot{x}_i &= Ax_i + Bw_i, \quad z_i = Cx_i + Dw_i \\ w_i &= \Delta z_i. \end{aligned} \tag{19.41}$$

To search for the condition of robust pole placement, first we look at the example of a disk region centered at $(-c, 0)$ and with a radius r. According to Example 19.1, the characteristic function of the disk is

$$f_D(z) = \begin{bmatrix} -r & c \\ c & -r \end{bmatrix} + z \begin{bmatrix} 0 & 1 \\ 0 & 0 \end{bmatrix} + \overline{z} \begin{bmatrix} 0 & 0 \\ 1 & 0 \end{bmatrix}.$$

If there is a positive definite matrix X satisfying the LMI

$$\begin{bmatrix} -rX & cX + A_\Delta X \\ cX + (A_\Delta X)^T & -rX \end{bmatrix} < 0,$$

then the poles of system (19.39) are located in the disk. To eliminate the uncertainty Δ in this inequality, we make use of the magnitude relation between input and output of the norm-bounded Δ. For this purpose, it is required to place A_Δ in the right side of the product. So we set $P = X^{-1}$ and multiply $\mathrm{diag}(P, P) = I_2 \otimes P$ to both sides of the inequality. Then, we obtain equivalently

$$N_D(A_\Delta, P) := \begin{bmatrix} -rP & cP + PA_\Delta \\ cP + A_\Delta^T P & -rP \end{bmatrix} < 0. \tag{19.42}$$

Furthermore, the condition is equivalent to

$$\begin{bmatrix} x_1 \\ x_2 \end{bmatrix}^T \begin{bmatrix} -rP & cP + PA_\Delta \\ cP + A_\Delta^T P & -rP \end{bmatrix} \begin{bmatrix} x_1 \\ x_2 \end{bmatrix} < 0 \quad \forall \begin{bmatrix} x_1 \\ x_2 \end{bmatrix} \neq 0.$$

As

$$A_\Delta x_2 = \dot{x}_2 = Ax_2 + Bw_2,$$

the preceding inequality can be written as

$$\begin{bmatrix} x_1 \\ x_2 \end{bmatrix}^T \begin{bmatrix} -rP & cP + PA \\ cP + A^T P & -rP \end{bmatrix} \begin{bmatrix} x_1 \\ x_2 \end{bmatrix} + x_1^T PBw_2 + w_2^T B^T Px_1 < 0$$

$$\Rightarrow \begin{bmatrix} x_1 \\ x_2 \\ w_2 \end{bmatrix}^T \begin{bmatrix} -rP & cP + PA & PB \\ cP + A^T P & -rP & 0 \\ B^T P & 0 & 0 \end{bmatrix} \begin{bmatrix} x_1 \\ x_2 \\ w_2 \end{bmatrix} < 0.$$

On the other hand, the norm condition of Δ implies that $z_2^T z_2 = (Cx_2 + Dw_2)^T(Cx_2 + Dw_2) \geq w_2^T w_2$, that is,

$$\begin{bmatrix} x_1 \\ x_2 \\ w_2 \end{bmatrix}^T \left\{ \begin{bmatrix} 0 \\ C^T \\ D^T \end{bmatrix} [0 \quad C \quad D] - \begin{bmatrix} 0 \\ 0 \\ I \end{bmatrix} [0 \quad 0 \quad I] \right\} \begin{bmatrix} x_1 \\ x_2 \\ w_2 \end{bmatrix} \geq 0$$

is always true. Like the proof of Theorem 13.2, an equivalent condition

$$\begin{bmatrix} -rP & cP + PA & PB \\ cP + A^T P & -rP & 0 \\ B^T P & 0 & -I \end{bmatrix} + \begin{bmatrix} 0 \\ C^T \\ D^T \end{bmatrix} [0 \quad C \quad D] < 0$$

is obtained by applying the S-procedure. This condition is independent of Δ. Applying Schur's lemma, we finally obtain a sufficient condition for robust pole placement in the disk region:

$$\begin{bmatrix} -rP & cP + PA & PB & 0 \\ cP + A^T P & -rP & 0 & C^T \\ B^T P & 0 & -I & D^T \\ 0 & C & D & -I \end{bmatrix} < 0. \tag{19.43}$$

Next, we generalize this result. An equivalent condition for $M_D(A, X) < 0$ is derived first. Set $P = X^{-1}$ and define a matrix

$$N_D(A, P) = L \otimes P + M \otimes (PA) + M^T \otimes (PA)^T. \tag{19.44}$$

By using relations such as $M \otimes (AX) = (I \otimes X)(M \otimes (PA))(I \otimes X)$, it is easy to know that

$$M_D(A, X) = (I \otimes X)N_D(A, P)(I \otimes X)$$

holds. So, $X > 0$ satisfying $M_D(A, X) < 0$ implies $P > 0$ satisfying $N_D(A, P) < 0$.

From the discussion in the preceding text, we see that a sufficient condition for robust pole placement is that there is a matrix $P > 0$ satisfying

$$N_D(A, P) = L \otimes P + M \otimes (PA_\Delta) + M^T \otimes (A_\Delta^T P) < 0 \quad \forall \|\Delta(t)\|_2 \leq 1. \tag{19.45}$$

The question now is how to find a condition independent of the uncertainty Δ. Again, the key is to utilize the relation between the norm of Δ and its input–output norms. Set three vectors made up of (x_i, w_i, z_i), respectively, as

$$\mathcal{X} = \begin{bmatrix} x_1 \\ \vdots \\ x_{n_f} \end{bmatrix}, \ \mathcal{W} = \begin{bmatrix} w_1 \\ \vdots \\ w_{n_f} \end{bmatrix}, \ \mathcal{Z} = \begin{bmatrix} z_1 \\ \vdots \\ z_{n_f} \end{bmatrix}; \quad n_f = \dim\ (f_D).$$

By the signal relationship (19.41), we have $A_\Delta x_j = \dot{x}_j = Ax_j + Bw_j$. So there is

$$x_i(m_{ij}PA_\Delta)x_j = x_i[m_{ij}P(Ax_j + Bw_j)] = x_i(m_{ij}PA)x_j + x_i(m_{ij}PB)w_j.$$

According to this equation, we surely have

$$\mathcal{X}^T N_D(A_\Delta, P)\mathcal{X} = \mathcal{X}^T N_D(A, P)\mathcal{X} + \mathcal{X}^T[M \otimes (PB)]\mathcal{W} + \mathcal{W}^T[M^T \otimes (B^T P)]\mathcal{X}.$$

When M does not have full rank, some signal w_i will be cancelled (e.g., w_1 in the disk region). In order to take out such signals from this equation, we perform a full-rank decomposition on matrix M

$$M = M_1 M_2^T \tag{19.46}$$

in which M_1, M_2 have full column rank. So, we obtain $M \otimes (PB) = (M_1 M_2^T) \otimes (PB) = (M_1 \otimes (PB))(M_2^T \otimes I)$. Define new signal vectors:

$$\mathcal{V} = (M_2^T \otimes I)\mathcal{W}, \ \mathcal{Y} = (M_2^T \otimes I)\mathcal{Z}.$$

Then a sufficient condition for the robust pole placement is given by

$$\begin{aligned}&\mathcal{X}^T N_D(A, P)\mathcal{X} + \mathcal{X}^T[M_1 \otimes (PB)]\mathcal{V} + \mathcal{V}^T[M_1^T \otimes (B^T P)]\mathcal{X}\\ &= \begin{bmatrix}\mathcal{X}\\ \mathcal{V}\end{bmatrix}^T \begin{bmatrix} N_D(A, P) & M_1 \otimes (PB)\\ M_1^T \otimes (B^T P) & 0\end{bmatrix}\begin{bmatrix}\mathcal{X}\\ \mathcal{V}\end{bmatrix} < 0 \quad \forall \mathcal{X} \neq 0.\end{aligned} \tag{19.47}$$

Since $z_i = Cx_i + Dw_i$, obviously there hold

$$\mathcal{Z} = (I \otimes C)\mathcal{X} + (I \otimes D)\mathcal{W}$$

and

$$\begin{aligned}\mathcal{Y} &= (M_2^T \otimes I)[(I \otimes C)\mathcal{X} + (I \otimes D)\mathcal{W}] = (M_2^T \otimes C)\mathcal{X} + (M_2^T \otimes D)\mathcal{W}\\ &= (M_2^T \otimes C)\mathcal{X} + (I \otimes D)(M_2^T \otimes I)\mathcal{W} = (M_2^T \otimes C)\mathcal{X} + (I \otimes D)\mathcal{V}.\end{aligned}$$

The remaining question is how to find the relationship between the magnitudes of signals $\mathcal{V}$ and $\mathcal{Z}$ and apply the S-procedure. For this purpose, we inspect the ith block v_i of the vector $\mathcal{V}$:

$$v_i = m_{2,1i}w_1 + \cdots + m_{2,n_f i}w_{n_f} = \Delta(m_{2,1i}z_1 + \cdots + m_{2,n_f i}z_{n_f}) = \Delta y_i.$$

From it we have

$$\mathcal{V} = (I_k \otimes \Delta)\mathcal{Y}, \quad k = \dim\ (M).$$

Hence, for any k-dimensional positive definite matrix Q, the inequality

$$\begin{aligned}\mathcal{V}^T(Q \otimes I)\mathcal{V} &= \mathcal{Y}^T(I \otimes \Delta^T)(Q \otimes I)(I \otimes \Delta)\mathcal{Y} = \mathcal{Y}^T(Q \otimes \Delta^T\Delta)\mathcal{Y}\\ &\leq \mathcal{Y}^T(Q \otimes I)\mathcal{Y}\end{aligned}$$

always holds (refer to Exercise 19.3 for the last inequality). So, expansion of this inequality leads to

$$\begin{bmatrix} \mathcal{X} \\ \mathcal{V} \end{bmatrix}^T \left\{ \begin{bmatrix} M_2 \otimes C^T \\ I \otimes D^T \end{bmatrix} (Q \otimes I)[M_2^T \otimes C \quad I \otimes D] - \begin{bmatrix} 0 & 0 \\ 0 & -Q \otimes I \end{bmatrix} \right\} \begin{bmatrix} \mathcal{X} \\ \mathcal{V} \end{bmatrix}$$
$$\leq 0 \quad \forall \mathcal{X} \neq 0. \tag{19.48}$$

Application of the S-procedure to (19.47) and (19.48) yields

$$\begin{bmatrix} N_D(A, P) & M_1 \otimes (PB) \\ M_1^T \otimes (B^T P) & 0 \end{bmatrix} + \begin{bmatrix} M_2 \otimes C^T \\ I \otimes D^T \end{bmatrix} (Q \otimes I)[M_2^T \otimes C \; I \otimes D]$$
$$- \begin{bmatrix} 0 & 0 \\ 0 & -Q \otimes I \end{bmatrix} < 0.$$

Finally, applying Schur's lemma and using $Q \otimes I = (Q \otimes I)(Q \otimes I)^{-1}(Q \otimes I)$, we can derive the following robust pole placement theorem.

Theorem 19.2 *If there exist matrices $P > 0, \ Q > 0$ satisfying LMI*

$$\begin{bmatrix} N_D(A, P) & M_1 \otimes (PB) & (M_2 Q) \otimes C^T \\ M_1^T \otimes (B^T P) & -Q \otimes I & Q \otimes D^T \\ (Q M_2^T) \otimes C & Q \otimes D & -Q \otimes I \end{bmatrix} < 0, \tag{19.49}$$

then poles of the uncertain system (19.39) are all located inside the region D.

19.6 Robust Design of Regional Pole Placement

19.6.1 On Polytopic Systems

Let us inspect whether we can do robust design of regional pole placement based on Corollary 19.3 of Section 19.5. For simplicity, consider a system with only one uncertain parameter:

$$\dot{x} = (\theta_1 A_1 + \theta_2 A_2)x + Bu, \ y = Cx; \quad \theta_1, \theta_2 \geq 0, \ \theta_1 + \theta_2 = 1. \tag{19.50}$$

For the controller

$$\dot{x}_K = A_K x_K + B_K y, \quad u = C_K x_K + D_K y, \tag{19.51}$$

application of the result of Section 19.4 shows that the condition to place the poles of closed-loop system in an LMI region D is that the matrix inequalities

$$L \otimes \begin{bmatrix} X & I \\ I & Y \end{bmatrix} + M \otimes \begin{bmatrix} AX + B\mathbb{C} & A + B\mathbb{D}C \\ \mathbb{A} & YA + \mathbb{B}C \end{bmatrix}$$
$$+ M^T \otimes \begin{bmatrix} AX + B\mathbb{C} & A + B\mathbb{D}C \\ \mathbb{A} & YA + \mathbb{B}C \end{bmatrix}^T < 0$$
$$\begin{bmatrix} X & I \\ I & Y \end{bmatrix} > 0$$

have a solution. Here,

$$\mathbb{A} = NA_K M^T + NB_K CX + YBC_K M^T + Y(\theta_1 A_1 + \theta_2 A_2 + BD_K C)X$$
$$\mathbb{B} = NB_K + YBD_K,\ \mathbb{C} = C_K M^T + D_K CX,\ \mathbb{D} = D_K.$$

Although the matrix $\mathbb{A}$ can be expressed as the convex combination $\mathbb{A} = \theta_1 \mathbb{A}_1 + \theta_2 \mathbb{A}_2$ of the vertex matrices in which

$$\mathbb{A}_1 = NA_K M^T + NB_K CX + YBC_K M^T + Y(A_1 + BD_K C)X$$
$$\mathbb{A}_2 = NA_K M^T + NB_K CX + YBC_K M^T + Y(A_2 + BD_K C)X,$$

it is not guaranteed that they have a common solution when calculating A_K from $(\mathbb{A}_1, \mathbb{A}_2)$ inversely. This means that the present variable change technique cannot solve this design problem.

However, when θ_1, θ_2 are known time-varying parameters, we may use a controller in a form of

$$A_K = \theta_1 A_{K1} + \theta_2 A_{K2}.$$

In this case,

$$\mathbb{A}_1 = NA_{K1} M^T + NB_K CX + YBC_K M^T + Y(A_1 + BD_K C)X$$
$$\mathbb{A}_2 = NA_{K2} M^T + NB_K CX + YBC_K M^T + Y(A_2 + BD_K C)X$$

and (A_{K1}, A_{K2}) can be inversely calculated from $(\mathbb{A}_1, \mathbb{A}_2)$. This is the so-called gain-scheduling method to be introduced in the next chapter.

19.6.2 Design for Norm-Bounded Parametric System

According to the discussion in Section 11.3.3, the general model of norm-bounded parametric system is

$$G \begin{cases} \dot{x} = Ax + B_1 w + B_2 u \\ z = C_1 x + D_{11} w + D_{12} u \\ y = C_2 x + D_{21} w \end{cases} \tag{19.52}$$

$$w = \Delta z, \quad \Delta_2 \leq 1. \tag{19.53}$$

We hope to design a dynamic output feedback controller K

$$\dot{x}_K = A_K x_K + B_K y$$
$$u = C_K x_K + D_K y \tag{19.54}$$

to robustly place the closed-loop poles inside an LMI region D. In this closed-loop system, the part except Δ is an LFT interconnection between the generalized plant G and the controller K

$$M(s) = \mathcal{F}_\ell(G, K) = (A_c, B_c, C_c, D_c) \tag{19.55}$$

in which the coefficient matrix is (refer to Section 7.4.1)

$$\begin{bmatrix} A_c & B_c \\ C_c & D_c \end{bmatrix} = \left[\begin{array}{cc|c} A + B_2 D_K C_2 & B_2 C_K & B_1 + B_2 D_K D_{21} \\ B_K C_2 & A_K & B_K D_{21} \\ \hline C_1 + D_{12} D_K C_2 & D_{12} C_K & D_{11} + D_{12} D_K D_{21} \end{array}\right]. \tag{19.56}$$

According to Theorem 19.2 in Section 19.5.2, a sufficient condition for the closed-loop poles being placed in the region D is that there exist matrices $P > 0, Q > 0$ satisfying LMI

$$\begin{bmatrix} N_D(A_c, P) & M_1 \otimes (PB_c) & (M_2 Q) \otimes C_c^T \\ M_1^T \otimes (B_c^T P) & -Q \otimes I & Q \otimes D_c^T \\ (Q M_2^T) \otimes C_c & Q \otimes D_c & -Q \otimes I \end{bmatrix} < 0. \tag{19.57}$$

In order to apply the variable change method of Section 3.2.4 to solve this problem, we need to fix the matrix Q. Here, we set $Q = I$. The corresponding condition of robust pole placement becomes

$$\begin{bmatrix} N_D(A_c, P) & M_1 \otimes (PB_c) & M_2 \otimes C_c^T \\ M_1^T \otimes (B_c^T P) & -I & I \otimes D_c^T \\ M_2^T \otimes C_c & I \otimes D_c & -I \end{bmatrix} < 0. \tag{19.58}$$

Factorize the positive definite matrix P as

$$P = \Pi_2 \Pi_1^{-1} = \Pi_1^{-T} \Pi_2^T, \quad \Pi_1 = \begin{bmatrix} X & I \\ M^T & 0 \end{bmatrix}, \quad \Pi_2 = \begin{bmatrix} I & Y \\ 0 & N^T \end{bmatrix}.$$

Multiplying this inequality from the left and right, respectively, with $\operatorname{diag}(I \otimes \Pi_1,\ I,\ I)$ and its transpose, we obtain equivalently

$$\begin{bmatrix} L \otimes (\Pi_1^T \Pi_2) + \operatorname{He}\{M \otimes (\Pi_2^T A_c \Pi_1)\} & M_1 \otimes (\Pi_2^T B_c) & M_2 \otimes (\Pi_1^T C_c^T) \\ M_1^T \otimes (B_c^T \Pi_2) & -I & I \otimes D_c^T \\ M_2^T \otimes (C_c \Pi_1) & I \otimes D_c & -I \end{bmatrix} < 0. \tag{19.59}$$

Define new variable matrices $\mathbb{A}, \mathbb{B}, \mathbb{C}, \mathbb{D}$ as

$$\begin{aligned} \mathbb{A} &= N A_K M^T + N B_K C_2 X + Y B_2 C_K M^T + Y(A + B_2 D_K C_2) X \\ \mathbb{B} &= N B_K + Y B_2 D_K, \quad \mathbb{C} = C_K M^T + D_K C_2 X, \quad \mathbb{D} = D_K, \end{aligned} \tag{19.60}$$

respectively. Then, substitution of

$$\Pi_1^T \Pi_2 = \begin{bmatrix} X & I \\ I & Y \end{bmatrix}, \quad \Pi_2^T A_c \Pi_1 = \begin{bmatrix} AX + B_2 \mathbb{C} & A + B_2 \mathbb{D} C_2 \\ \mathbb{A} & YA + \mathbb{B} C_2 \end{bmatrix}$$

$$\Pi_2^T B_c = \begin{bmatrix} B_1 + B_2 \mathbb{D} D_{21} \\ Y B_1 + \mathbb{B} D_{21} \end{bmatrix}, \quad C_c \Pi_1 = [C_1 X + D_{12} \mathbb{C} \quad C_1 + D_{12} \mathbb{D} C_2]$$

into (19.59) yields the design condition for the robust controller. Further, when using full-order controller, by the positive definiteness of P, there must be

$$\begin{bmatrix} X & I \\ I & Y \end{bmatrix} > 0. \tag{19.61}$$

In this case, coefficient matrices of the controller can be computed inversely from $\mathbb{A}, \mathbb{B}, \mathbb{C}, \mathbb{D}$.

19.6.3 *Robust Design Example: Mass–Spring–Damper System*

In the mass–spring–damper system, set the mass m as 1 kg, damping ratio as $b \in [0, 2]$ Ns/m, and spring constant as $k \in [80{,}120]$ N/m. Then, the state equation becomes

$$\dot{x} = \begin{bmatrix} 0 & 1 \\ -k & -b \end{bmatrix} x + \begin{bmatrix} 0 \\ 1 \end{bmatrix} u, \quad y = [1 \quad 0]x.$$

Describe each uncertainty parameter by

$$k = k_0(1 + w_1\delta_1), \quad b = b_0(1 + w_2\delta_2), \quad |\delta_i| \leq 1.$$

As Example 13.3, we take the nominal parameters as $k_0 = 100, b_0 = 1$ which are the averaged parameters. Then, the weights become

$$w_1 = \frac{k_{\max}}{k_0} - 1 = 0.2, \quad w_2 = \frac{b_{\max}}{b_0} - 1 = 1.$$

Normalizing the uncertain matrix $\Delta = [\delta_1 \ \delta_2]$ in the norm-bounded parametric system, each coefficient matrix becomes

$$A = \begin{bmatrix} 0 & 1 \\ -k_0 & -b_0 \end{bmatrix}, \quad B_1 = B_2 = \begin{bmatrix} 0 \\ 1 \end{bmatrix}, \quad C_1 = -\sqrt{2} \begin{bmatrix} k_0 w_1 & 0 \\ 0 & b_0 w_2 \end{bmatrix}$$

$$C_2 = [1 \quad 0], \quad D_{11} = D_{12} = \begin{bmatrix} 0 \\ 0 \end{bmatrix}, \quad D_{21} = 0,$$

respectively. Now, we design an output feedback controller to place the poles of closed-loop system inside the intersection of a disk ($c = 0, r = 15$) and a half-plane ($\sigma = 1$). Substituting into (19.59) the coefficient matrices (L, M) corresponding to this LMI region, we obtain a controller

$$K(s) = \frac{-66s^2 - 405s + 13315}{s^2 + 30.415s + 277.62}.$$

At the nominal value and four vertices of the parameter vector (k, b), poles of the closed-loop system are

$$\begin{aligned} &\text{Nominal}: \quad (-6.0189 \pm j8.717, \ -9.6886 \pm j5.9056) \\ &(-9.3103 \pm j7.3826, \ -5.8972 \pm j5.3125) \\ &(-9.2397 \pm j9.562, \ -6.9678 \pm j1.3258) \\ &(-3.9167 \pm j10.3592, \ -11.2458 \pm j6.0098) \\ &(-5.8157 \pm j10.5058, \ -10.3918 \pm j5.5414), \end{aligned}$$

respectively. They all are located in the given region.

It is remarked that this design is not so good because $K(\infty) = -66$ so that $K(s)$ is very sensitive to noise. Specification on noise reduction should better be supplemented, which leads to a multiobjective design. This topic will be touched on in the next chapter.

Exercises

19.1 Derive the description for the strip region in Figure 19.2 and the necessary and sufficient condition for the poles of system $\dot{x} = Ax$ being located inside this region. Finally, check whether poles of the system in Example 19.2 are all located in the strip region of $(h_1, h_2) = (2, 4)$ numerically.

19.2 In the LMI region of Figure 19.2, move the vertex of the sector region in Figure (c) to $(a, j0)$ and find the corresponding characteristic function $f_D(z)$.

19.3 Given $\|\Delta\|_2 \le 1, Q > 0$, prove that inequality $\mathcal{Y}^T(Q \otimes \Delta^T\Delta)\mathcal{Y} \le \mathcal{Y}^T(Q \otimes I)\mathcal{Y}$ holds for arbitrary vector $\mathcal{Y}$.

19.4 Suppose that LMI region D is in the left half-plane and its characteristic function is $f_D(z) = L + zM + \overline{z}M^T$. Prove that if the time-varying system

$$\dot{x} = A(t)x$$

satisfies

$$N_D(A(t), P) = L \otimes P + M \otimes (PA(t)) + M^T \otimes (PA(t))^T < 0 \quad \forall t,$$

then the quadratic function $V(x) = x^T Px$ satisfies

$$\frac{1}{2}\frac{\dot{V}(x)}{V(x)} \in D \cap \mathbb{R} = [-a, -b]$$

in which $a > b > 0$ ($(-a, j0), (-b, j0)$ are the two points of intersection of LMI region D and the real axis) [17]. Based on this, prove that the state $x(t)$ satisfies

$$c_1 e^{-at} \le \|x(t)\|_2 \le c_2 e^{-bt}, \quad c_1,\ c_2 > 0$$

and try to interpret the engineering meaning of this conclusion.

19.5 For the following LMI regions, concretely expand the condition (19.59) of robust pole placement:
1. disk: center $(-c, 0)$, radius r;
2. half-plane: $\Re(s) < -\sigma$;
3. sector: the vertex is the origin and the interior angle is 2θ.

Notes and References

Chilali and Gahinet [16] solved the regional pole placement problem and Ref. [17] treated the robust regional pole placement problem.

20

Gain-Scheduled Control

In Chapter 11, we have seen that in some occasions a nonlinear system can be described as an linear parameter-varying (LPV) model. Since the time-varying coefficients of an LPV model are functions of some states, they can be calculated so long as these states are measured online. Therefore, it is possible to change the controller gain according to the variation of coefficients in the LPV model so as to control the LPV system more effectively than controllers with fixed gains. This method is called the *gain-scheduled control.*

This chapter illustrates in detail the design methods of gain-scheduled control as well as two design examples.

20.1 General Structure

Here, let the LPV model be

$$\dot{x} = A(p(t))x + B(p(t))u \tag{20.1}$$

$$y = C(p(t))x. \tag{20.2}$$

The corresponding gain-scheduled controller is also set as the following LPV form

$$\dot{x}_K = A_K(p(t))x_K + B_K(p(t))y \tag{20.3}$$

$$u = C_K(p(t))x_K + D_K(p(t))y. \tag{20.4}$$

That is, the parameters of controller are changed together with that of the time-varying parameter vector $p(t)$ so as to realize high-performance control.

However, without specifying the relationship between the coefficient matrices and the parameter vector $p(t)$, concrete design method cannot be established. In addition, to determine the relationship between the controller and the parameter vector, we need to consider the relationship between the LPV model and the parameter vector. Due to such consideration, we will focus on two most typical LPV models.

Robust Control: Theory and Applications, First Edition. Kang-Zhi Liu and Yu Yao.

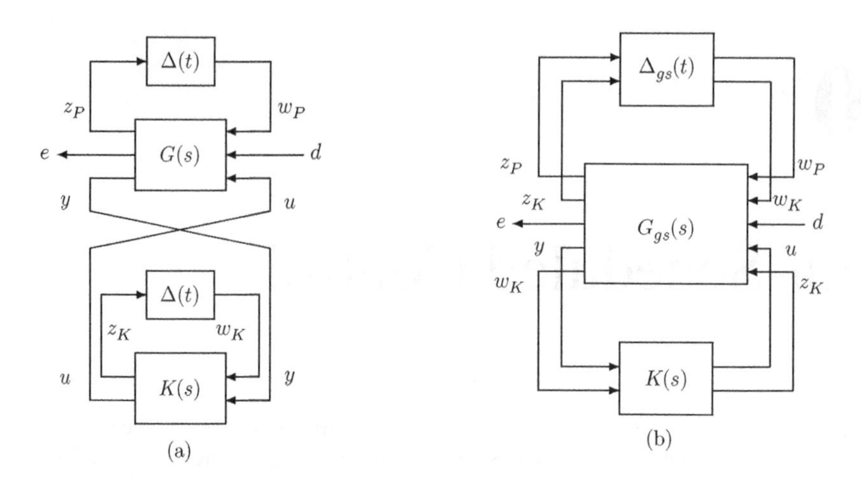

Figure 20.1 LFT-type gain-scheduled control system

20.2 LFT-Type Parametric Model

The system shown in Figure 20.1(a) is called the LFT model, in which the plant has a time-varying parameter uncertainty with a linear fractional transformation structure. Further, the uncertainty $\Delta(t)$ has the following diagonal scalar block structure:

$$\Delta(t) = \text{diag}(\delta_1(t)I_{r_1},\quad \delta_2(t)I_{r_2}, \dots, \delta_q(t)I_{r_q}),\quad |\delta_i(t)| \leq 1. \tag{20.5}$$

Its input–output relationship is described by

$$w_P = \Delta(t)z_P. \tag{20.6}$$

On the other hand, the state equation of the nominal transfer matrix $G(s)$ is given by

$$\dot{x} = Ax + B_1 w_P + B_2 d + B_3 u \tag{20.7}$$

$$z_P = C_1 x + D_{11} w_P + D_{12} d + D_{13} u \tag{20.8}$$

$$e = C_2 x + D_{21} w_P + D_{22} d + D_{23} u \tag{20.9}$$

$$y = C_3 x + D_{31} w_P + D_{32} d \tag{20.10}$$

in which d is the disturbance, e the performance output, y the measured output, and u the control input. Substitution of the equation of z_P into (20.6) yields

$$\begin{aligned} w_P &= (I - \Delta D_{11})^{-1}\Delta(C_1 x + D_{12} d + D_{13} u) \\ &= \Delta(I - D_{11}\Delta)^{-1}(C_1 x + D_{12} d + D_{13} u). \end{aligned}$$

Then substituting this equation back into the preceding state equation and eliminating w_P and z_P, we finally obtain the following state equation:

$$\begin{aligned} \dot{x} &= [A + B_1\Delta(I - D_{11}\Delta)^{-1}C_1]x + [B_2 + B_1\Delta(I - D_{11}\Delta)^{-1}D_{12}]d \\ &\quad + [B_3 + B_1\Delta(I - D_{11}\Delta)^{-1}D_{13}]u \end{aligned} \tag{20.11}$$

$$e = [C_2 + D_{21}\Delta(I - D_{11}\Delta)^{-1}C_1]x + [D_{22} + D_{21}\Delta(I - D_{11}\Delta)^{-1}D_{12}]d$$
$$+ [D_{23} + D_{21}\Delta(I - D_{11}\Delta)^{-1}D_{13}]u \tag{20.12}$$
$$y = [C_3 + D_{31}\Delta(I - D_{11}\Delta)^{-1}C_1]x + [D_{32} + D_{31}\Delta(I - D_{11}\Delta)^{-1}D_{12}]d$$
$$+ D_{31}\Delta(I - D_{11}\Delta)^{-1}D_{13}u. \tag{20.13}$$

In this state equation, all coefficient matrices are LFT functions about the time-varying coefficient matrix $\Delta(t)$.

Introducing the same LFT structure about $\Delta(t)$ into the gain-scheduled controller accordingly, Figure 20.1(a) is obtained. Concretely, the state equation of the coefficient matrix $K(s)$ is set as

$$\dot{x}_K = A_K\, x_K + B_{K1}\, w_K + B_{K2}\, y \tag{20.14}$$
$$z_K = C_{K1}\, x_K + D_{K11}\, w_K + D_{K12}\, y \tag{20.15}$$
$$u = C_{K2}\, x_K + D_{K21}\, w_K + D_{K22}\, y. \tag{20.16}$$

Here, the gain-scheduling signal w_K is

$$w_K = \Delta(t) z_K. \tag{20.17}$$

Since the signal w_K relies on the time-varying parameter $\Delta(t)$, it changes the gain of the controller online. This relationship can be read more clearly from the equation about the control input and the measured output:

$$\hat{u}(s) = \mathcal{F}_u(K(s), \Delta(t))\hat{y}(s) = [K_{22} + K_{21}\Delta(I - \Delta K_{11})^{-1}K_{12}]\hat{y}(s). \tag{20.18}$$

Here, $K_{ij}(s)$ denotes a block of the 2×2 block partition of the coefficient matrix $K(s)$.

In system design, these two time-varying parameter blocks can be merged into an augmented uncertainty block via block diagram transformation and written equivalently as Figure 20.1(b). Then, we may conduct the robust control design w.r.t. this closed-loop system. The design methods can be small-gain-based $\mathcal{H}_\infty$ control, μ synthesis, scaled $\mathcal{H}_\infty$ control, and so on. Here,

$$\Delta_{gs}(t) = \begin{bmatrix} \Delta(t) & \\ & \Delta(t) \end{bmatrix}. \tag{20.19}$$

Comparing the new generalized plant $G_{gs}(s)$ with $G(s)$, newly added are inputs (w_K, z_K) and outputs (w_K, z_K). So, putting these new input–output relations

$$z_K = z_K, \quad w_K = w_K$$

into the state equation of $G(s)$ leads to the state equation of $G_{gs}(s)$ (Figure 20.1(b)):

$$\begin{bmatrix} \dot{x} \\ \hline z_K \\ z_P \\ e \\ \hline w_K \\ y \end{bmatrix} = \left[\begin{array}{c|cccc|cc} A & 0 & B_1 & B_2 & 0 & B_3 \\ \hline 0 & 0 & 0 & 0 & I & 0 \\ C_1 & 0 & D_{11} & D_{12} & 0 & D_{13} \\ C_2 & 0 & D_{21} & D_{22} & 0 & D_{23} \\ \hline 0 & I & 0 & 0 & 0 & 0 \\ C_3 & 0 & D_{31} & D_{32} & 0 & 0 \end{array}\right] \begin{bmatrix} x \\ \hline w_K \\ w_P \\ d \\ \hline z_K \\ u \end{bmatrix}. \tag{20.20}$$

20.2.1 Gain-Scheduled $\mathcal{H}_\infty$ Control Design with Scaling

In the control design, it is necessary to treat the time-varying parameter matrix $\Delta_{gs}(t)$ as an uncertainty and apply methods such as the small-gain approach. But the small-gain approach is already conservative for real uncertainty, the diagonal structure of $\Delta_{gs}(t)$ further strengthens such conservatism. So, good performance cannot be expected from the small-gain condition only. More idea is needed. One is the introduction of scaling, another is the μ synthesis.

Next, a constant scaling method is introduced which is suitable for numerical computation. Assume that the control specification is to reduce the $\mathcal{H}_\infty$ norm of the closed-loop transfer matrix from the disturbance d to the performance output e:

$$\sup_{\|d\|_2 \neq 0} \frac{\|e\|_2}{\|d\|_2} = \|H_{ed}\|_\infty < 1. \tag{20.21}$$

According to the equivalence between the $\mathcal{H}_\infty$ norm performance and the robust stability (Section 12.3), the aforementioned performance specification is equivalent to the robust stability condition of the closed-loop system when a virtual norm-bounded uncertainty $\Delta_P(s)$ ($\|\Delta_P\|_\infty \leq 1$) is inserted between the signals d and e. Then, the overall uncertainty becomes

$$\mathrm{diag}(\Delta_{gs},\ \Delta_P) = \mathrm{diag}(\Delta,\ \Delta,\ \Delta_P).$$

To obtain a less conservative robust performance condition from the small-gain theorem, we shall look for a scaled small-gain condition according to the scaling method of Section 12.4.2. First of all, let us see what structure the scaling matrix should have. Due to the scalar block structure of $\Delta(s)$ in (20.5), any real matrix permutable with $\Delta(s)$ has the form of

$$L = \mathrm{diag}(L_1', L_2', \ldots, L_q'), \quad L_i' \in \mathbb{R}^{r_i \times r_i}.$$

As a scaling matrix, L must be invertible. So we may confine the scaling matrix in the set

$$\mathbb{L} = \{L > 0 | L = \mathrm{diag}(L_1', L_2', \ldots, L_q'), \quad L_i' \in \mathbb{R}^{r_i \times r_i}\}. \tag{20.22}$$

Obviously, a scaling matrix that is both positive definite and permutable with $\Delta_{gs}(s)$ has the following form:

$$L_{gs} = \begin{bmatrix} L_1 & L_2 \\ L_2^T & L_3 \end{bmatrix}, \quad L_1, L_3 \in \mathbb{L}, \quad L_2\Delta = \Delta L_2. \tag{20.23}$$

Secondly, since the virtual uncertainty $\Delta_P(s)$ is a full-block uncertainty, only a scalar block ℓI is permutable with it. So, the whole scaling matrix becomes

$$L_a = \begin{bmatrix} L_1 & L_2 & \\ L_2^T & L_3 & \\ & & \ell I \end{bmatrix} = \ell \begin{bmatrix} L_1/\ell & L_2/\ell & \\ L_2^T/\ell & L_3/\ell & \\ & & I \end{bmatrix}.$$

Since in the scaled small-gain condition both the scaling matrix L_a and its inverse matrix appear, we need only consider the case of $\ell = 1$. As such, we have shown that the scaled small-gain condition

$$\left\| \begin{bmatrix} L_{gs}^{1/2} & \\ & I \end{bmatrix} \mathcal{F}_\ell(G_{gs}, K) \begin{bmatrix} L_{gs}^{-1/2} & \\ & I \end{bmatrix} \right\|_\infty < 1 \tag{20.24}$$

guarantees the $\mathcal{H}_\infty$ performance of the gain-scheduled control system.

According to Exercise 16.6, the $\mathcal{H}_\infty$ norm specification with scaling matrix $L > 0$, that is,

$$\|L^{1/2} H_{zw} L^{-1/2}\|_\infty < 1$$

is achieved iff there exist matrices $X > 0$, $Y > 0$ and L, J satisfying

$$\begin{bmatrix} N_X^T & 0 \\ 0 & I_{n_w} \end{bmatrix} \begin{bmatrix} AX + XA^T & XC_1^T & B_1 \\ C_1 X & -J & D_{11} \\ B_1^T & D_{11}^T & -L \end{bmatrix} \begin{bmatrix} N_X & 0 \\ 0 & I_{n_w} \end{bmatrix} < 0 \tag{20.25}$$

$$\begin{bmatrix} N_Y^T & 0 \\ 0 & I_{n_z} \end{bmatrix} \begin{bmatrix} YA + A^T Y & YB_1 & C_1^T \\ B_1^T Y & -L & D_{11}^T \\ C_1 & D_{11} & -J \end{bmatrix} \begin{bmatrix} N_Y & 0 \\ 0 & I_{n_z} \end{bmatrix} < 0 \tag{20.26}$$

$$\begin{bmatrix} X & I \\ I & Y \end{bmatrix} \geq 0 \tag{20.27}$$

$$LJ = I. \tag{20.28}$$

The difficulty in applying this result is that the equation $LJ = I$ does not have convexity so that numerical computation is rather hard. What we should consider is if it is possible to avoid this nonconvexity by fully using the structure of the scheduling matrix (Δ appears twice). Here, we denote the inverse of L_{gs} by

$$J_{gs} = \begin{bmatrix} L_1 & L_2 \\ L_2^T & L_3 \end{bmatrix}^{-1} = \begin{bmatrix} J_1 & J_2 \\ J_2^T & J_3 \end{bmatrix}. \tag{20.29}$$

To apply the solvability condition of scaled $\mathcal{H}_\infty$ problem to the state space realization (20.20) of the generalized plant, we should make the following matrix replacements:

$$B_1 \to [0 \ \ B_1 \ \ B_2], \quad B_2 \to [0 \ \ B_3]$$

$$C_1 \to \begin{bmatrix} 0 \\ C_1 \\ C_2 \end{bmatrix}, \quad D_{11} \to \begin{bmatrix} 0 & 0 & 0 \\ 0 & D_{11} & D_{12} \\ 0 & D_{21} & D_{22} \end{bmatrix}, \quad D_{12} \to \begin{bmatrix} I & 0 \\ 0 & D_{13} \\ 0 & D_{23} \end{bmatrix}$$

$$C_2 \to \begin{bmatrix} 0 \\ C_3 \end{bmatrix}, \quad D_{21} \to \begin{bmatrix} I & 0 & 0 \\ 0 & D_{31} & D_{32} \end{bmatrix}$$

So, the matrix corresponding to orthogonal matrix N_Y is

$$\begin{bmatrix} 0 & I & 0 & 0 \\ C_3 & 0 & D_{31} & D_{32} \end{bmatrix} = \begin{bmatrix} I & 0 & 0 & 0 \\ 0 & C_3 & D_{31} & D_{32} \end{bmatrix} \begin{bmatrix} 0 & I & 0 & 0 \\ I & 0 & 0 & 0 \\ 0 & 0 & I & 0 \\ 0 & 0 & 0 & I \end{bmatrix}$$

Therefore,

$$N_Y = \begin{bmatrix} 0 & I & 0 & 0 \\ I & 0 & 0 & 0 \\ 0 & 0 & I & 0 \\ 0 & 0 & 0 & I \end{bmatrix} \begin{bmatrix} 0 \\ [C_3 \ \ D_{31} \ \ D_{32}]_\perp \end{bmatrix}.$$

Substituting it into the second LMI of the solvability condition (20.26) and removing the parts multiplied by zero blocks in N_Y, we obtain

$$\begin{bmatrix} N_Y^T & 0 \\ 0 & I \end{bmatrix} \begin{bmatrix} YA + A^TY & YB_1 & YB_2 & 0 & C_1^T & C_2^T \\ B_1^TY & -L_3 & 0 & 0 & D_{11}^T & D_{21}^T \\ B_2^TY & 0 & -I & 0 & D_{12}^T & D_{22}^T \\ 0 & 0 & 0 & -J_1 & -J_2 & 0 \\ C_1 & D_{11} & D_{12} & -J_2^T & -J_3 & 0 \\ C_2 & D_{21} & D_{22} & 0 & 0 & -I \end{bmatrix} \begin{bmatrix} N_Y & 0 \\ 0 & I \end{bmatrix} < 0 \quad (20.30)$$

in which N_Y has been redefined as

$$N_Y = [C_3 \;\; D_{31} \;\; D_{32}]_\perp \quad (20.31)$$

for convenience. However, this inequality still does not get rid of the nonconvexity of $LJ = I$. We note that in the (20.30) the elements of both rows and columns related with (J_1, J_2) are all zero. So, next we try to eliminate (J_1, J_2) using this property. If possible, then only (L_3, J_3) are left in the condition and they do not necessarily need to be the inverse of each other. So this removes the nonconvexity constraint.

It is easy to verify that

$$\begin{bmatrix} 0 & C_1^T & C_2^T \\ 0 & D_{11}^T & D_{21}^T \\ 0 & D_{12}^T & D_{22}^T \end{bmatrix} \begin{bmatrix} -J_1 & -J_2 & 0 \\ -J_2^T & -J_3 & 0 \\ 0 & 0 & -I \end{bmatrix}^{-1} \begin{bmatrix} 0 & 0 & 0 \\ C_1 & D_{11} & D_{12} \\ C_2 & D_{21} & D_{22} \end{bmatrix}$$

$$= -\begin{bmatrix} C_1^T & C_2^T \\ D_{11}^T & D_{21}^T \\ D_{12}^T & D_{22}^T \end{bmatrix} \begin{bmatrix} L_3 & 0 \\ 0 & I \end{bmatrix} \begin{bmatrix} C_1 & D_{11} & D_{12} \\ C_2 & D_{21} & D_{22} \end{bmatrix}.$$

Applying this equation and Schur's lemma to Eq. (20.30), we obtain an LMI

$$N_Y^T \left\{ \begin{bmatrix} YA + A^TY & YB_1 & YB_2 \\ B_1^TY & -L_3 & 0 \\ B_2^TY & 0 & -I \end{bmatrix} + \right.$$

$$\left. \begin{bmatrix} C_1^T & C_2^T \\ D_{11}^T & D_{21}^T \\ D_{12}^T & D_{22}^T \end{bmatrix} \begin{bmatrix} L_3 & 0 \\ 0 & I \end{bmatrix} \begin{bmatrix} C_1 & D_{11} & D_{12} \\ C_2 & D_{21} & D_{22} \end{bmatrix} \right\} N_Y < 0.$$

Similarly, the inequality about X is obtained as

$$N_X^T \left\{ \begin{bmatrix} AX + XA^T & XC_1^T & XC_2^T \\ C_1X & -J_3 & 0 \\ C_2X & 0 & -I \end{bmatrix} + \right.$$

$$\left. \begin{bmatrix} B_1 & B_2 \\ D_{11} & D_{12} \\ D_{21} & D_{22} \end{bmatrix} \begin{bmatrix} J_3 & 0 \\ 0 & I \end{bmatrix} \begin{bmatrix} B_1^T & D_{11}^T & D_{21}^T \\ B_2^T & D_{12}^T & D_{22}^T \end{bmatrix} \right\} N_X < 0$$

in which $N_X = [B_3^T \ D_{13}^T \ D_{23}^T]_\perp$. Finally, (L_3, J_3) should satisfy the following relationship

$$\begin{bmatrix} L_3 & I \\ I & J_3 \end{bmatrix} \geq 0, \quad L_3 > 0, J_3 > 0 \tag{20.32}$$

according to Lemma 3.1. These five LMIs, together with the relation between (X, Y)

$$\begin{bmatrix} X & I \\ I & Y \end{bmatrix} \geq 0 \tag{20.33}$$

become the solvability condition for the gain-scheduled problem with constant scaling matrix. So, we have succeeded in converting the original nonconvex condition to a convex condition by making full use of the structural properties of gain-scheduled problem.

Remark 20.1 *Note that the solvability condition*

$$\begin{bmatrix} N_Y^T & 0 \\ 0 & I \end{bmatrix} \begin{bmatrix} YA + A^TY & YB_1 & YB_2 & C_1^T & C_2^T \\ B_1^TY & -L_3 & 0 & D_{11}^T & D_{21}^T \\ B_2^TY & 0 & -I & D_{12}^T & D_{22}^T \\ C_1 & D_{11} & D_{12} & -J_3 & 0 \\ C_2 & D_{21} & D_{22} & 0 & -I \end{bmatrix} \begin{bmatrix} N_Y & 0 \\ 0 & I \end{bmatrix} < 0$$

given in Ref. [3] is not correct. Unless $J_2 = 0$, this inequality cannot guarantee (20.30). But when $J_2 = 0$, (L_3, J_3) must satisfy the nonconvex relation $L_3 J_3 = I$.

20.2.2 Computation of Controller

After obtaining matrices (X, Y, L_3, J_3), the coefficient matrices of controller

$$K = \begin{bmatrix} D_{K11} & D_{K12} & C_{K1} \\ D_{K21} & D_{K22} & C_{K2} \\ B_{K1} & B_{K2} & A_K \end{bmatrix} \tag{20.34}$$

can be computed as follows:

1. Matrix factorization:

$$MM^T = Y - X^{-1}, \quad N^TN = L_3 - J_3^{-1}. \tag{20.35}$$

2. Calculate the Lyapunov matrix P and scaling matrix L:

$$P = \begin{bmatrix} Y & M \\ M^T & I \end{bmatrix}, \quad L = \begin{bmatrix} I & N \\ N^T & L_3 \end{bmatrix},$$
$$L_a = \mathrm{diag}(L, I_{n_e}), \quad J_a = L_a^{-1}. \tag{20.36}$$

3. Solve the LMI:

$$Q + E^T \mathcal{K} F + F^T \mathcal{K}^T E < 0 \tag{20.37}$$

to get the coefficient matrix $\mathcal{K}$ of the controller. Here,

$$\begin{bmatrix} Q & E^T \\ F & \end{bmatrix} = \left[\begin{array}{ccc:c} \overline{A}^T P + P\overline{A} & P\overline{B}_1 & \overline{C}_1^T & P\overline{B}_2 \\ \overline{B}_1^T P & -L_a & \overline{D}_{11}^T & 0 \\ \overline{C}_1 & \overline{D}_{11} & -J_a & \overline{D}_{12} \\ \hdashline \overline{C}_2 & \overline{D}_{21} & 0 & \end{array}\right] \tag{20.38}$$

$$\begin{bmatrix} \overline{A} & \overline{B}_1 & \overline{B}_2 \\ \overline{C}_1 & \overline{D}_{11} & \overline{D}_{12} \\ \overline{C}_2 & \overline{D}_{21} & \end{bmatrix} = \left[\begin{array}{cc:ccc:ccc} A & 0 & 0 & B_1 & B_2 & 0 & B_3 & 0 \\ 0 & 0 & 0 & 0 & 0 & 0 & 0 & I \\ \hdashline 0 & 0 & 0 & 0 & 0 & I & 0 & 0 \\ C_1 & 0 & 0 & D_{11} & D_{12} & 0 & D_{13} & 0 \\ C_2 & 0 & 0 & D_{21} & D_{22} & 0 & D_{23} & 0 \\ \hdashline 0 & 0 & I & 0 & 0 & & & \\ C_3 & 0 & 0 & D_{31} & D_{32} & & & \\ 0 & I & 0 & 0 & 0 & & & \end{array}\right]. \tag{20.39}$$

20.3 Case Study: Stabilization of a Unicycle Robot

In this section, we will use the LFT model-based gain-scheduled method to design a posture stabilization and longitudinal motion controller for the unicycle robot [21] shown in Figure 20.2. This unicycle robot has two gyroscopes acting as an actuator for the lateral stabilization and a wheel as an actuator for the longitudinal motion.

20.3.1 Structure and Model

The definitions of physical variables are shown in Figure 20.3. The measured outputs are the rotational angles of the wheel (ϕ) and two gyroscopes (θ_3, θ_5), longitudinal and lateral angular rates of the body ($\dot{\theta}_1, \dot{\theta}_2$). Five sensors are used to measure these signals: one rotary encoder (ϕ), two potential meters (θ_3, θ_5), and two angular velocity sensors ($\dot{\theta}_1, \dot{\theta}_2$). The control inputs are the torques of the wheel motor (τ_1) and the outer motors of gyroscopes (τ_2, τ_2). Two inner motors are used to rotate the two gyroscopes at a fixed angular velocity of 8000 rpm.

20.3.1.1 Gyro Actuator

The mechanism of torque generation by gyroscope is as follows: when a flywheel rotating along z axis at speed ω_z is rotated along y axis at speed ω_y, a torque

$$\tau = -I_z \omega_z \omega_y \tag{20.40}$$

Figure 20.2 Unicycle robot in motion

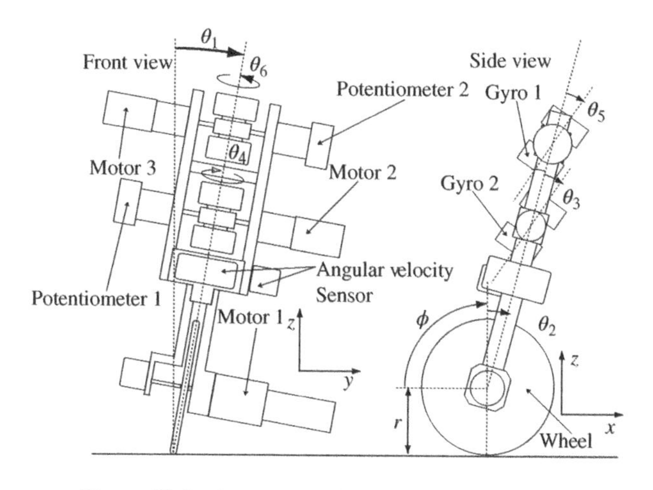

Figure 20.3 Front and side views of unicycle robot

is generated along x axis, as shown in Figure 20.4 (left-hand side). This torque τ is the so-called gyro-moment which contributes to the lateral stabilization. When the pitch angle along x axis is θ, the torque τ_{roll} in the lateral direction about x axis is given by

$$\tau_{\text{roll}} = -I_z \omega_z \cos\theta \omega_y = -R(\theta)\omega_y \tag{20.41}$$

in which $R(\theta) = I_z \omega_z \cos\theta$ is called the coefficient of gyro-moment.

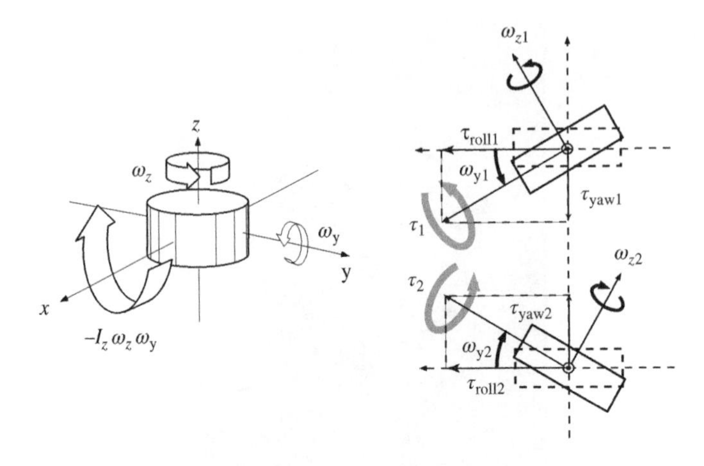

Figure 20.4 Generation of gyro-moment and cancellation of yawing torque

A large torque can be produced by raising the angular velocity ω_z of the flywheel. ω_y is used to control the torque τ. However, when the flywheel is at a tilted posture, the torque τ has an undesired yawing component. In order to cancel this yawing torque, two identical flywheels are used. The angular velocities ω_z and ω_y of each flywheel have the same amplitudes but opposite directions, respectively. When the initial angles are the same, the yawing components cancel each other and a purely lateral torque is obtained, as shown in Figure 20.4 (right-hand side).

20.3.1.2 Linearized Model

The nonlinear motion equation is derived based on Lagrangian dynamics [21]. Since the purpose is to achieve posture stabilization, the motion equation is linearized in the control design. Let the state vector be

$$x = [\phi, \theta_1, \theta_2, \theta_3, \theta_5, \dot{\phi}, \dot{\theta}_1, \dot{\theta}_2, \dot{\theta}_3, \dot{\theta}_5]^T.$$

The linearized motion equation in the neighborhood of the equilibrium $x = 0$ is given by the following descriptor form

$$E\dot{x} = Ax + Bu \tag{20.42}$$

$$y = Cx \tag{20.43}$$

where

$$E = \begin{bmatrix} I_{5\times 5} & & & O_{5\times 5} & & \\ & J_0 & 0 & J_{02} & 0 & 0 \\ & 0 & J_1 & 0 & 0 & 0 \\ O_{5\times 5} & J_{02} & 0 & J_2 & I_{y34} & I_{y56} \\ & 0 & 0 & I_{y34} & I_{y34} & 0 \\ & 0 & 0 & I_{y56} & 0 & I_{y56} \end{bmatrix}$$

$$A = \begin{bmatrix} & & O_{5\times5} & & & & I_{5\times2} & & I_{5\times3} & \\ 0 & 0 & 0 & 0 & 0 & -D_1 & 0 & D_1 & 0 & 0 \\ 0 & G_1 & 0 & 0 & 0 & 0 & 0 & -(R_1+R_2) & -R_1 & -R_2 \\ 0 & 0 & G_2 & 0 & 0 & D_1 & R_1+R_2 & -D_1 & 0 & 0 \\ 0 & 0 & 0 & 0 & 0 & 0 & R_1 & 0 & -D_2 & 0 \\ 0 & 0 & 0 & 0 & 0 & 0 & R_2 & 0 & 0 & -D_3 \end{bmatrix}$$

$$B = \begin{bmatrix} O_{3\times5} & \begin{matrix} 1 & 0 & -1 & 0 & 0 \\ 0 & 0 & 0 & 1 & 0 \\ 0 & 0 & 0 & 0 & 1 \end{matrix} \end{bmatrix}^T$$

$$C = \begin{bmatrix} 1 & 0 & -1 & 0 & 0 & 0 & 0 & 0 & 0 & 0 \\ 0 & 0 & 0 & 1 & 0 & 0 & 0 & 0 & 0 & 0 \\ 0 & 0 & 0 & 0 & 1 & 0 & 0 & 0 & 0 & 0 \\ 0 & 0 & 0 & 0 & 0 & 0 & 1 & 0 & 0 & 0 \\ 0 & 0 & 0 & 0 & 0 & 0 & 0 & 1 & 0 & 0 \end{bmatrix}$$

The parameters are obtained through a series of identification experiments [21].

20.3.1.3 LPV Model

In the posture stabilization and the longitudinal running control, posture angles θ_1, θ_2 are operated in the neighborhood of 0. Only θ_3 and θ_5, the rotation angles of gyroscopes are operated in a wide range ($\pm\pi/4$ rad). Since the coefficients of the gyro moments are given by

$$R_1(\theta_2, \theta_3) = R_1 \cos(\theta_2 + \theta_3), \quad R_2(\theta_2, \theta_5) = R_2 \cos(\theta_2 + \theta_5), \tag{20.44}$$

these two parameters change significantly during the motion, particularly when a lateral force disturbance is applied at the robot. Therefore, this parameter variation must be taken into consideration in the control design.

Noting that $\theta_2 \approx 0$, these two parameters can be written as

$$R_1(\theta_2, \theta_3) = R_1 + \Delta_{R_1}\delta_1(t), \quad R_2(\theta_2, \theta_5) = R_2 + \Delta_{R_2}\delta_2(t) \tag{20.45}$$

in which

$$\delta_1(t) \approx \frac{R_1}{\Delta_{R_1}}(\cos(\theta_3(t)) - 1), \quad \delta_2(t) \approx \frac{R_2}{\Delta_{R_2}}(\cos(\theta_5(t)) - 1) \tag{20.46}$$

and Δ_{R_1} and ΔR_2 denote the sizes of parameter uncertainties which can be determined once the ranges of θ_3, θ_5 are prescribed. Thus, $\delta_1(t), \delta_2(t)$ can be obtained online.

Replacing the constant parameters R_1, R_2 in the linearized model by $R_1(\theta_2, \theta_3), R_2(\theta_2, \theta_3)$ yields an LPV model as follows:

$$E\dot{x} = \left(A + \sum_{i=1}^{2} \delta_i(t) A_i\right) x + Bu \tag{20.47}$$

20.3.2 Control Design

The following disturbances exist in this unicycle robot: external force disturbance, persistent force disturbance due to the power cable, unbalance due to assembly error, and so on. These disturbances are modeled as two forces acting at $\ddot{\phi}$ (longitudinal) and $\ddot{\theta}_1$ (lateral), respectively. So a disturbance term Hd ($d \in \mathbb{R}^2$) is added to $E\dot{x}$:

$$E\dot{x} = \left(A + \sum_{i=1}^{2} \delta_i(t) A_i\right) x + Hd + Bu \tag{20.48}$$

in which

$$H = \begin{bmatrix} O_{2\times 5} & \begin{matrix} 1 & 0 & 0 & 0 & 0 \\ 0 & 1 & 0 & 0 & 0 \end{matrix} \end{bmatrix}^T.$$

The performance output is determined as follows. The tracking error $\phi - r$ is penalized so as to achieve the tracking of reference $r(t)$ by $\phi(t)$. To keep the balance of the unicycle robot, $\theta_1, \theta_2 \to 0$ needs to be guaranteed in principle. However, only the angular velocities $\dot{\theta}_1, \dot{\theta}_2$ are measured and the equilibrium state may not be $\theta_1, \theta_2 = 0$ due to manufacturing error. Therefore, the performance specification is set as that the body stands still at the equilibrium, that is, $\dot{\theta}_1, \dot{\theta}_2 \to 0$.

It follows from (20.40) and (20.44) that the rolling components of gyro-moments are $-R_1 \cos(\theta_3)\dot{\theta}_3$ and $-R_2 \cos(\theta_5)\dot{\theta}_5$ when θ_2 and its speed are small. So, large angular speeds $\dot{\theta}_3, \dot{\theta}_5$ are required which implies that the gyroscopes have to rotate over a quite wide range. Meanwhile, the gyro-moments decrease as the angles θ_3, θ_5 increase. For this reason, θ_3, θ_5 must be kept at zeros in the steady state.

Further, the angular velocity sensors used are piezoelectric vibrator type which have high-frequency noise due to the vibration of gyroscopes, so the controller gain must roll-off at high frequency. Also, the controller gain in the control band cannot be too high because of the limit of current. Based on these considerations, the performance output $\bar{e} \in \mathbb{R}^5$ is selected as

$$\bar{e} = [\phi, \theta_3, \theta_5, \dot{\phi}, \dot{\theta}_1, \dot{\theta}_2]^T = Mx$$

and the generalized plant $G_p(s)$ set as Figure 20.5, in which $w_p, z_p \in \mathbb{R}^4$ denote the output and input of the time-varying parameter block. All control inputs $e_u \in \mathbb{R}^3$ have also been put into the performance output, and the effect of sensor noise on $\dot{\theta}_1$ is taken into account. $r \in \mathbb{R}^1$ is the reference for the rotational angle ϕ of the wheel. According to Section 11.3.4, the generalized plant is given by

$$G(s) = \left[\begin{array}{c|ccc} E^{-1}A & E^{-1}L & E^{-1}H & E^{-1}B \\ \hline W & 0 & 0 & 0 \\ M & 0 & 0 & 0 \\ C & 0 & 0 & D \end{array}\right]. \tag{20.49}$$

Further, the weighting functions have the following structures:

$$W_e = \text{diag}(W_\phi, W_{\theta_3}, W_{\theta_5}, W_{\dot{\theta}_1}, W_{\dot{\theta}_2})$$
$$W_u = \text{diag}(W_{\tau 1}, W_{\tau 2}, W_{\tau 3})$$

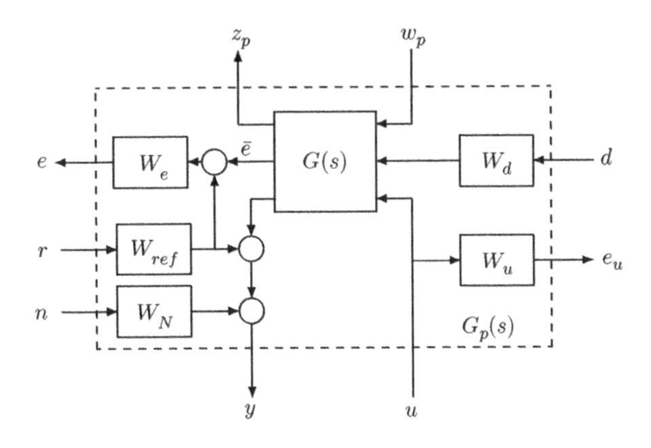

Figure 20.5 Block diagram of generalized plant

$$
\begin{aligned}
W_d &= \mathrm{diag}(W_{d_1}, W_{d_2}) \\
W_{ref} &= \begin{bmatrix} -W_r & 0 & 0 & 0 & 0 \end{bmatrix}^T \\
W_N &= \begin{bmatrix} 0 & 0 & 0 & W_{n_4} & 0 \end{bmatrix}^T
\end{aligned}
$$

in which the subscript of a weight denotes the corresponding signal.

The overall plant can be expressed as

$$
\begin{bmatrix} e_p \\ y \end{bmatrix} = \mathcal{F}_u(G(s), \Delta) \begin{bmatrix} d_p \\ u \end{bmatrix}, \quad e_p = \begin{bmatrix} e \\ e_u \end{bmatrix}, \quad d_p = \begin{bmatrix} d \\ r \\ n \end{bmatrix} \tag{20.50}
$$

where Δ denotes the diagonal matrix of time-varying parameters:

$$
\Delta = \mathrm{diag}(\delta_1 I_2, \delta_2 I_2), \quad \|\Delta\| \leq 1, \Delta \in \mathbb{R}^{4\times 4}. \tag{20.51}
$$

The performance is measured by the induced two-norm, that is,

$$
\sup_{d_p} \frac{\|e_p\|_2}{\|d_p\|_2} < \gamma \tag{20.52}
$$

for all $\Delta(t)$ in the prescribed domain. This performance is guaranteed by the following constant-scaled small-gain condition:

$$
\|S^{-1}\mathcal{F}_l(G_{gs}, K)SJ_\gamma\|_\infty < 1 \tag{20.53}
$$

in which $J_\gamma \equiv \mathrm{diag}(I, I/\gamma)$ and S is a constant scaling matrix which is permutable with $\mathrm{diag}(\Delta, \Delta, \Delta_p)$. Δ_p is a performance block representing the disturbance attenuation performance (20.52).

The tracking of step reference is presumed, so W_ϕ (model of the reference signal) is selected as an approximated integrator. In order to attenuate step disturbances, $W_{\theta_3}, W_{\theta_5}$ (models of mechanical disturbances) are also selected as approximated integrators. The input weighting functions are high-pass filters used to lower the high-frequency gain of controller. Other weights are chosen as constants and used as tuning gains in design. Concretely, the weighting functions are determined as

$$W_\phi = \frac{0.2s+4}{s+0.0001}, \quad W_{\theta_3} = W_{\theta_5} = \frac{0.3s+3}{s+0.0001}, \quad W_{\dot\theta_1} = W_{\dot\theta_2} = 0.01$$
$$W_{\tau_1} = \frac{0.5s+1}{s+1000} \times 10^3, \quad W_{\tau_2} = W_{\tau_3} = \frac{4s+40}{s+2000} \times 10^3$$
$$W_{d_1} = 0.02, \quad W_{d_2} = 0.03, \quad W_r = 0.1, \quad W_n = 0.001$$

by trial and error.

The working ranges of gyroscopes are presumed as $|\theta_3|, |\theta_5| \le \frac{\pi}{4}$, so the corresponding perturbation ranges are computed as

$$\Delta_{R_i} = |R_i \times (\cos\frac{\pi}{4} - 1)| \approx R_i \times 0.3, \quad i = 1, 2$$

A performance level of $\gamma = 0.961$ is achieved in the design. The order of the controller is 16. The Bode plots of the gain-scheduled controller $\mathcal{F}_u(K, \Delta)$ for $\theta_3 = \theta_5 = 0$ (solid line), $\theta_3 = \theta_5 = \frac{\pi}{6}$ (dashed line), $\theta_3 = \theta_5 = \frac{\pi}{4}$ $(\cdot -)$, and $\theta_3 = \theta_5 = \frac{\pi}{3}$ $(+)$ are investigated. Significant variation of gain is observed in the controllers between signals: $\dot\theta_1 \mapsto \tau_1$, $(\phi, \theta_3, \theta_5, \dot\theta_2) \mapsto \tau_2$, $(\phi, \theta_3, \theta_5, \dot\theta_2) \mapsto \tau_3$. As an example, the controller from $\dot\theta_1$ to τ_1 is shown in Figure 20.6.

20.3.3 Experiment Results

In the experiments, the LTI coefficient matrix is discretized by using Tustin method with a sampling period of 6 ms. The gain-scheduled controller is implemented using C program running on RTLinux. Three kinds of experiments are done.

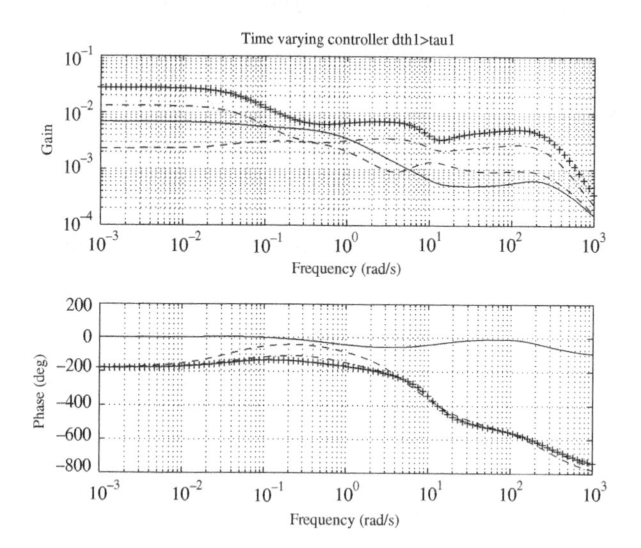

Figure 20.6 Bode plot of gain-scheduled controller

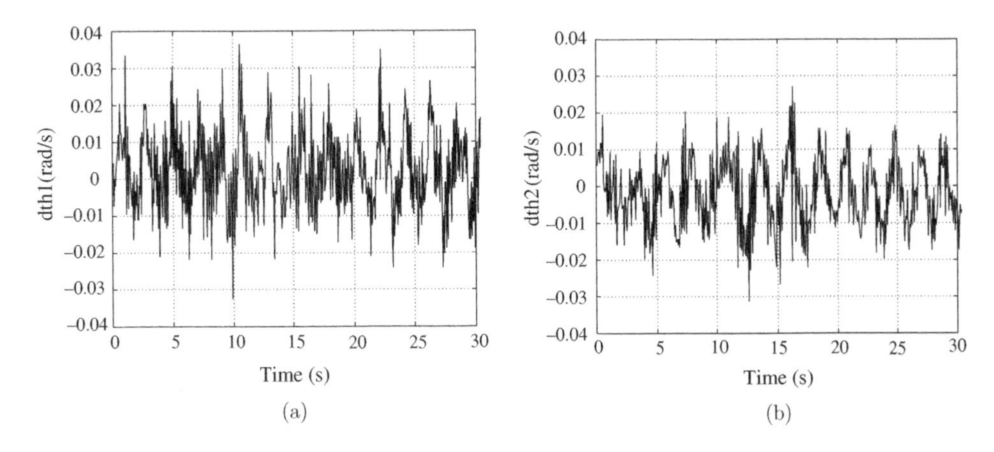

Figure 20.7 Posture stabilization (a) $\dot{\theta}_1$ (b) $\dot{\theta}_2$

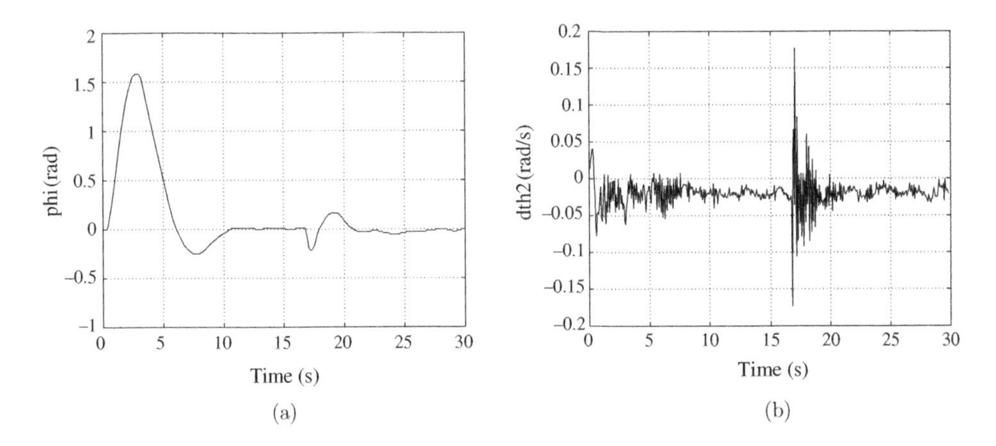

Figure 20.8 Responses to longitudinal force disturbance (a) ϕ (b) $\dot{\theta}_2$

20.3.3.1 Posture Stabilization

The responses of angular velocities are shown in Figure 20.7. As they are roughly symmetric to zero, the pitch angle θ_2 and roll angle θ_1 are nearly zero.

20.3.3.2 Disturbance Attenuation

Secondly, an external pulse force is applied by hand to test the disturbance attenuation performance. The force is applied in the longitudinal and lateral directions at 17 and 21 s, respectively. The results are shown separately in Figures 20.8 and 20.9. It is remarked that the two gyroscopes rotate in opposite directions when the lateral force is applied.

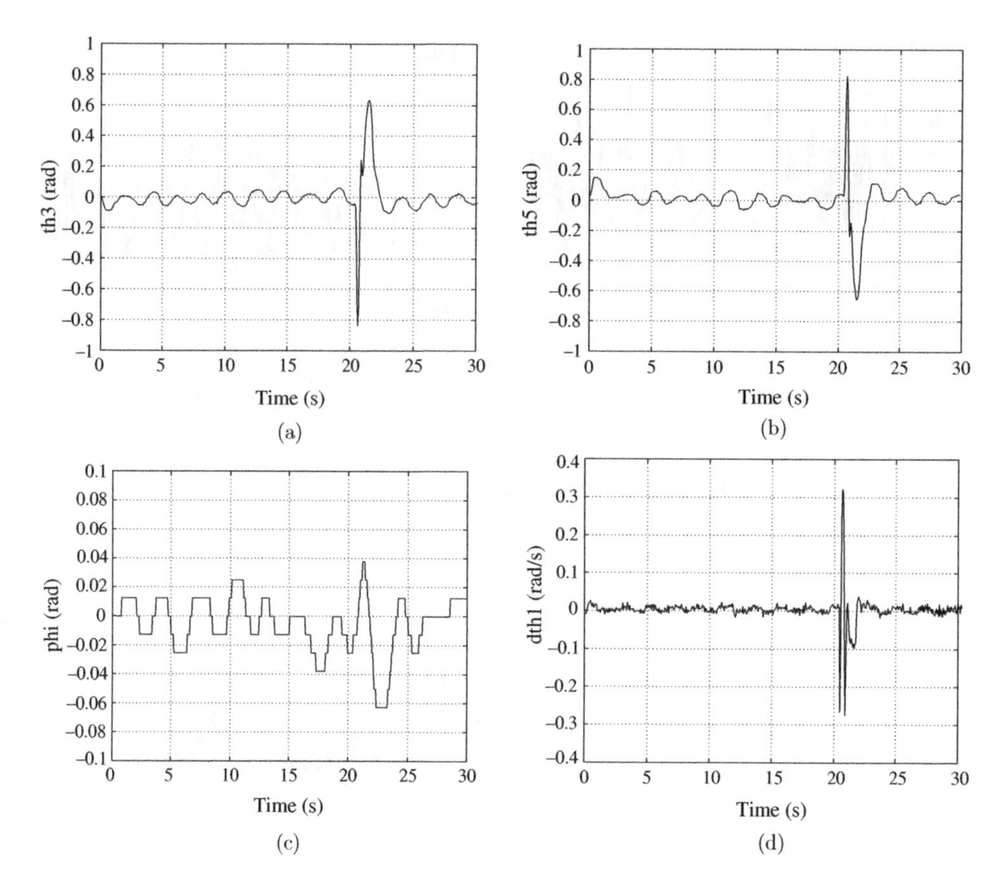

Figure 20.9 Responses to lateral force disturbance (a) θ_3 (b) θ_5 (c) ϕ (d) $\dot{\theta}_1$

20.3.3.3 Running Experiment

A reference $r(t)$ is applied to the rotational angle of the wheel, which corresponds to a distance of 40 cm. The response of wheel angle ϕ is shown in Figure 20.10.

These experiments indicate that the gain-scheduled control works well.

20.4 Affine LPV Model

From the LPV models of nonlinear systems[1] which we have encountered up to now, we see that compared with the LFT form it is more practical to describe the coefficient matrices of the state equation of plant as affine functions of time-varying parameters. So, we consider the following affine LPV model:

$$\dot{x} = A(p(t))x + B_1(p(t))d + B_2(p(t))u \tag{20.54}$$

[1] For example, the single-machine infinite-bus power system.

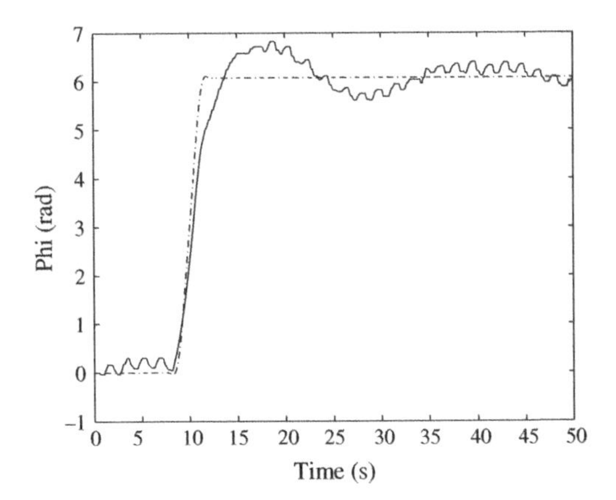

Figure 20.10 Longitudinal running

$$z = C_1(p(t))x + D_{11}d + D_{12}u \tag{20.55}$$

$$y = C_2(p(t))x + D_{21}d \tag{20.56}$$

in which (d, z, y) denote the disturbance, performance output, and measured output, respectively. Each coefficient has an affine structure about the time-varying parameter vector:

$$\begin{aligned} A(p) &= A_0 + \sum_{i=1}^{q} p_i(t)A_i \\ B_1(p) &= B_{10} + \sum_{i=1}^{q} p_i(t)B_{1i}, \quad B_2(p) = B_{20} + \sum_{i=1}^{q} p_i(t)B_{2i} \\ C_1(p) &= C_{10} + \sum_{i=1}^{q} p_i(t)C_{1i}, \quad C_2(p) = C_{20} + \sum_{i=1}^{q} p_i(t)C_{2i}. \end{aligned} \tag{20.57}$$

Here, all coefficient matrices are fixed except the time-varying parameter $p(t)$. Moreover, we assume that each time-varying parameter can be measured online and its range

$$p_i(\delta) \in [p_{im},\ p_{iM}], \quad i = 1, \ldots, q \tag{20.58}$$

can be estimated a priori. In this LPV model, all the DC terms are assumed to be constant. This is because the time-varying parameters are function of some states which usually do not appear in the DC terms in practice.

20.4.1 Easy-to-Design Structure of Gain-Scheduled Controller

We impose a similar affine structure on the gain-scheduled controller, that is, let the coefficient matrices of the controller

$$\dot{x}_K = A_K(p(t))x_K + B_K(p(t))y$$

$$u = C_K(p(t))x_K + D_K(p(t))y \tag{20.59}$$

be

$$A_K(p) = A_{K0} + \sum_{i=1}^{q} p_i(t)A_{Ki}, \quad B_K(p) = B_{K0} + \sum_{i=1}^{q} p_i(t)B_{Ki} \tag{20.60}$$
$$C_K(p) = C_{K0} + \sum_{i=1}^{q} p_i(t)C_{Ki}, \quad D_K(p) = D_{K0} + \sum_{i=1}^{q} p_i(t)D_{Ki}.$$

Each coefficient matrix is a constant matrix except $p(t)$. Let the state vector of the closed-loop system be

$$\xi = \begin{bmatrix} x \\ x_K \end{bmatrix}.$$

Then, the corresponding state equation becomes

$$\begin{aligned} \dot{\xi} &= A_c(p)\xi + B_c(p)d \\ z &= C_c(p)\xi + D_c(p)d \end{aligned} \tag{20.61}$$

where

$$\begin{aligned} A_c(p) &= \begin{bmatrix} A(p) + B_2(p)D_K(p)C_2(p) & B_2(p)C_K(p) \\ B_K(p)C_2(p) & A_K(p) \end{bmatrix} \\ B_c(p) &= \begin{bmatrix} B_1(p) + B_2(p)D_K(p)D_{21} \\ B_K(p)D_{21} \end{bmatrix} \\ C_c(p) &= [C_1(p) + D_{12}D_K(p)C_2(p) \quad D_{12}C_K(p)] \\ D_c(p) &= D_{11} + D_{12}D_K(p)D_{21}. \end{aligned} \tag{20.62}$$

It is clear from this equation that $A_c(p) \sim D_c(p)$ in general are not necessarily affine functions of the time-varying parameter vector $p(t)$. However, in order to facilitate the design, it is necessary to ensure that these coefficient matrices of closed-loop system are affine functions. In this way, we can reduce the design of gain-scheduled controller to vertex conditions. Basically, this is achieved by limiting the coefficient matrix form of the controller according to the structure of plant about $p(t)$.

Corresponding to the structure of LPV plant, the easy-to-design structures for gain-scheduled controller are summarized in the following:

- When matrices $(B_2(p), C_2(p))$ both depend on $p(t)$, (B_K, C_K) must be constant matrices besides $D_K = 0$.
- When (B_2, C_2) are both constant matrices, all coefficient matrices of the controller can be affine functions of the scheduling parameter.
- When only B_2 is a constant matrix, (B_K, D_K) must be constant matrices.
- When only C_2 is a constant matrix, (C_K, D_K) must be constant matrices.

It is easy to verify that under these conditions, all coefficient matrices of the closed-loop system are affine functions of the scheduling parameter vector $p(t)$.

In gain-scheduled control design, the affine model has the following merits:

1. It has a very good compatibility with practical systems. Since it is equivalent to the polytopic model, control specifications need to be satisfied only at all vertices, which greatly simplifies the numerical design.
2. Lyapunov method can be applied and the conservatism is weaker than the small-gain method for parametric uncertainty.
3. By the use of common Lyapunov function, it is easy to carry out *multiobjective control* design.

Its shortcoming is that all the time-varying parameters must be known, otherwise the design would be rather difficult.

20.4.2 Robust Multiobjective Control of Affine Systems

In the robust control design of affine systems, there are two methods: variable elimination and variable change. But the variable elimination method can only handle single objective. So, in this section we focus on the variable change method and study the multiobjective control problem of affine systems utilizing the feature of variable change. To this end, we need to first investigate what difficulty may occur and how to resolve it when applying this method to affine systems. Recall that in the variable change method, the new variables are

$$\begin{aligned}\mathbb{A} &= NA_KM^T + NB_KC_2X + YB_2C_KM^T + Y(A + B_2D_KC_2)X\\ \mathbb{B} &= NB_K + YB_2D_K, \quad \mathbb{C} = C_KM^T + D_KC_2X, \quad \mathbb{D} = D_K.\end{aligned}$$

When the coefficient matrices of the closed-loop system given in (20.62) are all affine functions of the scheduling parameter $p(t)$, these new variables automatically become affine functions of $p(t)$ and can be written as

$$\begin{aligned}\mathbb{A}(p) &= \mathbb{A}_0 + \sum_{i=1}^{q} p_i(t)\mathbb{A}_i, \quad \mathbb{B}(p) = \mathbb{B}_0 + \sum_{i=1}^{q} p_i(t)\mathbb{B}_i\\ \mathbb{C}(p) &= \mathbb{C}_0 + \sum_{i=1}^{q} p_i(t)\mathbb{C}_i, \quad \mathbb{D}(p) = \mathbb{D}_0 + \sum_{i=1}^{q} p_i(t)\mathbb{D}_i.\end{aligned} \tag{20.63}$$

Design based on the vertex conditions yields the constant matrices $(\mathbb{A}_i, \mathbb{B}_i, \mathbb{C}_i, \mathbb{D}_i)$ $(i = 0, 1, \ldots, q)$ in $(\mathbb{A}(p), \mathbb{B}(p), \mathbb{C}(p), \mathbb{D}(p))$. From these matrices we can compute the coefficient matrices of controller. For example, when (B_2, C_2) are both constant matrices, we have

$$\begin{aligned}D_{Ki} &= \mathbb{D}_i, \quad C_{Ki} = (\mathbb{C}_i - D_{Ki}C_2X)(M^\dagger)^T, \quad B_{Ki} = N^\dagger(\mathbb{B}_i - YB_2D_{Ki})\\ A_{Ki} &= N^\dagger(\mathbb{A}_i - NB_{Ki}C_2X - YB_2C_{Ki}M^T - Y(A_i + B_2D_{Ki}C_2)X)(M^\dagger)^T.\end{aligned} \tag{20.64}$$

(As for the other cases, refer to Exercise 20.2.)

Next, we concretely summarize the design conditions for various control specifications. The basic idea is to guarantee all performance specifications of the gain-scheduled control system with a common Lyapunov function (which corresponds to constant positive definite matrix P). That is to say, we apply the LMI performance conditions derived in the previous chapters to the gain-scheduled control system, then reduce the solvability condition to finite vertex conditions based on its polytopic structure. Assume that the vertices of the scheduling parameter vector polytope are θ_j $(j = 1, \ldots, N)$ hereafter.

20.4.2.1 $\mathcal{H}_\infty$ Norm Specification

For a given system, the necessary and sufficient condition for the $\mathcal{H}_\infty$ control problem $\|H_{zw}\|_\infty < 1$ is provided in Section 16.4. From this condition, we conclude that if there exist constant matrices $X = X^T, Y = Y^T$, and LPV matrices $\mathbb{A}(p), \mathbb{B}(p), \mathbb{C}(p), \mathbb{D}(p)$ satisfying

$$\text{He}\begin{bmatrix} A(p)X + (B_2\mathbb{C})(p) & A(p) + (B_2\mathbb{D}C_2)(p) & B_1(p) + (B_2\mathbb{D})(p)D_{21} & 0 \\ \mathbb{A}(p) & YA(p) + (\mathbb{B}C_2)(p) & YB_1(p) + \mathbb{B}(p)D_{21} & 0 \\ 0 & 0 & -\frac{1}{2}I & 0 \\ C_1(p)X + D_{12}\mathbb{C}(p) & C_1(p) + D_{12}(\mathbb{D}C_2)(p) & D_{11} + D_{12}\mathbb{D}D_{21} & -\frac{1}{2}I \end{bmatrix} < 0$$

$$\begin{bmatrix} X & I \\ I & Y \end{bmatrix} > 0,$$

then the robust $\mathcal{H}_\infty$ control specification is achieved even when the system parameters change. These conditions can be further reduced to the following vertex conditions $(j = 1, \ldots, N)$:

$$\text{He}\begin{bmatrix} A(\theta_j)X + (B_2\mathbb{C})(\theta_j) & A(\theta_j) + (B_2\mathbb{D}C_2)(\theta_j) & B_1(\theta_j) + (B_2\mathbb{D})(\theta_j)D_{21} & 0 \\ \mathbb{A}(\theta_j) & YA(\theta_j) + (\mathbb{B}C_2)(\theta_j) & YB_1(\theta_j) + \mathbb{B}(\theta_j)D_{21} & 0 \\ 0 & 0 & -\frac{1}{2}I & 0 \\ C_1(\theta_j)X + D_{12}\mathbb{C}(\theta_j) & C_1(\theta_j) + D_{12}(\mathbb{D}C_2)(\theta_j) & D_{11} + D_{12}\mathbb{D}D_{21} & -\frac{1}{2}I \end{bmatrix} < 0 \tag{20.65}$$

$$\begin{bmatrix} X & I \\ I & Y \end{bmatrix} > 0. \tag{20.66}$$

20.4.2.2 $\mathcal{H}_2$ Norm Specification

Here, we need to assume that $D_{11} = 0$ and $D_K = 0$. For a given system, the necessary and sufficient condition for the γ-optimal $\mathcal{H}_2$ control problem is given by Theorem 15.3. If there exist constant matrices $X = X^T, Y = Y^T, W = W^T$ and LPV matrices $\mathbb{A}(p), \mathbb{B}(p), \mathbb{C}(p)$ satisfying

$$\mathrm{He}\begin{bmatrix} A(p)X + (B_2\mathbb{C})(p) & A(p) & 0 \\ \mathbb{A}(p) & YA(p) + (\mathbb{B}C_2)(p) & 0 \\ C_1(p)X + D_{12}\mathbb{C}(p) & C_1(p) & -\frac{1}{2}I \end{bmatrix} < 0$$

$$\begin{bmatrix} W & B_1(p)^T & B_1(p)^TY \\ B_1(p) & X & I \\ YB_1(p) & I & Y \end{bmatrix} > 0$$

$$\mathrm{Tr}(W) < \gamma^2,$$

then the robust γ-optimal $\mathcal{H}_2$ control objective is guaranteed even when the system parameters change. This condition is equivalent to the following vertex conditions ($j = 1, \dots, N$):

$$\mathrm{He}\begin{bmatrix} A(\theta_j)X + (B_2\mathbb{C})(\theta_j) & A(\theta_j) & 0 \\ \mathbb{A}(\theta_j) & YA(\theta_j) + (\mathbb{B}C_2)(\theta_j) & 0 \\ C_1(\theta_j)X + D_{12}\mathbb{C}(\theta_j) & C_1(\theta_j) & -\frac{1}{2}I \end{bmatrix} < 0 \tag{20.67}$$

$$\begin{bmatrix} W & B_1(\theta_j)^T & B_1(\theta_j)^TY \\ B_1(\theta_j) & X & I \\ YB_1(\theta_j) & I & Y \end{bmatrix} > 0 \tag{20.68}$$

$$\mathrm{Tr}(W) < \gamma^2. \tag{20.69}$$

20.4.2.3 Regional Pole Placement

According to the result in Section 19.1.2, one of the sufficient conditions for placing the poles of LPV system to the LMI region

$$D = \{z \in \mathbb{C} : L + zM + \bar{z}M^T < 0\}$$

is that there are constant matrices $X = X^T, Y = Y^T$, and LPV matrices $\mathbb{A}(p), \mathbb{B}(p), \mathbb{C}(p)$, $\mathbb{D}(p)$ satisfying

$$L \otimes \begin{bmatrix} X & I \\ I & Y \end{bmatrix} + \mathrm{He}\left(M \otimes \begin{bmatrix} A(p)X + (B_2\mathbb{C})(p) & A(p) + (B_2\mathbb{D}C_2)(p) \\ \mathbb{A}(p) & YA(p) + (\mathbb{B}C_2)(p) \end{bmatrix}\right) < 0$$

$$\begin{bmatrix} X & I \\ I & Y \end{bmatrix} > 0.$$

This condition is equivalent to the following vertex conditions ($j = 1, \ldots, N$):

$$L \otimes \begin{bmatrix} X & I \\ I & Y \end{bmatrix} + \text{He}\left(M \otimes \begin{bmatrix} A(\theta_j)X + (B_2\mathbb{C})(\theta_j) & A(\theta_j) + (B_2\mathbb{D}C_2)(\theta_j) \\ \mathbb{A}(\theta_j) & YA(\theta_j) + (\mathbb{B}C_2)(\theta_j) \end{bmatrix} \right) < 0 \tag{20.70}$$

$$\begin{bmatrix} X & I \\ I & Y \end{bmatrix} > 0. \tag{20.71}$$

Moreover, concrete substitution of matrices L, M in the characteristic function of typical LMI regions leads to the following conditions:

Disk Centered at $(-c, 0)$ and with a Radius r

$$\text{He}\left(\begin{bmatrix} 0 & 1 \\ 0 & 0 \end{bmatrix} \otimes \begin{bmatrix} A(\theta_j)X + (B_2\mathbb{C})(\theta_j) & A(\theta_j) + (B_2\mathbb{D}C_2)(\theta_j) \\ \mathbb{A}(\theta_j) & YA(\theta_j) + (\mathbb{B}C_2)(\theta_j) \end{bmatrix} \right) + \begin{bmatrix} -r & c \\ c & -r \end{bmatrix} \otimes \begin{bmatrix} X & I \\ I & Y \end{bmatrix} < 0 \tag{20.72}$$

$$\begin{bmatrix} X & I \\ I & Y \end{bmatrix} > 0. \tag{20.73}$$

Half-Plane $\Re(z) < -\sigma$

$$2\sigma \begin{bmatrix} X & I \\ I & Y \end{bmatrix} + \text{He} \begin{bmatrix} A(\theta_j)X + (B_2\mathbb{C})(\theta_j) & A(\theta_j) + (B_2\mathbb{D}C_2)(\theta_j) \\ \mathbb{A}(\theta_j) & YA(\theta_j) + (\mathbb{B}C_2)(\theta_j) \end{bmatrix} < 0 \tag{20.74}$$

$$\begin{bmatrix} X & I \\ I & Y \end{bmatrix} > 0. \tag{20.75}$$

Sector $|\arg z - \pi| < \theta$

$$\text{He}\left(\begin{bmatrix} \sin\theta & \cos\theta \\ -\cos\theta & x\sin\theta \end{bmatrix} \otimes \begin{bmatrix} A(\theta_j)X + (B_2\mathbb{C})(\theta_j) & A(\theta_j) + (B_2\mathbb{D}C_2)(\theta_j) \\ \mathbb{A}(\theta_j) & YA(\theta_j) + (\mathbb{B}C_2)(\theta_j) \end{bmatrix} \right) < 0 \tag{20.76}$$

$$\begin{bmatrix} X & I \\ I & Y \end{bmatrix} > 0. \tag{20.77}$$

Refer to Ref. [46] for the application of this gain-scheduled method in the steering stabilization control of cars.

20.5 Case Study: Transient Stabilization of a Power System

As shown in Section 11.5.2, the nonlinear power system in Figure 20.11 can be equivalently converted into an LPV model with three time-varying parameters. However, in control design,

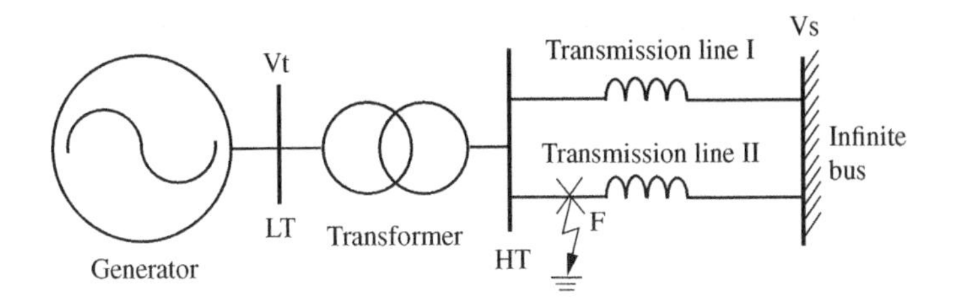

Figure 20.11 Single-machine infinite-bus power system

the time-varying parameters are treated as uncertainties, and the performance gets conservative as the number of uncertainties increases. For this reason, we derive an equivalent LPV model with only two time-varying parameters first, and then the gain-scheduled control design is carried out based on it.

20.5.1 LPV Model

The idea for the construction of two-parameter LPV model is to introduce a nonlinear feedback term in the field voltage:

$$V_f = \overline{V}_f - \frac{x_d - x_d'}{x_{d\Sigma}'} V_s \cos\delta \tag{20.78}$$

so that the nonlinear term in the dynamics of internal transient voltage E_q' is cancelled:

$$\dot{E}_q' = \frac{1}{T_{d0}}\left\{-\frac{x_{d\Sigma}}{x_{d\Sigma}'}E_q' + \frac{x_d - x_d'}{x_{d\Sigma}'}V_s\cos\delta + V_f\right\} = \frac{1}{T_{d0}}\left\{-\frac{x_{d\Sigma}}{x_{d\Sigma}'}E_q' + \overline{V}_f\right\} \tag{20.79}$$

Next, the deviation of the new voltage input $\overline{V}_f$ is defined as

$$u = \overline{V}_f - \overline{V}_{f0}.$$

As a result, the number of scheduling parameters is reduced to 2 in the new model. This two-parameter LPV model has an identical form as (11.48):

$$\begin{cases} \dot{x} = A(p)x + B_1 d + B_2 u \\ y = C_2 x. \end{cases} \tag{20.80}$$

The only difference is that it has only two time-varying parameters:

$$p_1(\delta) = \frac{k_1(\sin\delta - \sin\delta_0) - k_2(\sin 2\delta - \sin 2\delta_0)}{\delta - \delta_0}, \quad p_2(\delta) = \sin\delta$$

and $A(p)$ turns into

$$A(p) = \begin{bmatrix} 0 & 1 & 0 \\ c_1 p_1(\delta) & c_2 & c_3 p_2(\delta) \\ 0 & 0 & c_5 \end{bmatrix} = A_0 + p_1 A_1 + p_2 A_2. \tag{20.81}$$

It is assumed that the rotor angle δ is measured. So $p(\delta)$ can be computed online and used as scheduling parameters. Once the range of variation of δ is prescribed, the corresponding ranges $[p_{im}, p_{iM}]$ $(i = 1, 2)$ of p_1, p_2 can be computed offline.

20.5.2 Multiobjective Design

The goal of design is to ensure a good stability and to realize a satisfactory transient performance when large disturbance or fault occurs in the power system. Moreover, it should be kept in mind that the working range of the excitation voltage is rather narrow.

Since only the matrix $A(p)$ depends on $p(t)$, the parameter vector $p(t)$ may be put into all coefficient matrices of the controller (20.60).

In the swing equation (11.41) of synchronous generator, the damping is rather weak. To add damping to the power system, a good tool is to place the eigenvalues of the LPV system in some suitable region. Several kinds of region are tried, and it is found that the disk region yields the best response for the power system. So it is adopted here.

However, it is found via simulations that the swing of active power and rotor speed does not fade out fast enough. This is because the damping assigned by pole placement cannot be achieved due to the saturation of field voltage. In fact, the power system gets vulnerable to fault if the poles are placed too far away from the imaginary axis.

Further, it is observed that the rotor angle δ diverges first which causes the divergence of other variables. So, in addition to pole placement, the amplitude of δ should also be minimized. For this reason, we minimize the L_2 gain from the active power disturbance d to the rotor angle deviation

$$z = x_1 = C_1 x, \quad C_1 = [1 \quad 0 \quad 0] \tag{20.82}$$

in the design. That is, for a given $\gamma > 0$, ensure that

$$\sup_{\|d\|_2 \neq 0} \frac{\|z\|_2}{\|d\|_2} \leq \gamma. \tag{20.83}$$

For the multiobjective control under consideration, the generalized plant is given by

$$G(s) = \left[\begin{array}{c|cc} A(p) & B_1 & B_2 \\ \hline C_1 & 0 & 0 \\ C_2 & 0 & 0 \end{array}\right]. \tag{20.84}$$

In summary, the multiobjective optimization problem to be solved is to minimize the norm bound γ subject to the LMI constraints for pole placement in a disk and $\mathcal{H}_\infty$ norm as given in Section 20.4.2.

20.5.3 Simulation Results

In the simulations, the dynamics of the exciter is modeled as a first-order transfer function

$$\frac{K_A}{1 + sT_A}$$

with a limiter. All simulations are conducted w.r.t. the nonlinear model as illustrated in Figure 20.12.

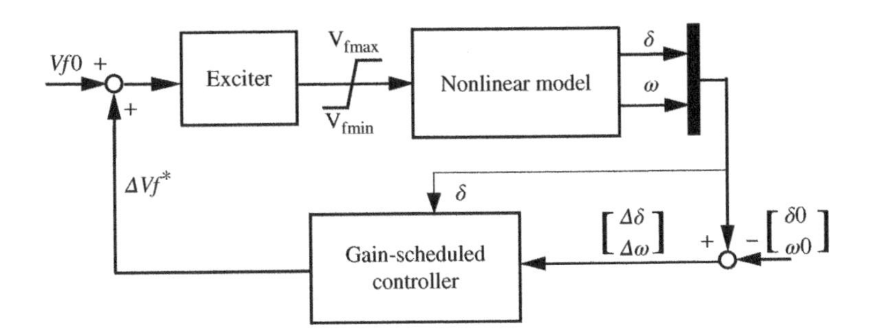

Figure 20.12 Block diagram of gain-scheduled control of power system

As a test of transient stability, a three-phase short circuit fault is applied behind the transformer as shown in Figure 20.11. To account for the influence of time delay in the measurement of rotor angle δ, a 2 ms pure time delay is set in the measurement of δ in all simulations. It is noted that the output voltage of the transformer is zero during the fault so that no power is transmitted to the infinite bus.

20.5.3.1 Fault Sequence (Permanent Fault)

- Step 1: A fault occurs at $t = 0.0$ s, that is, the system is in a prefault steady state before $t = 0.0$ s.
- Step 2: The fault is removed by opening the breakers of the faulted line at t_F s.
- Step 3: The system operates in a post-fault state.

  Parameters of the nonlinear power system are as follows (p. 864, Example 13.2 of [50]):

$$D = 0.15, \quad M = 7.00, \quad T_{d0} = 8.00, \quad V_s = 0.995$$

$$x_d = 1.81, \quad x_d^{'} = 0.30, \quad x_{l1} = 0.5, \quad x_{l2} = 0.93, \quad x_T = 0.15.$$

The operating point is

$$\delta_0 = 0.8807(\approx 50.5^\circ), \quad \omega_0 = 314, \quad E_{q0}^{'} = 1.3228, \quad V_{f0} = 2.6657.$$

Meanwhile, the parameters of the exciter are

$$T_A = 0.05, K_A = 50$$

and the limit on the field voltage is

$$0.0 \ [\text{pu}] \leq V_f(t) \leq 5.0 [\text{pu}].$$

Further, the range of δ is assumed to be $[40^\circ, 90^\circ]$ in the design. In all simulations, $t_F = 0.168$ [s] is used. Via some tuning, it is found that the best performance is achieved when the region of poles is placed in a disk centered at $(-8, j0)$ and with a radius $r = 6$.

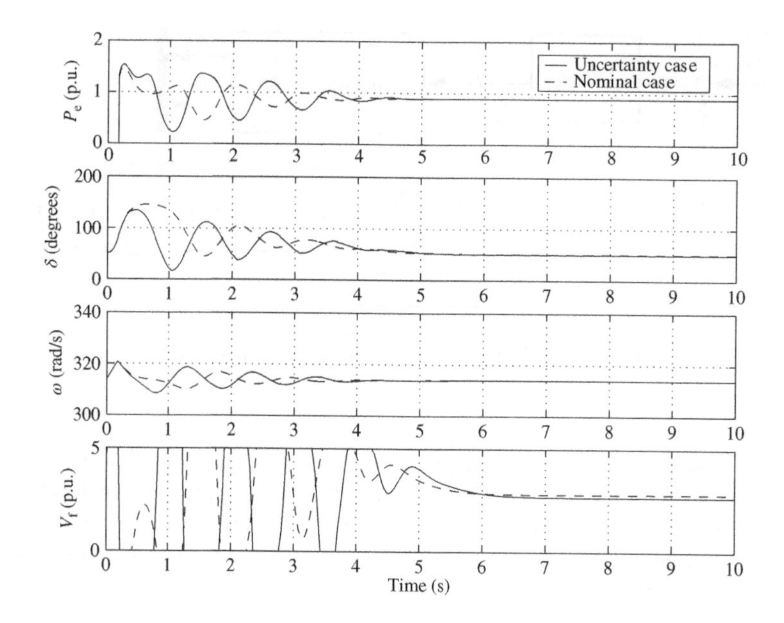

Figure 20.13 Robustness test ($\Delta V_s = 0.1V_s, \Delta x_L = 0.1x_L, \Delta x_T = 0.1x_T$)

20.5.4 Robustness

When a large fault occurs suddenly, both V_s and x_L change a lot, which have not been considered in the design for simplicity. So here to evaluate the robustness to these parameter changes, simulations about the changes of V_s, x_L, and x_T are performed, and the results are shown in Figures 20.13 and 20.14. A maximum variation of $\pm 10\%$ is considered here. ΔV_s, Δx_L, and Δx_T denote the uncertainties in V_s, x_L, and x_T. Although the quality of response worsens due to the change of parameters, the settling time remains roughly unchanged.

One may wonder why not to take these parameter uncertainties into account directly in the design. The answer is that the result would have been too conservative if these uncertainties had been considered.

20.5.5 Comparison with PSS

For comparison, we will first design a PSS (power system stabilizer) at the same operating point.

The philosophy of PSS is to add a damping signal (in phase with the oscillation) to the AVR (automatic voltage regulator) reference input through a phase-lead compensator so that the component of the electrical torque generated by this signal could enhance the damping of oscillation.

The PSS is designed following the design procedure described in Ref. [52]. Figure 20.15 depicts the configuration of the PSS with AVR and exciter. The terminal voltage V_t in the AVR is

$$V_t = \frac{1}{x_{d\Sigma}}\sqrt{E_q^2 x_s^2 + V_s^2 x_d^2 + 2x_s x_d E_q V_s \cos\delta}$$

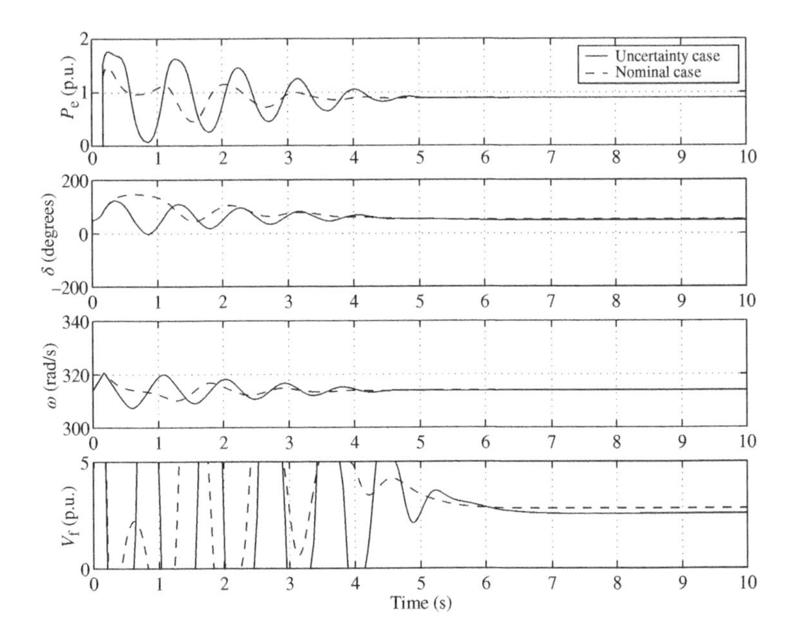

Figure 20.14 Robustness test ($\Delta V_s = 0.1V_s, \Delta x_L = -0.1x_L, \Delta x_T = -0.1x_T$)

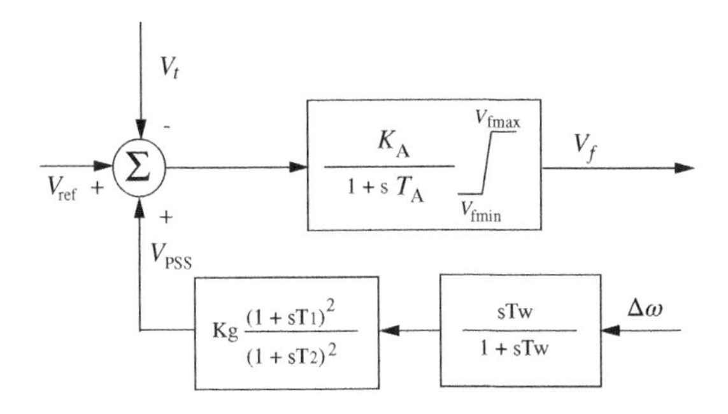

Figure 20.15 Block diagram of PSS

in which the internal voltage E_q is given by

$$E_q = \frac{x_{d\Sigma}}{x'_{d\Sigma}} E'_q - \frac{x_d - x'_d}{x'_{d\Sigma}} V_s \cos\delta.$$

The transfer function of PSS controller has the following form:

$$K_{\text{PSS}}(s) = K_g \left(\frac{sT_w}{1 + sT_w}\right) \left(\frac{1 + sT_1}{1 + sT_2}\right)^2 \tag{20.85}$$

and the parameters are tuned as

$$K_g = 0.3, \quad T_w = 0.1, \quad T_1 = 0.1, \quad T_2 = 0.05.$$

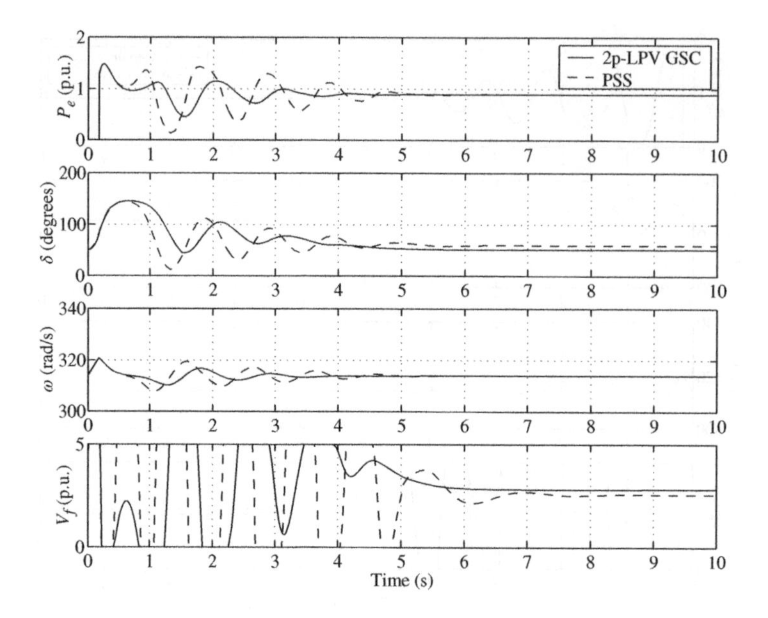

Figure 20.16 Comparisons of gain-scheduled control and PSS

Figure 20.16 shows the results of comparison between the gain-scheduled control (abbreviated as 2p-LPV GSC in the figure) and the PSS under the permanent fault. It is observed from the responses of active power P_e, rotor angle δ, and rotor speed ω that the GSC damps the oscillation faster than the PSS and the oscillation amplitude is smaller.

Exercise

20.1 Derive the algorithm of gain-scheduled control design given in Section 20.2.2.

20.2 Prove the easy-to-design controller structures summarized in Section 20.4:

- When matrices $(B_2(p), C_2(p))$ both depend on $p(t)$, (B_K, C_K) must be constant matrices besides $D_K = 0$.
- When (B_2, C_2) are both constant matrices, all coefficient matrices of the controller can be affine functions of the scheduling parameter.
- When only B_2 is constant matrix, (B_K, D_K) must be constant matrices.
- When only C_2 is constant matrix, (C_K, D_K) must be constant matrices.

And derive the formulas for controller coefficient matrices $(A_{Ki}, B_{Ki}, C_{Ki}, D_{Ki})$ corresponding to $(\mathbb{A}_i, \mathbb{B}_i, \mathbb{C}_i, \mathbb{D}_i)(i = 0, 1, \ldots, q)$ given in Section 20.4.

Description of Variables and Parameters

δ	rotor angle, in rad
ω	rotor speed, in rad/s

P_e	active electrical power, in p.u.
P_M	mechanical power, in p.u.
E_q	internal voltage, in p.u.
V_t	terminal voltage of generator, in p.u.
V_s	infinite bus voltage, in p.u.
E_q'	internal transient voltage, in p.u.
V_f	field voltage of generator, in p.u.
V_R	output voltage of exciter, in p.u.
V_{ref}	reference terminal voltage, in p.u.
D	damping constant, in p.u.
M	inertia coefficient of generator, in seconds
T_{d0}	field circuit time constant, in seconds
K_A	gain of the excitation system
T_A	time constant of the excitation system, in seconds
x_d	d-axis reactance, in p.u.
x_d'	d-axis transient reactance, in p.u.
x_T	reactance of transformer, in p.u.
x_{l1}	reactance of transmission line I, in p.u.
x_{l2}	reactance of transmission line II, in p.u.
x_L	reactance of transmission line, in p.u.
$x_{d\Sigma}$	$x_{d\Sigma} = x_d + x_T + x_L$, in p.u.
$x_{d\Sigma}'$	$x_{d\Sigma}' = x_d' + x_T + x_L$, in p.u.

Notes and References

For the LFT-type parametric systems, [76] gave a gain-scheduling method based on μ synthesis; and Refs [3, 42] provided a gain-scheduling method based on constant scaled $\mathcal{H}_\infty$ control. The former was applied to a unicycle robot [21] and the latter was applied to a missile [4]. The gain-scheduling method for affine systems was applied to automobiles [46] and power systems [38].

21

Positive Real Method

When the uncertainty of a system has the positive real property, a simple method to guarantee the stability of closed-loop system is to use the passivity method introduced in Chapter 13, that is, to design the nominal closed-loop system as a strongly positive real or strictly positive real transfer function. Its essence is to make use of the phase condition in the Nyquist stability criterion to guarantee the stability of closed-loop system. This chapter illustrates the positive real design method with an emphasis placed on the robust performance design, as well as its limit in some detail.

21.1 Structure of Uncertain Closed-Loop System

First, to figure out what structure the closed-loop system should have when dealing with a positive real uncertainty, let us look back at the flexible system discussed in Chapter 11:

$$\tilde{P}(s) = \frac{k_0 s}{s^2 + 2\zeta_0\omega_0 s + \omega_0^2} + \frac{k_1 s}{s^2 + 2\zeta_1\omega_1 s + \omega_1^2} + \cdots + \frac{k_n s}{s^2 + 2\zeta_n\omega_n s + \omega_n^2}$$

where all gains k_i are positive. Take the first-order resonant mode

$$P(s) = \frac{k_0 s}{s^2 + 2\zeta_0\omega_0 s + \omega_0^2} = (A, B, C, 0)$$

as the nominal model, and treat the sum of high-order resonant modes which are difficult to identify as a dynamic uncertainty:

$$\Delta(s) = \frac{k_1 s}{s^2 + 2\zeta_1\omega_1 s + \omega_1^2} + \cdots + \frac{k_n s}{s^2 + 2\zeta_n\omega_n s + \omega_n^2}.$$

Then, the uncertainty $\Delta(s)$ is positive real. The nominal closed-loop system, which is negatively feedback connected with the uncertainty, becomes

$$M(s) = \frac{K}{1 + PK}.$$

To make $M(s)$ strongly positive real, the relative degree of controller $K(s)$ must be zero. Obviously, such controller is sensitive to sensor noise. Hence, a better way is to make $M(s)$

Robust Control: Theory and Applications, First Edition. Kang-Zhi Liu and Yu Yao.

strictly positive real. Describing $M(s)$ as a generalized system about $K(s)$, the corresponding generalized plant becomes

$$\begin{bmatrix} z \\ y \end{bmatrix} = \begin{bmatrix} 0 & 1 \\ -1 & -P \end{bmatrix} \begin{bmatrix} w \\ u \end{bmatrix} = \left[\begin{array}{c|cc} A & 0 & B \\ \hline 0 & 0 & 1 \\ -C & -1 & 0 \end{array}\right] \begin{bmatrix} w \\ u \end{bmatrix}.$$

The second example is from Ref. [37]. In the vibration control of flexible systems, when the measured output is not velocity but displacement, the system model becomes

$$\tilde{P}(s) = \frac{k_0}{s^2 + 2\zeta_0\omega_0 s + \omega_0^2} + \frac{k_1}{s^2 + 2\zeta_1\omega_1 s + \omega_1^2} + \cdots + \frac{k_n}{s^2 + 2\zeta_n\omega_n s + \omega_n^2}.$$

In this case, the high-order resonant modes are no longer positive real. However, if we can identify the parameters (ζ_i, ω_i) of the high-order resonant modes to some extent, it is possible for us to extend the control bandwidth of closed-loop system. But due to manufacturing error the resonant frequencies usually vary in some ranges. The uncertainty of these resonant frequencies can be treated as a positive real function. The detail is as follows. The state-space realization of $\tilde{P}(s)$ is given by

$$\tilde{P}(s) = (A,\ B,\ C,\ 0)$$

where matrix A has the following block diagonal structure:

$$A = \mathrm{diag}\left(\begin{bmatrix} -\zeta_1 & \omega_1 \\ -\omega_1 & -\zeta_1 \end{bmatrix}, \ldots, \begin{bmatrix} -\zeta_n & \omega_n \\ -\omega_n & -\zeta_n \end{bmatrix} \right).$$

In vibration control this model is known as the *modal dynamics*. The uncertainty caused by the variation of resonant frequencies can be written as

$$\Delta A = \sum_{i=1}^{n} \delta_i A_i$$

where parameter δ_i corresponds to the uncertainty of resonant frequency ω_i, which changes in a certain range. The matrix A_i is given by

$$\begin{aligned} A_i &= \mathrm{diag}\left(0, \ldots, 0, \begin{bmatrix} 0 & 1 \\ -1 & 0 \end{bmatrix}, 0, \ldots, 0 \right) \\ &= [0, \ldots, I_2, \ldots, 0]^T \begin{bmatrix} 0 & 1 \\ -1 & 0 \end{bmatrix} [0, \ldots, I_2, \ldots, 0]. \end{aligned}$$

Apparently, $\Delta A + \Delta A^T = 0$. So ΔA can be treated as a positive real uncertainty.

In general, this kind of matrix uncertainty can be modeled as

$$\Delta A = -B_0 F C_0$$

where F is a positive real uncertainty satisfying $F + F^T \geq 0$. The corresponding state-space model for the uncertain system $\tilde{P}(s)$ is given by

$$\dot{x} = (A + \Delta A)x + Bu, \quad y = Cx. \tag{21.1}$$

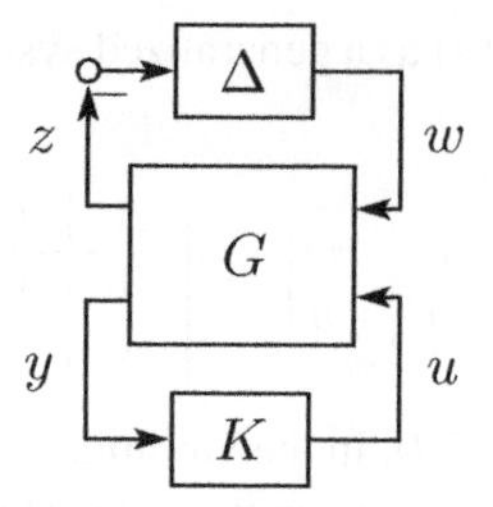

Figure 21.1 Closed-loop system with positive real uncertainty

Setting (z, w) as the input and output of uncertainty $-F$, we finally obtain the generalized plant as

$$G \begin{cases} \dot{x} = Ax + B_0 w + Bu \\ z = C_0 x \\ y = Cx. \end{cases}$$

It is negatively feedback connected with uncertainty F, that is,

$$w = -Fz.$$

Note that in these examples, the DC term from w to z is $D_{11} = 0$.

Summarizing these examples, in general a closed-loop system with positive real uncertainty can be expressed as the system shown in Figure 21.1. The generalized plant is given by

$$\begin{bmatrix} \dot{x} \\ z \\ y \end{bmatrix} = \begin{bmatrix} A & B_1 & B_2 \\ C_1 & D_{11} & D_{12} \\ C_2 & D_{21} & 0 \end{bmatrix} \begin{bmatrix} x \\ w \\ u \end{bmatrix}. \tag{21.2}$$

Note that to match the positive real theorem, Δ connects with the rest part in a negative feedback manner.

Setting $K(s) = (A_K, B_K, C_K, D_K)$, then the nominal closed-loop system $M(s)$ from w to z becomes

$$\begin{aligned} M(s) &= \left[\begin{array}{c|c} A_c & B_c \\ \hline C_c & D_c \end{array}\right] \\ &= \left[\begin{array}{cc|c} A + B_2 D_K C_2 & B_2 C_K & B_1 + B_2 D_K D_{21} \\ B_K C_2 & A_K & B_K D_{21} \\ \hline C_1 + D_{12} D_K C_2 & D_{12} C_K & D_{11} + D_{12} D_K D_{21} \end{array}\right]. \end{aligned} \tag{21.3}$$

21.2 Robust Stabilization Based on Strongly Positive Realness

According to the discussion in Section 13.4, as long as the nominal closed-loop system is strongly positive real, the closed-loop system in Figure 21.1 is asymptotically stable. Therefore, in this section we will discuss that under what condition $M(s)$ can be made strongly positive real by feedback control, as well as the concrete control design method. First of all, the strongly positive realness of $M(s)$ requires

$$D_c + D_c^T = D_{11} + D_{11}^T + D_{12}D_K D_{21} + (D_{12}D_K D_{21})^T > 0. \tag{21.4}$$

When $D_{11} + D_{11}^T > 0$, even if $D_{12} = 0$ or $D_{21} = 0$ this condition can be achieved. However, when $D_{11} = 0$, both D_{12} and D_{21} must be nonsingular in order to satisfy this condition. Hence, the strongly positive realness of $M(s)$ is a very strong condition.

According to the positive real lemma, $M(s)$ is strongly positive real iff there exists a positive definite matrix $P > 0$ satisfying

$$\begin{bmatrix} A_c^T P + PA_c & PB_c \\ B_c^T P & 0 \end{bmatrix} - \begin{bmatrix} 0 & C_c^T \\ C_c & D_c + D_c^T \end{bmatrix} < 0. \tag{21.5}$$

In the sequel, we design the controller using the methods of variable elimination and variable change, respectively.

21.2.1 *Variable Change*

Recalling the variable change method in Section 3.2.4, the positive definite matrix P in (21.5) can be factorized as

$$P = \Pi_2\Pi_1^{-1}, \quad \Pi_1 = \begin{bmatrix} X & I \\ M^T & 0 \end{bmatrix}, \quad \Pi_2 = \begin{bmatrix} I & Y \\ 0 & N^T \end{bmatrix}. \tag{21.6}$$

We set the new variable matrices as

$$\begin{aligned} \mathbb{A} &= NA_K M^T + NB_K C_2 X + YB_2 C_K M^T + Y(A + B_2 D_K C_2)X \\ \mathbb{B} &= NB_K + YB_2 D_K, \quad \mathbb{C} = C_K M^T + D_K C_2 X, \quad \mathbb{D} = D_K. \end{aligned} \tag{21.7}$$

The condition that ensures the positive realness of P and the one-to-one correspondence between the new variables and the controller parameters is

$$\begin{bmatrix} X & I \\ I & Y \end{bmatrix} > 0. \tag{21.8}$$

Multiplying (21.5) from both sides using a block diagonal matrix $\mathrm{diag}(\Pi_1, I)$ and its transpose, we obtain the following equivalent condition:

$$\mathrm{He}\begin{bmatrix} AX + B_2\mathbb{C} & A + B_2\mathbb{D}C_2 & B_1 + B_2\mathbb{D}D_{21} \\ \mathbb{A} & YA + \mathbb{B}C_2 & YB_1 + \mathbb{B}D_{21} \\ -(C_1X + D_{12}\mathbb{C}) & -(C_1 + D_{12}\mathbb{D}C_2) & -(D_{11} + D_{12}\mathbb{D}D_{21}) \end{bmatrix} < 0 \tag{21.9}$$

after the substitution of the new variables.

21.2.2 Variable Elimination

Here, we introduce the solution given in [89]. Arrange the coefficient matrices of controller as a variable matrix

$$\mathcal{K} = \begin{bmatrix} D_K & C_K \\ B_K & A_K \end{bmatrix}. \tag{21.10}$$

According to Section 7.4.1, the coefficient matrices in (21.3) can be written as an affine function of $\mathcal{K}$:

$$\begin{bmatrix} A_c & B_c \\ C_c & D_c \end{bmatrix} = \begin{bmatrix} \overline{A} & \overline{B}_1 \\ \overline{C}_1 & \overline{D}_{11} \end{bmatrix} + \begin{bmatrix} \overline{B}_2 \\ \overline{D}_{12} \end{bmatrix} \mathcal{K}[\overline{C}_2, \overline{D}_{21}] \tag{21.11}$$

where the coefficient matrices are given in (7.44). Substituting them into the strongly positive real condition (21.5), we obtain

$$Q + E^T \mathcal{K} F + F^T \mathcal{K}^T E < 0 \tag{21.12}$$

in which the coefficient matrices are

$$\begin{bmatrix} Q & E^T \\ F & \end{bmatrix} = \left[\begin{array}{cc|c} \overline{A}^T P + P\overline{A} & P\overline{B}_1 - \overline{C}_1^T & P\overline{B}_2 \\ \overline{B}_1^T P - \overline{C}_1 & -(\overline{D}_{11}^T + \overline{D}_{11}) & -\overline{D}_{12} \\ \hline \overline{C}_2 & \overline{D}_{21} & \end{array}\right].$$

By Theorem 3.1 about variable elimination, this inequality has a solution iff the following two inequalities

$$E_\perp^T Q E_\perp < 0, \quad F_\perp^T Q F_\perp < 0 \tag{21.13}$$

hold simultaneously. Substitution of the coefficient matrices given by (7.44) and a direct calculation lead to the orthogonal matrices of E, F

$$E_\perp = \begin{bmatrix} P^{-1} & 0 \\ 0 & -I_{n_z} \end{bmatrix} \begin{bmatrix} I_n & 0 \\ 0 & 0 \\ 0 & I_{n_z} \end{bmatrix} N_X, \quad F_\perp = \begin{bmatrix} I_n & 0 \\ 0 & 0 \\ 0 & I_{n_w} \end{bmatrix} N_Y$$

where

$$N_X = [B_2^T \ D_{12}^T]_\perp, \quad N_Y = [C_2 \ D_{21}]_\perp. \tag{21.14}$$

Partition the positive definite matrix P as

$$P = \begin{bmatrix} Y & * \\ * & * \end{bmatrix}, \quad P^{-1} = \begin{bmatrix} X & * \\ * & * \end{bmatrix}.$$

Substituting P^{-1} into the inequality $E_\perp^T Q E_\perp < 0$, after some calculation, we obtain

$$N_X^T \left\{ \begin{bmatrix} AX + XA^T & XC_1^T \\ C_1 X & 0 \end{bmatrix} - \begin{bmatrix} 0 & B_1 \\ B_1^T & D_{11} + D_{11}^T \end{bmatrix} \right\} N_X < 0. \tag{21.15}$$

Similarly, substitution of P into the inequality $F_\perp^T Q F_\perp < 0$ yields

$$N_Y^T \left\{ \begin{bmatrix} YA + A^T Y & YB_1 \\ B_1^T Y & 0 \end{bmatrix} - \begin{bmatrix} 0 & C_1^T \\ C_1 & D_{11} + D_{11}^T \end{bmatrix} \right\} N_Y < 0. \tag{21.16}$$

Finally, according to Lemma 3.1,the condition for the existence of such a positive definite matrix P is

$$\begin{bmatrix} X & I \\ I & Y \end{bmatrix} > 0. \tag{21.17}$$

These three simultaneous LMIs provide the necessary and sufficient condition for $M(s)$ to be strongly positive real. When they have a solution, we compute a nonsingular matrix M satisfying

$$MM^T = Y - X^{-1} > 0$$

first, then set P as

$$P = \begin{bmatrix} Y & M \\ M^T & I \end{bmatrix}, \tag{21.18}$$

and substitute it into (21.12). In this way, we can solve for the parameter matrix $\mathcal{K}$ of controller numerically.

21.3 Robust Stabilization Based on Strictly Positive Realness

Strongly positive real method requires the DC term of the nominal closed-loop transfer matrix to be nonsingular, which is an extremely strong requirement and difficult to satisfy in practical system designs. For instance, in the first example of Section 21.1, since $D_{11} = 0$, it requires the DC term of the controller to be nonzero which means that the high-frequency gain of controller cannot be rolled off. Therefore, it is sensitive to sensor noise. In the second example, as $D_{11} = D_{12} = D_{21} = 0$, it is impossible to satisfy $M(\infty) \neq 0$ at all. However, the essence of positive real method is to guarantee that the open-loop system satisfies the phase condition of stability theory, that is, the phase of $M(j\omega)\Delta(j\omega)$ is not 180 degree at any finite frequency. As for the infinite frequency, even if the phase condition fails, the closed-loop system will be stable so long as the open-loop gain is zero. This observation motivates us to make the nominal closed-loop system strictly positive real instead of strongly positive real.

Since Theorem 13.7 provides a method to ensure the asymptotic stability of closed-loop system by making the nominal closed-loop system $M(s)$ strictly positive real, we start from it. In this section, it is further assumed that

$$D_{11} = 0. \tag{21.19}$$

Then, $D_K = 0$ guarantees $D_c = 0$.

Next, we design a strictly proper controller $K(s) = (A_K, B_K, C_K, 0)$ such that the nominal closed-loop transfer matrix $M(s)$ is strictly positive real. Suppose that (A, B_2, C_2) in the realization of generalized plant $G(s)$ is both controllable and observable. As long as no zero–pole cancellation occurs between the controller and the generalized plant, which can be ensured by the controller design, the realization of nominal closed-loop transfer matrix $M(s)$ is both controllable and observable. Thus, the condition for the positive realness of $M(s - \epsilon)$ is that there exist an $\epsilon > 0$ and a positive definite matrix $P > 0$ satisfying

$$\begin{bmatrix} (A_c + \epsilon I)^T P + P(A_c + \epsilon I) & PB_c \\ B_c^T P & 0 \end{bmatrix} - \begin{bmatrix} 0 & C_c^T \\ C_c & 0 \end{bmatrix} \leq 0. \tag{21.20}$$

As our objective is to design a strictly proper controller, we should use the variable change method instead of the variable elimination method. The only difference between the current case and that in Section 21.2.1 is that there is one more term $2\epsilon P$, as well as $D_K = \mathbb{D} = 0$. Noting that

$$\Pi_1^T P \Pi_1 = \Pi_1^T \Pi_2 = \begin{bmatrix} X & I \\ I & Y \end{bmatrix},$$

via a similar calculation as Subsection 21.2.1, we get the design condition

$$\mathrm{He}\begin{bmatrix} AX + B_2\mathbb{C} & A & B_1 \\ \mathbb{A} & YA + \mathbb{B}C_2 & YB_1 + \mathbb{B}D_{21} \\ -(C_1X + D_{12}\mathbb{C}) & -C_1 & 0 \end{bmatrix} + 2\epsilon \begin{bmatrix} X & I & 0 \\ I & Y & 0 \\ 0 & 0 & 0 \end{bmatrix} \leq 0 \tag{21.21}$$

and

$$\begin{bmatrix} X & I \\ I & Y \end{bmatrix} > 0. \tag{21.22}$$

Here, the new variables are

$$\begin{aligned} \mathbb{A} &= NA_K M^T + NB_K C_2 X + YB_2 C_K M^T + YAX \\ \mathbb{B} &= NB_K, \quad \mathbb{C} = C_K M^T. \end{aligned} \tag{21.23}$$

The minimization of ϵ, that is,

$$\min\ \epsilon \tag{21.24}$$

subject to these conditions is a GEV problem and can be numerically solved. The coefficient matrices of the actual controller are given by

$$\begin{aligned} C_K &= \mathbb{C}(M^{-1})^T, \quad B_K = N^{-1}\mathbb{B} \\ A_K &= N^{-1}(\mathbb{A} - NB_K CX - YBC_K M^T - YAX)(M^{-1})^T. \end{aligned} \tag{21.25}$$

21.4 Robust Performance Design for Systems with Positive Real Uncertainty

In this section, we discuss how to carry out robust performance design for systems with positive real uncertainty. For example, in the first example of Section 21.1, the system $\tilde{P} = P + \Delta$ is affected by the input disturbance $d(t)$. Suppose that the dynamics of $d(t)$ is $W_D(s)$, that is, $\hat{d}(s) = W_D(s)\hat{w}(s)$. Then, in order to suppress the influence of disturbance on system output, we need to design a controller $K(s)$ such that the closed-loop transfer function $W_D\frac{\tilde{P}}{1+\tilde{P}K}$ from disturbance w to output y_P satisfies

$$\left\| W_D \frac{\tilde{P}}{1+\tilde{P}K} \right\|_\infty < 1.$$

According to the discussion in Section 12.3, this robust disturbance attenuation condition is equivalent to the robust stability when a virtual uncertainty $\Delta_g(s)$ ($\|\Delta_g\|_\infty \leq 1$) is added between disturbance w and system output y_P. To carry out the design in the framework of positive realness, we need to convert the norm-bounded virtual uncertainty into a positive real uncertainty. The essence of this transformation is to map the unit disk on the complex plane into the right half-plane, which can be achieved by a *conformal mapping*. Transforming a point g inside the unit disk $|g| \leq 1$ on the complex plane by

$$p = \frac{1+g}{1-g},$$

then all new variables p form the whole right half-plane of the complex plane as illustrated in Figure 21.2(a). This transformation is one to one. Thus, it is easy to know that the transformation converting a matrix $\Delta_g(s)$ in the unit ball to a positive real $\Delta_p(s)$ is

$$\begin{aligned} &\Delta_p(s) = (I - \Delta_g(s))^{-1}(I + \Delta_g(s)) \\ \Longleftrightarrow &\Delta_g(s) = (\Delta_p(s) - I)(I + \Delta_p(s))^{-1}. \end{aligned} \tag{21.26}$$

Note that the bounded real matrix $\Delta_g(s)$ can be written as

$$\Delta_g(s) = I - 2(I + \Delta_p(s))^{-1}. \tag{21.27}$$

Its relation with the positive real matrix $\Delta_p(s)$ is shown in Figure 21.2(b). From it we see that the relationship between the input–output (z_g, w_g) of $\Delta_g(s)$ and the input–output (z_p, w_p) of $-\Delta_p(s)$ is

$$z_p = z_g + w_p, \quad w_g = z_g - 2z_p = -2w_p - z_g. \tag{21.28}$$

Next, we consider the general case in Figure 21.3(a). Let the state realization of generalized plant $G_g(s)$ be

$$\begin{bmatrix} \dot{x} \\ z \\ z_g \\ y \end{bmatrix} = \begin{bmatrix} A & B_1 & B_2 & B_3 \\ C_1 & D_{11} & D_{12} & D_{13} \\ C_2 & D_{21} & 0 & D_{23} \\ C_3 & D_{31} & D_{32} & 0 \end{bmatrix} \begin{bmatrix} x \\ w \\ w_g \\ u \end{bmatrix}. \tag{21.29}$$

Here, for simplicity we have assumed that the DC term from w_g to z_g is zero. This condition holds in almost all practical applications. For example, this condition is satisfied in the tracking problem of Figure 21.4(a) as long as the weighting function $W_R(s)$, which is the model of the reference input, is strictly proper. Further, in the input disturbance control problem of Figure 21.4(b), this condition holds naturally since the plant $P(s)$ is always strictly proper.

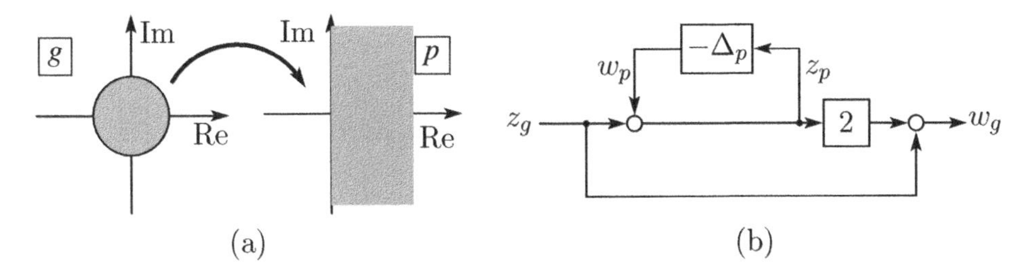

Figure 21.2 Relation between bounded real and positive real matrices

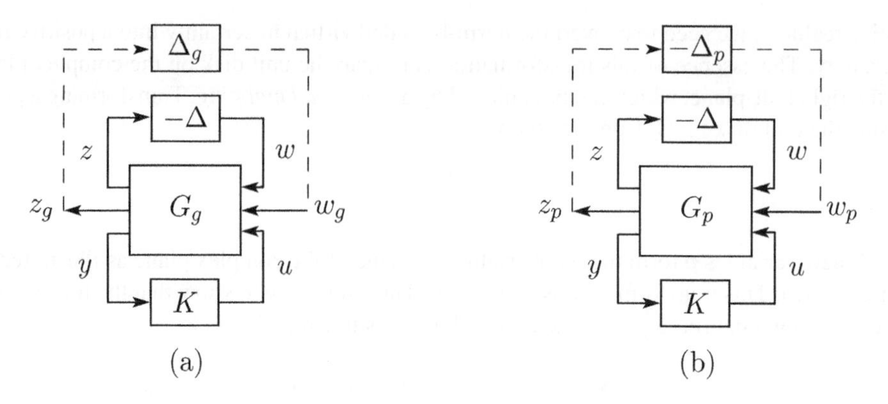

Figure 21.3 Robust performance problem in system with bounded real/positive real uncertainty

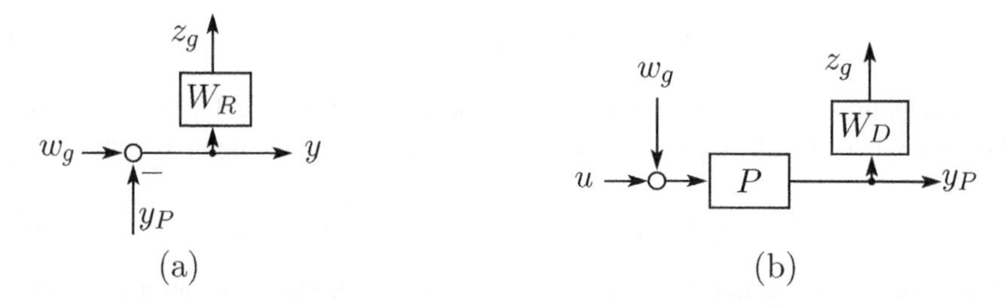

Figure 21.4 Examples about the relation between w_g and z_g

To do the robust performance design, we transform this closed-loop system into a generalized feedback system about positive real uncertainty $\text{diag}(\Delta, \Delta_p)$ shown in Figure 21.3(b). Now, we seek the corresponding generalized plant $G_p(s)$. First, substitution of $z_g = C_2x + D_{21}w + D_{23}u$ into $z_p = z_g + w_p$ and $w_g = -2w_p - z_g$ yields $z_p = C_2x + D_{21}w + w_p + D_{23}u$ and $w_g = -C_2x - D_{21}w - 2w_p - D_{23}u$. Further, substituting this expression of w_g into the state equation (21.29), we get the state equation of $G_p(s)$:

$$\begin{bmatrix} \dot{x} \\ z \\ z_p \\ y \end{bmatrix} = \left[\begin{array}{c|cc|c} A - B_2C_2 & B_1 - B_2D_{21} & -2B_2 & B_3 - B_2D_{23} \\ \hline C_1 - D_{12}C_2 & D_{11} - D_{12}D_{21} & -2D_{12} & D_{13} - D_{12}D_{23} \\ C_2 & D_{21} & I & D_{23} \\ \hline C_3 - D_{32}C_2 & D_{31} - D_{32}D_{21} & -2D_{32} & -D_{32}D_{23} \end{array}\right] \begin{bmatrix} x \\ w \\ w_p \\ u \end{bmatrix}. \tag{21.30}$$

The DC term between u and y can be removed by the method of Exercise 7.12. Hence, the controller can be designed using the methods given in the previous two sections. The detail is left to the reader.

21.5 Case Study

Consider a flexible system

$$\tilde{P}(s) = K_p\left(\frac{1}{s} + \frac{As}{s^2 + 2\zeta_1\omega_1 s + \omega_1^2}\right) \tag{21.31}$$

where $K_p = 3.74 \times 10^9, A = 0.4 \sim 1.0, \zeta_1 = 0.02 \sim 0.6, \omega_1 = 4000 \sim 4200$ Hz. Set the nominal model as

$$P(s) = \frac{K_p}{s}$$

and the resonant mode as a dynamic uncertainty:

$$\Delta(s) = \frac{K_p As}{s^2 + 2\zeta_1\omega_1 s + \omega_1^2}.$$

This uncertainty is positive real.

Let the controller be $K(s)$. Then, the nominal closed-loop system, which is negatively feedback connected with the positive real uncertainty Δ, becomes (refer to Figure 21.5)

$$M(s) = \frac{K(s)}{1 + P(s)K(s)} = \frac{sK(s)}{s + K_pK(s)}.$$

$M(j0) = 0$ when $\omega = 0$. So, $M(s)$ is neither strongly positive real nor strictly positive real. However, the gain of $M(s)$ is zero at $\omega = 0$. So, $M(s)\Delta(s)$ satisfies the gain condition of the Nyquist stability criterion at this point. This implies that there is no need of satisfying the phase condition at this point.

First, we consider the PI controller

$$K(s) = \frac{c_1 s + c_2}{s}, \quad c_1 > 0, c_2 > 0.$$

In this case, the frequency response of nominal closed-loop system

$$M(s) = \frac{s(c_1 s + c_2)}{s^2 + c_1 K_p s + c_2 K_p}$$

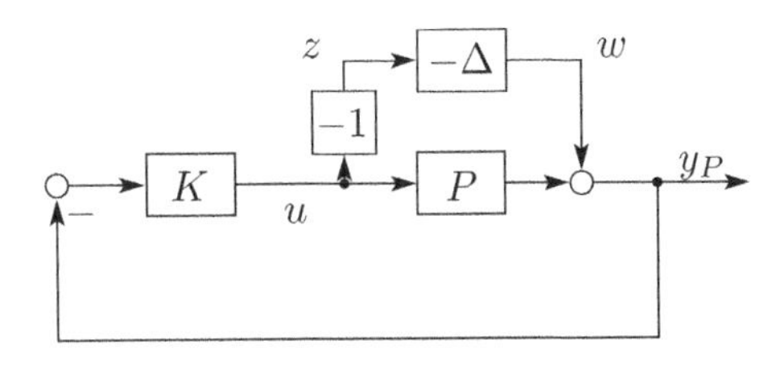

Figure 21.5 Robust stabilization design

is

$$M(j\omega) = \frac{j\omega(c_2 + jc_1\omega)}{c_2K_p - \omega^2 + jc_1K_p\omega}$$

$$= \frac{c_1\omega^4}{(c_2K_p - \omega^2)^2 + (c_1K_p\omega)^2} + j\frac{\omega[c_2(c_2K_p - \omega^2) + c_1^2K_p\omega^2]}{(c_2K_p - \omega^2)^2 + (c_1K_p\omega)^2}.$$

Obviously, the real part is positive at all frequencies except $\omega = 0$, thus satisfying the phase condition of the Nyquist stability criterion. Hence, the closed-loop system is robustly stable. In the determination of parameters (c_1, c_2), we should consider the response quality of nominal system. Let the damping ratio of the nominal closed-loop system be $3/4$ and the natural frequency be $\sqrt{40K_p}$ (which is set as such to obtain a response speed similar to the second controller). We obtain the following PI controller:

$$K_1(s) = \frac{3\sqrt{\frac{10}{K_p}}s + 40}{s}. \tag{21.32}$$

Simulations w.r.t. the 8 vertices of the uncertain parameter vector (A, ζ_1, ω_1) are carried out. The obtained step responses are shown in Figure 21.6.

Next, we do the robust disturbance control design. Noting that the pole of plant at the origin will be a zero of the nominal closed-loop transfer function, to carry out the robust performance design, we add a small perturbation to the pole of nominal plant at the origin so as to shift it to $-\epsilon(< 0)$. The perturbed nominal plant becomes

$$P(s) = \frac{K_p}{s + \epsilon}, \quad 1 \gg \epsilon > 0.$$

Choose the cost function as the closed-loop system shown in Figure 21.7. The purpose is to make the $\mathcal{H}_\infty$ norm of the transfer function from disturbance w_g to performance output z_g less

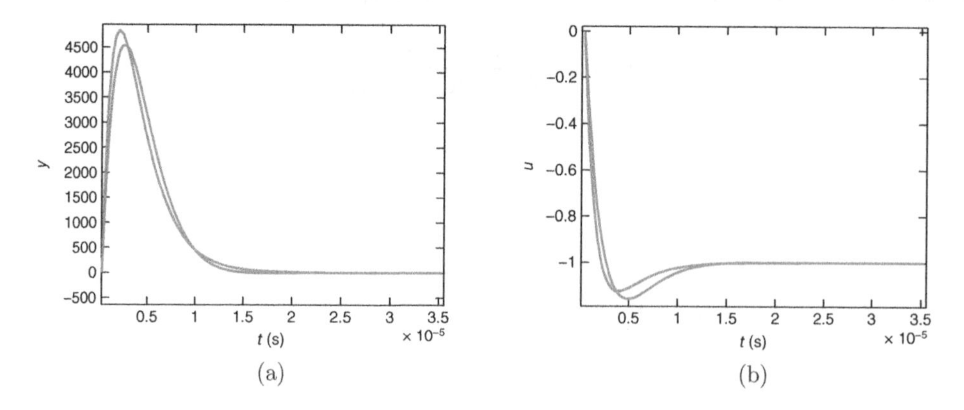

Figure 21.6 Responses of robust stabilizer (a) Output response (b) Input response

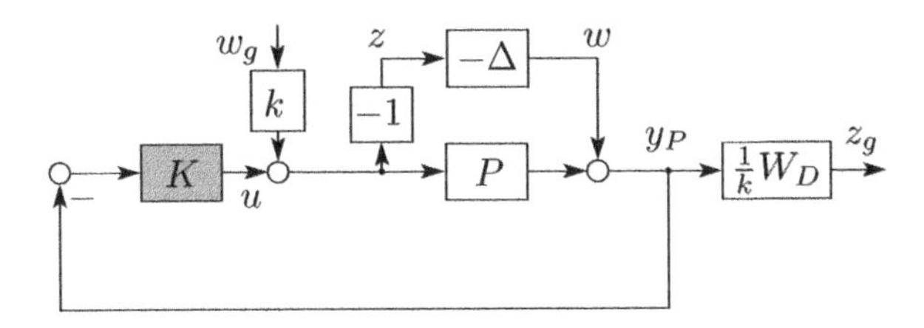

Figure 21.7 Robust performance design

than 1. The weighting function $W_G(s)$ representing the disturbance model is set as

$$W_D(s) = \frac{g}{s+\epsilon}, \quad g > 0. \tag{21.33}$$

The weighting gain $k(>0)$ functions as a scaling constant used for tuning the control performance. Since the gain K_p is extremely large and ϵ extremely small, the numerical error will be too big to yield the expected controller if we use directly the LMI method established earlier. To avoid this difficulty, we first change the time scale to millisecond which is equivalent to the following change of Laplace operator:

$$p = \frac{s}{10^3}. \tag{21.34}$$

Further, 10^2, a part of the gain K_p, is absorbed into the controller $K(s)$. Then, the transfer functions used in the design become

$$P(p) = \frac{3.47 \times 10^4}{p+\epsilon}, \quad W_D(s) = \frac{g}{p+\epsilon}.$$

After getting the controller $K(p)$, the true controller $K(s)$ is obtained by the substitution of $p = s/10^3$ and the multiplication of a gain 10^{-2}.

Via some gain tuning, we obtain a second-order controller

$$K(p) = \frac{0.04p^2 + 12.0p + 4.47}{p(p+2.02)}$$
$$\Rightarrow K_2(s) = \frac{4.0 \times 10^{-4}s^2 + 1.2 \times 10^2 s + 4.47 \times 10^4}{s(s+2.02 \times 10^3)} \tag{21.35}$$

w.r.t $g = 1.0, k = 0.2, \epsilon = 1.0 \times 10^{-5}$.

The step response of the corresponding closed-loop system is shown in Figure 21.8. Apparently, although the settling time and the magnitude of input of $K_1(s)$ and $K_2(s)$ are roughly the same, the disturbance suppression of the robust performance controller $K_2(s)$ is much better.

In the robust performance design based on the positive real method, dimensions of the disturbance and the performance output must be the same. This is a significant drawback of the positive real method. For instance, in the case study, as the input cannot be penalized together with the output, it causes trouble in tuning the response speed of control input. This goal can only be achieved by adjusting other weighting functions. Meanwhile, in the small-gain design,

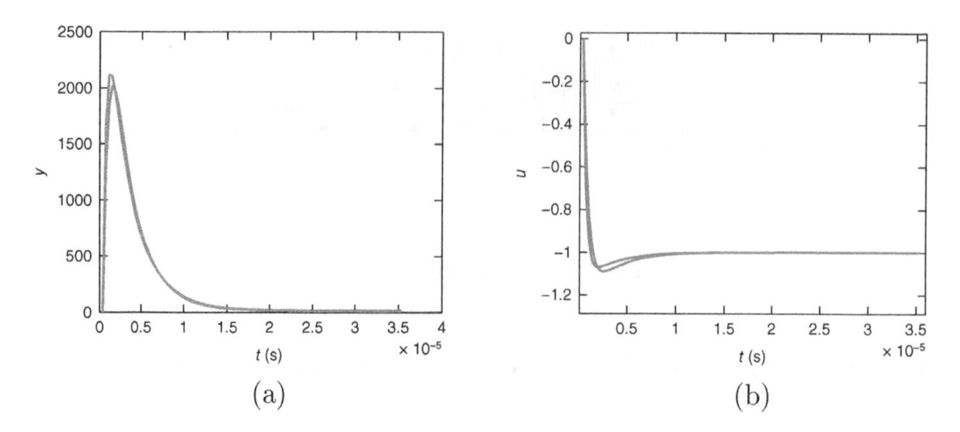

Figure 21.8 Responses of robust performance controller

the response speed of input can be easily changed by adjusting the crossover frequency of its weighting function.

Exercise

21.1 In industrial applications, even when the main characteristic of the uncertainty is represented by its phase, its gain cannot be arbitrarily large. Suppose that an uncertainty Δ is positive real and its largest gain over the whole frequency domain is less than a constant ρ (> 0). Prove that the mapping

$$\Delta_p = \frac{\rho^2 + \Delta^2}{\rho^2 - \Delta^2}$$

maps Δ into the whole right half-plane. Can we use this transformation to reduce the conservatism of the positive real design? Give the reason.

21.2 In the vibration control of flexible systems, when the measured output is the displacement, the transfer function becomes

$$\tilde{P}(s) = \frac{k_0}{s^2} + \frac{k_1}{s^2 + 2\zeta_1\omega_1 s + \omega_1^2} + \cdots + \frac{k_n}{s^2 + 2\zeta_n\omega_n s + \omega_n^2}.$$

Then, since the high-order resonant modes (uncertainty) are not positive real, the positive real method cannot be applied directly. In order to apply the positive real method to robust control design, a feasible approach is to absorb an integrator of the plant into the controller $K(s)$ first

$$\tilde{P}(s) = \frac{k_0}{s} + \frac{k_1 s}{s^2 + 2\zeta_1\omega_1 s + \omega_1^2} + \cdots + \frac{k_n s}{s^2 + 2\zeta_n\omega_n s + \omega_n^2}, \quad C(s) = \frac{1}{s}K(s),$$

then apply the positive real method. Prove that the coefficient matrices corresponding to the nominal closed-loop system in Section 21.2.2 are, respectively (n_K is the degree of $K(s)$),

$$\begin{bmatrix} \overline{A} & \overline{B}_1 & \overline{B}_2 \\ \overline{C}_1 & \overline{D}_{11} & \overline{D}_{12} \\ \overline{C}_2 & \overline{D}_{21} & \end{bmatrix} = \left[\begin{array}{ccc|c|cc} A & B_2 & 0 & B_1 & 0 & 0 \\ 0 & 0 & 0 & 0 & 1 & 0 \\ 0 & 0 & 0 & 0 & 0 & I_{n_K} \\ \hline C_1 & D_{12} & 0 & D_{11} & 0 & 0 \\ \hline C_2 & 0 & 0 & D_{21} & & \\ 0 & 0 & I_{n_K} & 0 & & \end{array}\right].$$

21.3 Repeat the design in Section 21.5 using the method of the preceding exercise. Here, suppose that the plant model is

$$\tilde{P}(s) = K_p \left(\frac{1}{s^2} + \frac{A}{s^2 + 2\zeta_1 \omega_1 s + \omega_1^2} \right).$$

The rest conditions are the same.

Notes and References

The contents of this chapter are taken from Refs [37, 89] and [64].

References

[1] Ahlfors, L.V. (1979) *Complex Analysis*, McGill-Hill Book Company.

[2] Anderson, B.D.O. and Moore, J.B. (1989) *Optimal Control: Linear Quadratic Methods*, Prentice Hall.

[3] Apkarian, P. and Gahinet, P. (1995) A convex characterization of gain-scheduled $\mathcal{H}_\infty$ controllers, *IEEE Transactions on Automatic Control*, **40**(5), 853–864.

[4] Apkarian, P., Gahinet, P. and Becker, G. (1995) Self-scheduled $\mathcal{H}_\infty$ control of linear parameter-varying systems: a design example. *Automatica*, **31**(9), 1251–1261.

[5] Balas, G., Doyle, J., Glover, K., Packard, A. and Smith, R. (1994) μ-Analysis and Synthesis Toolbox, MUSYN Inc. and The MathWorks Inc.

[6] Barmish, B.R. (1983) Stabilization of uncertain systems via linear control, *IEEE Transactions on Automatic Control*, **28**(8), 848–850.

[7] Bellman, R. (1995) *Introduction to Matrix Analysis*, SIAM.

[8] Boullion, T.L. and Odell, P.L. (1971) *Generalized Inverse Matrices*, Wiley-Interscience.

[9] Boyd, S.P. and Barratt, C.H. (1991) *Linear Controller Design: Limits of Performance*, Prentice Hall.

[10] Boyd, S.P., *et al.* (1994) *Linear Matrix Inequalities in System and Control*, SIAM.

[11] Boyd, S.P. and Vandenberghe, L. (2004) *Convex Optimization*, Cambridge.

[12] Chen, T. (1999) *Linear System Theory and Design*, Oxford University Press.

[13] Chen, M.J. and Desoer, C.A. (1982) Necessary and sufficient condition for robust stability of linear distributed feedback systems, *International Journal of Control*, **35**(2), 255–267.

[14] Chen, J., Hara, S. and Chen, G. (2003) Best tracking and regulation performance under control energy constraint, *IEEE Transactions on Automatic Control*, **48**(8), 1320–1336.

[15] Chen, J., Qiu, L. and Toker, O. (2000) Limitations on maximal tracking accuracy, *IEEE Transactions on Automatic Control*, **45**(2), 326–331.

[16] Chilali, M. and Gahinet, P. (1996) $\mathcal{H}_\infty$ design with pole placement constraints: an LMI approach, *IEEE Transactions on Automatic Control*, **41**(3), 358–367.

[17] Chilali, M., Gahinet, P. and Apkarian, P. (1999) Robust pole placement in LMI regions, *IEEE Transactions on Automatic Control*, **44**(12), 2257–2270.

[18] Corless, M. and Shorten, R. (2010) On the characterization of strict positive realness for general matrix transfer functions, *IEEE Transactions on Automatic Control*, **55**(8), 1899–1904.

[19] Cruz, J.B. and Perkins, W.R. (1964) A new approach to the sensitivity problem in multivariable feedback system design, *IEEE Transactions on Automatic Control*, **9**(3), 216–223.

Robust Control: Theory and Applications, First Edition. Kang-Zhi Liu and Yu Yao.

[20] Dahleh, M.A. and Diaz-Bobillo, I.J. (1995) *Control of Uncertain Systems—A Linear Programming Approach*, Prentice-Hall.

[21] Dao, M.Q. and Liu, K.Z. (2005) Gain-scheduled stabilization control of a unicycle robot, *JSME International Journal, Series C*, **48**(4), 649–656.

[22] Desoer, C.A. and Vidyasagar, M. (1975) *Feedback Systems: Input-Output Properties*, Academic Press.

[23] Doyle, J.C. (1978) Guaranteed margins for LQG regulators, *IEEE Transactions on Automatic Control*, **23**(4), 756–757.

[24] Doyle, J.C. (1982) Analysis of feedback systems with structured uncertainties, *IEE Proceedings, Part D*, **133**, 45–56.

[25] Doyle, J.C., Francis, B.A. and Tannenbaum, A.R. (1992) *Feedback Control Theory*, Prentice Hall.

[26] Doyle, J.C., Glover, K., Khargonekar, P.P. and Francis, B.A. (1989) State-space solutions to standard $\mathcal{H}_2$ and $\mathcal{H}_\infty$ control problems, *IEEE Transactions on Automatic Control*, **34**(8), 831–847.

[27] Doyle, J.C., Packard, A. and Zhou, K. (1991) Review of LFTs, LMIs and μ, *Proceedings of the 30th IEEE Conference on Decision and Control*, 1227–1232.

[28] Doyle, J.C. and Stein, G. (1981) Multivariable feedback design: concepts for a classical/modern synthesis, *IEEE Transactions on Automatic Control*, **26**(1), 4–16.

[29] Francis, B.A. (1987) A Course in $\mathcal{H}_\infty$ Control Theory, Vol. **88** of *Lecture Notes in Control and Information Sciences*, Springer-Verlag.

[30] Frankline, G.F., *et al.* (1991) *Feedback Control of Dynamic Systems*, Addison-Wesley.

[31] Freudenberg, J.S. and Looze, D.P. (1988) *Frequency Domain Properties of Scalar and Multivariable Feedback Systems*, Vol. **104** of *Lecture Notes in Control and Information Sciences*, Springer-Verlag.

[32] Gahinet, P. (1996) Explicit controller formulas for LMI-based $\mathcal{H}_\infty$ synthesis, *Automatica*, **32**(7), 1007–1014.

[33] Gahinet, P. and Apkarian, P. (1994) A linear matrix inequality approach to $\mathcal{H}_\infty$ control, *International Journal of Robust and Nonlinear Control*, **4**, 421–448.

[34] Gahinet, P., Apkarian, P. and Chilali, M. (1996) Affine parameter-dependent Lyapunov function and real parametric uncertainty, *IEEE Transactions on Automatic Control*, **41**(3), 436–442.

[35] Gahinet, P., Nemirovski, A., Laub, A.J. and Chilali, M. (1995) *LMI Control Toolbox*, The MathWorks Inc.

[36] Glover, K. and Doyle, J. (1988) State-space formulae for all stabilizing controllers that satisfy an $\mathcal{H}_\infty$ norm bound and relations to risk sensitivity, *Systems and Control Letters*, **11**(3), 167–172.

[37] Haddad, W.M. and Bernstein, D.S. (1991) Robust stabilization with positive real uncertainty: beyond the small gain theorem, *Systems and Control Letters*, **17**(3), 191–208.

[38] He, R., Liu, K.Z. and Mei, S. (2010) LPV modeling and gain-scheduled control approach for the transient stabilization of power systems, *IEEJ Transactions on Electrical and Electronic Engineering*, **5**(1), 87–95.

[39] Hirata, M. and Hasegawa, Y. (2007) High bandwidth design of track-following control system of hard disk drive using $\mathcal{H}_\infty$ control theory, *Proceedings of 2007 IEEE Conference on Control and Applications*, 114–117.

[40] Hirata, M., Liu, K.Z. and Mita, T. (1996) Active vibration control of a 2-mass system using μ-synthesis with a descriptor form representation, *Control Engineering Practice*, **4**(4), 545–552.

[41] Hirata, M., Liu, K.Z., Mita, T. and Yamaguchi, T. (1993) Heading positioning control of a hard disk drive using $\mathcal{H}_\infty$ control theory, *Transactions of SICE*, **29**(1), 71–77 (in Japanese). See also *Proceedings of the 31st IEEE Conference on Decision and Control*, 2460–2461, Tucson (1992).

[42] Iwasaki, T. (1997) *LMI and Control*, Shokodo Co., Ltd, Tokyo (in Japanese).

[43] Iwasaki, T. and Skelton, R.E. (1994) All controllers for the general $\mathcal{H}_\infty$ control problem: LMI existence conditions and state space formulas, *Automatica*, **30**(8), 1307–1317.

[44] Kailath, T. (1980) *Linear Systems*, Prentice Hall.

[45] Kalman, R.E. (1963) Lyapunov functions for the problem of Lur'e in automatic control, *Proceedings of the National Academy of Sciences of the United States of America*, **49**, 201–205.

[46] Kanehara, A. and Liu, K.Z. (2007) A gain scheduling control method for automotive vehicle stabilization, *Transactions of JSME*, **73**(729), 1285–1290 (in Japanese).

[47] Khalil, H.K. (1996) *Nonlinear Systems*, Prentice Hall.

[48] Kimura, H. (1984) Robust stabilization for a class of transfer functions, *IEEE Transactions on Automatic Control*, **29**(9), 788–793.

[49] Kodama, S. and Suda, N. (1998) *Matrix Theory for System Control*, Corona Publishing Co., Ltd, Tokyo (in Japanese).

[50] Kundur, P. (1994) *Power System Stability and Control*, McGraw-Hill, ERPI.

[51] Lancaster, P. and Rodman, L. (1995) *Algebraic Riccati Equations*, Oxford Science Publications.

[52] Larsen, E.V. and Swann, D.A. (1981) Applying Power System Stabilizers Part I, II, III, *IEEE Transactions on Power Apparatus and Systems*, **PAS-100**(6), 3017–3046.

[53] Liu, K.Z. (2008) Neo-robust control theory: how to make use of the phase information of uncertainty, *Proceedings of 2008 IFAC World Conference*, 5962–5968.

[54] Liu, K.Z. and He, R. (2006) A simple derivation of ARE solutions to the standard $\mathcal{H}_\infty$ control problem based on LMI solution, *Systems and Control Letters*, **55**(6), 487–493.

[55] Liu, K.Z., Kawatani, R. and Mita, T. (1990) Parameterization of state feedback $\mathcal{H}_\infty$ controllers, *International Journal of Control*, **51**(3), 535–551.

[56] Liu, K.Z. and Kobayashi, H. (2009) Neo-robust control theory (Part I): uncertainty model and robust stability analysis, *Proceedings of Asian Control Conference 2009*, 802–807.

[57] Liu, K.Z. and Kobayashi, H. (2009) Neo-robust control theory (Part II): analysis of robust sensitivity and robust bandwidth, *Proceedings of Asian Control Conference 2009*, 808–813.

[58] Liu, K.Z. and Mita, T. (1989) Conjugation and $\mathcal{H}_\infty$ control of discrete-time systems, *International Journal of Control*, **50**(4), 1435–1460.

[59] Liu, K.Z. and Mita, T. (1993) On robust steady state performance of feedback control systems, *Transactions of SICE*, **29**(3), 312–318 (in Japanese). See also *Proceedings of the 31st IEEE Conference on Decision and Control*, 225–230, Tucson (1992).

[60] Liu, K.Z. and Mita, T. (1995) A unified stability analysis for regulator and servomechanism problems, *Transactions of SICE*, **31**(7), 834–843 (in Japanese). See also *Proceedings of the 33rd IEEE Conference on Decision and Control*, 4198–4203, Orlando (1994).

[61] Liu, K.Z. and Mita, T. (1996) Parameterization of comprehensive stabilizing controllers and analysis of its structure, *Transactions of SICE*, **32**(3), 320–328 (in Japanese). See also *Proceedings of the 34th IEEE Conference on Decision and Control*, 4108–4113, New Orleans (1995).

[62] Liu, K.Z., Mita, T. and Luo, Z. (1993) Physical explanations of some $\mathcal{H}_p$ norms, *Proceedings of International Symposium MTNS1993*, 321–326.

[63] Liu, K.Z. and Shirnen, B. (2009) Neo-robust control theory for factorized uncertainty, *Proceedings of IEEE Conference on Decision and Control and Chinese Control Conference 2009*, 650–655.

[64] Liu, K.Z. and Yao, Y. (2012) Positive Real Method for Robust Control Problems, *Proceedings of Chinese Control Conference 2012*, Hefei.

[65] Liu, K.Z., Zhang, H. and Mita, T. (1997) Solution to nonsingular $\mathcal{H}_2$ optimal control problems with unstable weight, *Systems and Control Letters*, **32**(1), 1–10.

[66] Lu, Q., Wang, Z. and Han, Y. (1982) *Optimal Control of Power Systems*, Science Press, Beijing (in Chinese).

[67] Megretski, A. and Rantzer, A. (1997) System analysis via integral quadratic constraints, *IEEE Transactions on Automatic Control*, **42**(6), 819–830.

[68] Mei, S., Shen, T. and Liu, K.Z. (2003) *Modern Robust Control Theory and Application*, Tsinghua University Press, Beijing (in Chinese).

[69] Mita, T. (1977) On zeros and responses of linear regulators and linear observers, *IEEE Transactions on Automatic Control*, **22**(3), 423–428.

[70] Miyazaki, M., Liu, K.Z. and Saito, O. (2001) Robust decentralized control approach to the multi area frequency control problem of power systems, *IEEJ Transactions on Electronics, Information and Systems*, **121**(5), 953–960.

[71] Narendra, K. and Taylor, J.H. (1973) *Frequency Domain Criteria for Absolute Stability*, Academic Press, New York.

[72] Nesterov, Y. and Nemirovskii, A. (1994) *Interior-Point Polynomial Methods in Convex Programming*, SIAM.

[73] Okajima, K., Liu, K.Z. and Minowa, T. (1999) Improvement of the transient response of the automatic transmission of a car by $\mathcal{H}_\infty$ control, *Transactions of SICE*, **35**(1), 147–149 (in Japanese). See also Minowa, T., Ochi, T., Kuroiwa, H. and Liu, K.Z. (1999) Smooth gear shift control technology for clutch-to-clutch shifting, *SAE technical paper series*, 1999-01-1054.

[74] Ouchi, S., Liu, K.Z., Sato, S. and Mita, T. (1998) Mine train control system by $\mathcal{H}_\infty$ control, *Transactions of SICE*, **34**(3), 225–2312 (in Japanese). See also Takeuchi, H., Liu, K.Z. and Ouchi, S. (1996.12) Velocity control of a mine truck system using rationally scaled $\mathcal{H}_\infty$ control, *IEEE Conference on Decision and Control 1996*, 767–772.

[75] Paattilammi, J. and Mäkilä, P.M. (2000) Fragility and robustness: a case study on paper machine headbox control, *IEEE Control System Magazine*, **20**(1), 13–22.

[76] Packard, A. (1994) Gain scheduling via linear fractional transformations, *Systems and Control Letters*, **22**(2), 79–92.

[77] Packard, A. and Doyle, J.C. (1993) The complex structured singular value, *Automatica*, **29**(1), 71–109.

[78] Popov, V.M. (1962) Absolute stability of nonlinear systems of automatic control, *Automation and Remote Control*, **22**, 857–875.

[79] Qiu, L., *et al.* (1995) A formula for computation of real stability radius, *Automatica*, **31**(6), 879–890.

[80] Rantzer, A. (1996) On the Kalman-Yakubovich-Popov lemma, *Systems and Control Letters*, **28**(1), 7–10.

[81] Sabanovic, A. and Ohnishi, K. (2011) *Motion Control Systems*, Wiley-IEEE Press.

[82] Safonov, M.G. and Athans, M. (1977) Gain and phase margin for multiloop LQG regulators, *IEEE Transactions on Automatic Control*, **22**(2), 173–179.

[83] Sato, T. and Liu, K.Z. (1999) LMI solution to general $\mathcal{H}_2$ suboptimal control problems, *Systems and Control Letters*, **36**(4), 295–305.

[84] Seron, M.M., *et al.* (1997) *Fundamental Limitations in Filtering and Control*, Springer-Verlag.

[85] Takagi, R. and Liu, K.Z. (2011) Model based control for rapid thermal processing of semiconductor wafers, *IEEJ Transactions on Industry Applications*, **131-D**(2), 159–165 (in Japanese).

[86] Tao, G. and Ioannou, P.A. (1988) Strictly positive real matrices and the Lefschetz-Kalman-Yakubovich lemma, *IEEE Transactions on Automatic Control*, **33**(12), 1183–1185.

[87] Technical Committee on Servo Technology for the Next Generation Mass Storage System, IEEJ (2007) Hard disk drive benchmark problem, http://hflab.k.u-tokyo.ac.jp/nss/MSS_bench.htm.

[88] Tits, A.L., Balakrishnan, V. and Lee, L. (1999) Robustness under bounded uncertainty with phase information, *IEEE Transactions on Automatic Control*, **44**(1):50–65.

[89] Turan, L., Safonov, M.G. and Huang, C.H. (1997) Synthesis of positive real feedback systems: a simple derivation via Parrott's theorem, *IEEE Transactions on Automatic Control*, **42**(8), 1154–1157.

[90] Vidyasagar, M. (1985) *Control System Synthesis*, MIT Press.

[91] Wen, J.T. (1988) Time and frequency domain conditions for strictly positive real functions, *IEEE Transactions on Automatic Control*, **33**(10), 988–992.

[92] Willems, J.C. (1971) Least squares stationary optimal control and the algebraic Riccati equation, *IEEE Transactions on Automatic Control*, **16**(6), 621–634.

[93] Willems, J.C. (1972) Dissipative dynamical systems, Part I: General theory, *Archive for Rational Mechanics and Analysis*, **45**, 321–351.

[94] Wu, B.F. and Jonckeere, E.A. (1992) A simplified approach to Bode's theorem for continuous-time and discrete-time systems, *IEEE Transactions on Automatic Control*, **37**(11), 1797–1802.

[95] Yakubovich, V.A. (1962) Solution of certain matrix inequalities in the stability theory of nonlinear control systems, *Doklady Akademii Nauk SSSR*, **143**, 1304–1307.

[96] Yaw, K.J. and Lin, C.A. (1998) Lamp configuration design for rapid thermal processing systems, *IEEE Transactions on Semiconductor Manufacturing*, **11**(1), 75–84.

[97] Youla, D.C., Jabr, H.A. and Lu, C.N. (1974) Single-loop feedback stabilization of linear multivariable dynamical plants, *Automatica*, **10**, 159–173.

[98] Zames, G. (1966) On the input-output stability of nonlinear time-varying feedback systems, Parts I and II, *IEEE Transactions on Automatic Control*, **11**(2), 228–238 and (3), 465–476.

[99] Zames, G. (1981) Feedback and optimal sensitivity: Model reference transformations, multiplicative seminorms and approximate inverses, *IEEE Transactions on Automatic Control*, **26**(2), 301–320.

[100] Zhou, K., Doyle, J.C. and Glover, K. (1995) *Robust and Optimal Control*, Prentice Hall.

Index

absolute stability, 214, 297
actuator, 414
additive uncertainty, 250, 252
affine combination, 57
affine function, 68
affine LPV model, 422
affine set, 57
affine structure, 186, 187
affine system, 425
A-invariant, 30, 31
algebraic Riccati equation (ARE), 215
all-pass filter, 56
all-pass transfer function, 48, 223, 227, 228
analytical center, 81
analytic function, 226
antistability, 101, 125, 129, 335
antistable function, 50
ARE *see* algebraic Riccati equation
ARE solution, 348
asymptotic convergence, 123
asymptotic rejection, 128, 129
asymptotic tracking, 124, 125, 129
automatic voltage regulator (AVR), 432

balanced realization, 114
bandwidth, 139
barrier function, 71, 81
base matrix, 73
basis, 17, 22
BIBO stability *see* bounded-input bounded-output stability
bilinear matrix inequality (BMI), 75
blocking zero, 93
block triangular matrix, 10, 11, 31, 33, 183
BMI *see* bilinear matrix inequality
Bode phase formula, 234
Bode sensitivity integral relation, 232
boundary of the uncertainty, 5 (*see also* uncertainty bound)
bounded-input bounded-output (BIBO) stability, 100–103
bounded real lemma, 202, 203

cascade connection, 97, 98
Cauchy integral formula, 226
Cauchy theorem, 226
causality, 314
causal system, 314
Cayley–Hamilton theorem, 30
central path, 82
characteristic function, 387, 390–392, 399
characteristic polynomial, 28, 150
characteristic root, 28
circle criterion, 301, 303, 317
closed-loop control, 140, 142–145, 173
combined system, 167

Robust Control: Theory and Applications, First Edition. Kang-Zhi Liu and Yu Yao.

common Lyapunov function, 291, 318, 380, 426
complementary sensitivity, 225, 276
composite LMI region, 392
comprehensive stability, 180
concave function, 69, 71, 72
cone, 60
conformal mapping, 443
constant-scaled bounded real lemma, 214
constant-scaled small-gain condition, 419
controllability, 87
  canonical form, 152
  Gramian, 111–113, 327
  matrix, 87
controllable, 149, 156, 158
convex combination, 59
convex cone, 60
convex function, 68
convex hull, 59
convex set, 59
coordinate transformation, 18
coprime factorization, 185
coprime uncertainty, 286
critical point, 5

damping ratio, 132
descriptor form, 260
design of generalized plant, 352
detectability, 107
determinant, 10
diagonalization, 33, 34
dimension, 22
disk region, 386
dissipative system, 203
disturbance, 119
  observer, 146
  suppression, 119, 127, 144
D–K iteration, 369–371
domain, 23
dominant pole, 131
dual relation, 89
dynamic uncertainty, 247–249, 253, 265, 272, 353

eigenvalue, 28
eigenvalue problem (EVP), 74
eigenvector, 28
ellipsoid, 42, 62
energy, 36, 197, 199, 208, 288, 289, 346
  function, 35
  spectrum, 200
Euclidean distance, 19
EVP *see* eigenvalue problem
exciter, 430
expectation, 324

feasibility, 66, 73, 81
feedback connection, 97–99
feedback uncertainty, 250
first-order condition, 69
flexible space structure, 265
flexible system, 270, 436, 445
flywheel, 414
Fourier transform, 197
free parameter, 178, 180, 187, 332
full block uncertainty, 364, 366
full column rank, 26
full-order observer, 161, 167
full row rank, 26

gain-scheduled control, 407
  system, 426
gain-scheduled $\mathcal{H}_\infty$ control design, 410
gain-scheduled method, 414
gain scheduling, 7, 409
γ-optimal $\mathcal{H}_2$ control, 330
generalized eigenvalue problem (GEVP), 74
generalized eigenvector, 29
generalized feedback system, 174, 175
generalized plant, 174–177, 341, 352, 354, 372, 419
GEVP *see* generalized eigenvalue problem
Gopinath algorithm, 165
gyro-moment, 415
gyroscope, 414

$\mathcal{H}_\infty$ control, 348
$\mathcal{H}_\infty$ norm, 51, 52, 346
  specification, 426
$\mathcal{H}_\infty$ performance, 277
$\mathcal{H}_2$ control, 322
$\mathcal{H}_2$ norm, 51, 52, 322–328
  specification, 426

$\mathcal{H}_2$ optimal control, 330
half-space, 61, 62
Hamiltonian matrix, 216
hard disk drive (HDD), 1, 246, 354, 371
Hermitian matrix, 33, 55
high-order resonant mode, 246, 265, 437
hyperplane, 60

identification, 247
IEEJ benchmark, 246
image (range), 23
IMC *see* internal model control
implementation of 2-DOF, 191, 192
induced norm, 37
inertia ratio, 93
infinity-norm, 46
infinity zero, 97
inner function, 223
inner product, 20–22, 49
  matrices, 39, 40
  signals, 47–49, 199
  systems, 53
in-phase, 265
input blocking, 94
integral quadratic constraint (IQC), 312
  IQC theorem, 314
interior point method, 9, 81
internal model control (IMC), 193, 285
internal model principle, 125, 129
internal stability, 104
interpolation condition, 229
invariance, 56, 165, 230
invariant subspace, 30–32, 216–220
invariant zero, 95
IQC *see* integral quadratic constraint

kernel space (null space), 23
K–L iteration, 357
Kronecker product, 44, 108, 388
Kronecker sum, 44
KYP lemma, 201, 319

largest singular value, 40, 42
least quadratic Gaussian (LQG), 8
left eigenvector, 55
left singular vector, 41

LFT *see* linear fractional transformation
LFT-type gain-scheduled control, 408
limitation of reference tracking, 237
linear algebraic equation, 27
linear combination,15
linear dependence, 15
linear fractional transformation (LFT), 115, 116
  model, 408, 414
linear independence, 15
linear mapping, 23
linear matrix equation, 35
linear matrix inequality (LMI), 9, 72
  region, 385, 387, 390, 398, 406
  solution, 349
linear parameter-varying (LPV), 7, 266
  model, 266–269, 417, 429
  system, 7, 407
linear subspace, 22
linear transformation, 12
linear vector space, 14
LMI *see* linear matrix inequality
location of actuator, 337, 339
location of sensor, 337, 340
logarithmic function, 71
lower LFT, 116
LPV *see* linear parameter-varying
LQG *see* least quadratic Gaussian
Luenberger observer, 161
Lur'e system, 297, 298, 304
Lyapunov equation, 108, 327
Lyapunov function, 6, 289, 299
Lyapunov stability theory, 288

mapping restricted in invariant subspace, 32
mass–spring–damper system, 253, 258, 296, 396, 405
matrix decomposition, 11
matrix description, 12
matrix inverse, 12
matrix norm, 37, 38
matrix polytope, 257, 398
maximum modulus theorem, 230
measured disturbance, 145
measured output, 174
MIMO *see* multiple-input multiple-output

minimal order observer, 163, 168
minimal realization, 90
minimum phase, 140–142
  system, 234
  transfer function, 228, 251, 281, 371
modal dynamics, 437
model matching, 191
model reduction, 114
model uncertainty, 1, 142, 247
motor drive, 2
μ, 360
multiobjective control, 425, 430
multiple-input multiple-output (MIMO), 93
multiple-input system, 156
multiple uncertainty 360
multiplicative uncertainty, 249, 252

natural frequency, 132
nominal plant, 247
nonlinear time-varying block, 316
nonminimum-phase system, 145, 192
nonminimum-phase transfer function, 228
nonsingular $\mathcal{H}_2$ control, 331
nonsingularity condition, 330
nonsingular matrix, 26
normal rank, 93
norm-bounded parametric system, 259, 261, 282, 294, 379, 382, 398, 403
notch, 355
Nyquist plot, 201, 204, 265, 300–303
Nyquist stability criterion, 298, 373

observability, 89
  canonical form, 153
  Gramian, 111–113, 223, 327
  matrix, 89
observable, 162, 164
observer, 161
  gain, 161
  pole, 162
1-degree-of-freedom (1-DOF), 175
  system, 237
1-norm, 19, 38, 46
open-loop control, 140
orthogonal, 21, 50, 335
  matrix, 350
orthonormal matrix, 34
overshoot, 130, 135

parallel connection, 97, 98
parameter space, 253, 260
parameter uncertainty, 247, 261, 262
parameter vector, 255, 256
parametric system, 253, 258, 282, 375
parametrization, 178, 181
Parseval's theorem, 199
passive nonlinearity, 264
passive system, 208, 307
performance, 119
  criterion, 130
  limitation, 226
  output, 174
persistent disturbance, 121, 128
phase angle, 47, 48
phase information, 264
phase lead compensation, 355
plant set, 248
Poisson integral formula, 226
pole, 90
  location, 384
  placement, 149, 154, 156
pole–zero cancellation, 99, 106, 107
polygon, 254
polyhedron, 64
polytope, 64
  parameter uncertainty, 256
polytopic structure, 426
polytopic system, 258, 292, 318, 375, 396
Popov criterion, 304, 305
Popov plot, 306, 307
Popov-type IQC, 317
positive definite, 6
  function, 36
  matrix, 36
positive real, 6
  condition, 6
  function, 6, 204, 265
  uncertainty, 442
positive real lemma, 205, 208
positive semidefinite function, 36
positive semidefinite matrix, 36
positive-real function, 265

power system, 428
power system stabilizer (PSS), 432, 433
process control system, 285
proper, 96
prototype second-order system, 131, 132
pseudo-inverse, 34
PSS *see* power system stabilizer

Qiu's theorem, 283
quadratic form, 35
quadratic Lyapunov function, 6
quadratic stability, 291, 292, 294, 318
quadratic stabilization, 376, 377, 379

ramp signal, 121
rank, 26
rapid thermal processing (RTP), 337
realization, 90
reference input, 119
reference tracking, 119, 124
σ region, 386
regional pole placement, 387, 427
relative degree, 96
resonant mode, 3
right eigenvector, 55
right singular vector, 41
rigid body model, 3, 246
rise time, 130
robust control, 1
robust design of regional pole placement, 402
robust $\mathcal{H}_\infty$ control design, 382
robust $\mathcal{H}_\infty$ performance, 368
robust multiobjective control, 425
robust performance, 270, 279, 363
  design, 442
robust pole placement, 396
robust stability, 269
  condition, 276
robust stabilization, 438, 441
RTP *see* rapid thermal processing

scalar uncertainty, 364
scaled $\mathcal{H}_\infty$ control, 282, 355–357
scaled $\mathcal{H}_\infty$ problem, 282
scaling, 281
scheduling parameter, 424–426
Schur's lemma, 37
second-order condition,70
sector, 386
selection of weighting function, 353
sensitivity, 8, 229, 276
separating hyperplane, 64–66
separation principle, 168
settling time, 130
σ region, 386
signal norm, 45
similarity transformation, 25, 90
single-input single-output (SISO), 85
single-input system, 154
single-machine infinite-bus power system, 267, 429
singular $\mathcal{H}_2$ control, 335
singular value, 40
singular value decomposition (SVD), 34, 41
singular vector, 40
sinusoidal signal, 121
SISO *see* single-input single-output
smallest singular values, 40
small-gain approach, 5
small-gain condition, 5
small-gain theorem, 272, 315
specification, 130
spillover, 4
S-procedure, 295
stability radius, 283
stabilizability, 107
stabilization, 148
stabilizing controller, 179, 182, 184
stabilizing solution, 218
stable function, 50
state equation, 85
state feedback, 148
  gain, 148
  system, 148
static nonlinearity, 297
  in a sector, 316
steady-state response, 122
step signal, 120
storage function, 203
strictly concave function, 71
strictly convex function, 68

strictly positive real, 441
strictly positive real lemma, 207
strictly proper, 96
strongly positive real, 301, 438
strongly positive real lemma, 206
structured singular value, 360
structured uncertainty, 360
submultiplicative property, 39
subspace, 22–24, 30
supply rate, 208
SVD *see* singular value decomposition
symmetric matrix, 33
system matrix, 95
system norm, 50

terminal voltage, 432
test signal, 120
three-phase short circuit fault, 431
time delay, 56
time-varying parameter, 7, 269, 407
trace, 10
transfer function, 85
transfer matrix, 85
transient response, 130, 322
transient stabilization, 428
transmission zero, 93
truncation, 114, 314
2-degree-of-freedom (2-DOF), 175, 188, 191
  system, 244, 329
two-mass–spring system, 2, 86, 93, 271
2-norm, 19, 38, 46, 48, 49
type A undershoot, 137, 138
type B undershoot, 137

uncertain part, 246
uncertainty, 3, 245
  bound, 251, 252
  modeling, 251
uncontrollable mode, 91, 95
undershoot, 135
unicycle robot, 414, 415
unitary matrix, 34, 41
unmeasured disturbance, 144
unobservable mode, 91, 95
unstable zero, 238, 241, 244
upper LFT, 116

variable change, 79, 80, 351, 439
variable elimination, 74, 349, 440
variance, 324
vector norm, 19
vector space, 14
vertex, 254
  condition, 294, 426
  polytope, 292

water bed effect, 233
weighted closed-loop transfer function, 322
weighting function, 249, 251, 325, 347

Youla parametrization, 184, 229, 230

zero, 91–97, 185
  vector, 95, 185

Printed and bound by CPI Group (UK) Ltd, Croydon, CR0 4YY
16/04/2025
14658412-0001